MAMMAL SKIN

MAMMAL SKIN

V. E. Sokolov

UNIVERSITY OF CALIFORNIA PRESS
Berkeley   Los Angeles   London

University of California Press
Berkeley and Los Angeles, California

University of California Press, Ltd.
London, England

Library of Congress Cataloging in Publication Data

Sokolov, Vladimir Evgen'evich.
  Mammal skin.

  Includes bibliographical references and index.
  1. Mammals—Anatomy. 2. Mammals—Physiology.
3. Skin.   I. Title. [DNLM: 1. Mammals—Anatomy and
Histology. 2. Skin—Anatomy and Histology. WR 101
S683m]
QL739.S68   599.04'7   81-3037
ISBN 0-520-03198-9   AACR2

Printed in the United States of America

CONTENTS

ACKNOWLEDGMENTS

A large part of the material used for our investigation of comparative morphology was kindly given us by the Institute of Zoology's zoological museums, the USSR Academy, Moscow University, the British Natural History Museum, the Museum of Natural History in Paris, the Museum of the Institute of Zoology, the Cuban Academy of Sciences, and by L. M. Baskin, E. I. Betisheva, G. N. Bobrova, N. N. Vorontsov, V. V. Gerasimov, V. V. Dezkin, A. I. Ilyenko, N. I. Kalabukhov, S. K. Klumov, V. A. Komarov, P. G. Nikulin, M. V. Okhotina, T. B. Sablina, S. M. Uspensky, and E. S. Chuzhakina.

We obtained material on most of the species of *Ungulata, Pinnipedia, Cetacea, Carnivora,* and many *Rodentia* and *Insectivora* ourselves during expeditions to the Kuril islands, Tyuleny Island, Kamchatka, Sayany, Central Asia, and the European USSR.

Officers at the head office of the Frontier Service and the head office of Hunting and Reservations, RSFSR Council of Ministers, gave us a great deal of assistance.

I wish to express my sincere gratitude to the many colleagues who helped me with this book, especially to Professor B. A. Kuznetsov, who has contributed to the writing, and to Professor B. S. Matveyev, who undertook the difficult task of editing. Most of the specimens were processed by L. N. Skurat.

INTRODUCTION

As an ectosomatic system, the skin of mammals is in direct contact with an external, shaping environment that serves to determine its complex structure and functions. The study of mammalian skin involves both pure and applied sciences, for in different taxonomic groups it elucidates the relationship and evolution of mammals, and it demonstrates the peculiar adaptation of skin and organs to environments. This is of particular significance, for no other vertebrates have adapted to as varied climates or environments —surface, underground, water, and air. Because the skin manifests direct environmental effects, its adaptive features have special scientific import.

The peculiarities of skin adaptation in mammals can be simulated and used for various purposes by man. A knowledge of their skin structure is essential for breeding of furbearing animals and new domestic species. Studies of the formation of those skin glands with secretions having a specific odor are particularly interesting. A thorough investigation of glandular structure and the chemistry of glandular secretions, both in the laboratory and in natural conditions, makes it possible to ascertain their vital functions. Many scientists have been concerned with attractants and repellents in insects, yet the problem has barely been touched where mammals are concerned. To begin with, the morphology of mammalian glands merits thorough study, for the results may have commercial application to the problem of controlling mammal behavior.

Mammalian skin has long attracted the attention of scientists, though focus has primarily been on the skin of man and of domestic mammals, and most of the published work has dealt with the morphology, physiology, and biochemistry of human skin (Rothman, 1954; Kuno, 1959; Montagna,

1962; Kalantaevskaya, 1972; Parakkal and Alexander, 1972; Montagna and Parakkal, 1973; Spearman, 1973; and others). A few collections of papers on various aspects of structure and function have been published on man (Montagna and Ellis, 1958c; Montagna, 1960; Ellis and Silver, 1963; Montagna and Lobitz, 1964), and on domestic animals (Rumyantsev et al., 1933; Bezler, 1935; Braun et al., 1935; Tekhver, 1971). The skin structure of wild mammals is much more obscure. Gabe (1967) bases his work on data from human histology and fails to describe mammalian skin structure even at order level. This is true of the chapter devoted to mammalian skin in Weber (1928), and there are very few papers available on the skin structure of mammal orders (*Cetacea* and *Pinnipedia:* Sokolov, 1955, 1959, 1960 *a, b, c;* 1962, 1971; *Chiroptera:* Quay, 1970).

The skin structure of individual mammal species also calls for further investigation. Primate skin has received the most attention (thanks to studies in the laboratory headed by Montagna), but beyond the work of Montagna and his collaborators, only a few items on the specific glands of several primates have been published (Schaffer, 1940; Hill, 1956; Hill et al., 1958-1959) and on the skin and skin glands of some primate species (Straus, 1942, 1950; Hayasaki, 1959; Setogutti and Sakuma, 1959).

The skin of *Cetacea, Pinnipedia,* and of furbearing mammals has also been studied in considerable detail (Kuznetsov, 1927, 1928, 1945, 1952; Tserevitinov, 1951, 1958; Yeremeyeva, 1954, 1956; Leschinskaya, 1952a, b; Fadeev, 1955, 1958; Sokolov, 1955-1971; Belkovich, 1959-1964; Kogteva, 1963; and others). As far as the structure of individual parts of the skin in wild mammals is concerned, some works are available, such as Toldt (1906) (on hair); and Schaffer (1940) (on specific skin glands), though Schaffer's book is already dated.

There are few morphoecological works on mammalian skin, though some features of the hair of various ecological groups of furbearing mammals are described by Tserevitinov (1958), and some details on the adaptation of skin to water are given by Gudkova-Aksenova (1951); Sokolov (1955-1971); Belkovich (1959-1964); to the desert by Sokolov (1962); Quay (1965); underground by Sokolov (1970); to seasonal climatic changes by Leschinskaya (1952); Kogteva (1963); and others.

Clearly, the structure and adaptation of mammalian skin requires additional study. Furthermore, such knowledge as we have has not been available from a single source. This book is in three parts. Part one deals with the structure of mammalian skin and is based on the published work on man and laboratory animals. In addition to results obtained with light microscopy, histochemical and electron microscopic findings are briefly considered, though further studies of most mammals are needed. Part two presents a systematic description of the skin of various mammals, using published work and the author's own findings. Part three deals with the adaptive peculiarities of mammalian skin as they relate to varying conditions of life.

I

THE STRUCTURE OF DIFFERENT PARTS
OF MAMMALIAN SKIN

Mammalian skin is a multifunctional system that covers and protects the body of the animal. Since it is in direct contact with the environment, the skin must vary considerably with the mode of life of the species and so it is structurally one of the most diverse systems. Skin includes the skin proper, that is, the epidermis and dermis; plus subcutaneous fat tissue (fig. 1), hair, glands (sudoriferous, sebaceous, specific, mammary), and various modifications such as claws, nails, hooves, and horns. Nerves and nerve endings are distributed in the skin, blood vessels pass through it, and smooth and striated muscles are present.

The skin protects the organism from its environment. The dermis prevents loss of water and penetration of foreign microorganisms and substances, and the whole skin guards against mechanical damage. Hair and subcutaneous fat provide insulation, and the network of blood vessels and sudoriferous glands provides thermal regulation. The arrectores pilorum muscles can increase and decrease hair insulation. The skin is involved in the excretion of water and other metabolic products, such as urea, and may accumulate nutrients in the subcutaneous fat tissue. It also has certain essential sensory functions—it contains a vast number of nerve endings. Various glands in the skin have extra functions, such as producing milk to feed the offspring, and providing chemical signals.

EPIDERMIS

Mammalian epidermis is a layered, externally cornifying epithelium that

3

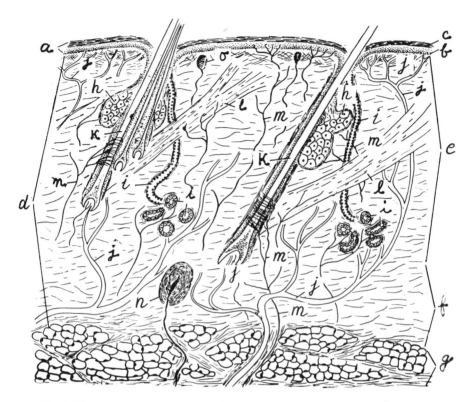

Fig. 1. Diagram of mammal skin: *a*, epidermis; *b*, stratum basale; *c*, stratum corneum; *d*, dermis; *e*, papillary layer; *f*, reticular layer; *g*, subcutaneous fat tissue; *h*, sebaceous gland; *i*, sweat gland; *j*, blood vessel; *k*, hair root; *l*, arrector pili muscle; *m*, nerve; *n*, Vater-Pacini body; *o*, Meissner's corpuscle.

covers the surface of the body. The epidermis varies a great deal in different mammals and on different parts of their bodies. It is most highly developed where the skin is hairless, especially on those areas subject to the most wear (the soles of the feet, the breast skin of goats and sheep) (fig. 2). Normally, in the epidermis of such areas, five layers are distinguishable. Working outwards, they are the basal and germinal layers (stratum basale or stratum germinativum), the spinous layer (stratum spinosum), the granular layer (stratum granulosum), the lucid layer (stratum lucidum), and the corneous layer (stratum corneum).

The cells of the basal and spinous layers are live, and these are frequently referred to as stratum malpighii or rete mucosum. The basal layer is commonly composed of cubiform and prismatic cells, the upper surface facing the spinous layer being conical or convex. The shape of the basal layer cells varies with the thickness of the epidermis. Where the epidermis cells of the basal layer are thick they are columnar or prismatic, where thin (in the skin

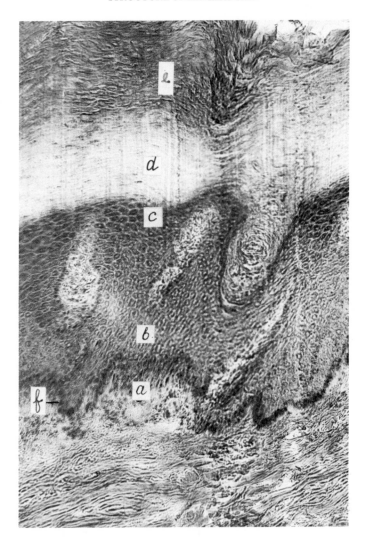

Fig. 2. *Capra sibirica*. Epidermal layers, breast: *a*, stratum basale; *b*, stratum spinosum; *c*, stratum granulosum; *d*, stratum lucidum; *e*, stratum corneum; *f*, dermis and dermal papilla.

with a well-developed pelage) they are flattened. In a thin epidermis, the basal layer may be patchy rather than solid. The surface of the epidermis facing the dermis has a complex network of ridges and (sometimes very deep) hollows or alveoli that link epidermis and dermis (Hoepke, 1927; Horstmann, 1952, 1957). Epidermal ridges tend to be radial around the ducts of the sweat gland and the hair bursae orifices or, where neither of these are found, in the direction of skin wear. In addition, from the lower surface of the basal layer cells, numerous outgrowths of protoplasm branch into the dermis and augment the cohesion between epidermis and dermis.

In between the epidermis and dermis is a basal membrane (or a basal plate). In guinea pig and man it attains a thickness of 200 to 500 Å (Snell, 1965). This consists of a network of collagen and reticular fibers (Selby, 1955). The reticular fibers may reach the epidermal cells, penetrate the cellular membrane, and fill the space between the cell outgrowths (Dick, 1947). Elastin fibers also strengthen the connection between dermis and epidermis. It was recently demonstrated that the narrow zone of a low electron density, about 300 Å thick, separates the basal plate from a plasmatic membrane of basal cells (Snell, 1965), running parallel to the full length of the basal plane (Odland and Reed, 1967; Stepanyan, 1969). The basal plate is attached to the dermis by many short fibrils, each with one end in the dermis and the other in the basal plate (Matoltsy, 1969). The basal plate is more compact where these fibrils end (Matoltsy, 1969).

The cells of the basal layer divide frequently, and the basal layer is the main producing layer of the epidermis. The basal layer cells contain mitochondria, especially in the perinuclear region. Numerous very small granules, possibly of protein origin, are found around the mitochondria. These are presumed to be the first precursors of keratin. The Golgi apparatus is located distally in the basal layer cells.

The basal cells are connected with each other and with cells in the spinous layer by interdigitating processes in the cell membranes. At the points of contact, small desmosomes 250 Å thick or less are recorded on the basal cells (Stepanyan, 1969), the number being largest on the distal and smallest on the lateral surfaces (Odland and Reed, 1967). The desmosome structure is very different on the plasmatic membrane of the basal cell bordering on the basal plate. These are called the hemidesmosomes (Odland and Reed, 1967).

The number of cell layers in the spinous layer of the epidermis (where there is one) varies. The cells that adhere to the basal layer are usually cubiform, though they may be prismatic. The upper cells are somewhat flattened and their shape approaches the polygonal. The cells of the spinous layer contain round nuclei and cytoplasmic organoids but, on the whole, they differ little from the basal layer cells, though mitochondria gradually disappear from the cells of stratum spinosum outer parts.

It should be noted that mitochondria play a part in many important biological syntheses, including keratinization. There are very pronounced intercellular bridges, desmosomes, and tonofibrils in the spinous layer. The polarity of location of the Golgi apparatus in the outer spinous layer cells disappears and, like the mitochondria, the Golgi apparatus is dispersed throughout the cell. There are fewer ribosomes than in the basal layer.

In the spinous layer, characteristic 90×140 mμ granules appear, ovoid or rod-shaped (Odland and Reed, 1967), whose inner structure consists of parallel membranes 60 to 70 Å thick embedded in a light-colored, homogenous substance, with the inner membranes parallel to the short axis of the ovoid granules (Matoltsy and Parakkal, 1965). Slits have been recorded between the cells of the spinous layer, crossed by epitheliofibrils or tonofibrils that interconnect to form intercellular bridges and a fibrillary tract through a great number of cells. Tonofibrils show up clearly in the intercellular bridges and also in the cytoplasm of the cells, even near the nucleus. In the granular layer tonofibrils are barely visible, and they disappear completely in the corneous layer. Electron microscopic studies have shown that these intercellular bridges are zones of cellular contact in which thickenings of the cytoplasmatic membrane correspond to the desmosomes. In these contact zones, series of solid and elongated granules connecting with fibril granules can be seen—the width of every fibril is approximately 100 Å. In the stratum spinosum many of the epithelial cells divide actively, though earlier researchers believed that mitosis could be observed only in the basal layer of the epidermis.

The mitotic activity of the basal and spinous layer cells varies in different parts of the body surface. Mitotic activity is high in the areas that need the most restoration of squamous epidermis. On the soles of white mice, the mitotic index (the number of mitoses per 100 cells) reaches 25.2 against 0.8 to 9.8 in the other parts of the body, and on the sole of the human foot the index reaches 74.0 (Chekuleyeva, 1969) against 0.0026 to 1.50 in other parts of the skin. The mitotic activity of epidermal cells also depends on several other factors. In particular, it varies with the time of day. In humans the peak is around 10 P.M. and the minimum at 10 A.M. In white mice, white rats, and domestic rabbits the peak is between noon and 2 P.M. There is a relationship between mitotic activity and motility in animals: maximum cell division in the epidermis is observed during rest.

Mitotic activity also varies with age. In man and in white rats the mitotic index goes steadily down from birth to a minimum at sexual maturity, then rises to turn downward again in old age. Experiments have revealed the effect of sex hormones on the mitotic rate in mammalian epidermis. In male white rats, mice, and guinea pigs, the mitotic process becomes less active after castration but returns to normal if androgenic sex hormones are injected. Inverse dependence between daily mitotic activity and the daily cycle of adrenalin production has been shown (Bullough and Lawrence, 1964).

The granular layer of the epidermis is not present in all species and, where present, it is not in all body areas. Well developed, it consists of four to five layers of polygonal cells with round nuclei and endoplasmic keratohyalin inclusions. The number and size of keratohyalin granules increases toward the surface of the granular layer. In a thin epidermis, the granular layer is not always continuous and it may be completely absent. A number of works deal with the process concurrent with changes in the morphology of the whole cell (Odland, 1964; Roth and Clark, 1964). In particular, the size of the cell increases, the tonofibril mechanism becomes less visible, chondriosomes and Golgi apparatus occur less frequently, and mitochondria are rare. Desmosomes appear to be essential to keratinization (Wilgram et al., 1964). The keratohyalin granules are fibrillar in structure. They contain calcium and are surrounded by a basophilous membrane. Keratohyalin has no proteins that contain sulfhydryl groups. This prevents keratohyalin from being regarded as a precursor of keratin. When keratohyalin granules appear, the epidermal cells are not capable of mitotic division and have lost all of their organelles (Snell, 1965; Odland and Reed, 1967).

In the lucid layer, which is pronounced only in thick epidermis, densely packed cells adhere to one another so that the borders between them become unclear and there is degeneration of nuclei. The keratohyalin granules in the cells merge and are transformed into eleidin.

The corneous layer of the epidermis is made up of flattened, squamous, polygonal (usually hexagonal), fully cornified dead cells arranged in peculiar columns, with the cells of the adjoining columns interdigitating (Christophers, 1971; Mackenzie, 1972). They lack nuclei and the main cytoplasmic organoids. The cells in the stratum corneum are arranged in columns (Christophers, 1971). The pattern of the peculiar outer surface of the corneous layer (Wolf, 1940) is affected by the lower epidermal layers and the dermis. There is continuous peeling of outer cornified squamous cells, singly or in clusters.

Epidermal cell proliferation and the peeling of corneous squamae are balanced by hormones (Bullough, 1962; Rothman, 1964). The renewal of the epidermis on the human palm takes 32 to 36 days; on the anterior surface of the forearm, 100 days; on the knee, 20 days; and on the elbow, 10 days (Fular, 1955). The epidermis on the sole of the guinea pig's foot is renewed in 6 to 7 weeks. The production of the corneous matter differs greatly in various parts of the body (Kligman, 1964).

The variability in water resistance of the corneous layer in different body areas is of interest (Kligman, 1964). For instance, the belly skin is one-hundred times as efficient at preventing water loss as an equal thickness of corneous layer on the foot sole (Blank, 1953). Attempts to single out a barrier layer in the stratum corneum (stratum compactum) (Szakall, 1955; Stupel and Szakall, 1957) appear to be erroneous (Kligman, 1964).

The surface of mammalian skin is covered with a thin film, half water and half fats (Rothman, 1954), which seems homogenous because it forms a fine emulsion. This surface film, which protects the skin from water losses, sharp temperature changes, and exogenous infection has a species and sex specific odor (Schaffer, 1930). Fats, peculiar fatty acids, higher alcohols, and carbohydrates are components of the superficial film (Rothman, 1954). Glycerides are rather poor here—they contain only traces of nitrogen and phosphorus. The fat mixture is hydrophilous.

In recent histochemical investigations of the epidermis, glucide tests have yielded positive results. Glycogen has been found in the upper spinous layer, mainly at the point of contact between hair follicles and sudoriferous gland ducts. No histochemically distinguishable glycogen has been found in the ectad spinous layer, and the relatively small amount of glycogen in the normal epidermis varies during embryonic development (decreasing sharply in the early stages of keratinization) and also with the time of day, parallel to changes in mitotic activity (Bullough and Eira, 1949). A different glucide has been found in the epidermis, which may be a glycoprotein conducive to the adhesion of epidermal cells. Mucopolysaccharides are present in the epidermis, their location indicating that they may be involved in keratinization (Hale, 1946; Meyer, 1947; Meyer and Rapport, 1951; Pearce and Watson, 1949; Wislocki et al., 1947 and others). RNA has been isolated from basal and spinous layers in large amounts (Brachet, 1942; Montagna and Noback, 1947; Sandritter, 1953).

Lipids are present in the epidermis of most mammals, and small, sudanophilous inclusions are noted in the basal, spinous, and granular layers. There are fewer lipids in the lucid and corneous layers. Fats may get into the latter through sebaceous and sudoriferous secretions. However, the corneous layer of the epidermis does retain sudanophily in animals lacking these glands, such as dolphins, or in instances where the sebaceous glands have been experimentally destroyed. The amount of lipids in the epidermis varies in different species and on different parts of the body. In fact, the level of lipids is rather higher in cetaceans than in other mammals, and in man it is highest in the armpits, the cochlea and other hairy areas. The stratum malpighii of the epidermis contains a considerable amount of ascorbic acid, while only traces are found in the corneous layer (Giroud et al., 1935a, b). Cytochrome oxidase is contained in the epidermis (Rogers, 1953; Steigleder, 1957). The basal and spinous layers have the highest enzyme activity. This drops off in the upper spinous layer and in the granular layer when the reaction is negative. With white mice, the maximum cytochrome oxidase corresponds to the peak of mitotic activity in the epidermis (Carruthers et al., 1959).

Succinedehydrogenase, nitrogen anhydrase, and monoamine oxidase are most active in the basal layer. This activity decreases in the spinous layer, decreases further (ceasing in some cases) in the granular layer, and sometimes disappears in the lucid and corneous layers. The degree of succinedehydrogenase change varies with the stages of hair growth. For example, with mice the epidermal enzyme activity is low in any area where the hair follicles are resting (Argyris, 1965), while succinedehydrogenase activity reaches a high level in the early stages of growth. When the hair follicles are fully differentiated and growing fast, succinedehydrogenase activity drops again.

Histochemical analysis has shown phosphorylase and transglucosidase activity in the spinous layer cells. This activity is negligible in the basal layer and absent in the other layers (Braun-Falco, 1956a; Takeuchi, 1958; Ellis and Montagna, 1958). There are more of these epidermal enzymes in the embryo epidermis than in the adult; the differences between species are considerable and the amount is always high in primates (Montagna and Ellis, 1959).

Opinions differ on aminopeptidase content. Some workers have found none in the epidermis (Burstone, 1958; Burstone and Folk, 1956). Others note low activity in the malpighian layer (Montagna, 1962). None have found basic phosphomonoesterase in the epidermis of healthy mammals—it appears only with irritation or skin damage. Acid phosphomonoesterase is found in considerable quantities in the spinous and granular layers, but to a much lesser extent in the basal layer (Moretti and Mescon, 1956a, b). In the corneous layer, the reaction is still positive. The cells of the granular layer of the rat show high phosphoamidase activity (Meyer and Weinmann, 1955, 1957). Beta-glucuronidase reaction has been proven positive in all the live layers of the epidermis, with the peak being recorded in the granular layer (Braun-Falco, 1956b; Montagna, 1957). In the cornified layers, the reaction is negative.

Carboxylesterase is frequently found in mammalian epidermis. Twin esterases are recorded in the stratum malpighii, but the reaction level is different in different parts of the body. Acetylnaphthylesterase is mainly localized in the keratogenic zone between the granular and corneous layers, but there is none in the human epidermis even though, like most mammalian epidermis, it contains cholinesterase. The distribution of enzymes in the epidermis varies qualitatively and quantitatively in different areas of the body (Ellis, 1964; Hershey, 1964).

The study of keratinization histochemistry (Spearman, 1966) has been much emphasized. It is known that the presence of large amounts of cystine, a sulphur-carrying amino acid with a disulfide bridge, is one of the basic chemical features of keratin. Cystine, an amino acid that contains a sulfhydryl group present in many proteins, is capable of forming a cysteine molecule of two of its own molecules, with the two disulfide groups yielding a disulfide bridge. This accounts for the interest in finding the sulfhydryl groups in the epidermis. Studies have demonstrated that the stratum basale in the mammalian epidermis contains a certain amount of proteins with free sulfhydryl groups (Giroud and Bulliard, 1930) and that this increases in the stratum spinosum to reach a maximum in the stratum granulosum. They are completely absent around the transition from the lucid to the corneous layer. The main feature of keratinization seems to be the oxidation of the sulfhydryl groups of cystine into the disulfide bridges characteristic of cysteine. This phenomenon begins in the outer parts of the spinous layer, reaches its peak in the granular layer, and is completed in the stratum corneum.

Recent research has, in principle, confirmed these data. Free sulfhydryl groups in the proteins of human and guinea pig skin have been found in the basal and spinous layers, but have not been recorded in the stratum corneum

(Chevremont and Frederic, 1943). They are also contained in the cytoplasm of the granular layer cells, where the keratohyalin granules have failed to react to them. On the border of the stratum corneum, the reaction level is high, while in the cornified cells themselves it is moderate. This is the main difference between Giroud and Bulliard's results (1930) and the results of present-day scientists. Recent research has confirmed a small cysteine content in the peeling squamae of the corneous epidermal layer (Rudall, 1946; Gustavson, 1956). The transformation of sulfhydryl groups into a disulfide bridge can be incomplete in the course of normal keratinization.

Thus, histochemical investigations of the epidermis show that basal and spinous layers contain not only the ribonucleic acid essential in the formation of intracellular proteins but the ascorbic acid involved in cell oxidation. Proteins with sulfhydryl groups occur in these layers, and their transformation into a disulfide bridge is one of the main reactions of keratinization. The high enzyme activity observed in the keratinous zone of the epidermis confirms the wide variety of chemical transformations in epidermal keratinization (Rothman, 1954, 1964).

Electron microscopy shows that keratohyalin granules are formed in the stratum spinosum. They develop in the cytoplasm as submicroscopic particles. Their number and size increase in the stratum granulosum (Parakkal and Alexander, 1972). They stop growing when the cells are fully developed. Keratohyalin granules are polymorphous and shapeless in humans but are spherical in rodents. The mechanism responsible for their formation and growth is unknown. Their growth may be caused by the accumulation of finely granulated material in the cytoplasm (Brody, 1959). The granules contain chromatin, RNA, mucopolysaccharides, lipids, and histidine-rich protein. Keratohyalin granules disappear as corneous cells are formed. The function of the granules is unknown.

A characteristic feature of keratinization is the thickened plasma membrane (Parakkal and Alexander, 1972). In the stratum malpighii it is only about 30 Å thick, but in the corneous layer cells it ranges from 150 to 200 Å. The corneous layer cells contain 80 Å filaments. In the final stage dehydration is essential to keratinization. As this occurs at some distance from the surface, and as common cells are found near the dehydrated cells, there is reason to believe that this is not only a result of water loss.

Melanin, melanoid, carotin, reduced hemoglobin, and oxyhemoglobin all play a part in coloring mammalian skin (Edwards and Duntley, 1939). The distribution of these pigments varies in different species and in different parts of the body. Of these, melanin, plentiful in the epidermis, is the most interesting. Melanin granules occur in the cells themselves or in specialized cells with arborizations. These are called melanoblasts, melanocytes, melanodendrocytes, or dendrite melanophores. To avoid confusion, we will use Fitzpatrick's and Lerner's terms (1953): (a) melanoblasts—cells that origi-

nate from the ridge and develop into cells forming melanin; (b) melanocytes
—differentiated melanin-forming cells that release granules of melanin
through their dendrites; (c) melanophores—pigment cells of lower vertebrates. The melanin is distributed in the melanocytes through the whole
cytoplasm. It may also be concentrated around the nucleus or in the outgrowths. Melanocytes themselves occur in the basal and, less frequently, in
the spinous layer. The number of melanocytes in different parts of the epidermal surface varies considerably from abundant to absent. For example,
there are 893 melanocytes per 1 mm² of guinea pig's ear epidermis and 406
on the entire body epidermis (Billingham and Medawar, 1953).

Skin color is less dependent on the density of melanocytes than on the
amount of melanin they produce (Staricco and Pinkus, 1957). The melanocyte dendrites may touch more than one of the surrounding common epidermal cells. The number of melanocytes in any area is related to the shape
and the length of the dendrites (Szabo, 1954; Staricco and Pinkus, 1957). At
around 1000 per 1 mm² the melanocytes become smaller and the dendrites
shorter. When the number is under 1000 per 1 mm² the melanocytes become
larger and angular, with long dendrites. Melanin synthesis occurs in melanocytes, not in the common epidermal cells. Melanoblast histogenesis is very
peculiar; it originates in the nerve ridge in common origin with spinal gan-
glia, some sensory nerves, and suprarenal glands (Harrison, 1910, 1938;
Holfreter, 1933; Rawles, 1953). It has been demonstrated on amphibian
embryos that the nerve plate itself can form melanoblasts (Niu, 1947;
Twitty, 1949). Migration of melanoblasts into the skin has been shown in
embryo mice (Reed and Henderson, 1940), embryo guinea pigs (Medawar,
1950; Meng, 1955), and human embryos (Chlopin, 1932; Zimmermann and
Cornblet, 1948, 1950).

In addition to melanocytes, mammalian epidermis contains other neuro-
genic elements, such as Langerhans cells. These are distinguished from
melanoblasts by the absence of melanin and certain subtle morphological
peculiarities (for instance, the outline of Langerhans cells is more angular
and there is no mitosis). The function of the Langerhans cells has not been
established.

As has been noted, melanin is formed only in melanocytes (Masson,
1926). However, as melanin occurs both in common epidermal cells and
in macrophages, one must assume not only the phagocytosis of melanin
granules but the migration of melanin granules from melanocytes to common epidermal cells (so-called cytochrony). Melanocyte processes are arranged over the epidermal cells (in the basal and inner spinous layer), and
melanin granules are run through into the epidermal cells where they are
dispersed or formed into a cap over the nucleus (Masson, 1926; Odland and
Reed, 1967). Some histologists suggest that there are no melanin granules in
common epidermal cells, accounting for what they see under the microscope by a close superimposition of melanin processes on epidermal cells

(Becker et al., 1952). A light microscopy study of melanogenesis has failed to confirm this view.

Electron microscopy produces some interesting data. Unlike common epidermal cells in the basal and spinosum strata, melanocytes have no fibrillar structures (Birbeck et al., 1956). Their dendrite processes are normally located in spaces between cells, and melanogenesis takes place in the cytoplasmic zone near the nucleus. There are numerous small vacuoles surrounded by a dense and sometimes plated ridge. From this first stage on, the thickness and density of the cortical layer increases, developing into melanin granules consisting of a row of plates superimposed on one another. The gaps between them are filled with small and very dense granules. This structure confirms the hypothesis that the melanin granule consists of a protein matrix, which is a plate with indol-5-6-quinone corresponding to the small granules. It is noteworthy that the melanocytes devoid of melanin found in some human albinos contain the same plate-structured grains as in normal people, though the dense granules are missing. The electron microscope reveals a large number of chondriomes clustered around the melanogenesis zone. Their role in this process is not yet known. There is a theory that melanin granules are produced by the Golgi apparatus.

DERMIS

The dermis is directly beneath the epidermis and is separated from it by a basal membrane. The dermis has two indistinct layers: the papillary which is right under the epidermis, and the reticular, which is deeper. The boundary between these layers is level with the hair bulbs or the bottom of the sudoriferous glands. The thickness of both layers of the dermis varies considerably in different mammals and in different body areas.

The dermis consists of connective tissue and contains collagen, reticulin and elastin fibers, all types of connective tissue cells, and nerves and blood vessels. Bundles of smooth muscle fibers, either attached to hairs or free (in the dermis of pinniped skin), are found in the human dermis around the nipples, anus, and in the genital skin. The dermis may also contain striated muscles that branch off from the subcutaneous muscles. The entire space between fibers and dermal cells is filled by the main substance of the dermis, a semiliquid, amorphous, nonfibrillar mass. This mass is greater in the papillary than in the reticular layer. The main substance of the dermis contains acid mucopolysaccharides, neutral heteropolysaccharides, proteins, soluble collagens, enzymes, immune bodies, and metabolites (Meyer and Rapport, 1951; Stearns, 1940; Schiller and Dorfman, 1957; Schiller, 1959; Gersch and Catchpole, 1960). It may also contain glycoproteins (Gersch and Catchpole, 1949). The total amount of the main substance and the ratio of constituents may vary with time and body area (Bunting, 1950; Duran-Reynolds et al.,

1950). The main substance of the dermis is multifunctional. In particular, the mucopolysaccharides may be involved in formation of collagen fibers during the healing of skin wounds (Bunting, 1950; Meyer, 1945; Sylvén, 1940).

There are many papers on collagen fibers (Bear, 1952; Gustavson, 1956; Schmitt, 1959; Eliseyev, 1961). Colorless fibers from 1 to 15μ thick and characteristically in thin, longitudinal bands, they occur in all animal tissue. Dark-field examination has shown that every collagen fiber is made up of a bundle of long, smooth, parallel fibrils. The fibrils are connected in a cement that forms a membrane at the surface of each collagen fiber. The fibrils probably contain glycoproteins like the main substance of the dermis (Gersch and Catchpole, 1949). Fibrillar thickness varies from 700 to 1400 Å (usually about 1000 Å) (Gross and Schmitt, 1948). The thickness of the fibrils increases with age. The fibrils, arranged in crosswise bands, are comprised of still finer filaments or protofibrils parallel to the long axis of the fibrils. The thickness of collagen protofibrils is comparable to the dimensions of separate polypeptide molecules (Bear, 1952).

Under the electron microscope collagenic fibers are seen to be surrounded by an amorphous substance (diminishing in quantity with age) made up of acid polysaccharides and protein (Gross, 1950a, b). The proteins of the collagen and unique in that they contain the amino acids hydroxyproline and hydroxylysine. About one-third of the amino acid is glycine; another third is proline and hydroxyproline. X-ray analysis reveals that the collagen macromolecule contains three polypeptide chains twisted around each other like a helix. The collagen molecule also contains cyclic structures of the diketopiperazine type (Mikhailov, 1949).

According to Grant and Prockop (1972) collagen bundles seen with the naked eye are called "fibers." A light microscopic examination reveals that these are in turn made up of numerous "fibrils," 10 to 15μ in diameter. As is seen under an electron microscope, the fibrils are composed of small "microfibrils" 300 to 600 Å. These microfibrils are formed by "filaments," 30 to 50 Å in diameter (Bouteille and Pease, 1971). A collagen molecule is 14μ in diameter, and there are 3 to 5 molecules in a filament.

Embryonic and children's skins contain numerous argyrophilous reticulin fibers. Electron microscopic investigations have failed to reveal marked morphological differences between reticulin and collagen fibers, though reticulin fibers are only one-half to one-third the thickness of collagen fibers (Gross and Schmitt, 1948; Gross, 1950a). With age there is a gradual increase in the number of thicker fibers. It is hard to say how collagen fibers and argyrophilous fibers are actually formed. However, collagen fibers are formed by joining tiny fibrils.

Kölliker (1868) assumed that the amorphous substance and the collagen fibers are influenced by connective tissue cells, and that the substance may be formed in the cells and be secreted by them. The inhibition of fibroblast growth in vitro arrests the growth of collagen (Hass and McDonald, 1940). Observations of a rabbit ear through the camera lucida have shown that when a wound is healing, a large number of granules appear in the fibro- blasts, or their surface may be covered by vesicles (Stearns, 1940a, b) prior to the formation of collagen fibers. The secretion of these vesicles may be involved in fiber formation. It has been shown that collagen fibrils are formed close to the surface of the fibroblasts (Porter and Pappas, 1959). Individual fibroblasts are formed or polymerized from material at the cell surface which is then excreted into the intercellular space. Once formed, individual fibrils increase in diameter through an accretion of material from the main substance.

Reticular fibers form a network through the whole connective tissue. They constitute about 0.38 percent of the dry weight of the skin. A dense network of these fibers is recorded in the upper part of the papillary layer, under the epidermis, or as the main membrane or part of it. Reticulin fibers are abundant around the sudoriferous glands and in the connective tissue of hair-follicle sheaths. In the reticular layer, reticulin fibers are plentiful but only around the blood vessels. In the subcutaneous fat tissue, reticulin fibers are well developed near blood vessels, forming basketlike capsules around every fat cell. Fibers vary in thickness, and the thickest appear to be bundles of thin fibrils. Reticulin fibers prove argyrophilous and Schick- positive.

Paper chromatography of reticulin fibers shows a great concentration of galactose, glucose, and mannose, and a small amount of fructose. Collagen also contains these carbohydrates but in considerably smaller quantities. Reticulin fibers are evidently "precollagen" rather than a different type of fiber. The physical and chemical properties of reticulin and collagen are similar, though Bowes and Kenton (1949) and Orlovskaya and Zaides (1952) believe that collagen and reticulin differ both physically and chemi- cally. Reticulin fibers often grow into collagen fibers. The embryo dermis contains only reticulin (Yuditskaya, 1949). Almost all the fibers in the der- mis of a two-day-old rat are argyrophilous, but as the rat grows they gradu- ally give way to collagen fibers (Gross, 1950a). When a wound begins to heal, the first fibers to form are argyrophilous, like reticulin fibers. They then give way to or grow into collagen fibers. Under the electron micro- scope, reticulin fibrils are very similar to collagen fibrils (Gross, 1950a), in- dicating that reticulin is a precursor of collagen and that the tissue contain- ing it is primitive.

Elastin fibers in the dermis appear in coarse, branching, prismatic, or flattened bands between collagen fibers. No fibrils are seen when they are stained. The diameters range from so small that they are barely visible under a light microscope to a few microns. The protein elastin is isolated from the elastin elements. It has a high glycine, proline, and valine content and a con- siderably smaller arginine content. About 2 percent of dry weight of the human skin is elastin. X-ray analysis of elastin has revealed a random arrangement of molecules, as in resin, which determines the elastic properties of the fibers. Elastin fibers, which are easily stained by the agents used to reveal the groups -S-S-, are not affected by alcohol, diluted acid, alkaline, or salt solutions. They can be decomposed by trypsin and, to a lesser extent, by pepsin. After treatment with acids and alkali, elastin fibers swell slightly in length and thickness. Elastase, an enzyme from the pancreas, appears specifically to affect elastin (Balo and Banga, 1949; Lansing et al., 1952).

Studies of elastin fibers under the electron microscope have yielded contradictory results. In contrast to collagen bundles, which are fibrillar in structure, elastin bundles are homogeneous (Montagna and Parakkal, 1974). These are formed by the inner, amorphous "medulla," which is elastin, and by the outer, cortical "cortex," comprised of nonelastin, proteinaceous microfibrils, 110 Å in diameter. When elastin fibers are partly decomposed by trypsin, intertwining fibers about 120 Å in diameter are revealed (Gross, 1949). When elastin fibers are heated in elastase, it is shown that twisted elastin fibrils are housed in a homogenous matrix with physical and chemical properties almost identical to those of the fibrils (Lansing et al., 1952). Opinions differ on the ontogenesis of elastin fibers. Some workers support intracellular origin; others believe that elastin fibers grow from the main substance.

Fibroblasts are widely distributed in the connective tissue. On the surface they look rather like 20μ amoebas with pseudopods of various sizes proing from the main cell. They have a spindle-shaped profile, but the shape of fibroblasts depends on their location. In the reticular layer they are larger and rather like mesenchymal cells. The large, oval nucleus of the fibroblast has very small chromatin particles in the meshes of a fine nuclear reticulum. It contains one of several large nucleoli. The cytoplasm of a live fibroblast appears amorphous under ordinary light, but phase-contrast and dark-field microscopy show innumerable mitochondria, granules, and vacuoles of various sizes. Thread- and rod-shaped mitochondria surround the nucleus almost completely, although only a few occur in the cytoplasmic processes. Almost all of the dermal fibroblasts have fine clusters of lipid sudanophilic granules in one or both cell poles corresponding to the location of the Golgi apparatus. All the dermal fibroblasts contain cytochrome oxidase, succindehydrogenase, and monoamine oxidase, beta-glucuronidase, and esterase. The functions of the fibroblast are not completely known, though there is indirect evidence that the formation of intracellular substance and of fibrillar structures of the connective tissue is connected with them.

Macrophages are similar to fibroblasts, but they are star- or spindle-shaped with a granular and commonly vacuolized cytoplasm and a nucleolus smaller and more compact than in fibroblasts. It is very difficult to identify macrophages in normal skin, but certain agents can be injected into the skin, making them easily identifiable due to their ability to incorporate such particles. They digest organic particles intracellularly, and retain inorganic particles within the cell or pass them on to the others. Two main types of macrophages, small and large, occur in the dermis. The large cells phagocytize both large and small particles, but the small cells manage only small particles. If a particle is too large for one cell to handle, several macrophages may get together around it and merge into a multicellular, syncytial formation. Phagocytic cells of normal, loose connective tissue are called stable macrophages, and the cells appearing where there is inflammation are called free macrophages. Macrophages originate from histogenic and hematogenic particles. The histogenic particles consist of connective and hemopoietic tissues, which are not too dissimilar. The hematogenic particles consist of monocytes and lymphocytes.

All connective tissue cells containing cytoplasmic granules that are stained metachromatically by agents of a thiazine group are termed mast cells. Thus, many cells that are not distinct from fibroblasts in size and appearance, and that contain varying quantities of metachromatic granules, are called mast cells. Typical mast cells are large and egg-shaped with one or two small nuclei, and their cytoplasm is filled with coarse basophilic granules stained metachromatically. The intergranular cytoplasm is commonly homogenous, although it may have vacuoles.

Mast cells very a great deal in different species (Riley, 1959). They occur in all parts of the dermis and subcutaneous fat tissue, but they are most numerous around the walls of small blood vessels. The number of fat cells increases in some skin diseases. Histochemical methods reveal that some of the granules and the intergranular cytoplasm of fat cells contain RNA. Mitochondria can be seen in mast cells, though only with great difficulty.

A very small Golgi apparatus has been found, but only in the mast cells of the hamster (Compton, 1952). Mast cells in the dermis of very young mice, under the electron microscope, show that metachromatic granules are electron-dense bodies appearing granular and reticulate (Rogers, 1956). Every granule has a solid periphery encircled by a cytoplasmic space, and mitochondria with a typical inner structure are scattered between the granules. A Golgi apparatus is present in the cytoplasm, but the endoplasmic reticulum is practically absent. The mast cells of hamster dermis reveal a structure in the intergranular cytoplasm identified as endoplasmic reticulum (Smith and Lewis, 1957). Histochemical tests of mast cells have yielded varying results for three reasons: (1) the mast cells in different tissues of the same animal are not identical; (2) their properties vary in different species; and (3) different methods have been used (Montagna, 1962).

Many factors suggest that mast cells are a source of heparin or some similar anticoagulant (Holmgren and Wilander, 1937; Jorpes et al., 1937). Mast cells also contain and produce histamine. It is thought possible that mast cells have both homo- and heteroplastic origins. Some workers have described mitotic and amitotic division of mast cells (Maximov, 1906; Dantschakowa, 1910; Michels, 1938; Lehner, 1924).

SUBCUTANEOUS FAT TISSUE

Skin is connected with the tissues beneath by its deepest layer, the subcutaneous fat tissue. This is a loose connective tissue in which fat cells frequently accumulate. Large groups of fat cells are separated by connective tissue layers formed by bundles of collagen, elastin fibers, and connective tissue cells (mainly fibrocytes). Blood vessels and nerves may be arranged in these layers. Fat cells are polygonal and have a marked membrane. Almost the whole cell is filled with fat, with a relatively small nucleus and cytoplasm at the periphery. Fat cells usually accumulate in the subcutaneous fat tissue of the obesity centers and merge into a continuous layer in fat animals only.

BLOOD SUPPLY

The network of blood vessels varies considerably with skin topography, so it is difficult to give a general picture of the skin's blood supply. Arteries are more abundant in the areas where the skin gets the most wear (Spalteholz, 1893). There are two types of skin arteries: skin arteries proper that branch only in the skin, and mixed-type skin arteries that branch off from arteries in the underlying tissues (Spalteholz, 1893). The branching of these vessels in man has been examined in great detail (Spalteholz, 1927; Satyukova et al., 1964). Deep in the skin the arteries form numerous anastomoses comprising the arterial network. Smaller vessels rise up from this and branch out longitudinally to form a subpapillary arterial network. Finally, there is a papillary (or subepidermal) arterial network in the papillary layer of the dermis. The vessels vary considerably in shape. Some arterioles appear to pass through the dermis and branch off in a tree shape just below the epidermis. There are usually anastomotic blood vessels along the hair follicles. The straight parts of the eccrine gland ducts are accompanied by interconnected arterioles and capillaries linked with the subepidermal arterial network in the subpapillary layer.

Large and small venules connect exactly like the network of large and small arterioles (Zweifach, 1959), though the venous network in the subcutaneous fat tissue is more pronounced than the related arterial network. Numerous anastomoses are arranged between the venous and arterial networks through which the blood may pass from arteries to veins, bypassing the capillaries. The venous capillary branches that project into the subpapillary venous network are connected with the second subpapillary venous network under the arterial plexus. The third venous plexus is in the middle of the dermis (Spalteholz, 1927; Comel, 1933).

It should be remembered that this pattern of blood vessels can be found only by the methods used to study the skin's blood supply system. It is formed by a continuous reticulum of vessels of varying diameter (Saunders et al., 1957; Saunders, 1961).

Two main, interconnected vascular structures can be distinguished in the skin. One includes all the vessels of the dermis (Moretti et al., 1959), and the other includes the vessels around the hair follicles. The former supplies all the parts of the skin with blood. The parafollicular reticulum is connected with the dermal blood vessel network through "vessel units" (Moretti et al., 1959). These networks surround single hair follicles or groups of follicles, and their diameter and depth vary with the size and position of the follicles. Adjoining vessel units are connected by branches, the superficial branches linking them with the vessels in the papillary layer. They are also connected with the networks that surround the sudoriferous glands (Moretti and Montagna, 1959).

The skin's lymphatic system is as well developed as its blood supply system, but the literature on this subject is scanty.

SKIN INNERVATION

The pattern of nerves in the skin is basically laid out as an inner network in the subcutaneous fat tissue. Twisted fibers branch out to the papillary layer of the dermis, where they form an outer network less complex than the inner one. Many of the fibers run straight from the inner network to the inner ridges of the epidermis. Directly under the epidermis, a third, subpapillary network has large, irregular alveoli. Nerves are not sparse in any zone between the inner and outer networks (Tamponi, 1940). The reticular layer of the dermis, once thought to be very poorly supplied with nerves, has been shown to contain an intricate neural network that becomes denser as it approaches the epidermis. The skin has an abundance and diversity of sensory nerve endings, and recent studies have tended to underline their similarity (Winkelmann, 1960). There is no substantial histological evidence to support the theory that pain, touch, cold, and heat are perceived through four different receptors (Weddel et al., 1955).

Most specialists on mammalian skin distinguish between free nerve endings, encapsulated endings, endings approaching the hairs, and free skin nerve endings subdivided into intraepidermal, intradermal, and perivascular, depending on their location. Free intraepidermal endings, formed by nonmyelinated fibers, are peculiar disks at the bases of special cells, which are possibly modified Schwann cells. Some authors believe that the disk may be a kind of artifact (Weddel et al., 1955). Free intraepidermal terminals are also formed by nonmyelinated fibers, fibers with free endings not in permanent contact with any part of the dermis. The arrangement of these terminals in the dermis varies. Perivascular nerve endings occur by the blood vessels, with more around the large vessels and fewer near the capillaries. The largest of these nerve fibers near the skin vessels retain their myelin sheath. In the adventitia and the muscular membrane of the blood vessels are large numbers of terminal arborizations of neural fibers, and they are always free and nonanastomotic.

There is no comprehensive, generally accepted classification of encapsulated nerve endings. The difficulty lies in the diversity and the multitude of intermediate forms. The most common in the mammal are Vater-Pacini bodies and Meissner's corpuscles.

SKIN GLANDS

A great deal of work has been done on the skin glands of man and of laboratory animals. Wild mammals have received much less attention, and the few works in this area are far from comprehensive (Brinkmann, 1912; Eggeling, 1931; Schaffer, 1924, 1930, 1940). There have been numerous attempts to classify skin glands, but none is fully satisfactory. The earliest were based on the differences between tubular and alveolar or lobate glands, morphological peculiarities that made it possible to distinguish between sebaceous and sweat glands. Later, the condition of sweat glands at the point of secretion was suggested as a further classification criterion (Ranvier, 1887).

Holocrine glands have a single secretion cycle resulting in irreversible changes. The end of secretion coincides with the death of the gland. Merocrine cells have several full secretion cycles—the merocrine secretion cell does not die when it secretes into the lumen of the gland.

An attempt has been made to classify mammalian skin glands by the presence or absence of myoepithelial cells (Brinkmann, 1912) based on the fact that tubular skin glands have myoepithelial cells and alveolar glands do not. However, the presence of myoepithelial cells can be very difficult to show (Eggeling, 1931). Schieffedeker (1917) returns to the Ranvier classification, adding a third type of secretion—apocrine—connected with the cell's losing the apical pole at the moment of secretion. Apocrine secretion is subdivided into macro- and microapocrine secretion (Kurosumi, 1961; Shubnikova, 1967). Macroapocrine secretion is characterized by part of the cytoplasm tearing off when the secretion exudes, while in microapocrine secretion, the dilated apexes of the microsetae are torn off the cell.

Schaffer (1924, 1927) puts forward a rather complicated classification of skin glands based on the number of cell layers in the gland. He distinguishes between monoptic (one-layer) and polyoptic glands. Monoptic glands can be merocrine or holocrine, and merocrine monoptic glands are subdivided into eccrine and apocrine. Schaffer goes on to divide monoptic eccrine glands into monocrine, dicrine, tricrine, and so on, by the quantity they secrete. The polyoptic glands he divides into holocrine, merocrine (hepatoid), and holomerocrine. In all these glands the secretory epithelium consists of only a few layers of glandular cells. Schaffer's classification involves too many factors—the number of cell layers, the way the cell functions, and the quantity secreted. The complexity of Schaffer's classification and its lack of unity accounts for its unpopularity. Few authors even mention it. Ranvier's classification, with Schieffedeker's additions, appears to be the most acceptable.

SEBACEOUS GLANDS

Although all sebaceous glands are similar in basic structure, they vary in shape and size in different parts of the body, and they differ considerably between species. In man, the largest and most numerous glands are in the skin of the head, forehead, cheeks, and chin—400 to 900 glands per 1 cm^2 as compared with less than 100 per 1 cm^2 in some parts of the body (Benfenati and Brillanti, 1939) and none at all on palms and soles. In other mammals, sebaceous glands are more equally distributed. The largest number occur in the external acoustic meatus, the perianal region, and one or two other places. Some mammals (e.g., cetaceans) have no sebaceous glands. Most of the sebaceous glands are appendages of the hair follicles and open into the hair-sebaceous gland canals (fig. 3). The size of the glands often varies inversely with the size of the hair follicles, with the largest glands occurring in very small hair follicles. Some sebaceous glands open on the skin surface. In man such glands are found on the eyelids (Meibomian glands), the mucous membrane of the cheeks, and the pigmented surface of the lips, the nipples, and the preputium (Tyson's glands). Other mammals have gigantic sebaceous glands which open on the skin surface in the anal and circumanal regions. Sebaceous glands are holocrine and usually multilobal. Their shape often depends on their position on the body and on the character of the surrounding dermis (Clara, 1929).

Regardless of size, shape, and position, all have a similar cell morphology. The lobes of every gland merge into a common excretory duct formed by squamous epithelium which runs onto the wall of the hair canal or the surface of the epidermis. The lobes of the sebaceous gland are surrounded by a rather thin basal membrane covered by a solid dermal connective tissue that forms a capsule. The peripheral layers of cells at the basal membrane are nondifferentiated and flattened, resembling cells in the epidermal germinal layer. The presence or absence of the central canal depends on the stage of the secretion cycle, for it is formed by the destruction of central cells. The adult gland cells and ducts contain some special secretion—a substance that consists of lipids and disintegrated cell pieces.

Different cells in the same gland are in different stages of development. In

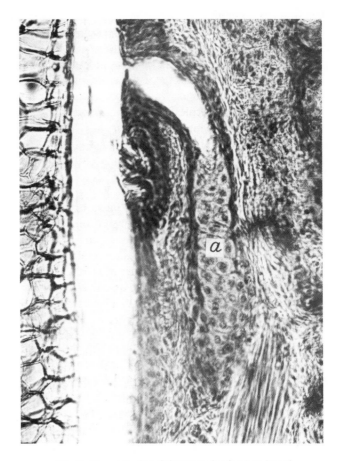

Fig. 3. *Cervus elaphus.* Sebaceous glands (*a*) on barrel.

some glands only the central cells are filled with fat droplets, in others the
peripheral cells are also filled with fat. The development of sebaceous
glands is closely connected with the development of hair follicles. Their rate
of development varies, depending on their position. In a human embryo of
three-and-a-half months, the glands on the face and scalp have already
become differentiated, while in other parts of the skin they appear rather
later. Two swellings on one side of the hair follicle develop before the full
differentiation of the follicle itself. The lower is the anlage of the swelling of
the hair follicle, to which the arrectores pilorum muscle will be attached.
The upper is the anlage of the sebaceous gland. The cells in the center of the
upper swelling lose glycogen and swell as they accumulate fat drops. Finally,
the largest cells in the center disintegrate. A similar cell differentiation and
subsequent disintegration occurs in the cells of the center of the continuous
strand above the sebaceous gland level. This column of the sebaceous glands

continues through the epidermis, forming the sebaceous canal that opens on the skin surface. In man, the sebaceous glands remain relatively large and continue to function from the moment of differentiation. They are even large in newborn babies, though they become much smaller soon after birth and remain small during childhood and early puberty, developing fully only with sexual maturity. Sebaceous glands are larger in males than in females.

In rats and mice some hair follicles and related sebaceous glands are developed in the first three or four days after birth. The glands first appear as large cellular swellings immediately above the swelling to which the arrectores pilorum muscle is attached. But differentiation of sebaceous glands does not occur until the inner sheath of the hair follicle is formed. Mitochondria are numerous in the fat cells of the sebaceous glands. At the early stages of sebaceous gland differentiation, the mitochondria are grouped in the perinuclear cytoplasm around the fat granules, which develop near but not directly from the mitochondria. Every fat drop is surrounded by mitochondria. With gradual fatty degeneration the size of the cells increases while the number of mitochondria is unchanged. In mature sebaceous gland cells, mitochondria form a dense network in the cytoplasm between the fat drops. They remain unchanged in number and appearance until the fat cells reach the late stages of maturity, after which they cease to be distinguishable. The mitochondria are not transformed into fat, though their role may be decisive in the synthesis of sebaceous gland lipids. This has been confirmed by electron microscopic studies of sebaceous glands (Palay, 1957, on rats' Meibomian glands; Charles, 1960, on man's sebaceous glands).

The Golgi apparatus is an important structure involved in the synthesis of fat drops in the differentiation of sebaceous gland cells. As the sebaceous gland cells mature, the Golgi apparatus virtually disappears.

Fat drops examined under the electron microscope at early stages of development look like a sparse network of fibers (Palay, 1957). As the cell grows older, the cytoplasm shrinks back to the spaces between the drops. The drops become larger and more numerous until they are separated only by a thin cytoplasmic plate containing mitochondria, fine granules, and tubules.

The chemical composition of sebaceous gland secretions differs considerably from the composition of fat tissue. Some substances are only synthesized by sebaceous glands (Lederer and Mercier, 1948). Human sebaceous glands are intensively sudanophilic. The duct cells, undifferentiated cells at the periphery of the sebaceous gland lobes and the cells between, have numerous perinuclear sudanophilic bodies. The accumulation of fat in the young lobes of sebaceous glands begins in the cells at the center and is subsequently distributed to the periphery (Montagna and Novack, 1946, 1947). As the cells store lipids, they grow bigger and bigger, increasing the volume of the lobes of the sebaceous glands. In cells that are not fully mature, the fat drops are spherical and about equal in size. Just before the disintegration of the cell, the fat drops merge to form globules that vary in size. There are some specific distinctions in the size of fat drops. They are large in the glands of man and very small in mice. The chemical composition of the secretion of the sebaceous glands in the ducts differs from the newly formed secretion and from the fat drops in the cells of the sebaceous glands (Montagna, 1949; Montagna and Hamilton, 1949).

The secretion of the sebaceous glands contains cholesterol esters, some phospholipids, and possibly some triglycerides. In polarized light, the sebaceous glands show a certain amount of anisotropic lipids. Except in human and higher primate skin, the sebaceous glands of the mammals examined contain virtually no glycogen (Montagna and Ellis, 1959a, 1960a, b). Almost all duct cells and all peripheral nondifferentiated glandular cells are rich in glycogen. In the cells in which fat transformation takes place, the glycogen content varies inversely with the fat content (Montagna et al., 1951, 1952). The glycogen content shows a significant correlation with fat content, and fat transformation evidently occurs through the transformation of carbohydrates into lipids.

The enzyme composition in the sebaceous glands varies considerably in different animals. Many of the enzymes occur in large concentrations. In fact, cytochrome oxidase in the sebaceous glands of man and other mammals shows an intensive reaction (Montagna, 1962). The sebaceous glands of mammals contain considerable quantities of succinodehydrogenase as granules in the cytoplasm of undifferentiated cells. The cells subject to fat transformation carry moderate quantities of enzymes; fully differentiated cells have none. Large concentrations of peroxidase are found only in the sebaceous glands of the preputium of rats and mice (Montagna and Novack, 1946). The sebaceous glands in other parts of their skin have mere traces of this enzyme (Montagna and Novack, 1947).

The sebaceous glands of mammals contain a large quantity of monoamine oxidase in the whole gland, especially in nondifferentiated and immature cells (Yasuda and Montagna, 1960). The sebaceous glands have a moderate amount of aminopeptidase. The presence of enzymes is noted in the newly formed secretion in the center of the gland and also in its mature cells, and considerable enzyme activity is recorded at the periphery of the gland. Considerable beta-glucuronidase activity is also noted in these glands (Braun-Falco, 1956b; Montagna, 1957) with a distribution similar to the monoamine oxidase.

The sebaceous glands of primates have a considerable phosphorylase activity, especially in undifferentiated cells and cells subject to fat transformation (Braun-Falco, 1956a; Ellis and Montagna, 1958). The distribution of alkaline phosphatase in the sebaceous glands varies tremendously, even in closely related species—the rat's sebaceous glands are very rich in this enzyme and in mice it is completely absent. The sebaceous glands of cats, raccoons, dogs, seals, weasels, lions, South American monkeys, and lemurs are very rich in alkaline phosphatase, which is concentrated in the undifferentiated cells at the periphery of the glands. Cells subject to fat transformation show a weaker reaction, and the enzyme is altogether absent in fully differentiated cells. The sebaceous glands of cows, pigs, rabbits, and guinea pigs have only a moderate alkaline phosphatase content and mice, primates, and man have no alkaline phosphatase in their sebaceous glands, except for weakly reacting granules in the secretion. The acid phosphatase content of mammalian sebaceous glands varies considerably (Moretti and Mescon, 1956a, b). This enzyme is widespread and has no species specificity. Acid phosphatase activity, highest at the periphery of the gland, decreases in the course of fat transformation of the cell. Some of this enzyme is even present in the secretion. All sebaceous glands contain a large quantity of esterase.

Because the differentiated cell of the sebaceous gland dies when it reaches maturity, a mechanism exists to replace the cells that die during cell secretion. There is some difference in opinion about mitotic activity in sebaceous glands. Some authorities believe that mitotic division takes place in the cells at the periphery of the lobes of the gland (Bizzozero and Vassale, 1887; Kyrle, 1925; Schaffer, 1927). Others suggest that it occurs almost entirely in the epithelium of the gland ducts at the point where they join the glandular lobes, from whence new cells pass into the body of the gland (Bab, 1904; Brinkmann, 1912; Clara, 1929). Since mitotic activity is found only in the cells of the ducts and bodies of the glands, the first hypothesis seems the more likely. In mice, mitotic activity in the sebaceous gland cells is characterized by a diurnal cyclic rhythm much like that of the epidermis. In female mice, mitotic activity in the sebaceous glands correlates with the sex cycle.

The viscosity of sebaceous secretion varies with skin temperature. This may have some influence on the distribution of surface secretion, though it is not the decisive factor, for the sebaceous glands appear to secrete continuously regardless of what happens at the skin surface. The quantity of lipids per square unit produced at any given moment is a function of the size of the gland, that is, of the number of differentiated cells it contains. Secretion is not regulated by the lipid surface layer. The stratum corneum of the epidermis functions like a wick, absorbing the sebaceous secretion from the glandular ducts by capillary action. The rate of the secretion of skin lipids is about 0.1 g/min per 1 cm^2 (Jones et al., 1951). The secretion of sebum is stimulated by androgens and inhibited by estrogens (Ebling, 1970). The contraction of arrectores pilorum muscles does not have any effect on sebaceous secretion.

ECCRINE SUDORIFEROUS GLANDS

Eccrine glands are most numerous and best developed in the higher primates. In other mammals they are found only on a thickened epidermis in

areas subject to wear (e.g., soles, the end of the grasping tail in some South American anthropoid apes). They are the most numerous in man. The chimpanzee and gorilla have more eccrine than apocrine glands. In man, the eccrine sudoriferous glands achieve the highest development. They differ sufficiently from such glands in other mammals to constitute a human characteristic. Man has two to five million eccrine glands on his body surface, or an average of 140 to 339 glands per 1 cm^2 (Kuno, 1959). In adult men the eccrine sudoriferous glands are most numerous on the soles and palms, then the head, then the trunk, with the smallest number on the limbs. The difference from one individual to another is considerable. The suggestion that Negroes have more eccrine sweat glands than Caucasians (Rose, 1948) has proven unfounded (Thomson, 1954). Caucasians average 130 glands per 1 cm^2, and Negroes average 127.

As no sudoriferous glands are formed after birth, the density is highest in newborn babies and decreases with growth (Szabo, 1958). The skin of a one-year-old child has eight to ten times as many sudoriferous glands per 1 cm^2 as the adult skin (Thomson, 1954), though the number of functioning sudoriferous glands is actually less because only glands fully developed anatomically are functioning. The first anlage of the eccrine sweat glands appears in the human embryo on the soles and palms. In the course of the first half of the fifth month, they appear in the armpits, then on the forehead and scalp, then gradually on other parts of the body. In human embryos at 16 to 19 weeks, cell buds grow downward from the epidermal ridges at the inner epidermal surface of the volar side of the sole and palm. In other body areas there are no gland anlagen, although the hair follicles have well-developed hair and differentiated sebaceous glands. The gland anlagen appear as epidermal buds, much like rudimental hair follicles but rather thinner, and they have characteristic dermal papillae at their base.

The cell rod gradually grows downward into the dermis—the swollen end of the rod is the growth part, where the mitotic figures are most numerous. Distally the rod gradually twists into a ball, with the glands developing at different speeds even in the same part of the body. Some glands are better developed than others. Toward the end of the fifth month, the cells in the center of the straight section of the rod become keratinized and separate to form a lumen. Later, during the seventh and eighth months, vacuoles appear in the center of the twisted part. Adjoining vacuoles merge to form elongated slits which, in turn, form a lumen that extends to the duct by the end of the seventh month. Thus the gland is formed. During the eighth month, the lumen expands and the first secretion cells develop some of the characteristics of adult gland cells (Tsuchiya, 1954). Myoepithelial cells, which in functioning glands resemble smooth muscle fibers and are found between epithelium and basal membrane, are not recorded in the embryo. In a nine-month fetus, their morphology approaches that of adults (Borsetto, 1951; Tsuchiya, 1954). During the first nine months of life, the glands gradually activate and become similar to adult glands.

The whole glomerular section of the eccrine sweat gland and the twisted duct are supplied by a large number of blood vessels which follow its outline (Ellis et al., 1958). These vessels often branch from a single arteriole, though those supplying larger glands may come from several.

The eccrine sudoriferous glands resemble common tubes stretching from the epidermis into the dermis or the subcutaneous fat tissue, so the glands comprise two groups: surface and deep. Each gland consists of an irregular, densely twisted basal section, a straight segment ascending to the epidermis, and another twisted segment running through the epidermis. Thus, the glands are twisted at both ends and are relatively straight in the middle.

The diameter of the duct is not as variable as that of the twisted secretion section. The diameter of the duct is usually less than one-third the diameter of the secretion cavity. The length and compactness of the main twisted section varies considerably. These differences are usually associated with the level of the gland within the dermis. Glands located in the middle of the dermis are usually smaller and more compact. The epithelium of sudoriferous glands has an irregular free border, and cytoplasmic processes sometimes jut into the lumen.

Some Japanese scientists believe that these processes are characteristic of apocrine secretion and that classifying sudoriferous glands into eccrine and apocrine on the basis of their secretion mechanism is not justified (Ito, 1943; Ito and Iwashiga, 1951; Ito et al., 1951; Iwashiga, 1952; Kuno, 1959). Others claim that the basophilic granules containing RNA which occur in the apexes of the eccrine cells are excreted into the duct (Ito et al., 1951). Vesicles and undulate epithelium are often noted on the edge of the duct in any serous secreting glands, though it would be wrong to claim this as the basis of the apocrine type of secretion (Montagna, 1962).

There are two distinct cell types in eccrine glands. The upper cells are pushed closer to the lumen and have cytoplasmic granules that stain dark in main dyes and are therefore known as dark cells (Montagna et al., 1953). The larger basal cells, which have sparsely distributed and slightly acidophilic cytoplasmic granules, are called light cells. The nuclei of dark cells are oval and very compact. The chromatin network is seldom clearly visible. The nuclei of light cells are round and filled with granules. The dark cells contain ribonucleic acid and other basophilic substances (Ito et al., 1951; Montagna et al., 1953) and also mucopolysaccharides (Formisano and Lobitz, 1957; Dobson and Lobitz, 1958; Dobson et al., 1958). The light cells have very small acidophilic granules and some basophilic granules evenly distributed in the cytoplasm. The ratio of dark and light cells varies, though it is more or less constant in the glands of each individual. Although in most mammals dark and light cells are very different, there are species in which they are almost indistinguishable. Most of their cells belong to the light type. Sometimes the secretion cells have a light cytoplasm surrounded by vacuoles and resemble sebaceous gland cells. These vacuoles do not, however, contain either glycogen or lipids. Such glands may occur in individuals of any age both in normal and pathological skin (Holyoke and Lobitz, 1952).

The electron microscope shows the cytoplasm of light cells to consist of fine microcilia with identical diameters that jut out into the lumen. In accordance with the functional state of the gland (or with the character of the fixing solution) the intercellular spaces differ in size. There are sometimes desmosomes between adjoining light cells. The light cells have numerous lumps of glycogen, mitochondria, and a Golgi apparatus. The supranuclear parts of dark cells examined under the electron microscope have a common endoplasmic reticulum, ergastoplasm, and a Golgi apparatus. A few small, membrane-bound vacuoles may be disposed in the apical cytoplasm. The nucleus is round or oval and irregularly dented at the periphery. During secretion, the apical cytoplasm of dark cells is filled by different-sized vacuoles. Vacuoles are separated by clear-cut membranes except where they are in contact with one another; in these zones they often appear to merge, with no boundaries between them. A solid layer of myoepithelial cells separates the thick hyaline membrane and the secretion cells.

The terminal layer of the secretory glomus, 33 to 80μ at the site of the transition into the duct, is called an ampule (Lowenthal, 1960, 1961). The ampule opens into a sphincter the same length as the ampule. The walls of the sphincter consist of the basal layer of large myoepithelial cells and cubical cells. It can be assumed that the ampule plays a part in sweat reabsorption and that the sphincter serves as a contracting unit.

As the secretion section of the gland has a unilayered epithelium beneath it and has two layers of cubical cells under the duct, the transition between these two segments is rather sharp. The cells under the lumen of the gland in the first part of the duct, however, contain granules not unlike the granules in the secreting cells. Thus, the transition from the gland segment to the duct is gradual. The existence of the transitional portion is shown not only by light microscopy but by electron microscopy (Ochi and Sano, 1970). This portion is composed of a layer of tall, columnar cells with occasional basal and myoepithelial cells. The basal cells of the duct's epithelium are square. The surface cells are also square or slightly flattened. Under the electron microscope, the surface facing the lumen has low microcilia surrounded by plasmic membrane (Ellis and Montagna, 1962).

The eccrine gland ducts open onto the surface of the palms and soles through the ridges in the epidermis, and the apertures are distributed very regularly in the center of the epidermal ridges (Hambrick and Blank, 1954). The excretion duct becomes helical when it enters the epidermal ridge. Most of the twists turn clockwise (Takagi, 1952). The lumen of the twist increases nearer the surface (Hambrick and Blank, 1954). Epidermal thickness determines the shape of the intraepidermal part of the sudoriferous gland lumen (Pinkus, 1939). Some researchers believe that the duct of the gland in the epidermis is a canal between the epidermal cells. Others believe that it has walls made up of square cells (two layers). As the cells of the intraepidermal segment of the duct are evidently cytologically different from the epidermal cells around it, the intraepidermal duct cells and the surrounding cell layer are termed "the common epidermal sudoriferous duct" (Lobitz et al., 1954a).

Mitoses are recorded only in the basal cells where the duct connects with the epidermis and in no other part of the helical intraepidermal duct (Lobitz et al., 1954b). These basal cells are thus the germinal layer of the helical duct, and the cells formed move up along the helix. The duct cells have no pigment. The cells of the intraepidermal section of the duct are subject to keratinization, which is more rapid here than in the epidermal cells around the duct. The keratinization in the duct cells in the middle of the stratum malpighii, for instance, corresponds to the keratinization in the epidermal cells of the stratum granulosum.

Mitotic activity seems to be rare in eccrine gland secretion cavity cells (Bunting et al., 1948; Holyoke and Lobitz, 1952) because there is little need for cell replacement. However, mitotic division can occur both in light and dark cells. Duct cells, which have no mitotic activity, normally retain their capacity for rapid proliferation when the need arises, and mitotic activity is recorded only in the basal duct cells where the loss of cells from the side of the lumen stimulates replacement.

The cytoplasm of the eccrine gland's light secretion cells is filled with fine lipid granules—dark cells contain small quantities of these. There are variations in the number of lipids in each gland, and the glands of individuals vary. Children's glands contain fewer lipid granules than the glands of adults. The glands of old people may fill completely (Kano, 1951).

Most of the eccrine cells have a diffuse yellow pigment in the cytoplasm which is evidently carotene (Bunting et al., 1948). The eccrine sudoriferous glands contain a large amount of glycogen (Tsukagoshi, 1951; Montagna et al., 1952), and they have more light than dark cells. The dark cells contain a substance that appears to be a mucopolysaccharide (Formisano and Lobitz, 1957). The epithelium of the duct usually contains glycogen, mostly concentrated in the basal cells (Montagna, 1955; Lobitz et al., 1955). The eccrine glands have more cytochrome oxidase. The cytoplasm of the duct and the secretion cells has been found to contain large quantities of reactive granules. The glands contain large amounts of succinodehydrogenase (Braun-Falco and Rathjens, 1954; Serri, 1955; Montagna and Formisano, 1955). The reactive granules are concentrated around the nucleus in the duct cells and in a thin band parallel to the surface of the gland lumen. A large quantity of carbonic anhydrase has been recorded in human eccrine glands (Braun-Falco and Rathjens, 1955). The reaction is pronounced both in the duct cells and in the secretion section, where the concentration is greater in the light cells than the dark.

Some scientists believe that the monoamine oxidase content of the eccrine sudoriferous glands is slight (Hellmann, 1955; Shelley et al., 1955); others (Montagna, 1962) believe that it is very high. The eccrine sweat glands contain a larger amount of phosphorylase than other skin structures (Braun-Falco, 1956a; Ellis and Montagna, 1958; Yasuda et al., 1958). The cytoplasm of both light and dark cells is very rich in phosphorylase. The reaction is especially pronounced in the apical sectors of the cells. This beta-glucuronidase activity in man (Braun-Falco, 1956b; Montagna, 1957), and aminopeptidase activity is pronounced in the secretion section of the eccrine sweat glands. As a result of the reaction, rough granules along the whole length of the secretion section are visible. Alkaline phosphatase activity is found only in the deep, twisted parts of the gland glomerule, and both acid phosphatase and esterase activity is low or moderate.

APOCRINE SUDORIFEROUS GLANDS

Most mammals have apocrine sudoriferous glands. In humans, apocrine glands are recorded in the armpit region, the mons pubis, meatus acousticus externus, around the anus, the papillae, labia minora, preputium, scrotum, and umbilicus. Human apocrine glands are common, twisted tubular glands whose secretion cavities are very tortuous with adjoining loops sometimes interconnected. The secretion cavities of smaller glands are in the dermis; those of the larger glands reach deep into the subcutaneous fat tissue. The apocrine gland duct runs closely parallel to the hair follicle, usually opening inside the hair canal above the sebaceous gland duct. Occasionally two glands open into one hair canal (Hurley and Shelley, 1954), and some ducts may open on the skin surface.

The diameter of the duct is considerably narrower than that of the secretion cavity ducts, and the transition between the two segments is abrupt. The epithelium of the secretion cavity is common and irregularly columnar. The terminal sections of some glands are elongated and jut into the lumen, giving its border an irregular edge. In the duct, the epithelium is represented by two layers of square cells. All the epithelial cells are above the underlying myoepithelial cells, outside of which is a thick hyalin membrane. The gland is surrounded by a loose connective tissue rich in blood vessels.

In the basal section of the glandular cells are one or sometimes two round nuclei, each of which contains one or two large nucleoli rich in RNA. The electron microscope reveals that the nucleus of the secretion cells is surrounded by a thick inner and thin outer membrane.

The cytoplasm at the base of every cell of the apocrine gland is usually filled with fine basophilic granules. In the excretion ducts both cell layers are highly basophilic. Except for its characteristic granules, the cytoplasm of the secretion cells resembles the cytoplasm of other gland cells under the electron microscope. There are not many mitochondria. They are most numerous in the cells with few or no secretion granules; there are none in cells filled with secretion granules (Minamitani, 1941*a, b;* Ota, 1950). Many researchers believe that mitochondria change into secretion granules. The Golgi apparatus is situated in the apocrine gland cells over the nucleus. Most researchers regard it as contributing to the production of secretion granules.

The apocrine glandular cells contain large quantities of four categories of lipids (Bunting et al., 1948). The ferrum content of the cells varies among individuals, from a large amount to none. The pigmentation of the apocrine glands differs widely among individuals and depends on the size, number, and color of pigment granules. The size of pigment granules varies from the threshold of visibility to the diameter of the nucleus, and the color ranges from light yellow to dark brown. Unlike the eccrine glands, man's apocrine glands have only a moderate amount of succinodehydrogenase (Montagna and Formisano, 1955).

Cytochrome oxidase activity is moderate in the ducts and intense in secretion sections. The duct cells and the glandular epithelium contain a large concentration of monoamine oxidase, but amylophosphorylase activity is found only in the ducts. Unlike the human eccrine glands, the apocrine glands contain a considerable amount of esterase (Montagna and Ellis, 1958). The apexes of high epithelial cells and myoepithelial cells contain varying amounts of alkaline phosphatase, which is completely absent in the duct cells. Man's apocrine glands have very little phosphatase activity in the duct cells and moderate activity in the apexes of the secretion cells. The glands are rich in beta-glucuronidase. Apocrine sudoriferous gland ducts have virtually no aminopeptidase, the content of which is fairly high in the secretion cells, though not as high as in the eccrine glands (Montagna, 1962). Blood supply to the apocrine glands is like that to the eccrine glands, and cholinesterase-rich nerves are found around the apocrine glands of dark-skinned humans (Montagna, 1962).

The anlage of the apocrine sweat gland is a lateral outgrowth of the hair follicle above the sebaceous gland. This outgrowth is buried in the connective tissue and gives rise to the glandular glomerule and the duct. The lumen first appears in the unformed duct with partial keratinization. Later, slits appear in the twisted section of the duct which then grow and merge to form a duct in the secretion canal. The apocrine glands are not fully formed at birth, as the myoepithelial cells are not yet visible. Man's apocrine glands grow from the age of seven into the teens and gradually acquire the structure and histochemical properties of functioning glands.

HAIR

Mammals are the only vertebrates that have hair. The morphological properties of hair vary in different mammals and the hair of the same animal may vary considerably in different areas of the body. The hair can be spine-shaped, rough, soft, long, short, thick, thin, colored, or colorless. Even in a small skin area the hair may vary in length, density, and color. A hair is a filament of keratinized cells solidly fixed together. The free part of the hair above the skin surface is the shaft; the part buried in the skin is the root. The root is surrounded by the outgrowth of the epidermis, and the whole complex, called a hair follicle, passes below the dermis into the subcutaneous fat tissue. The lower bulbous part of the hair follicle is called the bulb.

Hair follicles are one of the most primitive systems in the organism in which growth and differentiation are observed. Despite the considerable size of the hair, only a small number of cells in the bulb of the hair follicle, the matrix, are responsible for its production. Every hair follicle in the skin of a human head produces about 0.35 mm of hair per day, so the daily metabolic activity of the germinal tissue with mitoses and syntheses of the whole protein complex is very high. Periodically the follicles stop producing hair. A large part of the bulb disappears and the remaining cells of the follicle enter into a rest stage. At different intervals the resting follicles become active again, pass into the period of organogenesis, and begin to produce new hair. The growing hair follicles are said to be in the stage of anagen and the resting follicles in the stage of telogen; the transition period between the two stages is called catagen (Dry, 1926). Hair varies widely in diameter and shape. Most hair is oval or round in section, though some is flat or bean-shaped.

The hair shafts may be straight, bent, wavy, curly, spiral, or crimped. The curvature of the hair appears to be determined by a deviation of the hair bulb from the straight line, by the eccentric position of the hair in the follicle, and by asymmetric keratinization. Hair shafts usually consist of a cuticle layer around the medulla with a cortical layer in between. The cuticle is a single layer of tile-patterned squamae with a free edge pointing to the apex or end. Although the cuticle is made up of a single layer of cells, the cells are elongated so that many of them overlap.

The squamae of the cuticle are semitransparent and contain no pigment. Inside the hair follicle, the cuticle cells are connected with the cells of the so-called inner root sheath. This structure fastens the hair shaft tightly to the follicle. The cuticle of the hair protects the cortical layer, which peels and is quickly destroyed without the cuticle. A thin layer of fat and carbohydrates covers the cuticle to protect the hair from physical and chemical damage. Two main types of cuticle cell disposition are distinguished: ring-shaped and others (fig. 4). Where the cells are ring-shaped each cell goes right around the hair; in others there are several cuticle cells around the hair. The free edge of the cuticle cells may be common, dentate, or serrate. The cuticle cells may be subdivided into elongated, pointed, oval, or flattened.

Under the cuticle layer lies the cortical layer, which is composed of elongated, spindlelike, keratinized cells tightly linked to one another. In

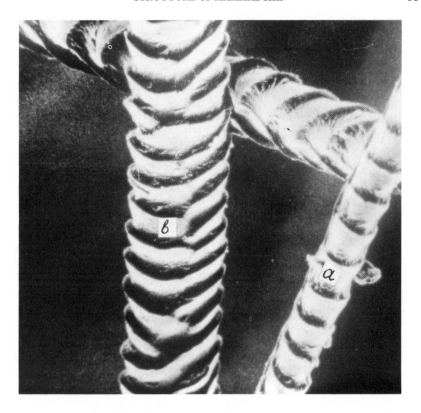

Fig. 4. *Nyctalus noctula:* ring-shaped cells (*a*) of fur hair cuticle and
(*b*) cuticle cells of pile hair (not ring-shaped).

pigmented hair there are melanin granules in the cortical cells. Where there is no pigment, the cortical layer appears transparent. There are small air cavities between the cortical cells which are filled with liquid in the live parts of the hair (Hausman, 1932; 1944). When the hair grows, it dries and the liquid is replaced by air. The thickness of the cortical layer varies greatly in different species. The durability of the shaft depends on the solidity of the cortical layer. In the center of the hair there is a continuous or discontinuous medulla. In thin hair the medulla is usually absent. The medulla type may vary even within the same hair. The medulla consists of large, loosely-connected, keratinized cells. Large intra- and intercellular air cavities in the medulla determine the luster and tint of hair color by light reflection.

The hair of most mammals is categorized as vibrissae, guard, pile, intermediate, and fur. Vibrissae or tactile hairs have long, thick, conical, straight, or slightly bent shafts. Cuticle that is not ring-shaped is formed by polyhedral or round cells with edges that touch but do not overlap. In vibrissae, blood vessel lacunae surround the follicle, which has abundant innervation. Vibrissae hairs are few and are usually found on the head or the feet.

The guard, pile, intermediate, and fur hairs form the pelage of mammals. The guard hairs are longest but least numerous. The shafts of the guard hairs are thick and usually spindle-shaped, thinning gradually at the tip. The cuticular layer is not ring-shaped and has a multiple-row medulla. Pile hairs are usually lancet-shaped. The thicker and usually flattened upper part of the axis is called a granna. There is a necklike contraction before the granna in which the axis forms a small kink or fracture. The thin, basal part of the pile hair may be crimped. The cuticle is not ring-shaped. The medulla in the thin, basal part of the axis consists of two to three rows of cells. In the granna it numbers five to six rows and more. Sometimes the guard and pile hairs are called overhairs.

The intermediate hairs share characteristics of both pile and fur hairs. They are shorter and thinner than the pile hairs, but longer and thicker than fur. Unlike the fur, intermediate hairs have a granna, though it is not pronounced. The cuticle of this granna is not ring-shaped. There are two to three layers of medulla cells, while the main part of the hair has a ring-shaped cuticle and a medulla one cell deep.

The fur hairs are usually numerous, short, and thin. Their shaft is sinuate and the thickness is even along the length of the hair. The cuticle of fur is usually ring-shaped, and there is no medulla. Sometimes pile, intermediate, and fur hairs are subdivided into size orders. Hair may grow singly or in tufts (fig. 5). Where it grows in tufts, several follicles open into a hair funnel through which these hairs protrude from the skin. Each tuft is made up of one pile and several fur hairs, or of fur hairs only.

Hair is usually in groups. Most commonly, groups are formed by one guard or pile hair with some fur hairs on each side. Complex hair groups consist of two or three hair tufts around a single guard or pile hair. The hair tips are oriented to give direction to the pile. The pile varies considerably in different mammals. In many furbearing animals, the tips of the back hair point toward the tail, diagonally backward on the sides and downward and toward the middle on the belly.

The hair follicles are positioned obliquely in the skin with a small diameter at the base that dilates into a bulbous swelling. The cavity inside the bulb is filled with loose connective tissue forming the dermal papilla. The upper part of the follicle, stretching from the end of the sebaceous gland duct to the skin surface, is the hair canal. The bundles of smooth muscle fibers of the arrectores pilorum muscle branch off at an oblique angle to the surface of a small swelling on the lateral side of the hair follicle right below the sebaceous gland. These muscles, innervated by a large number of adrenergic nerves, pull the hair into a vertical position when they contract, raising the skin around the follicle into "goose pimples." In most mammals these muscles are well developed on the tail.

Hair follicles consist of an outer root sheath and an inner root sheath

Fig. 5. Tufts of hairs in polar bear skin (middle of back).
Hair tips are cut off.

with the hair in the center. The thickness of the outer sheath varies with the
size of the follicle. The inner root sheath is made up of three concentric
layers. The outer Henle layer is one cell deep. The Huxley layer, which is
intermediate, is two or more cells deep. The inner layer of the deep inner
sheath is a single cuticle layer of tile-patterned, squamous, keratinized cells
whose free edge points downward, overlapped by the hair's cuticle cells.
The hair follicles are surrounded by a connective tissue sheath lined with an
inner circle of fibers; the fibers of the external layer are longitudinal. The
outer root sheath is separated from the connective tissue row by a noncellular hyalin membrane.

A resting hair follicle has no bulb and is from one-third to one-half the
length of the growing follicle. Even the deepest resting follicles are wholly in
the dermis and their hairs have an ampullaceous swelling at the base. The
resting follicle consists of an ampullar hair root enveloped by an epithelial
sac. A strand of small cells stretches from the sac to the dermal papilla. A
partially keratinized capsule, one or two cells deep, lies between the cells of
the epithelial sac and the ampullaceous swelling of the hair. The sac consists
of cells retained from the external sheath. The hair anlage at the base of the
epithelial sac, at the lower level of the ampullaceous swelling, is also a rem-
nant of the external sheath.

Hair cuticle cells are distinguishable in the upper part of the bulb, where they branch upward from the matrix. About halfway up the bulb these cells are cubical, the cytoplasm being highly basophilic and filled with basophilic granules. Unlike the cells of the cortical layer, these cells do not contain melanin. In the upper part of the bulb the cells become columnar, the long axis being oriented radially. The cells remain columnar within a short segment of the bulb, then the outer layer begins to stretch upward. Possibly the cells of the internal root sheath (which is lateral) grow faster than the hair cuticle cells, which turn their edges upward and look like tiles. The cuticular cells are flattened and look like acuminate squamae in vertical section.

The inner sheath is formed by peripheral and central masses of matrix cells. The cells, which move almost upward in relation to the matrix, are surrounded by concentric layers in the upper part of the bulb. In the hair follicles of most mammals, each of the three layers of the inner root sheath is only one cell deep, but in larger follicles the Huxley layer is two cells deep. The cells in all three layers of the inner root sheath contain trichohyalin granules. None of the layers of the inner root sheath contain pigment, even in follicles that produce highly pigmented hair.

The Henle layer cells have trichohyalin granules right after they originate from the matrix. The cells start roughly square, then elongate vertically toward the upper parts of the bulb. The trichohyalin granules, very small at first, presently merge into large homogenous globules. These changes are accompanied by cell hyalinization, and their nuclei become indistinguishable and ultimately disappear. Unlike the cells in other keratinized tissues, the cells of the inner sheath do not decrease in volume and do not shrink, possibly owing to the absorption of the liquid by trichohyalin (Auber, 1952).

The Huxley layer cells differentiate more slowly than the Henle layer cells. In the Huxley layer the trichohyalin granules first appear in the cells on the bulb apex, where the Henle layer cells are fully hyalinized. At first the granules are small, spherical, and evenly distributed in the cytoplasm. Higher up, they merge and become large and irregular. Some cells without trichohyalin granules extend lateral cytoplasmic outgrowths across the layer of the external sheath. The live cytoplasm bridges around the dead Henle layer. The food or energy stores from the external sheath may pass across these bridges to the live cells of the Huxley layer. The Huxley layer cells are hyalinized about halfway down the hair follicle.

The cells of the inner layer cuticle do not acquire trichohyalin granules until their size is equal to about half the length of the follicle. They are the smallest cells in the follicle and can be identified even in the lower part of the bulb. The cuticle cells remain small and flat even to the level of half the follicle, where very small, irregular trichohyalin granules can be seen in them. At this level the cells are rather flat with their proximal edges turned toward the axis so that they overlap the distal edges of the preceding cells. Soon after this the cells acquire hyalin granules. The cuticle cells become hyalinized and their nuclei gradually disappear. Above the middle of the follicle, the cuticle cells and the Huxley and Henle layers merge into a solid hyalin layer.

In the hair follicles of most mammals, the inner root sheath forms horizontal folds or wrinkles at the base of the hair canal near the sebaceous gland ducts. In human hair follicles these folds do not occur. As the inner sheath connects with the hair, both grow and move outward at about the same rate. The outer section of the Henle layer glides over the surface of the outer root sheath, which is not mobile. Both these layers have smooth surfaces conducive to movement over the outer root sheath. The Henle layer cells become keratinized immediately after they appear out of the matrix and move easily, gliding on the partially keratinized cells of the outer root sheath. The thickness of the outer sheath of large follicles usually varies because the hair's position in the follicle is eccentric. The two-layer outer sheath stretches to the apex of the bulb around which it is formed. The outer cells are slightly elongated, the inner cells more flattened.

Immediately above the bulb the sheath has three layers. It gradually becomes multilayered, gaining greater length in the upper third of the hair follicle. Here almost all of its cells, except those on the axial surface, are filled by vacuoles. The tall, columnar peripheral cells are perpendicular to the follicle axis. In the upper third of the follicle, where the cells are not highly vacuolized, the outer cells are square.

Along the whole length of the sheath the peripheral cells are more strongly vacuolized than the cells on the axial surface, which have a relatively continuous cytoplasm. The intercellular bridges, desmosomes, and tonofibrils are especially well developed in the cells closer to the axis. The cells facing the Henle layer are rich in tonofibrils. About halfway up the follicle, their cytoplasm becomes hyalinized and is subject to partial keratinization (Gibbs, 1938).

The mitotic activity of the cells is relatively normal in the upper part of the hair follicle, where the external root sheath turns into the outer epidermis. This part of the external sheath forms a keratinized, superficial layer along the hair canal which peels so that cell replacement is necessary.

The upper and lower halves of active hair follicles have different cycles. The lower half is a relatively unstable structure appearing at every growth cycle. During the resting stage the larger part of the external sheath perishes. The remaining part forms the hair rudiment and the epithelial sac around the ampullar hair. The upper part of the external sheath is a constant element of the follicle.

The hyalin, noncellular, two-layered membrane is situated between the external root sheath and the connective tissue membrane. It is barely visible around the upper follicle, relatively thick around the middle, and thickest around the thickest part of the bulb (Schaffer, 1933). The ampullar membrane is thin around the matrix and almost invisible in the papillary cavity.

The thickness of the hyalin membrane in different parts of the follicle is proportional to the size of the hair follicle.

There is one layer of ampullar membrane level with the upper follicle, and there are two layers where the follicle reaches its greatest diameter in its lower third or half. Only the outer layer, continuing into the basal membrane of the whole epidermis, encircles the whole follicle. It has a hyalin structure mainly composed of fine collagen fibers.

Under the electron microscope two layers of collagen fibers can be seen in the ampullar membrane in mice and guinea pigs (Rogers, 1957). The fibers of the inner plate are parallel and the fibers of the outer plate are perpendicular to the long axis of the follicle.

The term "dermal papilla" should be used only for the connective tissue incorporated into the follicle bulb in the course of anagen; it forms a compact globule of cells under the rudiment of the hair in the course of telogen. In active follicles the dermal papilla is fixed to the connective tissue sheath by a basal foot. The degree of vascularization of the dermal papilla depends on the size of the follicle (Durward and Rudall, 1958).

The dermal papilla is subject to considerable morphological change during the hair growth cycle. In resting follicles it becomes a compact cell globule with solid, dense nuclei and hardly visible cytoplasm. A small number of pigment granules are found between the cells. In the growing follicles, the papilla is larger and the cells are farther apart. The nuclei, large and ovoid, stain white with major stains.

It is generally accepted that during catagen there is a reduction in the number of cells in the papilla and that in early anagen the cells divide and increase in number (Wolbach, 1951). However, the degradation of papillary cells is actually very low. Changes in the size of the dermal papillae mostly result from changes in the size of cells, a reduction or increase in the size of capillary networks, and changes in the intercellular substance. The number of cells remains relatively stable.

Mitochondria are hardly distinguishable in the follicular cells. Very small, threadlike mitochondria are visible around the nuclei of the matrix cells, becoming slightly more numerous in the cells of the upper part of the bulb. There are very few mitochondria in the small cells of the hair rudiment in resting hair follicles.

During the hair anlage stage, the Golgi apparatus is seen at the apical pole of the nuclei in the vertically elongated cells above the bulb. As soon as keratin fibers are formed in these cells, the osmiophilic bodies become less and less distinguishable and finally disappear. In the internal sheath, the Golgi apparatus is found at the apex of the cells, but it disappears when trichohyalin granules are formed. In the cells of the external sheath the Golgi apparatus is not so distinct.

The hair follicles contain sudanophilic lipids (Nicolau, 1911). They are represented by different-sized granules in nearly all the nonkeratinized cells.

The cells of the matrix have accumulations of two or more lipid bodies. Thin, flattened cells in the internal root sheath around the lower part of the bulb have more lipid bodies than any of the other cells in the bulb. The cuticle cells of the internal root sheath and the hair cells lose the perinuclear sudanophilic bodies completely as soon as they move upward from the matrix. In the hair root, the cells of the cortical layer and the medulla are relatively free of sudanophilic lipids, but fine, clearly visible granules are found in the apexes of the dermal papilla cells. In resting follicles, almost all the epithelial sheath cells have lipid granules. The hair rudiment and dermal papilla cells have no lipids. During catagen, the degenerating follicular cells frequently contain large quantities of lipids. The dermal papilla also contains increased quantities of lipids.

The active hair follicles are always rich in glycogen. In the connective tissue sheath, certain quantities of glycogen granules are found in the cytoplasm of the fibroblasts and extracellularly along the fibers. The outer root sheath is also rich in glycogen, but the cells in the inner sheath have none. The cuticle cells over the bulb are rich in glycogen, which they lose as soon as they cornify around the middle section of the follicle. Above the bulb, some cells of the cornifying cortical layer may have traces of glycogen, and the large medullary cells have glycogen from the level right above the dermal papilla to about halfway up the follicle, where they are shriveled and cornified. The dermal papilla does not usually have glycogen.

The resting hair follicles of most mammals do not contain glycogen (Bollinger and McDonald, 1949; Johnson and Bevelander, 1946; Shipman et al., 1955). But the resting hair follicles of almost all primates and of man contain ample glycogen in the cells of the epithelial sac (Montagna and Ellis, 1959). The cells of the dermal papilla, which almost invariably contain glycogen during catagen, are free of it in telogen.

Mucopolysaccharides of the hair follicle are mostly localized to the dermal papilla. The metachromasis is low in the connective tissue membrane at the bulb level and more intensive in the hair follicle's middle third. Metachromatic diffusion staining is observed in most cells in the outer sheath.

Cytochrome oxidase is concentrated in the hair canal of active follicles, especially in the cells of the matrix. In resting follicles, a positive reaction is obtained in the cells of the hair canal and in the epithelial cells, though it is virtually absent in hair rudiment cells.

All the live cells of the hair follicle contain varying quantities of succinodehydrogenase (Montagna and Formisano, 1955). In the active follicles, the highest enzyme activity is found in the cells of the matrix and the upper part of the bulb. All the cells of the resting follicles have a lower enzyme content than the active follicles.

The active and resting follicles contain small quantities of monoamine oxidase (Yasuda and Montagna, 1960). In the outer root sheath of the active follicles, high enzyme activity is recorded around the keratogenic zone.

The active and resting hair follicles show intensive amylophosphorylase activity (Braun-Falco, 1956a; Ellis and Montagna, 1958). The highest concentrations of this enzyme are found in the outer root sheath of the upper part of the active follicle. The hair rudiment shows a low activity. The presence of this enzyme in the resting follicles is specific to the skin of primates. In most other mammals the resting follicles do not have phosphorylase.

A large amount of aminopeptidase is present in the outer root sheath around the keratogenic zone and around the bulbs of active follicles, the largest concentration being in the dermal papilla. The hair follicles also contain beta-glucuronidase (Braun-Falco, 1956b; Montagna, 1957).

Esterase is widely distributed in the hair follicles (Braun-Falco, 1958; Montagna and Ellis, 1958a). Alkaline phosphatase activity is species specific. In most cases only the dermal papilla and the lower part of the outer sheath react to this enzyme. The activity of acid phosphatase especially manifests itself in the inner root sheath and the keratogenic zone of the hair bulb.

Sulfhydryl proteins are well represented in the hair follicle. The reaction to these proteins is moderate and homogenous along the whole length of the outer root sheath of the active follicles. In the inner root sheath the Henle layer is rich in sulfhydryls. The cuticle of the internal root sheath is moderately active, and the cuticle of the hair is highly active from bulb to skin surface. The skin of the matrix shows only moderate activity. The upper part of the bulb is rich in sylfydryls.

Blood is supplied to the hair follicles through collateral branches of arteries arranged in a candlestick pattern. Every follicle is fed by several arterioles, which are usually close to the lower third of the follicular membrane. Descending from them, the capillary network branches and penetrates into the dermal papilla. Straight branches, parallel to the longitudinal axis of the follicle and connected by lateral branches, run upward from the arterioles. Level with the sebaceous gland, a network envelops both gland and hair follicle. The capillaries of this plexus merge with the postepidermal papillary layer network in the dermis and form a ring around the hair funnel. Larger follicles have a more complex system of blood vessels than smaller follicles. The blood supply to the vibrissae is peculiar, featuring a rather large network of veins in the connective tissue membrane between longitudinal and transverse layers. The vascular network around the hair follicle is very species specific.

The innervation of the hair varies with species, hair category, and body area. The innervation of the vibrissae is extensive. The hair follicle is often innervated by several nerve fibers coming from different directions and forming two networks (Szymonowicz, 1909; Weddel et al., 1954), an inner network of transverse fibers in the connective tissue membrane, and an outer network, also mainly of transverse fibers, touching the hyalin membrane. The nerve fibers in the hair follicle have free and independent terminals (Weddel et al., 1954). The outer nerve fiber networks terminate freely between the cells of the middle section of the connective membrane. The internal plexus fibers cross the hyalin membrane and terminate freely on the border with the outer root sheath cells. The follicular nerves have a high acetylcholinesterase content (Montagna and Ellis, 1958*b*).

Hair anlagen in the skin appear at different stages of ontogenesis in different species and asynchronously in different parts of the body. The vibrissa anlagen usually appear first and others come later. Hair follicles may even appear at different times in the same body area. The hair follicles are the first skin derivates formed in the relatively undifferentiated embryo epidermis. The first sign of hair follicles is a grouping of cells in the basal layer of the epidermis forming a small swelling on its inner side (Pinkus, 1958). The cells in the nuclei of these primitive hair rudiments become smaller and darker than the surrounding cells. Later the basal cells of the hair rudiment and their nuclei elongate perpendicular to the epidermis as the hair rudiment gradually swells into the dermis. The fibroblasts from the dermis form a small cup around the ampullar swelling of the hair rudiment. This cup is the dermal papilla, which shows an intensive reaction to alkaline phosphatase (Achten, 1959).

As the hair follicle pushes into the dermis, it forms a peglike accumulation of cells dorsally leveled. Tall, columnar cells at the periphery of this peg are radial to the center, with the central cells being longitudinal. The free end of this peg becomes cleft. As it grows downward, a notch develops in the middle of the free end of the bulb occupied by the dermal papilla. As the follicle becomes longer, the distal end expands and grows around the dermal papilla so that the latter is completely incorporated into the bulb.

Meanwhile, the central and then the peripheral cells of the follicle fill with glycogen. Later, two processes are formed on the back of the follicle. The lower process has sloping sides forming the so-called *pincus colliculus,* and the upper process is spherical. The upper process is a sebaceous gland anlage; the lower process is the point at which the arrectores pilorum muscle will be fastened. The cells in the center of this strand reveal keratohyalin at the moment when the differentiation of sebaceous glands begins. Sudan black staining shows that some of the cells in the center of this strand are also subject to fat transformation, forming the hair-gland canal before the hair begins to grow in the follicle.

The dilated distal part of the follicle is differentiated into a bulb that overgrows the dermal pipilla. The papilla remains fastened to the basal plate of the dermal cells with a narrow strand. The pigment cells, which before this stage of development are found along the entire bulb, shift to its upper part, leaving the lower part unpigmented. The cells below the pigment line form the follicular matrix. The cells of the bulb around the dermal papilla are subsequently arranged longitudinally. Trichohyalin granules appear in the cells at the center of the upper part of the bulb over the apex of the dermal papilla. In other cells trichohyalin granules are also formed above them, creating a strand in the center of the full length of the follicle. The cells of this strand later become keratinized and slightly wrinkled, forming the first inner root sheath. The differentiation of the apex of the first hair takes place over the dermal papilla in the upper part of the bulb. The parallel rows of the proliferative matrix cells advance upward, forming a keratin cylinder in the center (i.e., the hair encased in a keratin tubule), which is the inner root sheath. The differentiation of the inner root sheath occurs soon after birth. In the external Henle layer, trichohyalin is first formed in the upper part of the bulb. Keratohyalin is later formed in the Huxley layer, and the cuticle of the outer root sheath is the last to be keratinized. The hair is keratinized over the bulb neck. The end of the hair is invariably free of pigment and the first hairs have no medulla. The hyalin membrane is formed around the lower part of the follicle below the *Pincus colliculus* while the first hairs begin to differentiate. Two connective tissue membranes start to form around the follicle. Finally, the cells of the connective tissue (between the *Pincus colliculus* and the surface) elongate, grow larger, and differentiate into a strand of smooth muscle, the arrectores pilorum.

II

COMPARATIVE MORPHOLOGY OF SKIN
OF DIFFERENT ORDERS

Having detailed the structure of various parts of the skin, we will now consider the skin of individual members of various orders of mammals, describing members of different taxonomic groups and species that are taxonomically related but that follow different modes of life. To avoid repetition, the general description of the skin of the whole order will sometimes be followed by familial descriptions.

MATERIALS AND METHODS

Skin samples were taken from 194 mammals (18 orders). In some of the species, specimens were taken at different seasons. Samples from two or three animals of each species were used, sometimes from different areas. Samples were cut from the withers and from the breast between the forelimbs. In *Pinnipedia, Cetacea, Artiodactyla,* and *Perissodactyla,* skin from many parts of the body was studied to elucidate the skin layer topography. In almost all the mammals, samples were taken from the foot soles to examine specific glands.

The skin was fixed in 10 percent formalin solution and was subsequently included in celloidin or celloidin-paraffin. Perpendicular and horizontal sections were taken. From thick skin they were made by Christeller's method and stained with hematoxylin-eosin after Van-Gieson. For staining elastin fibers, elastin "H" and resorcinfuctisin were used. Some samples were made by cryomicrotome and the sections were stained with Sudan III.

The nerves in the skin of some *Pinnipedia* were stained after Compass. Electron microscopic studies were made with a HEM-7A microscope. The thickness of skin, epidermis and layers, dermis and layers, and subcutaneous fat tissue were measured. The disposition of sebaceous and sweat glands and arrectores pilorum muscles was described and measured on perpendicular sections. On sebaceous glands, the smallest and largest diameters were measured, and on sweat glands, the external diameter of the secretion cavity was estimated.

Hair was divided into categories and size orders by length and thickness. The thickness of the medulla (as a percentage of the hair thickness) and the length of the granna (in pile hairs) were used, taking the average of no fewer than ten hairs. The number of hairs was estimated and the pattern of distribution was described from horizontal sections. Hair cuticles were photographed under an electron-scan microscope. Skin pellucidity was recorded with an auto-spectrophotometer (Sokolov, 1962*a,* 1962*b*).

We experimented with skin fixed in formalin and rinsed in running water. The pellucidity of skin and hair of a newly killed great gerbil (November) used for control proved almost identical to skin simply fixed in formalin, though pellucidity was higher in the fresh hairless than in the fixed skin. Our data should be valid for comparative descriptions of species. *Spermatophilopsis leptodactylus, C. pygmaeus, C. suslicus, Rhombomys opimus,* and *Meriones meridianus* were taken for investigation, using *Sciurus vulgaris* for comparison.

ORDO MONOTREMATA

The skin of *Monotremata* has not received sufficient attention. Specific and common glands of *Tachyglossus* and *Ornithorhynchus* were described (Schaffer, 1940; Montagna and Ellis, 1960*a;* de Meijere, 1894; Hausman, 1920; Wildmann and Manby, 1938). We studied the skin of *Tachyglossus* sp. *and Ornithorhynchus paradoxus.*

The skin of the *Monotremata* body is thin to medium. The stratum corneum is very well developed in the epidermis, which is not pigmented. In the dermis, bundles of collagen fibers run in various directions to form a dense plexus. The papillary layer is thicker than the reticular layer, on which it borders level with the hair bulbs. Sweat glands are either absent or rather well developed, and both apocrine and eccrine glands occur. The sebaceous glands are small, with a single lobe. The absence of specific glands in the sole skin is characteristic (according to Klaatsch [1888] these glands are present in *Monotremata*). A peculiar gland formed of alveolar glands is found in the femoral region of *Monotremata*. Its duct extends to the foot and pierces the horny spur in the calcaneal region to reach the skin surface. In females this gland is only slightly developed. There is an anal gland formed by sebaceous and tubular glands. The pelage is either well developed or very sparse;

in species where it is sparse, spines are present and the arrectores pilorum muscles are absent.

The skin is medium thick. The back and the sides are covered by long, thick spines with a very well developed cortical layer and thin-walled medulla cells. The pelage is sparse. The hairs are coarse, thick, and without medulla. The femoral gland is rather poorly developed.

Tachyglossus sp.

Material. Skin from withers, sole of hind foot of adult (body length 57 cm) taken in Australia. The withers skin is in poor condition with the epidermis not preserved.

The papillary layer of the withers dermis takes up the full 6.3 mm of the skin (there is no reticular layer). In the skin of the sacrum, the epidermis is 200μ, the stratum corneum is 170μ. There is a peculiar dermal zone 3.7 mm from the outer surface (withers) where horizontal bundles of cross-striated muscles alternate with bundles of collagen fiber. The spine roots pass through this zone to a depth of 6.3 mm. Since there are no arrectores pilorum muscles, the spiny anteater raises his spines by contracting cross-striated skin muscles. The hair bulbs are 0.9 mm below the outer surface. Elastin fibers occur through the whole dermis and are especially abundant in the inner layers. Sweat glands are absent and no sebaceous glands are found at the spine roots. Posterior and anterior to the hair roots is a small one-lobe gland (up to $44 \times 123\mu$).

In the meati acoustici are small sebaceous and large, abundant tubular glands (Burne after Schaffer, 1940). The femoral gland grows to 26 mm long, 20 mm wide, and 12 mm thick (Schaffer, 1940). The anal gland is formed of superficial sebaceous glands and underlying tubular gland glomeruli (Schaffer, 1940).

Both hair and spines are round. The spines, up to 60 mm long, gradually decrease in thickness from the distal to the proximal third, their thickness uniform in the middle third, their average diameter 4 mm. In the upper third they gradually taper to a sharp tip. Oval in section, they are outwardly smooth. The cortical layer is thick; the medulla accounts for 79% of the thickness (fig. 6). The cuticle scales have peculiar, irregular edges (fig. 7). The medulla is comprised of thin-walled cells with none of the fortifying "beams" found in hedgehog spines. The medulla cells are smaller in the middle of the spine than on the periphery. Coarse, flattened hairs in the base are of an even thickness for some of the length, after which they thicken, retaining their thickness down most of the shaft before tapering toward the tip. The medulla is either absent or is present in isolated fragments.

Hair of this category subdivides into two orders (Hausman, 1920). The

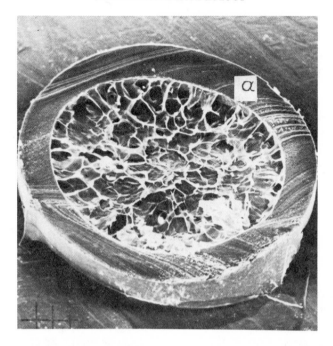

Fig. 6. *Tachyglossus* sp. Section of spine: *a*, cortical layer (x 45).

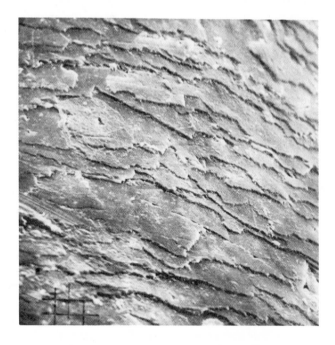

Fig. 7. *Tachyglossus* sp. Cuticle of spine (x 750).

first order averages 15.9 mm in length. The second order averages 14.2 mm. Their thicknesses are 127 and 78μ, respectively. Brunner and Coman (1972) report the maximum hair thickness to be 350μ. The hair is oval in cross section. There are more hairs of the second order than of the first. Some peculiar hairs are found among the spines. They do not grow in the skin and we cannot positively attribute them to our specimen (possibly these are the hairs described by Toldt [1906], who called them fur hairs). They have a long, twisted club and a straight granna; the granna is only slightly thicker than the club. The medulla extends along the whole hair, though it is not always continuous.

The abdominal hairs fall into three orders: bristle-like, flattened; wavy, flattened; less flattened. The wavy hairs predominate (Hausman, 1920).

FAMILIA ORNITHORHYNCHIDAE

The skin is thin. There are sweat glands of eccrine and apocrine types with a tubular secretion cavity. There are no spines. The pelage is dense. The hairs divide into pile and fur. Granna and neck are very pronounced in the pile hairs. The medulla of the granna is thin. The fur hairs are sinuate with medulla. The femoral gland is well developed.

Ornithorhynchus anatinus

Material. Skin from breast and foot soles and pelage from withers of adult (body length 46 cm).

The chest hair is growing. The thickness of the skin is 1.87 mm. The epidermis (without the outer layer, which was not preserved) is 34μ. The dermis is 1.83 mm, the papillary layer is 1.5 mm; the reticular layer is 0.33 mm (fig. 8). The sebaceous glands are very small (evidently due to the growth of the hair). One anterior, one posterior to certain hair tufts, the sebaceous glands reach $34 \times 112\mu$. The whole outer surface of the flattened bill of the platypus is covered with small pores connected with the sweat gland ducts and with some special sensitive bodies (Poulton, 1895).

A transparent liquid is secreted through these pores which slightly irritates the skin of the hands (Montagna and Ellis, 1960). The sweat glands are considerably larger on the bill than on the body skin. (Poulton [1894] noted the presence of one or two sinuate sweat glands at each hair group in the body skin.) In the skin of the duck-billed platypus under investigation, apocrine sweat glands have long tubular secretion cavities reaching 45μ in diameter and stretching along the hair roots (fig. 9). On the head are two types of common, sinuate tubular sweat glands (Montagna and Ellis, 1960).

There are eccrine glands in the surface skin of the throat and frontal shields, and apocrine glands in the skin of the inner surface of the bill and the bases of the frontal and throat shields. The secretion cavities of the eccrine glands of the duck-billed platypus contain glycogen as well as a

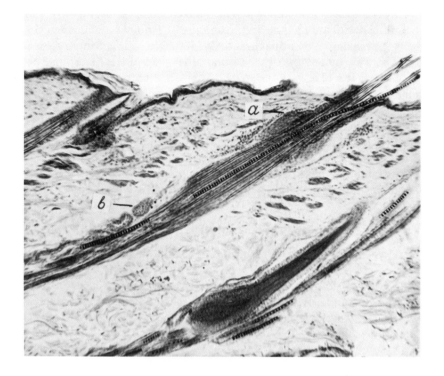

Fig. 8. *Ornithorhynchus anatinus.* General view of the outer skin layers:
a, sebaceous gland; *b,* sweat gland.

small number of PAS-positive, saliva-resistant particles. The glands, which react actively with succinate dehydrogenase and monoamine oxidase, do not contain alkaline phosphatase. The gland ducts react readily with non-specific esterase. In the secretion cavities some cells are and some are not reactive. The nerves around the sinuate section of each of these glands form a loose, tightly adhering network. They react with cholinesterase. The presence of a PAS-positive, saliva-resistant substance is characteristic of the apocrine glands on the bill. Like the eccrine glands, these glands react strongly to succinate dehydrogenase and monoamine oxidase, but do not react strongly to nonspecific esterase (though their secretion cavities do). The apocrine glands do not contain cholinesterase nor are they surrounded by cholinesterase-reactive nerves. The secretion of the skin glands appears to keep the thickened epidermis soft and elastic (Montagna and Ellis, 1960).

The femoral gland is 40 mm long, 20 mm wide, and 15 mm thick in males. Its size varies with the sex cycle. During rut it may reach 44 mm in length. The secretion is toxic (Pawlowsky, 1927).

There are only two categories of hair on the withers; pile and fur (Blumenbach, 1802). The pile hairs have a pronounced, flattened granna

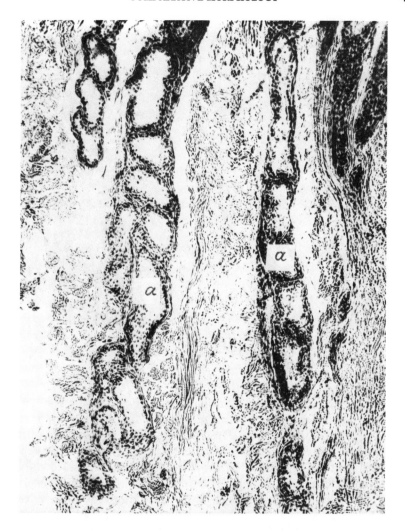

Fig. 9. *Ornithorhynchus anatinus. a,* Apocrine sweat glands.

(fig. 10) and a long thin club separated from the granna by a still thinner neck. The medulla is well developed in the club but is absent from the neck, and it runs through the upper third of the granna in a thin strip. When there is pigment in the granna, it is concentrated on its inner and lateral sides (Wildman and Manby, 1938). The hairs are 10.2 mm long (the granna 3.1 mm) and up to 146μ thick (up to 180μ according to Brunner and Coman, 1974). The medulla comprises 15% of the granna width and 58.3% of the shaft thickness in the club. In the granna, the medulla is not in the middle of the hair but is closer to one of the flattened sides of the granna (Brunner and Coman, 1974).

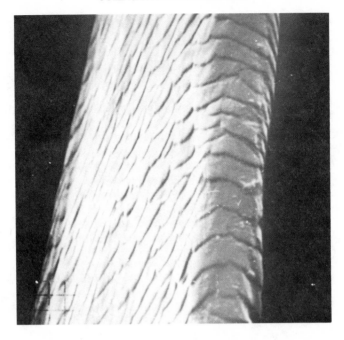

Fig. 10. *Ornithorhynchus anatinus*. Granna of pile hair.
In foreground, narrow side of granna.

Wildman and Manby (1938) distinguish three types of pile hairs in the duck-billed platypus, with two being further subdivided into two variants and the third into three variants. However, the differences in these variants and types are, in my opinion, negligible and not supported by measurement data. Hausman (1920) gives six parts on the body of a duck-billed platypus as distinguishable by the pile hairs. Wavy fur hairs (the number of the waves being from seven to nine) are the same thickness down their whole length (13 mm), and are 8.5 mm long. The cuticle is not ring-shaped. The medulla (69.3% of hair thickness) is absent only at the base. The hairs grow in tufts and groups (Leydig, 1859). Each tuft has twelve to twenty fur hairs, with four to five tufts beside and behind a pile hair forming a group. The pile hairs form rows divided by tufts of fur hairs (de Meijere, 1894). There are 750 pile hairs and 56,250 other hairs per 2 cm² of chest skin.

ORDO MARSUPIALIA

Practically no research has been done on the skin of *Ordo Marsupialia,* though there are some descriptions of the structure of sweat glands (Green, 1961), certain other glands (Ford, 1934; Schaffer, 1940; Schulze-Westrum,

1965) and of the pelage (Gibbs, 1938; Boardman, 1943-1952; Lyne, 1951, 1959).

We studied the skin of *Chironectes minimum, Dasyurus quoll, Perameles nasuta, Phalanger orientalis,* and *Petaurus breviceps.* The rarity of the marsupial species investigated makes it impossible to define skin structure at the level of families, so an account of the skin of the order is required.

The skin is thin. The epidermis of the body is thin, without stratum granulosum or stratum lucidum. There is no pigment. The stratum malpighii is one to two cells deep, and stratum corneum is relatively thick. The border between the papillary and reticular layers is level with the hair bulbs, or at the bottom of the sweat glands. In the reticular layer, there are horizontal bundles of collagen fibers (fig. 11). Sebaceous glands are unilobal, one anterior, one posterior to every hair tuft. There are sweat gland secretion cavities in several shapes: saccular, sausage-shaped and tubular, straight, or slightly sinuate. Specific glands are not abundant. Evidently all the marsupials have tubular glands on the soles, and anal and circumanal glands formed by sebaceous glands. The arrectores pilorum muscles are absent.

The pelage varies among the different families. There are guard, pile, intermediate, and fur hairs (*Familia Didelphidae*), or guard, pile, and fur

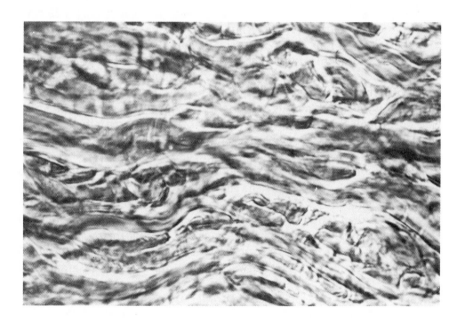

Fig. 11. *Dasyurus quoll.* Horizontal arrangement of collagen bundles in reticular layer in withers skin.

hairs (*Familia Phalangeridae*), guard and pile (*Familia Peramelidae*), and pile and fur (*Familia Dasyuridae*). The medulla is usually not very well developed. The hairs grow in tufts of two to seven. They fluoresce under UV light—this appears to be a property of some of the skin gland secretions. Fluorescence in blue light is typical of *Trichosurus vulpecula* (Bollinger, 1944). The direction of the pile varies, starting from a primitive pattern as in *Myrmecobius* and *Dasycercus,* whose body hairs are directed caudally and ventrally (Boardman, 1943, 1945, 1949, 1952; Lyne, 1951). The pile is made more complex by additional convergent and divergent centers.

Vibrissae are well developed in all the marsupials, except in *Notoryctes* where they are absent (Lyne, 1959). Vibrissae are best developed in mobile, arboreal marsupials, rather less in slow-climbing arboreal and terrestrial forms, and poorly in *Myrmecobius fasciatus* (Lyne, 1959). Unlike placental mammals, where the vibrissae grow in ontogenesis, the marsupials develop vibrissae only after the whole embryonic body is covered with hair (Gibbs, 1938).

FAMILIA DIDELPHIDAE

Chironectes minimum

Material. Skin from withers and pelage from chest (body length 23 cm).

The thickness of the skin is 0.66 mm, the epidermis 45μ, stratum corneum 34μ, dermis 0.615 mm, papillary layer 0.515 mm, reticular layer 0.100 mm, The stratum corneum is loose. The border between the papillary and reticular layers is level with the bottom of the sweat glands. The plexus of collagen fibers in the dermis is not solid; the bundles in the upper and middle papillary layer run at various angles to the surface—those in the lower part are horizontal. Sebaceous glands ($22 \times 84\mu$) are elongated along the hair bursa (fig. 12). Sweat glands have almost horizontal botuliform secretion cavities up to 50μ in diameter beneath the hair bulbs (fig. 12).

The hair of the withers divides into four categories: guard, pile (two orders), intermediate, and fur. The guard hairs (length 12.9 mm, width 73μ) have a straight shaft, which gradually thickens from the base to the upper quarter and then tapers to a thin tip. The medulla, rather well developed (56.2%), is missing only from base and tip. The pile hairs (first order) (length 16.4 mm, thickness 8μ, medulla 58.1%) have a long thin (16μ) medulla-free club with up to ten waves and a much-thickened granna 7.4 mm long. Their cuticle cells have sinuate edges. The shaft bends before the granna. The pile hairs of the second order are distinct, length 10.6 mm, thickness 63μ, medulla 49.3%. Their granna extends a little over half of the shaft (6.0 mm). There are six waves in the neck, which has no medulla. The intermediate hairs (length 10.5 mm) have up to ten waves and are of even thickness along nearly the whole length (21μ). Only the distal section has a

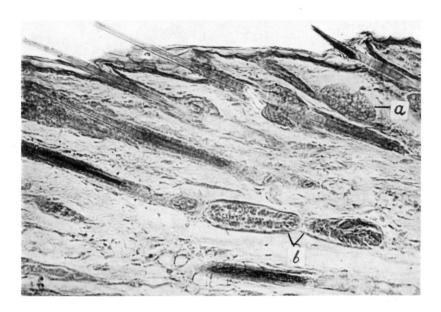

Fig. 12. *Chironectes minimus.* Structural view of withers skin:
a, sebaceous gland; *b,* sweat glands.

type of granna—a slightly thickened, straight piece of the shaft (mean length 1.9 mm). The medulla (62.0%) extends the whole length, except tip and base. The fur hairs (length 10.2 mm) are the same thickness throughout their length but are thinner toward the tip. These hairs are wavy with up to fourteen waves. Usually they do not have a medulla (in which case they are 11μ thick). Fur hairs (with a medulla) are somewhat thicker (15μ). The cuticle cells of the fur hairs may be ring-shaped or not. Guard hairs grow singly, the others in tufts (two to four in each). We failed to distinguish hair categories on the cross-section parallel to the skin surface. A count of 333 guard and 39,333 other hairs is recorded per 1 cm² of skin.

Hair categories on the chest are similar to the withers. Guard hairs (length 15.4 mm, thickness 82μ, medulla 48.7%) have a granna 6.5 mm long. The width of the neck is 39μ, that of the medulla in the neck, 19μ. In some hairs the medulla becomes narrower just before the granna. In the pile hairs of the first order (length 16.5 mm, thickness 57μ, medulla 33.3%), the length of the granna is 6.2 mm. The club of the hair has up to ten waves. In the pile hairs of the second order (length 10.8 mm, thickness 47μ, medulla 36.1%) the granna length is 5.0 mm. The intermediate hairs (length 11.9 mm, thickness 22μ) are few in number. There is no medulla in the granna and it is rather poorly developed in the club. The fur hairs (length 10.5 mm, thickness 11μ) have no medulla and have up to fourteen waves.

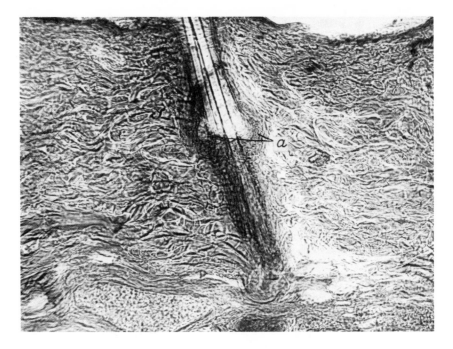

Fig. 13. *Dasyurus quoll.* Withers skin: *a,* sebaceous glands.

FAMILIA DASYURIDAE

Dasyurus quoll

Material. Skin from withers and chest of adult male (body length 32 cm).

The skin from the withers is rather thick (1.65 mm owing to hair growth; epidermis 39μ, stratum corneum 22μ, dermis 1.611 mm, papillary layer 1.281 mm, reticular layer 0.33 mm). The chest hairs are not growing and the skin is much thinner (skin 0.44 mm, epidermis 22μ, stratum corneum 12μ, dermis 0.418 mm, papillary layer 0.165 mm, reticular layer 0.253 mm). The outer surface of the skin is very uneven with large folds and indentations. In the withers skin, the stratum corneum is rather loose. In the dermis, the papillary and reticular layers meet at the hair bulbs. In the papillary layer of the withers, the bundles of collagen fibers run in various directions and form a compact plexus. Elastin fibers in the dermis are few in number; a network is recorded only at the border with the subcutaneous muscles. In the upper reticular layer are small groups of little fat cells (34 x 39; 45 x 67μ). The sebaceous glands are tiny (34 x 62; 22 x 56μ) and not visible at most of the hairs (fig. 13). This is evidently due to hair growth. At some of the hair roots only the ducts of sebaceous glands are seen. Apocrine sweat glands are found only in the sacrum skin, where there is a small, tubular sweat gland behind every hair tuft. The secretion cavities of the apparently inactive glands are almost vertical along the hair roots. They are slightly sinuate and the ducts open into hair funnels. *Sminthopsis crassicaudata* (*Dasyuridae*) has apocrine glands in the body skin and eccrine glands in the soles (Green, 1961).

The withers hairs divide into two categories: pile and fur. The hair is 21.4 mm long, 77μ thick, and not abundant. The granna at the distal end of the shaft is 11.0 mm (first order). The thickness of the club is 33μ. The hairs are straight with a medulla (63.7%) along the whole length. The second order of pile hairs (length 22.2 mm, thickness 64μ, medulla 59.4%) has a bent club 24μ thick; the granna is shorter (7.8 mm). In the third order (length 17.4 mm, thickness 45μ, medulla 60.0%), the granna is less developed (5.1 mm). The thickness of the club is 20μ. The fur hairs (length 12.7 mm, thickness 22μ, medulla 45.4%), with no granna, have up to six waves. The hair grows in tufts of one large and four to seven small hairs (we failed to distinguish hair categories and orders on the horizontal section). There are 1,000 large and 8,000 small hairs per 1 cm².

FAMILIA PERAMELIDAE

Perameles nasuta

Material. Skin from withers and chest of adult (body length 26 cm) taken March 28.

The hair is not growing. The thickness of the withers skin is 1.325 mm, chest 0.66 mm, epidermis both on withers and on chest 8μ. The corneous layer is not available. The dermis is 1.324 and 0.652 mm thick, respectively. The papillary layer is 0.246 and 0.234 mm. The reticular layer is 1.078 and 0.418 mm. The boundary between the papillary and reticular layers is level with the bottom of the sweat glands. The plexus of the collagen fibers in the dermis is compact (especially in the withers). In the papillary layer horizontal bundles prevail, though numerous bundles rise to the surface at various angles (fig. 14). Horizontal elastin fibers are abundant in the papillary layer of the withers dermis. There are small sebaceous glands (17×39; $28 \times 50\mu$), but not at every hair tuft. Sweat glands, poorly developed in the withers, are present at every hair tuft. Their secretion cavity is tubular, 28μ in diameter, and slightly sinuate. They are well below the hair bulbs and open into the hair funnel. In the chest the sweat glands are longer, more sinuate, with larger secretion cavities (50μ). Apocrine glands are found in *Isodon obesulus* (*Peramelidae*) in the skin of the body, and in the eccrine glands in the sole (Green, 1961).

The withers hairs divide into two categories: guard and pile. The straight guard hairs thicken gradually from the base to the distal quarter of the shaft and taper toward the tip. The medulla is peculiarly divided into halves. The

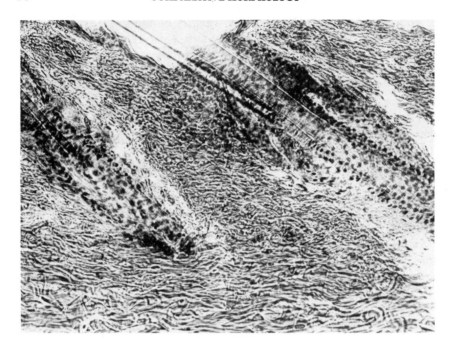

Fig. 14. *Perameles nasuta.* Network of collagen fiber bundles in withers skin.

guard hairs are in three orders which differ in size (length 20.5, 18.1, 15.2 mm; thickness 206, 137, 61μ; medulla 90.3; 88.4; 83.3%). The hairs are slightly convex on one side and strongly concave on the other in cross section, with two lateral, semicircular thickenings (Brunner and Coman, 1974). There are two orders of pile hairs. The first order (length 13.4 mm, thickness 35μ) is slightly wavy in the lower section (three to five waves). The second order (length 10.2 mm, thickness 25μ) is wavy down the whole length (four to seven waves), with the shaft thickening slightly in the distal section. The medulla (no halving, as in guard hairs) is more poorly developed (65.8 and 52.1%).

Familia Phalangeridae

Phalanger orientalis

Material. Skin from withers and chest (body length 32 cm) taken in New Guinea.

The hair is growing, so the skin is thick (0.9 mm on the withers and 1.0 mm on the chest); the epidermis is 17 and 22μ, respectively, the stratum corneum is 11μ, the dermis is 0.743 and 0.648 mm, the papillary layer is 0.603 and 0.318 mm, and the reticular layer is 0.140 and 0.330 mm. The stratum

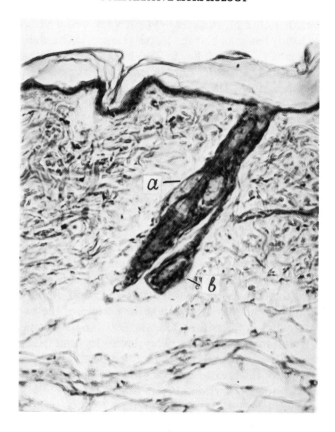

Fig. 15. *Phalanger orientalis.* General structure of dermis papillary
layer, breast: *a,* sebaceous gland; *b,* sweat gland.

corneum is compact. The papillary layer is thicker than the reticular layer
where the hairs are growing and thinner where they are not. The plexus of
collagen fibers is compact; most of the bundles in the papillary layer are at
acute angles to the surface. Elastin fibers are numerous in the papillary and
lower reticular layer. Sebaceous glands behind the hair tufts are larger than
those in front (posterior 50 x 140, anterior 39 x 123μ). Sweat glands occur
rarely, probably because the hair is growing. Their tubular, twisted secretion cavities measure 80 to 90μ in diameter (fig. 15). The ducts open into the
hair funnels. In *Trichosurus vulpecula* (*Phalangeridae*), apocrine sweat
glands are recorded in the body skin, eccrine glands in the foot sole (Green,
1961).

The hair on the withers divides into three categories: guard, pile, and fur
(two orders). The guard hairs (length 19.0 mm, thickness 64μ) are straight,
barely thickening from the base to the distal three quarters of the shaft. The
medulla (45.1%) extends through the whole hair, except tip and base. The

pile hairs (length 17.0 mm, thickness 49μ) have a well-developed granna that extends slightly less than half the length (7.3 mm). The club of the hair has up to eight waves. The medulla (55.2%) is missing only in base and tip. The two orders of fur hair differ only in length (17.8 and 14.8 mm, thickness 21μ, medulla 62.0%). The hair shaft has up to eleven waves. The guard hairs grow singly; the other hairs grow in tufts of one pile and two to three fur hairs, or two to three fur hairs only. There are 1,333 guard, 3,666 pile, and 29,000 fur hairs per 1 cm². The number on the chest (11,333 per 1 cm²) is much smaller.

Petaurus breviceps

Material. Skin from withers, chest, and flying membrane of male (body length 15.5 cm) taken in New Guinea.

The withers skin is 0.5 mm thick; the chest, 0.55 mm; epidermis, respectively, 22 and 39 mm; stratum corneum, 11 and 28μ; dermis, 0.313 and 0.331 mm; papillary layer, 0.268 and 0.286 mm; reticular layer, 0.045 and 0.035 mm; subcutaneous fat tissue, 0.165 and 0.145 mm. In the dermis, the papillary and the reticular layers adjoin at the bottom of the sweat glands. The bundles of collagen fibers form a dense plexus in the dermis. Horizontal collagen fibers prevail in the papillary layer. In some places they are crossed by vertically rising bundles. Elastin fibers occur rarely in the dermis. The reticular layer of the breast has rows of fat cells parallel to the surface, and there are two to five horizontal rows of fat cells (34 × 45; 34 × 62μ) under the reticular layer of the withers dermis. As there are few collagen fiber bundles, I regard it as subcutaneous fat tissue (thickness 145 to 165μ). On the chest only small (28 × 84μ) elongated sebaceous glands are pressed against the hair roots, but not at every tuft. In the skin of the withers there are apocrine sweat glands with saccular, slightly twisted secretion cavities up to 40μ in diameter, one at every tuft (fig. 16), their ducts opening posteriorly into the hair funnel. Sweat glands resemble *Insectivora* glands in shape. In the chest skin, sweat glands have twisted secretion sections (up to 34μ) in diameter) and long ducts. The frontal gland is on the line connecting right ear and left eye where it crosses the line from left ear to right eye (Schultze-Westrum, 1965). In males, the glandular field reaches 16 × 18 mm. The gland is formed by a combination of tubular and sebaceous glands. Sweat and sebaceous glands and anal glandular sacs are well developed in the perianal area (Ortmann, 1960; Schultze-Westrum, 1965).

A specific gland on the chest is mentioned by Leche (1900, quoted after Schaffer, 1940). *Myrmecobius fasciatus* also has a thoracic gland at the base of the neck (Ford, 1934). The hair roots around the gland are 440μ long. There is a gradual increase in the size of sebaceous and sweat glands from the periphery to the center of the glandular formation. The glands range from normal to very large. Sebaceous glands have more than twelve lobes

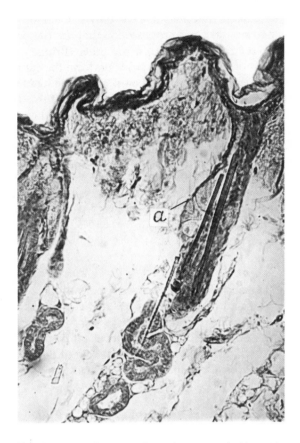

Fig. 16. *Petaurus breviceps*. General structure of withers skin:
a, sebaceous gland.

and are up to 1.0 × 0.55 mm in size. They take up the whole thickness of the skin, no other tissues being present.

Gland ducts open into the hair sheaths. It appears that only the glands posterior to the hairs are transformed into these huge glands. The diameter of the sweat glands increases up to 100μ; their secretion cavity is much longer. The glomi into which these glands are twisted (220 × 290μ) are deeper than the sebaceous glands. Long ducts open into the hair funnels. Abundant fat cells (28 × 39; 34 × 39μ) are found at the periphery of the glandular formation. Our data on the thoracic gland are similar to Schultze-Westrum's description (1965).

The hair of the withers divides into three categories: guard, pile, and fur. The guard hairs are straight, of almost uniform thickness (18μ) the entire length (9.6 mm), and with no granna (medulla 90%). The pile hairs (length 9.4 mm, thickness 24μ, medulla 66.7%) have many similarities to *Insecti-*

vora, the difference being that the shaft has only one narrowing. The club of the hair has up to six waves. In cross section they are flat or slightly convex on one side and strongly convex on the other (Brunner and Coman, 1974). The thickness of the fur hairs (18μ) is uniform the entire length (9.2 mm) but there is a weak granna near the tip above the ligature. The shaft has six to eight waves. The medulla takes up to 77.8% of the thickness. The hairs grow in tufts of two to four (mostly three) on the withers, and three to five on the breast. There are some poorly defined groups of three to four tufts. An average of 16,000 hairs per 1 cm² grows on the withers and 22,500 on the breast.

The flying membrane ($250\text{-}300\mu$ thick) has pelage on both sides with groups of cross-striated muscle bundles between. There are small sebaceous and sweat glands (diameter of secretion cavity 28μ). The hair of the flying membrane grows in tufts of two to four.

FAMILIA MACROPODIDAE

Apocrine sweat glands are found in the body skin of *Macropus major* and *Setonyx brachyurus,* and eccrine glands are found in the skin of the feet (Green, 1961). In *S. brachyurus,* the twisted section of the apocrine glands is deep in the dermis or in the upper subcutaneous fat tissue. Each gland is a spiral tube, which straightens higher up, narrowing into a duct that opens into the hair's sebaceous gland channel.

ORDO INSECTIVORA

The skin of *Insectivora* has not received sufficient attention. We examined the members of all the families in this order. Their skin characteristics are in some cases diametrically opposed because of extreme specialization, so it is very difficult to use skin to define this order.

Insectivora skin is either thin or medium. The epidermis and stratum corneum are always thin (though their relative thickness may be considerable). The stratum granulosum and the stratum lucidum are absent in the epidermis, which is rarely pigmented. In most *Insectivora,* the papillary layer of the dermis is considerably thicker than the reticular (the opposite is noted in moles). In the dermis, horizontal bundles of collagen fibers invariably predominate (in the mole breast dermis, the bundles run in various directions). Sebaceous glands are usually large, with a single lobe. The secretion cavities of the sweat glands are saccular, sausage-shaped, twisted slightly or not at all (in the desman they are very twisted). Eccrine sweat glands are found in the soles of most *Insectivora.* The arrectores pilorum muscles are invariably absent. The pelage is varied and often highly specialized. Ligatures in the pile and fur hair shafts are common to the *Insectivora.* The hair is twisted at an obtuse angle at the ligatures, making a slow, smooth 180° turn along the longitudinal axis. The edge of the hair is back in its original position at the next turn so that the turns are not spiral. In most species the hairs grow singly.

Familia Erinaceidae

Descriptions are available of the specific skin glands in *Hylomys suillus* (Weber, 1928), in the European hedgehog (Schaffer, 1940; Murariu, 1972, 1974, 1975, 1976) and its spines (Lvov, 1884). We studied the skins of *Hylomys suillus, Erinaceus europaeus, Hemiechinus auritus,* and *H. hypomelas.*

Familia Erinaceidae divides into two subfamilies: *Echinosorecinae* and *Erinaceinae.* The members of the first subfamily, *Hylomys suillus,* have a number of features of skin structure in common with other *Insectivora.* They have no spines. The skin, the epidermis, and the stratum corneum are thin with no pigment in the body skin. The subcutaneous tissue contains no fat cells. The arrectores pilorum muscles are absent. The sweat glands are large with highly twisted botuliform and saccular secretion cavities. The hairs have ligatures and longitudinal turns. The skin of *Erinaceinae* is different from that of other *Insectivora* because it has spines with extra cavities in the stratum corneum. The absolute thickness of the skin is greater than in any other *Insectivora* under investigation. The epidermis and the stratum corneum are thickened. The epidermis may contain pigment. The subcutaneous fat tissue is usually well developed. There are no sweat glands in the body skin and the sebaceous glands are small. Powerful smooth muscles adhere to the spine roots. The hairs have no ligatures with longitudinal turns. The breast hairs grow in tufts. There are no specific glands in the sole or anal area.

Hylomys suillus

Material. Skin from withers and breast (table 1) of specimen (body and head length 13.5 cm) taken in June on Kalimantan Island.

The stratum malpighii of the body epidermis contains one to two layers of cells. The epidermis is up to 84μ thick, and stratum corneum is 22μ in the hind sole skin. The stratum granulosum is pronounced. There are no alveoli in the epidermis. The epidermis (especially stratum basale) is heavily pigmented.

The boundary between the papillary and reticular layers in the body skin is level with the sweat glands. In the papillary layer, bundles of collagen fibers form a rather dense plexus. The bundles, which run in various directions, are predominantly horizontal. In the reticular layer, the collagen bundles are horizontal, forming a loose plexus. Sebaceous glands are small ($34 \times 78\mu$), elongated oval. There are two glands at each hair root on the

TABLE 1

ERINACEIDAE: SKIN MEASUREMENTS (μ)

	Skin	Epidermis	Stratum corneum	Dermis	Papillary layer	Reticular layer
Hylomys suillus						
withers	420	9	5	411	355	56
breast	230	7	4	233	161	62
Erinaceus europaeus						
(2) withers	3,190	196	185	2,994	2,389	605
(1) breast	935	56		879	604	.275
(2) breast	506	22	14	484	384	100
Hemiechinus auritus						
withers*	2,090	28	17	1,402	1,402	Absent
breast*	1,518	29	15	254	1,254	Absent
H. hypomelas						
withers*	2,600	56	39	1,600	1,600	"-"
breast	550	45	39	505	285	220

*Subcutaneous fat tissue measurements are given in the text.

withers. On the breast, the glands are somewhat larger (34 × 118μ) and are singly posterior to the hair roots. There are two sebaceous glands at each hair of the sole (67 × 84μ; 67 × 101μ). The apocrine glands are rather far apart. Their secretion cavities lie deeper in the skin than the hair bulbs. In the skin of the withers, the secretion cavities are botuliform and strongly twisted, 50μ in diameter. In the breast they are saccular and straight, 45μ in diameter, and their long axis is horizontal. The gland ducts open into the hair funnels. The sole of the hind foot has no tubular glands. An anal gland is present (Weber, 1928).

The hair is very interesting because it has properties in common with that of the shrews. The hair of the breast divides into two categories. Guard hairs (length 7.8 mm, thickness 77μ) are straight, thicken gradually from base to distal third, and taper at the tip. The medulla (35.0%) extends the whole length except base and tip. The pile hairs divide into three orders. The first and longest (6.9 mm) has a long granna (1.3 mm) that extends almost half the hair (thickness 52μ, medulla 34.6%). As in the guard hairs, the medulla is missing only at base and tip. The club of the hair has up to three ligatures where the shaft is bent. The pile hairs of the second group are shorter (6.6 mm) and thinner (42μ), with granna only a third of the length. The medulla is thinner (28.5%). The club has up to two ligatures. These hairs are not abundant. The third order is 5.8 mm long and 25μ thick (granna just over 36.0%). The club has up to three ligatures.

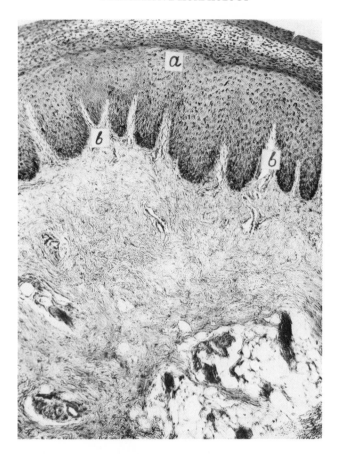

Fig. 17. *Erinaceus europaeus.* Skin of front foot sole.
a, epidermis; *b,* dermal papillae.

Erinaceus europaeus

Material. Skin from withers (only specimen 2), breast (table 1), and soles of front and hind feet of two specimens (body lengths 26 cm and 24.5 cm). Specimen (2) taken in June, Moscow region.

The stratum malpighii is two to three cells deep; the stratum corneum is rather compact. Epidermis and stratum corneum are thickest in the hind soles—280 and 56μ on the front feet and 440 and 100μ on hind feet (fig. 17). The inner surface of the epidermis is alveolar. There are pigment cells in the basal layer of the epidermis. The border between the papillary and reticular layers is level with the lower section of the arrectores pilorum muscles. The plexus of collagen fiber bundles in the dermis, running in various directions

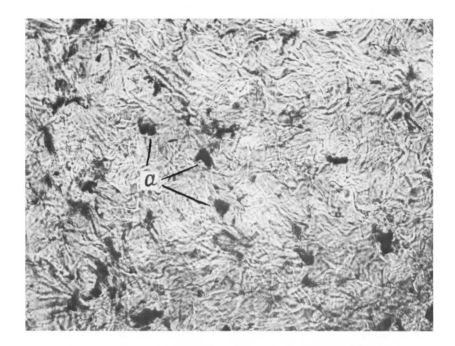

Fig. 18. *Erinaceus europaeus.* Withers skin, network of collagen fiber bundles, papillary layer: *a,* melanocytes.

in the outer papillary layer, is very dense (fig. 18). In the middle of this layer, near the spine bursae, are many large, horizontal bundles, though by the spine bulbs the bundles are mainly vertical. Many oblique bundles are at the lower end of the arrectores pilorum muscle level. In the reticular layer, the collagen bundles are mainly horizontal. Their plexus is looser than in the papillary layer. Groups of fat cells (28×34; $34 \times 39\mu$) are in the lower papillary layer, and large melanocytes are abundant in the upper papillary layer (fig. 18). The border between the papillary and reticular layers of the breast dermis is at the hair bulb level. Here the papillary layer is thicker than the reticular layer, and the bundles of collagen fibers are mostly horizontal, forming a solid plexus in both.

In some areas of the upper papillary layer of specimen (1), there are groups of fat cells (28×45; $34 \times 50\mu$) near the hair bulbs. No fat cells are in the subcutaneous fat tissue. A pair of sebaceous glands (17×123; $28 \times 40\mu$) is at every spine and hair. The secretion cavities of the glands at the spines are elongated and perpendicular to the spine bursae. The glands near the hair roots in the withers skin are smaller and round. The secretion cavities of the sebaceous glands in the breast skin are 55×112; $101 \times 224\mu$. They are elongated along the hair roots, with perpendicular ducts. The lower end of the sebaceous glands almost reaches the hair bulbs.

Most glands are one-lobed, but some are multilobed. According to Murariu (1972), there are no sebaceous or sweat glands on the back. Both gland types occur occasionally on the belly. On the belly, the sebaceous glands are small and round. On the lips, there are multilobed sebaceous glands and eccrine sweat glands. Most of the glands have one lobe, though some have several.

In hedgehog specimens (1) and (2), no sweat glands are found in the samples under study. In the skin of the withers and lower lip of an adult male, which died during hibernation (January) in the Moscow region, sparse, slightly twisted sweat glands are revealed. The secretion tube has an outer diameter of 57μ in the withers skin and 22μ on the lower lip. No sweat glands are found on the breast and sacrum. An anal gland is made up of strongly twisted glands with dilated ducts opening into the hair sheaths (Schaffer, 1940). There are also some strongly twisted tubular glands.

According to Murariu (1972), there are sebaceous glands in the perianal region of the hedgehog which do not open into the hair follicles but into the skin surface of peculiar inpouchings; sweat glands are not described by the author. Murariu reports (1975b, 1976) sebaceous and apocrine sweat glands in the circumanal region of the hedgehog. The acinus number in sebaceous glands may be up to ten. The sweat glands are poorly developed. The anal glands are large sebaceous glands and eccrine sweat glands. In the dermis of the sole are glomi of tubular glands (usually surrounded by fat cells) 0.7 to 0.8 mm from the outer surface. These glomi are usually larger on the hind sole than on the front. The diameter of the gland secretion cavities is 17μ (front foot) to 22μ (hind foot). Eccrine sweat glands in hedgehog soles are reported by Murariu (1974, 1976); there are more glands in the hind foot soles than in the front. The arrectores pilorum muscles in the withers are very well developed (fig. 19). They embrace the lower section of the spine bulb, retrace a semicircle and get lost in the mass of collagen bundles in the middle papillary layer. We failed to find any near the epidermis. There are no muscles in the breast skin.

Hair on the withers is comprised of spines and fur. The spines in *E. europaeus* are described in detail by Lvov (1884). They are 20.5 mm long, 0.8 mm thick, with a smooth surface. The spines give way to a thin, short club, which thickens sharply into the shaft. Longitudinal outgrowths numbering 18 to 25, do not reach the center but divide the peripheral part into minor sections, and jut into the medulla from the thickened cortical layer (fig. 20). The 57% medulla does not extend into the tip of the spine.

The inner, longitudinal outgrowths stretch continuously through the entire spine. They are approximately equal in height. They are isosceles triangles in cross section with a short base and very elongated sides. In addition to the longitudinal outgrowths of the cortical layer, the central spinal cavity has transverse medullar membranes fixed to the inner surface of the cortical layer, both to the outgrowth and interoutgrowth spaces (fig. 21).

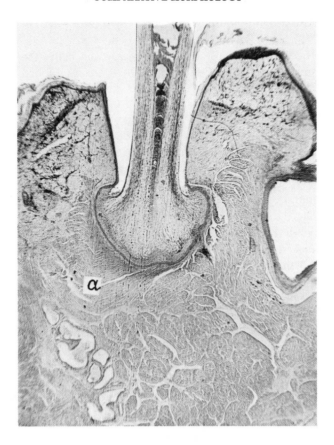

Fig. 19. *Erinaceus europaeus*. Withers skin:
a, arrector pili muscle at spine root.

The spaces between the transverse membranes are approximately equal.
Occasionally they are connected by diagonal septa (fig. 21). The longitudinal membranes of the medulla are fixed to the cortical layer by numerous
"clubs"* merging at some distance from the cortical layer into a membrane
retaining the layered pattern. The surface of the medullar lengthwise membranes is irregular with folds and hollows.

The spines are shed in autumn and in spring, with no less than 50% of the
total spine number remaining (Niethammer, 1970). They can be replaced in
any area of the back. Fur hairs, 16.5 mm long and 30μ thick, with neither
granna nor medulla, twine between the spines.

There are two types of hair on the breast: guard and fur. The guard hairs

---

*They are referred to as "clubs" only conventionally. They look like clubs in
spinal cross section. Actually, they are layers or membranes.

Fig. 20. *Erinaceus europaeus.* Section of spine.

fall into two orders. Both are straight, gradually thicken from base to middle and taper at the tip, and differ only in size (length 16 and 11 mm, thickness 60μ). The medulla is well developed in both. There is no granna. The fur hairs have no granna or medulla. They are twisted and of even thickness (15μ) down most of the length (9 mm).

Erinaceus concolor roumanicus demonstrates that hedgehogs develop three categories of spiny coat during their life (Kratochvil, 1974). The first develops during uterine life, its growth being completed during the first ten days of postnatal development (average length is 9.8-10.8 mm, the total number reaching 300). These spines begin falling out (first from the back) between forty and fifty days of life, but may remain until the animal is two years old.

The second spiny category, forming the juvenile coat, begins to grow at one to two days of postnatal development, completing growth at the beginning of the second ten-day-period of life (average length is 9.6-11.9 mm, the total number reaching 1000). The spines of this second category begin falling out between forty and fifty days of life, but some remain in adult animals.

The subadult spiny coat originates only when the skin is well pigmented (about the fifth day of life). The growth of these spines is complete in the

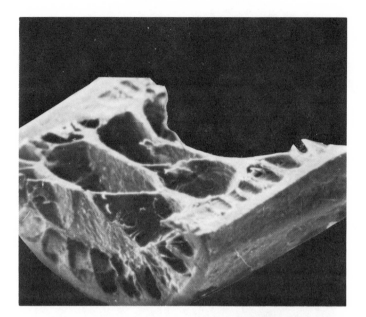

(a)

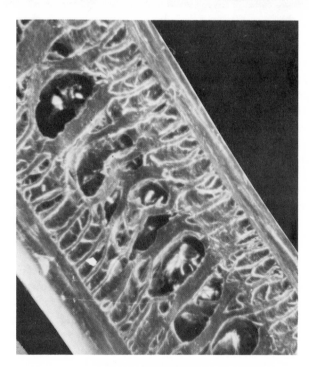

(b)

Fig. 21. *Erinaceus europaeus*. Inner structure of spine:
a, cross-section; b, longitudinal section.

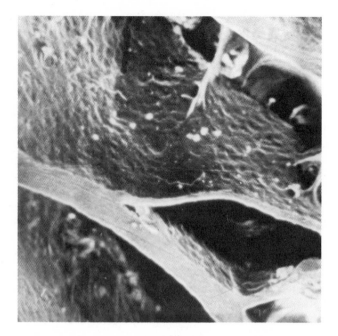

(c)

Fig. 21. c, enlarged view.

final third of the first month (average length is 144-154 mm, the total number reaching 1000-3000). The spines of this category fall out slowly. They may be found even in two-year-old hedgehogs. The adult spiny coat originates between twenty and thirty days of life, its growth proceeding through the entire life of the animal (average length is 17.4-21.9 mm, the total number reaching 6,500 ± 300).

Hemiechinus auritus

Materials. Skin from withers, breast (table 1), and front and hind soles (body length 20 cm), taken in Repetec (Kara Kum), July.

The stratum malpighii is three to four cells deep and contains pigment cells. In the withers skin, the reticular layer gives way to large accumulations of fat cells directly beneath the arrectores pilorum muscles, 2 mm from the epidermis. The plexus of mainly horizontal collagen fiber bundles in the dermis is rather compact. A sparse network of elastin fibers running in various directions is recorded only in the upper dermal layer. The dermis abounds in pigment cells. In the dermis of the breast only, the upper layer, 60-300μ thick, is composed of a dense plexus of collagen fibers, mostly parallel to the surface. The entire skin from under this zone down to the subcutaneous muscles is filled with fat cells 45 x 118; 34 x 73μ. No bundles of collagen fibers are visible among them. The hair roots penetrate deep into the fat plastid (the hair bulbs are 1254μ from the epidermis), and the reticular layer is consequently absent. There are no pigment cells in the breast dermis.

Subcutaneous fat tissue is well developed, 660μ thick on the withers and 235μ on the breast. The size of the fat cells is 39 × 84; 61 × 90μ. A sebaceous gland measures 39 × 84; 61 × 90μ. There is a sebaceous gland (39 × 67μ) at each hair with two larger glands (34 × 169μ) at each spine. The breast sebaceous glands are 35 × 39μ. No sweat glands are revealed on the withers and breast of the individual under study. In a subadult specimen taken in June, sweat glands are found on the withers, sacrum, front, and in the anogenital region (not found on the breast). They are also found in the withers skin of two 5-day-old newborns. In the skin of the subadult, sweat glands form glomi 350 × 700μ (withers), 57 × 226μ (sacrum), 226 × 565μ (front). The outer diameter of secretion tubes is 22 to 79μ. The duct diameter is 11 to 28μ. In young animals, sweat glands form no glomi. The diameter of their secretion cavities is 57μ, that of the ducts is 11 to 22μ. On the front and hind soles, tubular glands form rather large glomi. The diameter of the secretion cavity of the gland reaches 28μ. The arrectores pilorum muscles are positioned in the skin much like in *E. europaeus*. Most of the muscles in the withers miss the epidermis by 200 to 240μ. Small bundles of muscles adhere to the hairs and reach up into the epidermis from these muscles.

The withers include both spines and hair. Spines are 21.3 mm long and 0.9 mm thick. Unlike *E. europaeus, H. auritus* has no spines on the sides of the body. The spines have longitudinal crests or ridges with grooves between. At the tip of each crest is a row of fairly pronounced tubercules. Their cross section is oval with a sinuate outer edge. The structure of the spines of the longeared hedgehog is similar to that of the European hedgehog. The continuous cortical layer of the spine is 55μ thick. The medulla is 67.3%. The sides and the lower part of the body are covered with soft, relatively short (0.8 mm), thin (60μ), straight hairs, which are thickest in the upper third or half. The medulla (90%) is well developed. The two orders of hair on the breast have pronounced differences. The first is 39μ thick with a better developed medulla (48.7%). The second is thinner (28μ) with a less developed medulla (35.7%). The tips of both are broken.

Hemichinus hypomelas

Material. Skin from withers and breast (table 1) (body length 30 cm).

The stratum malpighii is two to three cells deep. The epidermis is not pigmented. In withers dermis, the spine bulbs penetrate as far as 1.1 to 1.3 mm from the outer surface. Powerful arrectors pilorum muscles reach still deeper (to 1.6 mm), with a fat layer directly beneath so there is no reticular layer in the withers. The bundles of collagen fibers in the dermis form a dense plexus of mainly vertical bundles in the upper part and horizontal bundles in the lower part. There are a few elastin fibers. The breast dermis has papillary and reticular layers, their border level with their bulbs. Most of the collagen fiber bundles in the dermis run parallel to the surface, with the plexus firm in the papillary and lax in the reticular layer. The few fat cells in the reticular layer are 28 x 56; 30 x 56μ. Subcutaneous fat tissue in the withers skin reaches considerable thickness (944μ). It is less pronounced in the breast and does not contain fat cells. The sebaceous glands in the withers skin near the hair roots are very small (45 x 50μ), but are rather larger at the spines. There are two glands at every hair or at the spine. The ducts are perpendicular to the hair and spine roots. There is a single, round sebaceous gland (62 x 84μ) at every hair of the breast skin. The arrectores pilorum muscles in the withers are very well developed. Each of them has several parts. The muscles stretch backward and downward from the spine bulb. There is also a connection with the posterior spine.

The withers skin has spines and hairs proper. The spines are 39.7 mm long and 1.4 mm thick, with surface crests, ridges, and a thin, short club. They are almost round but have an inward curve on one side. The cortical layer has numerous small, oval cavities perpendicular to the length of the spine. The continuous outer cortical layer is 70μ thick (medulla 50%). There are small spines as well as large. Only one small spine is found in the sample, length 11.4 mm, thickness 0.5 mm.

The hair divides into two categories. The first (length 27.7 mm), slightly curved, is almost evenly thick (54μ) the whole length, thickening gradually from the base toward the distal quarter to taper gently at the tip. The medulla reaches down the whole tip. The second category is much shorter (9.9 mm) but almost equally thick (53μ). It is straight, thickening gradually from base to middle, tapering at the tip. The poorly developed medulla (32%) is found only in the middle section of the hair. The breast hairs grow in tufts of two to five hairs (mostly two). Six to ten tufts form a group. There are 3,500 hairs per 1 cm^2 of breast skin.

FAMILIA TALPIDAE

More work has been done on this family than on other *Insectivora*. There are descriptions of *Desmana moschata* (Belkovich, 1962), of *Talpa europaea* (Kogteva, 1963), of desman and mole glands (Weber, 1928; Schaffer, 1940), and of the pelage of *D. moschata* (Gudkova-Aksenova, 1951).

We studied *Uropsilus soriceps, Desmana moschata, Galemys pyrenaicus, Talpa europaea, Talpa altaica, Talpa caeca, Talpa micrura wogura, Talpa m. robusta,* and *Condylura cristata.*

The family includes highly varied, specialized animals. *Uropsilus soriceps* (usually regarded as a subfamily of *Uropsilinae*) lives on land, *Desmana moschata* (subfamily *Desmaninae*) is aquatic, and moles (subfamily *Scalo-*

pinae, Talpinae, Condylurinae) live underground. Their pelage varies with their mode of life.

The skin of primitive *Uropsilus soriceps* has much in common with the shrew. Skin, epidermis, and stratum corneum are thin, with the reticular layer of the dermis only slightly developed. Sebaceous glands are small, sweat glands saccular and straight, and the hairs grow singly. Three categories are distinguished: guard, pile, and fur. The pile hair shafts have ligatures and longitudinal waves. All the hairs have a thick medulla.

Desman skin is rather thick, though the epidermis and stratum corneum are thin. The reticular layer, rather well developed, adjoins the papillary layer at the bottom of the sweat glands. The sebaceous glands are large and very twisted. There are no specific glands in the sole, but there is a large subcaudal gland. The hairs are markedly differentiated into distinguishable categories. Shaft ligatures are not recorded. In the pile hairs, the granna is pronounced but is without a medulla. The fur hairs grow in tufts. There are no arrectores pilorum muscles. The hair is denser on the belly than the back. Molting is prolonged.

Moles are subterranean and their skin is thick, especially on the breast, with the epidermis and stratum corneum being thicker there than on the back. The reticular layer is very well developed, and the sebaceous glands are very large. Sweat glands have slightly twisted, sausage-shaped secretion cavities. There are no specific glands in the sole. Categories are pronounced. The pile is not marked. The medulla is very thin. The pile hair shafts have ligatures and longitudinal waves. In some mole species the hair grows in tufts. Compensatory molts occur.

Uropsilus soriceps

Material. Skin from withers (table 2), body length 11 cm) taken in China.

The stratum malpighii is one cell deep. The stratum corneum is compact. The sebaceous glands are small ($34 \times 67\mu$) with one lobe, and sweat glands are separate. We found only one sweat gland, which was saccular, straight, 34μ in diameter, and stretched horizontally beneath the hair bulbs.

The hair divides into three categories: guard (length 6.0 mm, thickness 95μ, medulla 71.5%), pile (5.7 mm, 28μ, 71.5%), and fur (4.9 mm, 15μ, 86.7%). Pile hairs have three ligatures with longitudinal turns and a 1.7 mm granna. The hairs grow singly.

Desmana moschata

Material. Skin from withers, breast (table 2), hind soles and base of tail of three adults (body lengths 16-18 cm), taken in Khoper Reservation, September.

The epidermis measures $15\text{-}20\mu$, the stratum corneum $2\text{-}5\mu$. The indexes of the epidermis reported by Belkovich (1962) are rather different from ours (table 2). The stratum malpighii complex is two to four cell layers. The stratum corneum is lax, composed of numerous corneous plastids. According to Belkovich (1962), the thickness of epidermis and stratum corneum increases from back to belly. In the dermis, bundles of collagen fibers run in different directions in an intricate plexus in the upper parts of the papillary layer only. The other parts of the dermis have mainly horizontal bundles. The reticular layer is rather well developed (75% of the dermal thickness according to Belkovich, 1962). There are fat cells (28×50; $34 \times 56\mu$) on the withers between the hair roots in the middle and deeper papillary and upper reticular layers. The upper half of the specimen (1) reticular layer in the breast skin is filled by continuous accumulation of fat cells (39×56, $45 \times 50\mu$), which the roots of growing hairs reach in due course. There are few fat cells in the breast skin of specimen (2). The sebaceous glands are small (39×84; $34 \times 146\mu$) and oval, two (or even three, according to Belkovich, 1962) at every hair tuft. They are clearly visible even at the growing hairs (fig. 22). The sweat glands are unevenly developed. They are relatively small (34 and 56μ in diameter) and slightly sinuate in the withers of *D. moschata,* specimens (1) and (2). In specimen (3) the glands are more sinuate and their greatly elongated secretion cavities reach deep into the skin beyond the bulbs of the growing hairs; the diameter of specimen (3) is considerably larger (112μ). The sweat glands are fewer and smaller on the breast (22 to 39μ) (fig. 22) than on the withers. No glands are found in the hind sole. An anal gland is noted (Weber, 1928).

The ventral tail surface has a subcaudal gland, 1 to 1.5 cm from the tail base. The relative length of the glandular complex is 20 to 26.6% of the tail length. The complex comprises sebaceous glands varying in shape from a common to an alveolar sac (Sokolov et al., 1977) (fig. 23). The gland numbers range from 13 to 20. Their size is 5.24×4.07 mm in adult females, and 5.75×2.63 mm in adult males. Alternating ducts open on both sides of the tail. There are hypertrophied sebaceous glands on the dorsal tail surface over the subcaudal gland. These are similar in structure to those of the subcaudal complex, being lesser in size than the latter.

The withers hair divides into three categories: guard, pile (first and second orders), and fur. Gudkova-Aksenova (1951) also distinguishes three categories in desman sacrum skin, but does not divide pile hairs into orders. The guard hairs, few in number, are straight (length 18.5 mm, thickness 125μ), with a rather thin, long club gradually thickening into a pronounced granna in the distal third of the hair. The medulla is present only in the club. The pile hairs, first order, are 17.1 mm long and 124μ thick. Unlike the guard hairs, their wide granna is separated from the club by a sharp neck. The granna extends along two-thirds of the upper hair shaft while the medulla is pronounced only in the club. In the pile hair, the club is characteristically wavy (five to six waves).

TABLE 2

TALPIDAE: SKIN MEASUREMENTS (μ)

	Skin	Epidermis	Stratum corneum	Dermis	Papillary layer	Reticular layer
Uropsilus soriceps						
withers	118	7	5	111	100	11
Desmana moschata						
(1) withers	2,090	45	27	2,045	1,825	220
(2) withers	1,350	27	18	1,323	938	385
(3) withers	2,527	27	16	2,500	1,950	550
(1) breast	1,407	27	16	1,380	698	682
(2) breast	1,925	25	14	1,900	1,680	220
Galemys pyrenaicus						
(1) withers	912	32	14	880	640	240
(2) withers	770	16	11	754	534	220
(1) breast	880	38	24	842	512	330
(2) breast	715	31	20	684	464	220
Talpa europea						
withers						
Trans-Carpathian region						
(1)	392	23	14	369	201	168
(2)	320	14	7	306	144	162
Moscow region	693	14	7	679	239	440
Komi ASSR	1,243	18	13	1,225	620	605
breast						
Trans-Carpathian region						
(1)	1,650	34	23	1,616	1,066	550
(2)	913	21	16	892	232	660
Moscow region	890	28	13	862	224	638
Talpa europea altaica						
(1) withers	451	14	9	437	151	286
(2) withers	528	18	13	510	224	286
(1) breast	627	20	15	607	176	431
(2) breast	784	27	18	757	337	420
Talpa caeca						
(1) withers	285	9	4	276	242	34
(2) withers	224	16	11	208	152	56
(3) withers	235	23	18	212	139	73
(1) breast	2,035	12	4	2,023	1,088	935
(2) breast	1,456	18	9	1,438	1,102	336
(3) breast	1,960	27	18	1,933	813	1,120
Talpa micrura wogura						
withers	1,133	11	4	1,122	374	748
breast	1,749	50	28	1,699	907	792

TABLE 2—*continued*

	Skin	Epidermis	Stratum corneum	Dermis	Papillary layer	Reticular layer
T. m. robusta						
(1) withers	1,210	9	5	1,201	321	880
(2) withers	1,485	–	–	1,485	990	495
(1) breast	2,310	25	11	2,285	745	1,540
(2) breast	1,980	21	7	1,959	1,189	770
Condylura cristata						
withers	360	–	–	360	280	80

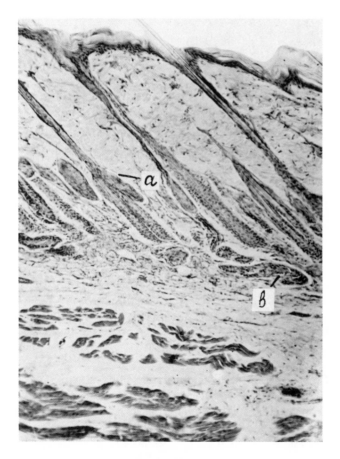

Fig. 22. *Desmana moschata*. Breast: General structure of skin;
a, sebaceous glands; *b*, sweat glands.

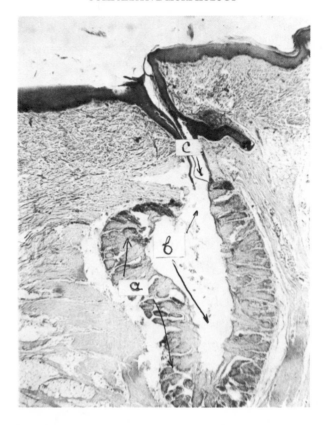

Fig. 23. *Desmana moschata.* Subcaudal complex gland: *a,* alveoles;
b, glandular cavity; *c,* duct.

The second-order pile hairs differ from the first only in size (length 16.2 mm, thickness 93μ). The granna extends only along one-fourth of the shaft. The fur hairs have no granna, are wavy with up to fourteen waves (length 13.7 mm, thickness 13μ, medulla 23.0%). The pile hairs grow singly; the fur hairs grow in tufts of two to four, mostly four (up to seven, according to Belkovich, 1962). There are 333 pile and 22,333 fur hairs per 1 cm^2 (according to Belkovich [1962], 993.% are fur hairs).

Like the withers, the breast has guard, pile, and fur hairs. The guard hairs (length 15.0 mm, thickness 115μ) are similar to the withers hairs, with the granna running almost halfway along the hair. The first-order pile hairs (length 13.6 mm, thickness 110μ) like the withers, have a wavy club with up to five waves and a neck before the granna. The pile hairs, second order, differ in size (length 12.9 mm, thickness 62μ). The granna is considerably shorter (average length 3.4 mm). The number of waves in the club is seven to nine. The fur hairs (length 10.2 mm, thickness 10μ) do not have a granna.

The medulla (60%) is developed down the whole hair except tip and base. There are up to nine waves.

Galemys pyrenaicus

Material. Skin from withers and breast (table 2) from two specimens (body lengths 12 and 13 cm) taken in French Pyrenees.

The lower border of the breast skin is rather ordinary with fairly thick interlayers of cross-striated muscle fiber bundles, as in the middle of the reticular layer. The pattern and the density of the collagen fiber plexus over and under them is identical. Skin thickness is measured to the cross-striated muscles. The stratum malpighii is two to three cells deep; the stratum corneum is lax. In the withers dermis, bundles of collagen fibers, mostly horizontal, form a dense plexus. In the breast dermis, collagen bundles run in various directions, horizontal bundles being found only in the upper papillary layer. The single-lobe sebaceous glands (50 x 95; 123 x 224 mm) (fig. 24) are seen in horizontal section as anterior and posterior to each hair tuft. As in *D. moschata,* there is a specific skin gland, though it is not as well developed and forms no swelling on the tail (Schaffer, 1940).

The sweat glands are well developed (56 to 78μ in diameter) (fig. 24). Horizontally, slightly elongated glomeruli reach down deeper than the hair bulbs. The adjoining sweat glands form an almost continuous glandular layer. The sweat gland glomeruli are larger in the withers than the breast. Anal glands are formed by tubular and sebaceous glands (Schaffer, 1940) but no glands are found in the sole callosities (Schaffer, 1940). A subcaudal gland at the base of the tail consists of sebaceous glands of a complex, arborizing, alveolar type (Sokolov et al., 1977). Their size is 1.35 x 1.03 mm. The ducts open into hair follicles.

The withers pelage divides into three categories: guard, pile (first, second, and third orders) and fur. The guard hairs (length 11.9 mm, thickness 113μ) have a long, rather thin club gradually thickening into a granna. The medulla is pronounced only in the club, and the hair tips are broken. First-order pile hairs (length 12.6 mm, thickness 124μ) have a wavy club (four waves), a neck before the 3.7 mm granna, stretching slightly over a third down the hair. The medulla is present only in the club. The pile hairs of the second and third orders are smaller (length 12.6 and 11.6 mm, thickness 98 and 46μ). The second order has a club with five waves; the third order has eight. The granna length reaches 2.8 mm; in second-order pile hairs, just over one-fourth of the hair length; in third-order, 1.5 mm, about one-sixth of the length. The fur hairs (length 9.9 mm, thickness 13μ) have no granna. There are up to nine waves in the shaft. The medulla (69.3%) is well developed in the whole hair length except in the tip and base. Both guard and pile hairs grow singly, and fur hairs grow in tufts of two to five (mostly four) hairs. There are 1,000 pile and 28,333 fur hairs per 1 cm², but no groups.

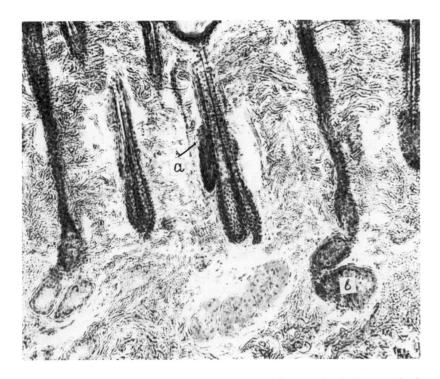

Fig. 24. *Galemys pyrenaicus*. Papillary layer (breast): *a*, sebaceous glands; *b*, sweat glands.

The same categories are found on the breast as on the withers, and they do not differ appreciably in morphology. The guard hairs are 11.6 mm long, 93 μ thick; the pile hairs, first order, are 11.5 mm long, 124 μ thick; the pile hairs, second order, are 11.4 mm long, 84 μ thick; the pile hairs, third order, are 10.5 mm long, 30 μ thick; the fur hairs are 9.8 mm long, 11 μ thick, medulla 56.0%. The granna in the first-order pile hairs averages 3.1 mm (a little less than one-third of the hair); the second order averages 2.5 mm (a little over one-fifth of the hair); the third order averages 1.8 mm (about one-sixth of the hair). The guard and pile hairs grow singly, and the fur hairs grow in tufts of mainly four, occasionally three, and rarely two hairs. There are 2,333 pile and 53,666 fur hairs per 1 cm² of skin.

Talpa europaea

Material. Skin from withers and breast (table 2). Four specimens (body length (1) 11.7 cm, (2) 11.5 cm, (3) 11 cm, (4) 12 cm); specimens (1), (2) taken in Porechye (Trans-Carpathia) July. Specimen (3) near Zvenigorod (Moscow region), July, (4) in Pechora (Komi ASSR), September.

The Komi mole's hair is growing. The stratum malpighii is one to two cells deep in all specimens. The sacrum skin of specimens taken in Leningrad and Novgorod is 480 to 580μ in winter, 705μ in summer (Kogteva, 1963). The thickness of the epidermis varies from 9μ in winter to 12μ in summer. The stratum corneum, consisting of solid rows of cornified cells, is well developed in these specimens. In spring it becomes lax and exfoliates. The stratum malpighii is simple, except in the pelage replacement season, when it is complex, three to four cells. The dermis thickness ranges from 300 to 365μ in winter and from 423 to 495μ in summer. In the sole of the June Zvenigorod specimen, the epidermis is very thick, 230μ on the front and 130μ on the hind feet. The stratum corneum takes up 100μ on front and 35μ on hind feet. The papillary and reticular layers are level with the hair bulbs. Sweat glands, deeper than the hair bulbs, are very sparse. The upper papillary layer is compact, the middle of the papillary layer is a lax collagen fiber plexus, becoming compact at hair bulb level. The reticular layer is pronounced, often thicker than the papillary layer (according to Kogteva, 1963, 47 to 48% of the sacrum dermis thickness). The upper reticular layer has a compact collagen fiber plexus; the lower layer plexus is loose. In the dermis, the majority of bundles run parallel to the surface. No fat cells are recorded in the dermis or the subcutaneous fat tissue. The small area of hairless skin on the mole's muzzle has a large number of nerve endings where "Eimer organs" are formed by well-innervated epidermal papillae that jut into the dermis (Eimer, 1871). Encapsulated nerve endings are near the deep parts of the Eimer organs (Ranvier, 1887; Bielschowsky, 1907; Boeke, 1926). Armstrong and Quilliam (1961) refute the Cauna and Alberti (1961) claim that there are no encapsulated endings in the mole muzzle dermis. They have found differentiated capsules with very lightly pronounced inner bulbs (Armstrong and Quilliam, 1961). The nerve endings are not twisted but almost straight where they enter the capsule, with no visible modifications in the terminal. The encapsulated nerve endings in the mole muzzle appear to be capable of perceiving high-frequency pressure changes.

Single-lobe sebaceous glands in the withers are mostly well developed (62 ×196; 39 × 90μ) (fig. 25). They are rather smaller in the breast (44 × 84; 28 × 101μ). Adjoining glands adhere closely to one another, forming a continuous glandular layer. There is a gland at every hair embracing the hair sheath on almost all sides (according to Kogteva, 1963, there are two glands at every hair). Large vacuoles are noted in a number of glands. The sebaceous glands in the sacrum skin are larger in winter (112 to 152μ) than in summer (83 to 90μ) (Kogteva, 1963).

The sweat glands are rare (but not absent, as Kogteva reported in 1963). Their secretion cavities are botuliform, 22 to 45μ in diameter, and slightly twisted (fig. 26). A specific gland may be found in the middle of the back, both in males and females (in the mating season). It is made up of large sebaceous glands (V. Lehman, 1969). The anal gland is made up of sebace-

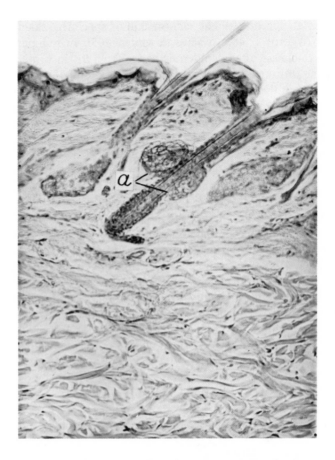

Fig. 25. *Talpa europaea*. Skin of breast: *a*, sebaceous glands.

ous and tubular glands around the anus (Schaffer, 1940). There are large
preputial glands—rather large sebaceous glands connected with the hairs
(Schaffer, 1940). The meatus acousticus externus has large sebaceous glands
opening as ducts at the transition of the acoustic duct into the tympanic
cavity (Schaffer, 1940).

According to Murariu (1971, 1974, 1975b, 1976), sebaceous glands occasionally open into the trunk skin surface; in the lips there are hypertrophied
sebaceous glands and occasional apocrine sweat glands; in foot soles there
are rare eccrine sweat glands; in the circumanal region, there are sebaceous
apocrine sweat glands 50μ in diameter; the anal glands are formed by sebaceous and eccrine sweat glands; the preputial glands are small, sebaceous
type; and abundant, short, apocrine sweat type.

The number of molts per year is not certain. Some researchers believe
that *T. europaea* molts three times a year in spring, summer, and autumn

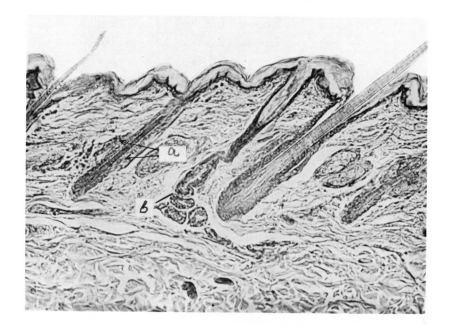

Fig. 26. *Talpa europaea.* Breast: *a,* sebaceous gland; *b,* sweat gland.

(Bashkirov and Zharkov, 1934; Kuznetsov, 1952; Stein, 1954; Stroganov, 1957), while others report four molts (Tserevetinov et al., 1948; Deparma, 1951). Kogteva (1963) showed substantial evidence that in spring the mole has a complete change of coat. In summer the first pelage is replaced by a similar but new one. The summer molt is protracted. A complete replacement of pelage is characteristic of the autumn molt, and a compensatory molt in certain areas of the body has been noted in some moles in winter. There are three clear zones for pelage length (Tserevetinov, 1958). The first (hair length 8 mm) takes in almost the whole body. The second (hair length 7 mm) includes a short strip on the head, shoulder blades, and belly. At the forefeet and flank the hair is 6 mm long. Kogteva (1963) classifies four zones with different hair lengths on *T. europaea,* the longest on the sacrum, rather shorter on the back, still shorter on the belly, and shortest at the limbs.

We studied the pelage of a mole taken in the Komi ASSR (withers). The hair divides into three categories: guard, pile, and fur (most researchers distinguish three categories of hair [Toldt, 1928; Kogteva, 1963]). The guard hairs are straight (length 8.0 mm, thickness 45μ), gradually thicken into a granna in the upper third of the shaft, then taper to a thin tip. The medulla is well developed along the whole hair (48.9%), except for base and tip. The granna of the pile hairs is less distinct. The club usually has four narrowings with longitudinal turns where the hair is bent at an obtuse angle (fig. 27).

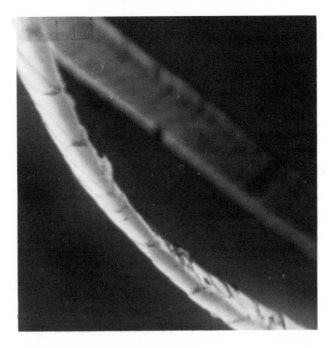

Fig. 27. *Talpa europaea.* Longitudinal turn of shaft of pile hair.

The pile hairs divide into two orders, which vary in the degree of granna development. The first-order hairs are 7.7 mm long, 37μ thick, medulla 73%; the second-order are 8.1 mm long, 32μ thick, medulla 57.3%. The granna is shorter in the second order. The fur hairs (length 6.9 mm, thickness 18μ) are slightly twisted without a granna. Their medulla is pronounced (66.6%). The hairs grow singly, 29.3 to 34.6 thousand hairs per 1 cm² on the withers. On the breast the hairs also divide into three orders, 27.6 thousand per 1 cm². The length varies with seasonal changes (Bashkirov and Zharkov, 1934), as do thickness, medulla development, and the number of hairs (Kogteva, 1963). In the winter the hairs are longer and thinner (table 3), the medulla thickens (table 4), their number being very much larger. In June the mean number of the hairs per 1 cm² of the sacrum is 12,150; in December it is 17,380 (Kogteva, 1963).

T. e. altaica

Material. Skin from withers and breast (table 2) of two animals (body length 10.3 cm, 10 cm) taken in Altai near Shebalino, one on June 22, one on July 3.

The stratum malpighii of the epidermis is one to two cells deep. The stratum corneum is rather compact. In the dermis, the outer, more compact

TABLE 3

TALPA EUROPAEA: WINTER AND SUMMER PELAGE, WITHERS (KOGTEVA, 1963)

Hair categories	n	Summer pelage		n	Winter pelage	
		M±m	±σ		M±m	±σ
Length (mm)						
Guard hair	36	6.6±0.09	0.54	25	9.7±0.06	0.31
Pile hair	75	6.6±0.05	0.46	25	9.5±0.04	0.2
Fur hair	74	5.9±0.05	0.43	25	8.5±0.08	0.42
Thickness (μ)						
Guard hair	64	40.5±0.03	0.25	50	34.5±0.04	0.29
Pile hair	25	41.4±0.07	0.31	36	35.1±0.03	0.18
Fur hair	25	12.9±0.04	0.20	25	10.5±0.07	0.31

TABLE 4

TALPA EUROPAEA: THICKNESS OF THE MEDULLA ON SACRUM, SUMMER AND WINTER PELAGE
(KOGTEVA, 1963)

Hair category	Medulla			
			% of hair thickness	
	summer	winter	summer	winter
Guard hair				
granna	21.6	21.6	54	64
base	17.6	12.8	50	66
Pile hair				
granna	19.2	23.5	48	69
base	17.6	11.2	50	70
Fur hair				
distal section	—	—	—	—
base	11.5	12	60	66

part of the papillary layer is 55 to $58\,\mu$ on the withers, and 85 to $110\,\mu$ on the breast. The border between the papillary and reticular layers is level with the bottom of the sweat glands. In the dermis the bundles of collagen fibers are mainly horizontal. The sebaceous glands have a single lobe, 28×67; $50 \times 196\,\mu$. On the withers of specimen (1) the sweat glands are barely visible ($22\,\mu$ in diameter). On specimen (2) they are well developed ($45\,\mu$). They are sausage-shaped, slightly twisted, and lie horizontally under the hair bulbs. The sweat glands on the breast are more pronounced (45 to $67\,\mu$ in diameter) and have strongly twisted secretion sections.

The withers hair is not growing on specimen (1), but on specimen (2) it is in the first stage—the hair bulbs have begun to swell. The hair roots form an angle of $40°$ with the surface. The Altai mole has three molts—spring, summer, and autumn (Deparma, 1951). The hairs divide into three categories. On the withers the guard hairs are 7.2 mm long, $36\,\mu$ thick, medulla 44.4%, the breast guard hairs are 6.0 mm long, $40\,\mu$ thick, medulla 45%. There is only one order of pile hairs. On the withers it is 7.3 mm long, $33\,\mu$ thick, medulla 63.7%, and 5.8 mm, $33\,\mu$ and 63.7% on the breast. The pile hair club has up to three necks with longitudinal turns. The withers fur hairs are 6.3 mm, $15\,\mu$, medulla 73.4%; the breast fur hairs 4.8 mm, $15\,\mu$ and 63.4%. There are 23,000 hairs per 1 cm^2 on the withers. All hairs grow singly, the breast hairs growing in not very distinct rows, 20,000 per 1 cm^2.

Talpa caeca

Material. Skin from withers and breast (table 2) of three specimens (body length 10 cm, 10.3 cm, 10 cm). Specimens (1) and (2) taken near Psebay (Caucasus), August 14, and specimen (3) near river Belaya.

The structure of the epidermis and dermis are similar to that of *T. europaea*. Single-lobed sebaceous glands are well developed (39×90; $112 \times 168\,\mu$). No sweat glands are found in specimen (1), though specimen (2) does have sweat glands on the withers ($28\,\mu$ wide) with tubular, twisted secretion sections. Three typical categories of hair are similar in shape to *T. europaea*. On the withers, guard hairs are 7.1 mm long, $34\,\mu$ thick, medulla 47.1%; pile hairs 7.5 mm, $32\,\mu$ and 68.8%; and fur hairs 6.4 mm, $14\,\mu$, and 71.5%. On the breast the measurements are: guard hairs 6.0 mm, $40\,\mu$, and 51.4%; pile hairs 6.8 mm, $38\,\mu$, and 57.9%; fur hairs 5.8 mm, $16\,\mu$, and 75%. The hairs grow singly, an estimated 29,000 per 1 cm^2 on the withers and 23,000 on the breast.

The figures and the skin structure of the December mole differ mainly in that the stratum corneum is much thicker. In the dermis of the withers, the outer, firmer section of the papillary layer is 45 to $70\,\mu$ thick. The breast dermis has a characteristic thickened reticular layer. The pelage is preparing for growth, the hair roots have lengthened, yet the papillary layer is thinner than in summer. Single-lobe sebaceous glands, rather large in the withers

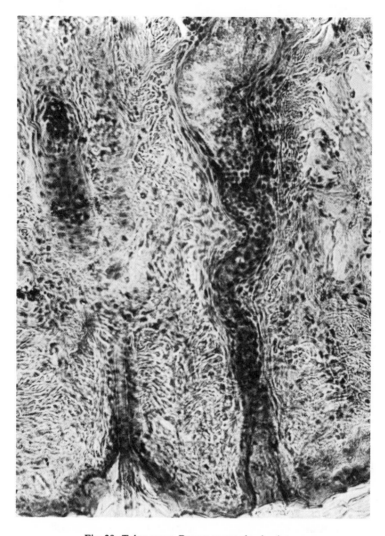

Fig. 28. *Talpa caeca*. Breast: sweat gland, winter.

$(67 \times 140\mu)$, are smaller in the breast $(45 \times 73\mu)$. The sweat glands, which are sparse and slightly twisted, are considerably larger in the breast skin (withers 34μ in diameter, breast 73μ) (fig. 28).

The hair divides into three categories, differing from the summer specimen in size only—it is longer and the medulla is better developed. The withers guard hairs are 9.3 mm long, 35μ thick, medulla 47.1%; the pile hairs 9.3 mm, 34μ, 77%; the fur hairs 8.0 mm, 15μ, 75%. The breast guard hairs

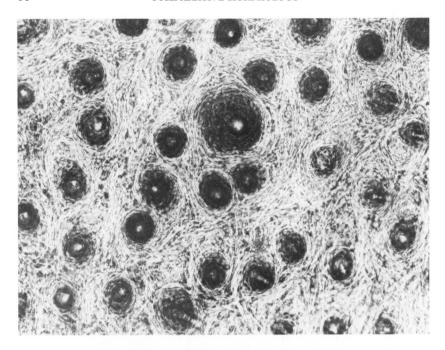

Fig. 29. *Talpa caeca.* Single arrangement of hairs in skin.

are 7.8 mm, 45μ, 33.3%; the pile hairs 7.0 mm, 32μ, 68.8%; the fur hairs 6.2 mm, 16μ, 75%. There are 23,000 single-growing hairs per 1 cm^2 on the withers (fig. 29), and 29,333 on the breast.

Talpa micrura wogura

Material. Skin from withers and breast (table 2) of summer specimen from the Primorye territory.

The withers hairs are in early anagen (the hair bulbs are slightly swollen), so the papillary layer of the dermis is thinner than the reticular. The breast hairs are growing, so the papillary layer is thickened. The outer papillary layer is not pronounced. In the reticular layer, the bundles of collagen fibers run in various directions. There is a pair of rather large sebaceous glands at every hair tuft, single-lobed on the withers (50 × 196μ), with two to three lobes on the breast skin (112 × 224μ). Pigment cells are concentrated around the sebaceous glands. Numerous strongly twisted sweat glands have vertically lengthened secretion cavities. In the withers skin they reach no deeper than the hair bulbs and are 56μ in diameter. In the breast skin there are sweat glands at almost every hair tuft, but only where the hair is in anagen. They are strongly twisted and penetrate deeper into the skin than the hair bulbs, the secretion cavity diameter reaching 95μ. In the skin regions with very long hair bursae, sweat glands are greatly reduced or absent.

The hairs grow in tufts but not in groups, two to four hairs (breast) and four to six hairs (withers) to a tuft. There are 16,333 hairs per 1 cm² (withers) and 14,333 (breast).

Talpa micrura robusta

Material. Skin from the withers and breast (table 2) of male (body length 182 mm) and female (body length 173 mm) both taken in the Sikhote Alin Mountains, Soviet Far East, September.

The stratum malpighii is one to three cells deep. The epidermis is highly developed on front and hind soles, from 220 to 275μ on the front and 106 to 154μ on the hind feet. The stratum corneum is also thick—132μ on the front and 45 to 50μ on the hind feet. The male's hair is not growing, and the papillary layer is thinner than the reticular. The female's hair is growing, so the papillary layer is thicker. In the outer papillary and the inner reticular layers, bundles of collagen fibers are mainly horizontal (fig. 30). In the other parts of the dermis, more bundles are arranged in other directions. There is a pair of sebaceous glands at every tuft of hair (anterior and posterior). They are large (56 × 196; 34 × 112μ) and single-lobed. The sweat glands are rather weakly developed. In the withers, their slightly twisted or straight secretion cavity (diameter 34 to 35μ) along the hair roots is almost

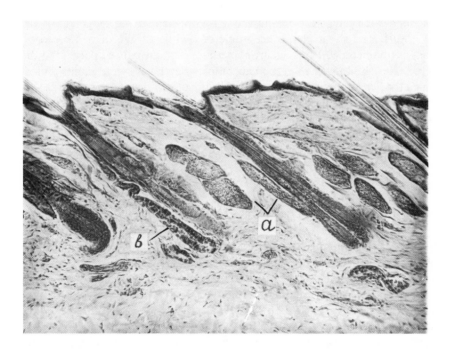

Fig. 30. *Talpa micrura robusta.* Breast, papillary layer: *a,* sebaceous gland; *b,* sweat gland.

vertical. The breast sweat glands ($22\,\mu$ in diameter) have a slightly twisted, sausage-shaped secretion cavity that runs horizontally under the hair bulbs.

The pelage of the withers divides into three categories. The guard hairs are 6.8 mm long, 41μ thick, medulla 36.5%; the pile hairs are 7.8 mm, 47μ, 48.9%; the fur hairs are 6.2 mm, 17μ and 47%. The granna takes up 1.9 mm, about a quarter of the pile hair length. The club of the pile hair has up to five ligatures with longitudinal turns. The fur hairs have up to five waves. The hairs grow in tufts. There are three categories of hair on the breast: guard hairs 6.0 mm, 54μ, 25.9%; pile hairs 5.1 mm, 33μ, 51.5%; fur hairs 6.0 mm, 17μ, 52.9%. The length of the granna of the pile hairs is 1.9 mm, more than one-third the length. The club of the pile hairs has up to three ligatures. The shaft of the fur hairs has up to four waves. The hairs grow in tufts.

Condylura cristata

Material. Skin from withers (table 2) and hind sole skin of adult star-nosed mole. Epidermis destroyed by poor fixing.

The epidermis of the hind sole is $64\,\mu$ thick, the compact stratum corneum 26μ. The epidermis does not have real alveoli; it has protuberances and recesses on the inner surface. The withers hair is not growing. In the reticular layer the collagen bundles are horizontal. In the papillary layer they are at various angles. The arrectores pilorum muscles are very thin (about 4μ). The sebaceous glands at every hair root are not strongly developed (25×70; $35 \times 80\mu$). The sweat glands are numerous with sausage-shaped secretion cavities (diameter 60μ), which are sometimes strongly twisted parallel to the skin surface and lower than the hair bulbs. Individual sweat glands almost touch the subcutaneous muscles. On the dorsal surface of the head (also on the mentum, carpus, perianal, and inguinal regions) larger, highly twisted sweat glands form peculiar glandular complexes in the inner dermis. They produce a secretion imparting to the fur a brownish or yellowish tint noted during the reproduction season (April, May, June in north eastern America) (Eadie, 1948, 1954). A similar phenomenon is recorded in the members of the genera *Parascalops* (Eadie, 1939), *Scalops u. scapanus* (Eadie, 1954). The odor of the secretion appears to attract the opposite sex. The hairs grow singly, 16,600 hairs per 1 cm² on the withers.

At the end of the muzzle of *C. cristata* are twenty-two sensory rays formed of thick dermis and epidermis (Van Vleck, 1965). Sinuslike arterioles and a relatively large nerve pass inside the rays. On the outside, the epidermis forms numerous papillae; each contains 1 to 3 tactile organs called Eimer organs. These organs are found in high concentrations predominantly on the medial surface of the ray, although some are found occasionally on the lateral sides. The function of the organ is thought to be primarily tactile, although pain and temperature may also be served (Cauna and

Alberti, 1961). Each Eimer organ consists of 6 to 9 nerve fibers that extend up through a column of epithelial cells of the stratum corneum.

FAMILIA TENRECIDAE

The skin of tenrecs has not been examined.

We studied *Tenrec ecaudatus, Hemicentetes semispinosus, Echinops telfairi, Microgale pusilla, Limnogale mergulus, Potamogale velox, Micropotamogale lamottei.*

The skin may reach a considerable thickness because some of the hairs are modified into spines. The dorsal skin is considerably thicker than the belly skin. Cross-striated muscles in twos or threes reach high into the skin (fig. 34). The sebaceous glands are small, with one lobe. The sweat glands have tubular, twisted secretion cavities which penetrate deeper into the skin than the hair bulbs. The soles have no specific glands. The arrectores pilorum muscles are well developed. Spines sometimes occur in the skin, developed to a different extent in various tenrecs. The hair is mainly coarse and sparse, but sometimes is soft and abundant.

Tenrec ecaudatus

Material. Skin from withers, breast (table 5), and soles of two specimens (body length 24 and 26 cm).

The stratum malpighii is two to three cells deep in the withers epidermis and three to five cells deep in the breast. The stratum corneum is rather compact, consisting of many corneous layers. In the soles, the epidermis is rather thick with slight alveoli on the inner surface; in the epidermis, stratum granulosum, and stratum lucidum are distinguished.

The lower border of the body dermis coincides with the zone of sparse, horizontal bundles of cross-striated muscle fibers in groups of two to three (fig. 31). Some of these groups are close to the sweat glands, some are nearer to the epidermis than to the hair bulbs. The connective tissue under the cross-striated muscles is laxer than that above. The smaller collagen fiber bundles in the upper dermis (50 to 60μ thick) lie parallel to the skin surface in a dense plexus. The bundles of cross-striated muscles lying in the same direction are much more loosely packed.

In the remaining part of the dermis, the collagen bundles lie at various angles but are predominantly horizontal. The plexus of bundles is rather dense. The border between the papillary and reticular layers is level with the lower sections of the sweat glands. There are fat cells (22 x 50, 28 x 62μ) in the withers skin in the reticular layer of the dermis near groups of horizontal, cross-striated muscles, but there are few fat cells in the dermis of the breast. Their size is 39 x 62; 45 x 56μ. There are sensory cells and a network of nerve endings that adhere to the epidermis in the dermis of two small

TABLE 5

TENRECIDAE AND CHRYSOCHLORIDAE: SKIN MEASUREMENTS (μ)

	Skin	Epidermis	Stratum corneum	Dermis	Papillary layer	Reticular layer
Tenrec ecaudatus						
(1) withers	1,320	32	14	1,288	1,068	220
(2) withers	1,679	29	14	1,650	990	660
(1) breast	1,244	56	22	1,188	693	495
(2) breast	715	30	18	685	465	220
Hemicentetes semispinosus						
withers	726	22	13	704	704	—
breast	660	28	17	632	522	110
Microgale pusilla						
withers	144	—	16	128	108	20
breast	172-400	—	16	156-384	136-364	20
Limnogale mergulus						
withers	608	30	8	578	426	152
breast	456	15	7	441	320	121
Potamogale velox						
withers	680	38	30	642	342	300
breast	1,030	20	12	1,010	804	206
Micropotomogale lamottei						
side of body at scapula	380	25	12	355	275	80
Chrysochloris sp.						
(1) withers	600	45	38	555	479	76
(2) withers	608	38	23	570	464	106
(1) breast	570	45	38	525	420	105
(2) breast	520	53	35	467	362	105

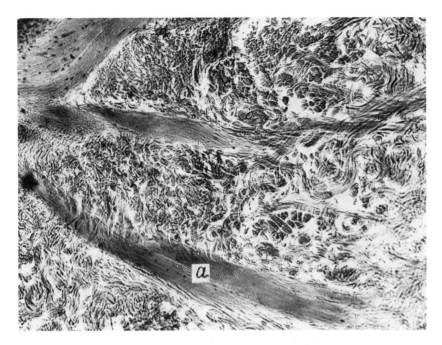

Fig. 31. *Tenrec ecaudatus.* Withers skin: *a,* cross-striated muscles.

swellings on the muzzle near the nostrils (Bielschowsky, 1907). There are sebaceous glands (72 × 140; 84 to 207μ) behind each hair (fig. 32). Some breast hairs have two glands. Sweat glands (diameter of secretion cavity 45 to 73μ) are present at each hair (fig. 33). The sweat glands are better developed on the breast than on the withers because of the larger glomeruli.

The body is covered by coarse, dense hair, with hairs half spine, half bristle on the dorsal side. The hair roots have arrectores pilorum muscles 90 to 165μ in diameter. The withers hairs fall into three categories and are so individual that the accepted names are less suitable than simple enumeration.

The first category (half spine, half bristle) are 22.1 mm long. They are too soft for spines and too hard for bristles. They are slightly wavy (up to sixteen waves); they thicken gradually toward the distal in the upper two-thirds to three-quarters of the shaft. There is a dark ring of pigment at the thickest part. Base and tip are white. The tip is sharpened and slightly split, almost oval in cross-section, with a thick cortical layer. The 70% medulla is absent only in the base and tip.

The three sizes of true hairs fall into the second category. The first are 23.8 mm long and 16.5 thick. They have a similar profile to the bristle-spine and have twenty-two waves (various transitions between the first and second categories can be found). The medulla (83.4%) is developed along the whole length except base and tip. The second order differs from the first

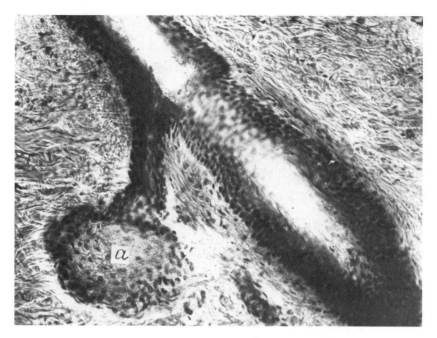

Fig. 32. *Tenrec ecaudatus.* Withers skin: *a,* sebaceous gland.

only in size (length 16.1 mm, thickness 101μ, medulla 38.5%) and in the number of waves (eighteen). The third order hairs are still shorter and thinner with up to sixteen waves, an average length of 12.1 mm, and a thickness of 73μ. The medulla is lightly developed. The lower quarter of the hair has only a narrow strip of medulla (39.7%) at the middle of the hair. Some hairs have a broken line of medulla. There are a few third category hairs and they are very short (4 mm), thin (34μ), without waves or medulla. They gradually thicken toward the middle of the shaft and thin near the tip. There is an average of 250 first category hairs and 833 others per 1 cm².

The single category of breast hairs subdivides into three orders with various transitional forms. The first order is thick and long (length 13.9 mm, thickness 152μ), straight (some very slightly twisted), thickening gradually from the base to middle, and then tapering toward the tip. The medulla (67.2%) is well developed along the whole shaft except at the base and tip. The second order differs from the first in size (length 11.0 mm, thickness 86μ, medulla 63.7%), and in some a broken line of medulla 25% is present only in the lower part. Third order hairs are sparse, still thinner (68μ), and shorter (7.1 mm), with a very poor medulla (12.5%) or, more frequently, none at all. An average of 750 hairs is recorded per 1 cm².

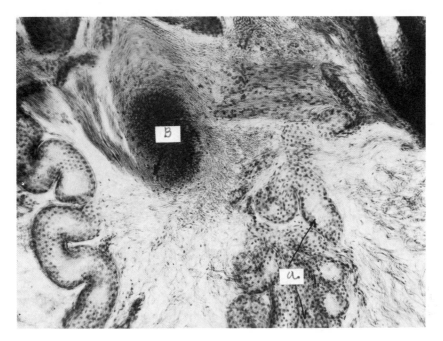

Fig. 33. *Tenrec ecaudatus.* Withers skin: *a,* sweat glands; *b,* follicle hair bulb.

Hemicentetes semispinosus

Material. Skin from withers and soles (body length 18 cm).

The lower skin border as in *T. caudatus* is indistinct. Individual bundles of cross-striated muscles lie at various depths, some rising higher toward the epidermis than the hair bulbs. The stratum malpighii is two to four cells deep. The stratum corneum is rather firm. There is no reticular layer in the withers dermis, as some hair bulbs penetrate the cross-striated muscle zone. The plexus of mostly horizontal collagen fiber bundles is not very dense. Fat cells (17 x 34; 28 x 28μ), which are abundant in the withers dermis, reach beyond the 0.4 to 0.5 mm upper layer. There are no fat cells in the breast skin. There is a single sebaceous gland (50 x 101; 34 x 84μ) behind every hair root. Sweat glands (secretion section 39 x 45μ) occur rarely, as most are reduced to ducts near many of the hair roots; their secretion cavities are absent. The arrectores pilorum muscles vary in size. Powerful bundles of smooth muscle fibers adhere to the large spine roots deep in the skin, while the muscles at the hairs are small. The cross-section of the muscle is 112μ on the withers and 45μ on the breast.

The pelage on the withers divides into three categories. The spines, first order, are thick (0.63 mm), relatively short (10.8 mm), and white. They have a thin club and a sharp tip. A considerable thickening, sharp at first, then gradual, starts from the base, after which the spine tapers toward the tip. Round in section, their 88.9% medulla is developed down the whole length except club and tip. The second-order spines, 11.0 mm long and 366μ thick, are often slightly wavy (medulla 85.3%). The first-category hairs are long (20.8 mm), thin (121μ), and bristly. Their thickness does not change down the length of the shaft, and they have no medulla. The second-category hairs (first size order) are bow-shaped and relatively short (7.6 mm). They thicken gradually from base to middle to 13μ, then taper gently toward the tip. In some, the medulla is well developed (53.9%). In some, it is a broken line (23.0%). In a few, it is altogether absent.

The second-category (second-order) hairs, shorter than the first order (5.1 mm long and 13μ thick), have no medulla . There is an average of 83 spines (first order), 166 spines (second order), and 46,666 hairs per 1 cm². The first category of hairs (length 8.0 mm, thickness 76μ) has a thick base, thickens further in the lower third, and then tapers very gradually into a long whip-like tip. These hairs are straight and, except for base and tip, have a developed medulla (73.7%). The second category of hairs is short (6.7 mm), thin (49μ), and arched backward. The medulla (50%) is found only in the lower half or two-thirds. The third category of hairs, 4.3 mm long and 31μ thick, has no medulla.

Echinops telfairi

Material. Pelage from withers and breast of one specimen (body length 11 cm) taken in June.

More spines are found among *E. telfairi* hairs than are found among the hairs of the other tenrecs under investigation. There are three categories of hair on the withers. The spines of the first category, up to 11.8 mm long and 622μ thick, have a club above which they gradually taper to a sharp tip. They are round in section. Exteriorly, the spines have peculiar, round notches situated in irregular, lengthwise rows (fig. 34). The cortical layer is thick, without lengthwise processes like those in hedgehogs. At great magnification, the cortical layer inner surface reveals transverse folds (fig. 34). The spine inner cavity is divided into "stories" by longitudinal membranes (fig. 34). The transverse membranes are at various distances from one another. Some of the membranes merge into one another. Every membrane of the medulla consists of numerous layers stemming from the cortical layer to fuse into a compact formation. The membrane surface is irregular with notches and thickenings.

The second category, 6.6 mm long, 27μ thick, is similar to shrew hairs with the same two or three ligatures, some with a slightly developed granna. The medulla (63.1%) is well developed. The tiny, thin, white third category hairs have not been measured. There are 83 spines and 83 hairs per 1 cm².

The breast hairs divide into two categories. Straight hairs (6.4 mm, and 58μ) thicken in the bottom third of the shaft then taper gradually toward the tip (medulla 41.3%). The second category hairs are white and so tiny that only the tips show on the skin surface. These hairs have not been measured. There are 249 hairs per 1 cm².

Microgale pusilla

Material. Skin from withers and breast (table 5) of an adult *M. pusilla*.
The breast pelage is at various stages of growth so the skin thickness varies. Inadequate preservation prevents structural examination of the epidermis, which is not pigmented. In the withers dermis, the upper papillary layer 20 to 30μ thick has a fairly dense plexus of collagen fiber bundles. The collagen plexus is loose in the rest of the papillary layer and in the reticular layer, the bundles being mainly horizontal. The reticular layer contains occasional fat cells. The breast dermis is loose and abounds in fat cells, except for the solid outer section of the 20μ thick papillary layer. The sebaceous glands on the withers (one or two at every hair) are large (41 × 84μ) and oval, their lower ends almost reaching the hair bulbs. In horizontal section they are bean-shaped and nearly encircle the hair. They take up a considerable part of the papillary layer. On the breast, where the hair is growing, they are considerably smaller (24 × 60μ). There are few skin glands in the withers. Their saccular-shaped secretion sections (20μ in diameter) are level with the hair bulbs, their longitudinal axis being horizontal. No sweat glands are found in the breast. No arrectores pilorum muscles are recorded in these samples. The hairs grow singly with 40,145 hairs per 1 cm² on the withers.

Limnogale mergulus

Material. Skin from withers and breast (table 5) of one specimen (body length 12 cm).
The stratum malpighii is one to two cells deep. The epidermis of the foot sole is 150μ thick, of which 40μ is stratum corneum. There are small alveoli on the inner epidermal surface, and a few pigment cells occur in the epidermis. Pigment granules can be seen in some epidermal cells. The upper papillary layer of the body dermis (thickness about 150μ) has a dense plexus of collagen fiber bundles, the network of bundles being loose in the rest of the dermis. The collagen bundles are mainly horizontal. Pigment cells are numerous in the dermis of the foot sole. The sebaceous glands (single-lobed, 30 × 76; 38 × 76μ) are in front of the hair roots, to which they adhere closely. Horizontal sections show that some of the pile hairs have two glands (or one with twin lobes). The sweat glands (secretion cavity, 30 to 35μ in diameter), sausage-shaped and slightly twisted, are right under the hair bulbs. The

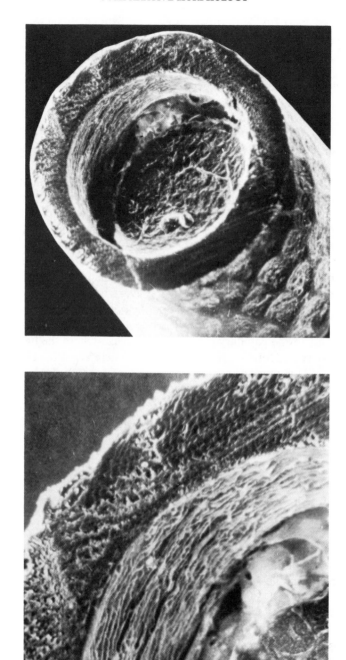

Fig. 34. *Echinops telfari.* Structure of (*a*) inner and (*b*) outer part of spine.

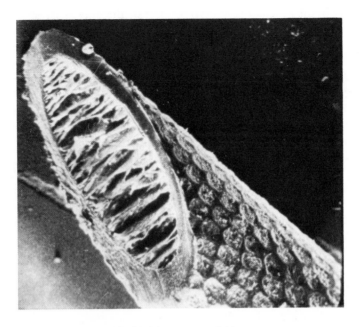

(c)

Fig. 34. (c) outer part of spine.

hairs grow singly, not in groups. The withers skin has 1,100 pile and 31,000 fur hairs per 1 cm²; the breast has 1,100 and 33,300, respectively.

Potamogale velox

Materials. Skin from withers, breast (table 5), crest of the tail, and the hind sole.

The skin is thicker on the breast than on the withers, probably because the breast hairs are in full growth while the withers hairs are incipient. The stratum malpighii is one to two cells deep. There is no pigment in the epidermis, which is thick on the tail crest and hind sole. The stratum malpighii is four to five cells deep on the tail crest, though the stratum lucidum and the stratum granulosum are not very pronounced, one to two outer layers being formed of flattened cells with spindle-shaped nuclei. The hind sole has a lightly developed stratum granulosum but no stratum lucidum. Only the epidermis of the tail crest has a small amount of pigment. The stratum corneum is thin on the tail crest and is rather thick on the sole. The inner surface of the sole epidermis has abundant alveoli. In the upper withers dermis an intricate plexus of collagen fiber is 100 to 120μ thick.

The collagen bundles, loose in the remainder of the papillary layer, become dense again in the reticular layer. Predominantly but not entirely horizontal in the papillary layer, the collagen bundles are almost exclusively horizontal in the reticular layer. On the breast, the outer part of the papillary layer is about 100μ thick, forming a rather dense horizontal collagen fiber plexus. The remainder of the papillary layer contains the roots of growing hairs. The reticular layer has rather a loose plexus of horizontal collagen bundles. In the dermis, elastin fibers are numerous (especially in the reticular and outer papillary layers) and mostly horizontal.

The skin on the tail crest is peculiar in that its middle layers have great accumulations of fat cells (40×60; $30 \times 50\mu$) with bundles of collagen fibers 30 to 50μ thick stretching between the inner dermal layers from one side of the crest to the other. The tail crest skin (its thickness reaches 500μ) has no reticular layer. The collagen bundles lie in the dermis at various angles. The withers sebaceous glands, which are large (45×120; $75 \times 170\mu$) and oval, are set lengthwise down the hair root, two at every tuft. In the breast skin, sebaceous glands are small (16×48; $16 \times 60\mu$) because the hair is growing. The tail crest has a large sebaceous gland ($80 \times 130\mu$) at every hair, and the few sausage-shaped or saccular sweat glands ($60 \times 70\mu$) have straight secretion cavities. On the withers they are level with the hair bulbs. There are no sweat glands on the tail crest. There is an anal gland (Weber, 1928), but the hind foot has no specific glands. Arrectores pilorum muscles are absent. The withers hair grows in tufts of five to seven fur hairs or one pile and six or seven fur hairs. There are about 1,100 pile and 48,890 fur hairs per 1 cm² (as the hairs are in the process of growth, this figure is not exact). The pile hairs grow singly. The fur hairs grow in tufts of four to eight. On the tail crest the hairs grow singly, sometimes in pairs.

Micropotamogale lamottei

Material. Skin from the side, near the shoulder blades (table 6).

The hair is not in growth. The stratum malpighii is one to two cell layers. A dense network of mainly horizontal collagen fibers comprises the outer papillary layer, the middle and inner papillary layers being laxer with bundles rising toward the surface at all angles. The reticular layer is loose; the collagen bundles are horizontal. The few elastin fibers are mostly in the papillary layer. The sebaceous glands (40×90; $45 \times 110\mu$) are oval at the sagittal section, longitudinally perpendicular or set at an angle. The gland excretion ducts open into the hair sheaths about halfway down the hair root, with two glands at every hair tuft. The horizontal section shows that the sebaceous glands almost completely surround the hair tuft. The apocrine sweat glands are few and far between. Their saccular secretion sections, 38μ in diameter, are below the hair bulb level. The glands are longitudinally horizontal. The excretion ducts open into the hair follicle. The arrectores pilorum muscles are very thin—only 8 to 10μ. Usually the pile hairs grow singly and the fur hairs grow in tufts of four or five, but some tufts have one pile and four or five fur hairs. There are about 1,330 pile and 35,550 fur hairs per 1 cm².

Familia Chrysochloridae

The skin of *Chrysochloridae* has not been studied. We examined *Chrysochloris* sp.

The skin is rather thin, and though in absolute terms the epidermis and stratum corneum are also thin, their relative thickness is considerable. The papillary layer is much thicker than the reticular, which is made up of an intricate, mainly horizontal network of collagen fiber bundles. The sebaceous glands are large and single-lobed. The hairs grow in tufts of one pile and several fur hairs.

Chrysochloris sp.

Materials. Skin from withers and breast (table 5) of two golden moles (body lengths 10 and 11 cm).

The thickness of the epidermis is 6 to 10% that of the skin; that of the stratum corneum is 60 to 80% of the epidermis. The outer papillary layer, about 150μ thick in specimen (1) and 350 to 580μ in specimen (2), is an intricate plexus of collagen fibers. The reticular layer is made up of abundant fat cells (38×70; $35 \times 76\mu$) between sparse horizontal bundles of collagen fiber. There are fewer fat cells in the breast than in the withers. Specimen (2) has a few individual cross-striated muscle bundles in the reticular layer (breast). The few elastin fibers in the dermis are mostly in the papillary layer, mostly horizontal.

The somewhat peculiar sebaceous glands, much the same size in withers and breast, are large (80×190; $140 \times 160\mu$) and oval, with the long axis parallel to the surface. The glands themselves are deep in the papillary layer, level with the hair bulbs. The ducts open into the hair sheaths halfway to two-thirds down the roots. There is one gland at every hair tuft. The anal gland is a single formation divided into two parts made up of sebaceous glands (Schaffer, 1940). No sweat glands are found. Arrectores pilorum are very thin (about 7μ). The hairs grow in tufts of one pile and eight to thirteen fur hairs (withers) and one pile and six to eight fur hairs (breast). There are 3,300 pile and 37,700 to 38,800 fur hairs per 1 cm² (withers) and 2,200 pile and 15,500 to 17,700 fur hairs (breast).

Familia Solenodontidae

The skin of *Solenodontidae* has not been studied. We investigated the skin of *Solenodon cubanus*.

The skin is thick. In the dermis the bundles of collagen fiber, mainly horizontal, form a loose plexus. Elastin fibers are abundant in the outer dermis. The sebaceous glands have single lobes. The apocrine sweat glands behind the hair tufts have tubular and slightly twisted secretion cavities. The ducts

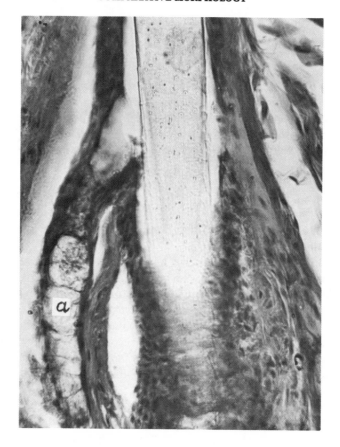

Fig. 35. *Solenodon cubanus. a,* sebaceous gland.

open into the hair sheaths near the hair funnels. There are arrectores pilorum muscles. Pile and fur hairs grow in tufts.

Solenodon cubanus

Material. Skin from the belly (body length 25 cm).

The hairs are growing; the roots pierce the whole skin almost down to the subcutaneous muscles. The skin is 2.32 mm thick; the epidermis is 29μ; the stratum corneum is 13μ; and the stratum malpighii comprises three to four rows. The dermis, not divided into layers, is 0.87 mm thick. The subcutaneous fat tissue is strongly developed (1.42 mm). The sebaceous glands are sacciform ($75 \times 240\mu$) with two near every hair tuft (fig. 35). Their ducts open into the hair sheaths near the epidermis. The secretion cavities of the apocrine sweat glands are 59μ in diameter. The arrectores pilorum muscles are long and thin (21μ), but are not found at every hair tuft. The hair tufts consist of one or two pile hairs and one to three fur hairs (fig. 36). There are 320 hairs per 1 cm² of skin.

FAMILIA SORICIDAE

Descriptions of the pelage of some shrew species are available (Smith, 1933; Gudkova-Aksenova, 1951; Borowski, 1952, 1959). A fair proportion of the literature is devoted to the specific glands (see review by Schaffer, 1940). A brief account of the pelage of shrews in the USSR was given by Sokolov and Sumina (1962).

We studied the skin of *Sorex minutus, S. araneus, Neomys fodiens, Crocidura suaveolens, Crocidura russula guldenstaedti, Diplomesodon pulchellum.*

On the body, the skin, epidermis, and stratum corneum are thin. The stratum malpighii is one to two cells deep. The stratum granulosum and stratum lucidum are absent. The epidermis is not pigmented. In the dermis, the bundles of collagen fibers are mostly horizontal. The sebaceous glands are well developed; the sweat glands are saccular. The lateral specific glands formed by sweat and sebaceous glands are characteristic. Arrector pili

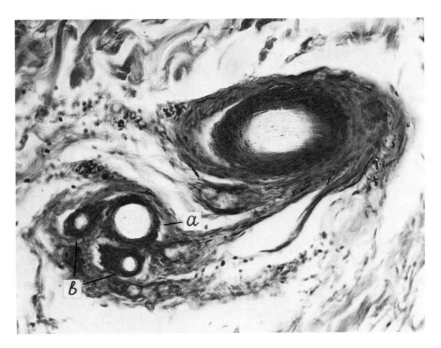

Fig. 36. *Solenodon cubanus.* Hair tuft in belly skin: *a,* pile hair; *b,* fur hair.

TABLE 6

SKIN MEASUREMENTS OF SORICIDAE (SOREX AND NEOMYS FODIENS) (μ)

	Skin	Epidermis	Stratum corneum	Dermis	Papillary layer	Reticular layer
Sorex minutus						
Withers						
Tjumen	101	11	3	90	184	6
(1) Chashnikovo	106	11	4	95	95	Not
(2) Chashnikovo	78	7	3	71	71	available
Semirechie	112	9	2	103	103	absent
(1) Pechora	112	13	6	99	99	absent
(2) Pechora	228	17	6	211	211	absent
Breast						
Crimea	129	17	6	112	112	absent
(1) Chashnikovo	84	9	4	75	68	7
(2) Chashnikovo	84	9	4	75	75	absent
Sorex araneus						
Withers						
(2) Chashnikovo	106	16	9	90	90	—
Switzerland	162	14	5	148	131	17
(2) Switzerland	106	17	7	89	81	8
Sayany	179	17	8	162	154	8
(1) Pechora	179	11	5	168	151	17
(2) Pechora	157	11	5	146	129	17
Semipalatinsk region	129	17	8	112	101	11
Breast						
Chashnikovo	101	32	25	69	52	17
Neomys fodiens						
Withers						
Semirechie	364	13	9	351	267	84
(1) Chashnikovo	123	16	9	107	107	—
(2) Chashnikovo	168	18	9	150	133	17
Pechora, July	230	6	4	224	196	28
Pechora, October	140	28	23	112	106	6
Altai, August	280	13	4	267	250	17
Altai, September	157	7	4	150	139	11
(1) Sayany	146	27	20	119	108	11
(2) Sayany	95	11	9	84	84	—
Belovezh, January	146	27	23	119	97	22
Breast						
(1) Chashnikovo	134	13	9	121	106	15
(2) Chashnikovo	185	9	5	176	159	17
Belovezh, January	146	18	15	128	100	28

muscles are absent. The hairs divide into three fairly pronounced categories (guard, pile, and fur) with marked ligatures. The hairs grow singly.

Sorex minutus

Material. Skin from withers, breast (table 6), and the front sole taken June 25 in Tumen region (body length 4.3 cm); July in Chashnikovo, Moscow region (two adult shrews); July in Semirechinsk region (body length 5.7 cm); August 27 and September 6 in the Pechora-Ilych reservation (body length 4.7 cm and 4.8 cm); September 29 in the Crimea on Chernaya Rechka (body length 5.2 cm).

Although the dates on which these *S. minutus* were taken range from June to September, the pelage is in the resting stage in most of them so the skin is suitable for comparison. Only in the two September shrews (from the Pechora-Ilych reservation and from the Crimea), is the pelage growing—the autumn molt has started and their skin is thickest. No peculiarities in the indexes and structure of the skin of pygmy shrews are recorded related to area. In the skin of the front foot sole (only one *S. minutus* from the Moscow region was studied), the epidermis reaches a considerable thickness (140μ); the stratum granulosum is pronounced; the thickness of the very solid stratum corneum is 84μ; the epidermis (specifically its inner part) is heavily pigmented. The outer compact part of the papillary layer is more pronounced than in other shrews. The hair bulbs are nearer the subcutaneous muscles, with the result that there is no reticular layer. The sebaceous glands are large (22×56; $39 \times 78\mu$), oval or bean-shaped, two at each hair. Neighboring glands adhere to each other very closely. The sweat glands are well developed (diameter 22 to 39μ) with horizontal, saccular, slightly twisted secretion cavities beneath the hair bulbs. There are tubular glands in the dermis of the foot skin glomi, and the diameter of the secretion sections is 17μ. The Semirechye shrew has 37,500 hairs per 1 cm^2. Borowski (1973) reports four molts in *S. minutus:* the autumn molt, two spring molts, and the senile molt. The longest hair is characteristic of winter animals (see table 7).

Sorex araneus

Material. Skin from withers, breast (table 6), and soles of adult male taken June in Chashnikovo, Moscow region; July 19 and September 3 in Switzerland (body length of both, 5 cm); August 11, Sayany (body length 5.8 cm); September 3 and 4 in the Pechora-Ilych reservation (Northern Urals) (body length of both, 6.5 cm), and September 14 in the Semipalatinsk region.

Differences in skin thickness bear no relation to where they were taken.

TABLE 7

HAIR LENGTH IN SOREX MINUTUS FROM BIALOWIEZA NATIONAL PARK

Season	N	Min.-Max. (mm)	$\overline{X} \pm$ S. D.
Summer, juvenile	51	2.5-3.5	2.97 ± 0.32
Summer, adult	77	2.5-3.7	3.00 ± 0.29
Spring	11	3.4-4.2	3.74 ± 0.22
Winter	68	4.5-5.9	5.03 ± 0.30

In fact, both shrews from Switzerland are similar in size, with nongrowing pelage. They were taken at about the same time, yet the thickness of their skin differs (table 6). The sole epidermis and stratum corneum are very thick (epidermis 72 to 84μ, stratum corneum 34μ). In the body skin the papillary layer is well developed (fig. 37). The reticular layer is thin (completely absent in the Chashnikovo specimen), and the hair roots run through the whole dermis. The sebaceous glands are large (22 x 56; 34 x 97μ). The development level of the sweat glands varies (diameter 28 to 61μ), but no relationship to area and climate is shown. The presence of a great number of specific glands is characteristic.

We have studied only the sole glands, the large glomi of eccrine sweat glands with 17 to 22 secretion cavities. The presence of sweat glands in the foot soles of the common shrew is also reported by Murariu (1974, 1976), refuting Leydig's claim (1876) that the shrew has no glands. Lateral glands are found on both sides (Schaffer, 1940; Murariu, 1973, 1976 et al.) in both males and females. They vary seasonally in size. In the female the gland is 2.86 mm long, 1.7 mm wide, and 0.4 mm thick in August. In September the size increases, with the thickness reaching 0.7 mm. The lateral gland is a complex of sweat and sebaceous glands. A few layers of apocrine sweat glands are at the base and the larger sebaceous glands are higher up. The ducts of both gland types open into the hair bursae. The secretion of the sweat glands has an odor that depends on the stage in life, in particular, on breeding. This odor may give the shrew some protection against its enemies (Schaffer, 1940). The anal glands are a combination of sebaceous and sweat glands (Hamperl, 1923) Murariu 1975a, 1976). The sweat glands are apocrine type. There are three categories of hair: guard, pile, and fur (table 8). Gudkoka-Aksenova, 1951, gives only the latter two categories.

The two orders of pile hairs have well-developed grannas and a characteristic ligature with longitudinal bends in the shaft (table 8). The first order has three such ligatures and the second order has four, the hairs being obtusely bent at each. The fur hairs are of even thickness down the whole length; each has four ligatures and longitudinal bends. It should be noted that in the northern Pechora-Ilych animal, the guard hairs are longer than

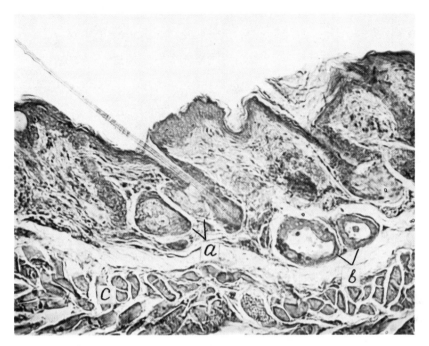

Fig. 37. *Sorex araneus.* Withers skin: *a,* sebaceous gland; *b,* sweat gland;
c, subcutaneous muscles.

in the southern Semipalatinsk animal (both were taken in early September). The Semipalatinsk animal has 27,500 hairs per 1 cm^2 on the withers (according to Gudkova-Aksenova, 1951, there are 9,800 hairs on the sacrum). The hair density doubles in winter (Borowski, 1959). According to Borowski (1968) *S. araneus* has four molts: the autumn molt, two spring molts, and the senile molt.

Neomys fodiens

Material. Skin from withers, breast (table 6), and hind sole of *N. fodiens,* taken in May in Semirechye region (body length 8 cm); in June in Chashnikovo, Moscow region, male and female (body lengths 7 and 6.5, respectively); in July and October, Pechora-Ilych reservation, Northern Urals (body lengths 8.5 and 6.5 cm); in August and September, Altai (body lengths 6.8 and 8 cm), in August, Sayan mountains (body lengths 7 and 6 cm); in January, Belovezhskaya Puscha, Western Byelorussia (body length 11.5 cm).

The hair is not growing on any of these *N. fodiens,* yet the skin thickness varies greatly, but apparently this is not geographically related. The thickness of the epidermis and stratum corneum is also highly variable. The

TABLE 8

SORICIDAE: HAIR MEASUREMENTS

Indexes	Guard hair			Pile hair							Fur hair		
				1st order			2d order						
	Length (mm)	Thickness (μ)	Medulla (%)	Length (mm)	Thickness (μ)	Medulla (%)	Length (mm)	Thickness (μ)	Medulla (%)	Length (mm)	Thickness (μ)	Medulla (%)	
Sorex araneus													
Pechora	6.3	40	65.0	5.9	34	70.6	5.9	27	70.4	3.0	15	73.4	
Semipalatinsk region	5.2	41	65.9	5.2	33	69.7	5.5	24	75.0	4.3	16	75.0	
Neomys fodiens													
Withers													
Semirechie, May	7.4	41	46.4	6.4	29	58.7	6.5	27	48.2	5.6	16	75.0	
Pechora, July	7.4	50	60.0	6.9	30	60.0	6.3	25	52.0	5.7	17	76.5	
Belovezh, January	9.9	45	64.5	8.7	33	57.6	7.9	27	33.4	7.8	18	77.8	
Breast													
Belovezh, January	—	—	—	7.4	28	57.2	6.3	24	25.0	6.5	15	63.4	
Crocidura suaveolens													
Withers													
Tula	4.9	37	73.0	3.7	32	75.0	3.9	21	62.0	3.5	18	77.8	
Bokhara	5.0	31	74.2	4.1	28	71.5	4.0	20	70.0	3.5	16	75.0	
C. r. güldenstaedti													
Withers													
(1) August	4.1	30	66.7	3.3	27	70.4	3.4	19	68.6	2.9	16	75.0	
(3) January	7.3	37	62.2	6.4	27	72.5	—	—	—	5.4	18	77.8	
Diplomesodon pulchellum													
(1) Kazakhstan	5.8	20	60.0	5.1	17	64.8	—	—	—	4.4	9	55.5	

stratum corneum is predominantly solid. In the skin of the hind sole the epidermis is thick, up to 120μ; the solid stratum corneum measures 45μ. The stratum granulosum and stratum lucidum are recorded. In the papillary layer of the dermis a 40μ to 50μ subepidermal layer is recorded, consisting of an intricate plexus of horizontal collagen fibers (fig. 38). The remainder of the papillary layer contains a small number of collagen fibers between large sebaceous glands and hair sheaths. A thin reticular layer is formed by a dense network of horizontal collagen bundles.

The sebaceous glands are large (39 x 56; 56 x 140μ) and oval with a vertical long axis. There are two glands anterior and posterior to each hair (fig. 38). In horizontal section through the sebaceous gland zone, the glands, separated only by thin collagen interlayers, fill the microscope's field of vision. The massive development of the sebaceous glands appears to be associated with the aquatic mode of life. The sweat glands have saccular, slightly twisted secretion sections, the long axis being horizontal beneath the hair bulbs (fig. 38). The size varies considerably, the diameter ranging from 22 to 50μ. It has proven impossible to relate their level of development to climate factors. According to Murariu (1972) the sweat glands on the belly are smaller in size than in other body areas. As in *Sorex,* there are eccrine sweat

Fig. 38. *Neomys fodiens.* Withers skin, papillary layer: *a,* sebaceous gland; *b,* sweat gland; *c,* duct of sweat gland.

glands in the hind soles. Their large glomi are embedded in the dermis. The diameter of the secretion cavities is not large (22μ). Murariu (1974, 1976) has reported sweat glands in water shrew foot soles. There are lateral glands (in a male taken in August) reaching 6 mm long and 4 mm wide (Schaffer, 1940). The structure of the gland is similar to that of *Sorex araneus,* the apocrine sweat glands at the base with multilobal sebaceous glands above. In males, the lateral glands develop concurrently with the gonads (Dehnel, 1950). Active functioning of lateral glands is observed in young males as early as a few months after they leave the burrow. The anal glands are a complex of sebaceous and sweat glands (Hamperl, 1923; Murariu, 1972, 1975*a,* 1976). Sweat glands are less abundant than sebaceous.

Dehnel (1950) and Borowski (1952) record spring and autumn molts: Borowski holds that the hair breaks in summer and thus becomes shorter. In Borowski (1973) five molts (juvenile molt, autumn molt, two spring molts, and the senile molt) are reported, no mention being made of the breaking of the hairs in water shrews.

The withers hairs are studied in the May Semirechye and July Pechora specimens. The pelage falls into three categories (also indicated by Gudkova-Aksenova, 1951) (table 8). The guard hairs are few, straight, thickening gradually halfway to two-thirds from proximal to distal, tapering toward the tip with a medulla extending the length except for base and tip. The pile hairs of the first order have a long, thin club and three ligatures with obtuse, longitudinal turns bent at about 45° before the granna. The granna reaches about halfway along the shaft. The medulla extends the length of the hair except for base and tip. The pile hairs of the second order have a longer club with three to four ligatures, longitudinal turns, a granna down one-third of the hair, and a medulla extending the length except for base and tip. The fur hairs have no granna but have four ligatures with longitudinal turns. The medulla is developed similarly to other categories. The pelage density is much the same on all specimens—25,000 hairs per 1 cm² on the Semirechye specimen; 27,500 to 28,000 or 30,000 on the Altai, 30,000 on the Pechora, and 32,500 on the Sayan specimen. This is probably associated with fairly similar water temperatures in those areas.

Although we cannot establish a geographic relation of skin structure in *N. fodiens* with so few specimens, some seasonal variations in skin characteristics can be clearly demonstrated. The epidermis of the January Belovezhskaya Puscha specimen is thicker owing to a strongly developed, loose stratum corneum creating an additional layer of thermoinsulation (similarly thickened epidermis and stratum corneum are found in the late autumn, October, Pechora specimen). There is a well developed, loose reticular layer, 22μ thick. The sebaceous glands are well developed in the winter animal (34×50; $28 \times 39\mu$) but no more so than in summer. The sweat glands show a sharp seasonal variation; they are inactive in winter and the diameter of the secretion cavity is much smaller (11 to 28μ).

All categories of hair are considerably longer in winter than in summer animals (table 8). The greatest length of the hair in winter water shrews from the Bialowieza National Park (hair categories not being indicated) is also reported by Borowski (1973) (table 8a).

TABLE 8a

CHANGES IN HAIR LENGTH IN NEOMYS FODIENS (BOROWSKI, 1973)

Season	N	Min.-Max. (mm)	$\overline{X} \pm$ S. D.
Summer, juvenile	726	3.6-6.7	5.40±0.33
Summer, adult	333	4.0-6.8	5.34±0.58
Spring	95	5.2-7.6	6.39±0.48
Winter	308	6.9-9.5	7.99±0.55

The density is also usually greater, reaching 40,000 per 1 cm². In the October animal from the north (Pechora-Ilych reservation), the density is still greater—42,500 per 1 cm². The hairs on the breast have been studied only in the January animal. No guard hairs are found, though they appear to exist, being absent in our sample only by chance as they are rather sparse. All the hair categories of the breast are shorter and thinner than their equivalents on the withers (table 8). The shafts of the pile hairs have four to five ligatures with longitudinal turns; the fur hairs have five ligatures. In the pile hairs of the first order, the grannae reach less than one-third of the hair length. In the pile hairs of the second order, the granna reach one-fourth. No functional explanation has been found for the much lower pelage density on the breast than on the withers: there are 22,500 hairs per 1 cm² of skin.

Crocidura suaveolens

Material. Skin from withers (table 9) taken in June in Kazakhstan in Aral Kara Kum, two males (body lengths 6.1 and 5.8 cm); July 30 and 31 in Tula region (body lengths 4.8 and 6 cm); July 4 in Askania Nova (body length 6 cm); July 8 near Bukhara (body length 4.5 cm); June 20 in Krasnodar region (body length 4.8 cm); September 19 near Suifun River, Primorye territory, two specimens (4.6 and 5.8 cm).

The pelage is in a state of autumn molt on specimen (1) from the Primorye territory. In the other animal, the pelage is in a state of rest suitable for comparison. Sebaceous glands, large (17 x 34; 38 x 84μ), are single-lobed and placed behind every hair follicle. Sweat glands have saccular secretion cavities, with their long axis horizontally arranged beneath the hair bulbs. However, in specimen (1) from Tula, in the inner skin, an

TABLE 9

SORICIDAE (CROCIDURA, DIPLOMESODON PULCHELLUM): SKIN MEASUREMENTS (μ)

Indexes	Skin	Epidermis	Stratum corneum	Dermis	Papillary layer	Reticular layer
Crocidura suaveolens						
Withers						
(1) Kazakhstan	112	9	4	103	90	13
(2) Kazakhstan	123	9	4	114	106	8
(1) Tula region	252	13	4	239	127	112
(2) Tula region	118	6	2	112	101	11
Askania Nova	145	6	2	139	128	11
Bokhara	106	7	2	99	88	11
Zaagdan	118	6	2	112	101	11
(1) Primorie	436	8	4	428	411	17
(2) Primorie	156	6	2	150	139	11
Crocidura russula güldenstaedti						
Withers						
(1) August	263	6	1	257	207	50
(2) September	224	11	3	213	168	45
(3) January	157	13	6	144	133	11
Breast						
(1) (skin thickening), August	269	9	3	260	198	62
(1) (skin thinning), August	146	9	3	137	137	—
(2) September	90	6	1	84	73	11
(3) January	95	18	11	77	71	6
Diplomesodon pulchellum						
Withers						
(1) Kazakhstan	147	16	9	131	109	22
(2) Kazakhstan	119	13	5	106	84	22
(3) Turkmenia	770	27	18	743	743	—
Breast						
(1) Kazakhstan	123	16	8	107	107	—
(3) Turkmenia	178	16	9	162	140	22

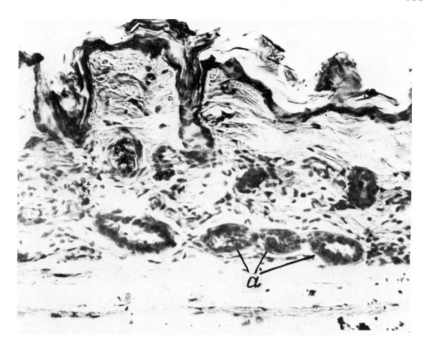

Fig. 39. *Crocidura suaveolens.* Withers skin: *a,* sweat gland.

unusual exception is found: glomeruli of sweat glands (fig. 39). This is the only case of a sweat gland twisted into a glomus recorded by us in Soricidae. The diameter of the secretion cavities varies considerably (22 to 45μ). However, we do not relate this to habitat.

Withers hair divides into three categories: guard, pile (two orders), and fur (table 9). The Tula specimen differs from the Bukhara only in its lighter color. Size categories, orders, and medulla are much the same. The pile hairs (first order) have granna over half the length of the hair (2.2 to 2.3 mm). The shaft has two ligatures with longitudinal turns. The Bukhara specimen has 25,000 hairs per 1 cm² on the withers; the Kazakhstan specimen has 27,500. In the breast skin the pelage is less dense: 17,500 to 20,000 hairs (Kazakhstan specimen) per 1 cm².

Crocidura russula güldenstaedti

Material. Skin from withers and breast (table 9), (body lengths 5 cm, 6 cm, and 5 cm) taken in August, September, and January, near Tbilisi, specimens (1), (2), and (3), respectively.

In specimen (1) the pelage on part of the breast area is growing, the skin being thicker here. Only the outer zone of the papillary layer of the dermis is

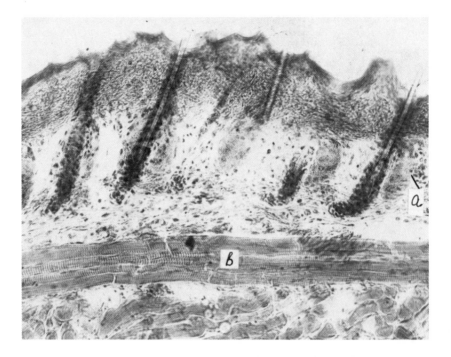

Fig. 40. *Crocidura russula güldenstaedti. a,* sebaceous glands; *b,* subcutaneous muscles.

formed by a dense plexus of collagen fibers (fig. 40). The reticular layer is loose. The sebaceous glands (22 × 45; 34 × 62μ) are unilobal and oval, arranged singly or in pairs at every hair root. The secretion cavities of the sweat glands in the withers skin are saccular, 40 to 45μ in diameter, horizontally slightly or strongly bent.

As with most shrews, the hair divides into three categories (table 7). The guard hairs are thickest in the middle of the shaft. In the pile hairs, first order, the granna is 22 mm, more than half the hair length. The shaft has two ligatures with longitudinal turns. In the second order, the granna is considerably shorter (1.1 mm, about one-third of the length). The pile hair has up to three ligatures with longitudinal turns, as does the fur hair.

In winter the skin is thinner and the stratum corneum is looser than in summer. In the dermis there is a denser outer zone and a looser mesial zone (at the level of sebaceous glands). The collagen bundles in the dermis are mostly horizontal. The reticular layer is rather loose. No fat cells are recorded in the dermis or subcutaneous tissue. In the withers most of the sebaceous glands are well developed, oval or round (34 × 45; 45 × 45μ), with a pair at every hair bursa. In the breast skin the sebaceous glands are more poorly developed, with one per hair. The secretion cavities of the sweat glands (withers), with a smaller diameter than in summer (34μ), are still well developed and are horizontally firmly bent, the skin around them being somewhat thicker. There are no sweat glands in the breast skin. The winter pelage (withers) has three typical categories (table 7) and is longer than the summer pelage. The pile hairs have only one order. The shaft of the pile hairs has three ligatures, and that of the fur hairs has two ligatures with longitudinal turns. There are 19, 333 hairs per 1 cm².

Diplomesodon pulchellum

Material. Skin from withers and breast (table 9) of three specimens taken in June in Kazakhstan and the Aral Kara Kum (body length, female 68, male 65 mm) and in April in Turkmenistan (body length 70 cm).

In the two June specimens from Kazakhstan, the hair is in a dormant state and the withers skin is thin. The stratum corneum is loose. Between the hair bulbs and the subcutaneous muscles is a thin (22μ) connective tissue layer. The secretion sections of the sweat glands are in this layer, and there is no reticular layer in the dermis. However, as sweat glands are rare, and the collagen bundles beneath the hair bulbs form a denser plexus, one might regard the dermis zone under the hair bulbs as a reticular layer. Sebaceous glands, only visible at some hairs, are small $(17 \times 34; 18 \times 40\mu)$ and sideways posterior to the hair roots. The sweat glands, with saccular, straight secretion cavities 20 to 34μ in diameter, are under the hair bulbs, their long axis horizontal to the skin.

The skin of the *D. pulchellum* (April, Turkmenistan) is very thick—the hair is in a state of growth. The epidermis and stratum corneum are thicker than in the June specimens. The outer surface of the epidermis is uneven and covered by small crests. The stratum corneum consists of multilayered corneous scales. Large hair roots containing a medulla run through the dermis. The roots are peculiarly twisted; the upper part lies at an angle of about 30° to the skin, the lower is almost perpendicular. The dermis has an outer, more compact layer about 170μ thick beneath which there is a loose tissue with fat cells $(28 \times 50; 20 \times 39\mu)$ between the hair roots. In some places the hair bulbs adhere directly to subcutaneous muscles, in others a thin, subcutaneous fat tissue containing one to three layers of fat cells $(22 \times 40; 22 \times 28\mu)$ lies between the bulbs and subcutaneous muscles. The reticular layer is absent. Sebaceous glands $(18 \times 36; 39 \times 73\mu)$ are anterior, posterior, or lateral to the hair root. The sweat glands are better developed than in June specimens. Their secretion sections of 23 to 67μ in diameter are almost sausage-shaped with a long axis along the hair roots. The glands are straight (fig. 41).

The June specimen has three categories of hairs (table 7). In the pile (one order only) the granna extends less than half the shaft. The club of the hair has three ligatures with longitudinal turns. Fur hairs have up to three such turns. There are 47,000 hairs recorded per 1 cm² on the withers of the April

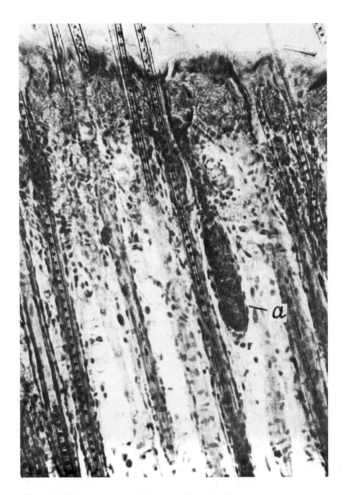

Fig. 41. *Diplomesodon pulchellum*. Skin of withers: *a*, sweat gland.

specimen from Turkmenistan, and 50,000 on the June specimen from Kazakhstan. There are 31,333 hairs per 1 cm² on the breast of the April specimen and 32,500 on the June specimen.

FAMILIA MACROSCELIDIDAE

A description of specific skin glands of *Macroscelididae* is available (Wagner, 1834; Peters, 1852; Dobson 1883-1890). In the skin of *Elephantulus rozeti,* the skin, epidermis, and stratum corneum are thin; the epidermis is not pigmented. The sweat glands have saccular secretion cavities. Long, medulla-free tips are characteristic of the guard hairs. The hair shafts have ligatures with longitudinal turns and the hairs grow singly. Arrector pili muscles are absent. A caudal gland is recorded.

Elephantulus rozeti

Material. Skin from withers and breast (table 10) of three specimens (body lengths of 9.5 cm, 10 cm, and 9 cm) taken in Algiers.

TABLE 10

ELEPHANTULUS ROZETI: SKIN MEASUREMENTS (μ)

Indexes	Skin	Epidermis	Stratum corneum	Dermis	Papillary layer	Reticular layer
Withers						
(1)	106	7	5	99	77	22
(2)	448	9	4	439	400	39
(3)	224	7	4	217	200	17
Breast						
(1)	157	—	—	157	123	34
(2)	373	9	2	364	336	28
(3)	168	14	4	154	126	28

The skin varies in thickness with the stages of hair growth. The stratum malpighii is one to two cells deep. The stratum corneum (thickness is 30 to 56μ on withers and 110 to 170μ on breast) is solid. The outer part of the papillary layer has a dense plexus of collagen fibers. The reticular layer of the dermis, which has horizontal collagen bundles, is clearly distinguishable. In specimen (2), considerable deposits of fat cells are found in the middle of the papillary layer of the withers dermis because of hair growth. The cells are uniformly large (28 x 56; 28 x 50; 28 x 67μ) throughout the dermis, mostly horizontally elongated (fig. 42).

A continuous mass of round fat cells 45 to 50μ in diameter is in the lower part of the dermis beneath the outer solid zone of the papillary layer (breast and the collagen bundles are almost invisible). The reticular layer is replaced by one to two layers of fat cells between the hair bulbs and in the subcutaneous muscles. The sebaceous glands (17 x 73μ, 28 x 67μ) are closely pressed against the hair bursa and are present only in nongrowing hairs. Apocrine sweat glands with saccular secretion cavities thickened at the lower end (diameter 34 x 56μ) are recorded (fig. 43). The long axis of the secretion cavity is parallel to the hair roots. The sweat glands penetrate 140 to 200μ down from the skin surface. The ducts open into the hair funnels behind the hairs.

No sweat glands are recorded in specimen (1), which has a subcaudal

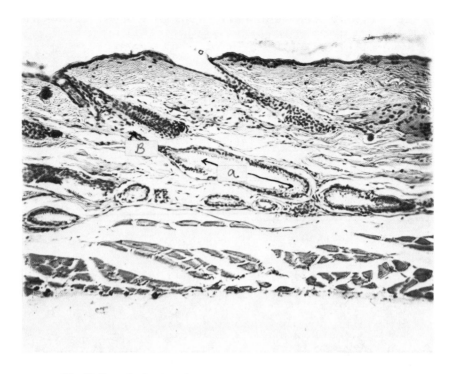

Fig. 42. *Petrodromus tetradactylus. a,* sweat gland; *b,* duct of sweat gland.

gland, as does the desman (Wagner, 1834). The gland forms a small extrusion (1.3 cm long) on the tail, about 2.5 cm from the anus, with excretion ducts opening onto its surface. It appears to be present only in mature males (Peters, 1852). The subcaudal gland is formed by expanded alveolar and tubular glands (Weber, 1928). The presence of this complex of glands distinguishes it clearly from the subcaudal gland in the desmans. *E. rozeti* also has anal glands (Dobson, 1883, 1890).

The hair of the withers divides into two categories: guard and fur. The guard hairs grow to 15.0 mm long, 48μ thick. These straight hairs gradually thicken in the proximal two-thirds, after which they taper into a long whip-like tip. The medulla is developed along the whole of the shaft, comprising 75% at the widest part. The fur hairs grow to 10.2 mm long, 21μ thick. Two ligatures are halfway and three-fourths of the way up the hair. Above the upper ligature the hair becomes somewhat thicker. The medulla is well developed along the hair length (62.0% in the granna). There are an estimated 44,333 hairs per 1 cm² growing singly. The categories in the breast hair are also guard and fur. The guard hairs are straight, without ligatures or turns, and have a long, thin club that thickens into a granna in the distal third. The tip is elongated into a long, nonmedulla "whip." The average length of the guard hairs is 11.4 mm, their thickness is 31μ. The medulla

(87.1%) is well developed along the whole hair. The fur hairs have four or five ligatures with longitudinal turns. No granna proper is recorded, as the thickness of the shaft is almost uniform. Nevertheless, the upper segment separated by the ligature is different, distinguishing this category from typical *Insectivora* fur hairs. The fur hairs, which average 9.9 mm in length and 19μ in thickness, grow singly, not in groups or tufts.

ORDO DERMOPTERA

FAMILIA CYNOCEPHALIDAE

The skin of *Dermoptera* has hardly been studied. A small number of short vibrissae are reported (Pocock, 1926), as are the presence of sebaceous and sweat glands (Grasse, 1955) and the absence of anal glands (Weber, 1928).

We studied the common integument of *C. temminckii.* The order includes one family, so the skin description below accounts for both. The presence of the flying membrane is peculiar. The body skin is thin. The stratum malpighii is one to two cells thick, and there is no stratum granulosum or stratum lucidum. The body epidermis is

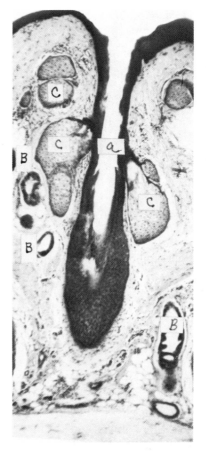

Fig. 43. *Petrodromus tetradactylus.* Skin of ventral part of tail: *a,* hair bursa; *b,* sweat gland; *c,* sebaceous gland.

not pigmented; the sole epidermis is. The dermal bundles of collagen fibers are horizontal; the sebaceous glands are unilobal, singly posterior to the hairs. Their ducts open into hair sheaths somewhat lower than the hair funnels. There are eccrine sweat glands in the sole skin. Arrector pili muscles are absent. The hairs are not divided into categories and grow singly.

Cynocephalus temminckii

Material. Skin from withers and breast of specimens (2) and (3), from the medial part of the flying membrane, level with the middle of the body side in specimens (1) and (3), and from the hind sole of specimens (1) and (2) (body lengths 18.5 cm, 18 cm, and 20 cm), taken in Sumatra.

The hairs are at different growth stages in the three animals. The hair on

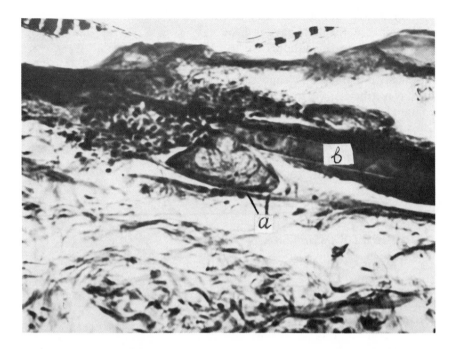

Fig. 44. *Cynocephalus temminckii.* Withers skin: *a,* sebaceous gland; *b,* hair follicle.

the withers of specimen (1) is at a comparatively late stage of growth, the skin being thicker (930μ) than in the other specimens (517 and 660μ). The breast skin (550 and 990μ) is thicker on the withers, apparently because most of the withers hairs are in the process of fast growth. The thickness of the epidermis is 27 to 34μ. The thickness of the stratum corneum is 18 to 28μ (withers), and 20 to 22μ (breast). The epidermis is thickest on the hind soles, its inner surface having numerous alveoli. The thickness of the withers dermis in specimen (1) is 903μ, in (2) 517μ, and in (3) 626μ; the papillary layers are 628, 417, and 494μ, and the reticular layers are 275, 100, and 132μ, respectively. On the breast in (2) and (3), the dermis is 522 and 956μ, the papillary layer is 442 and 681μ, and the reticular layer is 80 and 275μ.

The boundary between papillary and reticular layers is level with the hair bulbs. The upper zone of the papillary layer, 60 to 90μ thick, is a dense plexus (fig. 44). The rest of the papillary layer is made up of tightly adherent hair bursae with a small number of mainly rising bundles of collagen fibers. The reticular layer is of loose collagen bundles. Elastin fibers are very few except deep in the reticular layer. In (1) and (3), small fat cells occur rather frequently in the papillary layer of the withers between the hair bursae. In the skin of specimen (2), fat cells are found in the middle and inner papillary layer. In specimen (1), the outer reticular layer of the withers (about 190μ thick) is a continuous accumulation of fat cells. Sebaceous glands are small (22 x 56; 39 x 56μ) and not present at all hairs (fig. 44).

By the few apocrine sweat glands (withers, diameter 22μ; breast, 34μ), the hairs are virtually nongrowing. The glands are almost straight (though more twisted in the breast than in the withers). The long axes of the secretion cavities run parallel to the hair bursae.

In the sole skin, small glomeruli of eccrine sweat glands are surrounded by fat cells (secretion section diameter 28μ). These data do not support Grasse's claim (1955) for the absence of specific glands in the sole skin.

There is one category of withers hair, but there are three different orders. All are uniformly thick through their length, narrowing only toward the apex into a long, thin tip. The medulla is well developed through the whole shaft. The hairs of the first order are few in number, straight, 12.1 mm long, 36μ thick, medulla 77.8%. The second order, of which there are many more, are slightly wavy. They are 6.6 mm long and 28μ thick, medulla 71.5%. The third order (5.9 mm, 21μ, medulla 71.5%) has more waves (up to three). The hairs grow in rows. Per 1 cm^2, on the withers there are 22,330 hairs in specimens (2) and (3) and 27,000 in specimen (1). On the breast there are 18,333 hairs.

The skin of the upper and lower surfaces of the flying membrane has a structure similar to the body skin, with rare, sausage-like sweat and small sebaceous glands. Thick networks of elastin fibers run through the middle of the flying membrane, in the same places as bundles of cross-striated muscles.

ORDO CHIROPTERA

Ordo Chiroptera is one of the few mammal orders the skin of which has been described in a special review (Quay, 1970). On the basis of the published data as well as his own evidence, Quay (1970) describes all the parts of *Chiroptera* skin with emphasis on specific glands. However, the number of *Chiroptera* species whose skin (except pelage) has been described are few. The work by Benedicts (1957) devoted to the hair of *Chiroptera* should be noted. This author presents much evidence demonstrating hair and cuticle scale characteristics and indexes (a similar work, though on several species, has been done by Volzhina, 1951).

We studied the skin of *Pteropus melanotus, P. giganteus, Eidolon helvum, Rhinolophus bocharicus, Myotis mystacinus, Plecotus auritus, Barbastella barbastella, Nystatus noctula, Pipistrellus kuhli, Vespertilio murinus, Miniopterus scheibersi.*

The characteristic feature of *Chiroptera* is the cutaneous flying membrane. The ratio of the area of fur-covered skin on the body and wing varies from 4:1 to 8:1 in neoarctic bats to 12:1 in tropical fruit bats (Cowles, 1947). The flying membrane is two-layered and its structure replicates that of the body skin. In the flying membrane the epidermis and the stratum corneum are rather thin. The stratum lucidum and stratum granulosum are absent.

The epidermis of the flying membrane is highly pigmented both on the outer and inner sides. There is an intricate plexus of elastin fibers and bundles in the dermis. The bundles of collagen fibers in the dermis are mostly parallel to the skin surface. There is no subcutaneous fat tissue in the flying membrane. Rarefied hairs have a few sebaceous glands and one sweat gland apiece. In a number of *Chiroptera* species sweat glands have not been recorded. The blood vessels of the flying membrane have characteristic venous valves and its veins and arteries pulsate independently (recent works: Nicoll and Webb, 1946; Mislin and Kaufmann, 1947; Cowles, 1947; Reeder and Cowles, 1951; Huggel, 1959 and others).

The *Chiroptera* body skin is thin or medium thick, the epidermis is thin or relatively thick, and there is no stratum granulosum or stratum lucidum. In or immediately under the epidermis there may be melanocytes. Langerhans cells are recorded in the epidermis (Bourland et al., 1967). In the hind sole the epidermis attains considerable thickness, strata granulosum and lucidum being present. In the body dermis, bundles of collagen fibers are mostly horizontal. Elastin fibers are numerous, and cross-striated muscle fibers may occur. The content of cellular elements in *Chiroptera* dermis is relatively higher than in other mammals of similar size (Quay, 1970). In some *Chiroptera,* fat accumulations in the subcutaneous tissue can vary with the season. In fact, in *Taphozous kachhensis* in India, large fat deposits are accumulated in the tail base and in the intercostal membrane by the dry season (Brosset, 1962). The sebaceous glands have one or two lobes. Sweat glands are absent (*Megachiroptera*) or well developed (*Microchiroptera*) (Quay, 1970). There are no specific glands in the sole, though they are recorded in the face, on the body, on the wings, and in the preputial and anal areas. In most of the species the hairs grow singly and usually divide into two very similar categories apart from the vibrissae, pile, and fur. The cuticle of the pile hairs is sometimes ring-shaped, sometimes not. The medulla in the pile hair is either absent or rather well developed. Hairs different from the body hairs are peculiar to the specific skin glands, which are either longer and different in color or rather atrophied. Most *Chiroptera* appear to have only one molt per year, with only *Rhinolophus* from India having two (Andersen, 1917).

SUBORDO MEGACHIROPTERA

The skin of *Megachiroptera* (which is rather primitive) is very different from the more specialized *Microchiroptera* bats. The absolute thickness of body skin is considerable. The epidermis is thick and usually pigmented. The stratum spinosum is two to three cells deep. In the dermis, bundles of collagen fibers form a dense plexus, which is less solid in the reticular than in the papillary layer. The boundary between papillary and reticular layers is

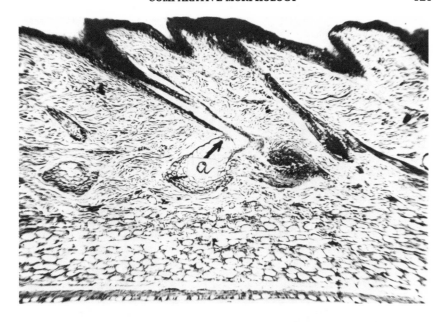

Fig. 45a. *Pteropus sp.* Back skin: *a,* common sebaceous glands.

level with the hair bulbs. The horizontal elastin fibers are numerous in the papillary layer of the dermis. Deep in the reticular layer, elastin fibers form a solid horizontal plexus, and fat cells occur in most cases in the reticular layer. Sebaceous glands are large and oval or round, with one, two, or sometimes three at every hair, rather deep in the dermis almost level with the hair bulbs (fig. 45a). Sweat glands are absent. Specific glands may occur in the skin of the body (fig. 45b). In fact, on the sides of the occiput or on the shoulders, some species have pouchlike, specific glands.

The males of genera *Rousettus, Epomophorus, Micropteropus, Epomops, Myonycteris, Pteropus* and some others have glandular sacs in the brachial region from which hair tufts grow. In *Myonycteris* males, the specific gland is situated on the upper part of the neck. Paired, bean-shaped glands have been found on the sides of the anus in *Bonycteris* (postanal or perianal, Schaffer, 1940, Quay, 1970). Arrector pili muscles are sometimes well developed (Quay, 1969). The hairs are usually divided into pile and fur (Volzhina, 1951, distinguishes guard hairs in some members of the suborder, such as *Pteropus pselophon*).

In some fruit bats, such as *Aethalops aequalis, Pentetor lucasi* (Benedicts, 1957), the hairs are not divided into categories. The pile hairs range from 6 to 24 mm long and from 14 to 121μ thick, the fur hairs range from 5 to 20 mm and 19 to 54μ. Benedicts (1957) compiled detailed keys for the hair and cuticle scale indexes of numerous species of fruit bats. The pile

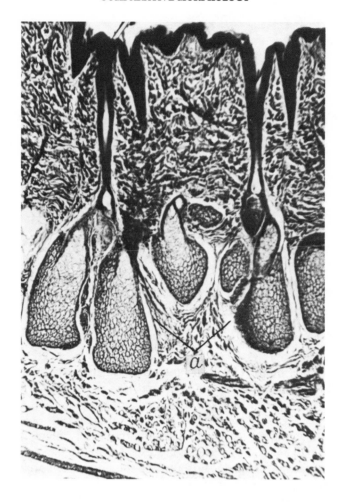

Fig. 45b. *Pteropus sp.* Withers skin: *a,* hypertrophied specific
sebaceous glands.

hairs have a weak granna in the upper part of the shaft. The lower part of
the pile hair is wavy. The fur hairs are uniformly thick the whole length.
The medulla is either absent or well developed. The cuticle cells of the pile
hairs are usually ring-shaped. The cuticle cells of the fur hairs are invariably
ring-shaped. The hairs may form groups.

In the skin of the flying membrane, the epidermis is pigmented more
heavily on the outer side. The sebaceous glands, smaller than in the body
skin, have a single lobe. Sweat glands and arrector pili muscles are absent.
Intensive pigmentation of the epidermis of the hind foot soles is characteristic. Alveoli on the inner side of the epidermis interdigitate with the dermis.
The specific tubular glands in the foot sole skin are absent.

Pteropus melanotus

Material. Skin from withers, flying membrane, and hind foot sole; adult specimen taken in Java.

The withers skin is thick (913μ) due to hair growth. The epidermis (34μ) is not pigmented. The stratum corneum (17μ) is rather solid. The thickness of the dermis is 879μ, the papillary layer is 615μ, and the reticular layer is 264μ. The reticular layer of the dermis has many fat cells (40 x 50 and 33 x 57μ). The sebaceous glands have one or two lobes, with one or two glands at every hair.

There are two categories of withers hair: pile and fur (Volzhina, 1951). The pile hairs are 20.8 mm long and are thickest (86μ) in the upper third of the shaft. The medulla starts from the distal third. The cuticle, which is not ring-shaped, consists of irregularly shaped cells. The fur hairs are 15 mm long, 29μ thick, and are thickest in the lower third of the shaft. The medulla is absent. The cuticle is similar to the pile hair cuticle.

On the sacrum and belly, in addition to pile hairs and fur, there are straight and spindle-shaped guard hairs. There are 1,580 hairs per 1 cm² of skin. The flying membrane epidermis is 27μ thick, the stratum corneum is 11μ. The sebaceous glands are unilobate, oval, and smaller than on the body (110 x 165μ). Arrector pili muscles are absent. The hind sole epidermis attains the greatest thickness (176μ). Here the stratum corneum is 44μ thick, and the inner layers of the epidermis are pigmented. The epidermal alveoli are 77μ high.

Pteropus giganteus

Material. Skin from withers, breast, flying membrane, and hind soles; two specimens (body length 26 cm and 25 cm) taken in Sri Lanka.

The hairs are growing. The roots are much longer in specimen (1), which has a thicker skin. On the withers the skin is 1,430μ and 770μ thick in specimens (1) and (2). The thickness of the epidermis is 56μ, of the stratum corneum 39μ, dermis 1,374 and 714μ, papillary layer 1,100 and 527μ, reticular layer 274 and 187μ. In the breast, skin thickness is 1,078 and 660μ, epidermis (without stratum corneum) 22 and 17μ, dermis 1,056 and 643μ, papillary layer 856 and 593μ, reticular layer 220 and 50μ. The stratum corneum is loose. Melanocytes and pigment granules are found in the epidermal cells.

The hind sole epidermis is thick (154μ); the stratum granulosum contains two or three rows of cells; the stratum lucidum is absent; the stratum corneum is 28μ thick; the inner layers of the epidermis are densely pigmented; the height of the dermal protuberances reaches 56μ. In the inner papillary layer and the middle and outer reticular layer of the body skin there are many fat cells (40 x 45; 45 x 45μ). The bundles of collagen fibers are mostly horizontal, though there are many bundles at various angles. The elastin fibers are rather numerous. The withers sebaceous glands are very large

(220 x 440; 220 x 390μ). In the breast skin, sebaceous glands are smaller (154 x 330; 165 x 290μ) and are somewhat higher in the middle layers than in the withers.

The pile hairs are up to 14.0 mm long and 182μ thick with a weak granna in the upper third of the shaft. The main part of the hair is wavy (up to ten waves). There is no medulla. The fur hairs are only 5.1 mm long, even, and 25μ thick. They are wavy (up to five waves), without medulla. The hairs grow singly, with 2,632 hairs per 1 cm².

The thickness of the flying membrane epidermis is 22μ and of the stratum corneum 11μ. Elastin fibers are numerous, forming bundles up to 60μ in diameter in places. The bulbs of the growing hairs almost adhere to the subcutaneous muscles, so a reticular layer is not distinguishable in the dermis. The sebaceous glands, which are oval, are smaller than the body glands (67 x 112; 73 x 112).

Eidolon helvum

Material. Skin from withers, breast, flying membrane, and hind foot soles of two adults taken in Zanzibar, March 6 (specimen 1) and April 22 (specimen 2).

In specimen (2) the withers skin is thickened by numerous fat cells in the inner reticular layer. In both specimens the hair has completed growth. In specimens (1) and (2), the thickness of the withers skin is 518μ and 638μ, the epidermis 67 and 11μ, the stratum corneum 55μ in both, the dermis 451 and 627μ, the papillary layer 275μ and 242μ, and the reticular layer 176μ and 357μ. The breast skin of specimen (1) is 495μ thick, the epidermis (without stratum corneum) 11μ, dermis 484μ, papillary layer 264μ, reticular layer 220μ. The stratum corneum is loose. Numerous pigment granules concentrate mostly around the nucleus of the epidermis cells. The stratum corneum is also pigmented. The sole skin has a relatively thick epidermis (up to 150μ) with a solid stratum corneum up to 28μ thick and a one to two cell stratum granulosum. The dermal projections are 50 to 70μ high. In specimen (2), the base of the reticular layer (160 to 180μ thick) has a large number of fat cells (24 x 45; 30 x 36μ). The sebaceous glands are oval (62 x 65μ; 78 x 84μ). Most hairs have one, two, and sometimes three glands whose short ducts open in the lower third of the hair roots. The sebaceous glands, surrounded by numerous densely adhering pigment cells, are larger in the breast.

The withers skin has pile (15 mm, 48μ) and fur (7 mm, 26μ) (Benedicts, 1957). In the pile hairs, the length of the cuticle scale in the middle of the hair is 10 mm, the width 46μ; on the fur hairs it is 12 and 26μ. Long (16 mm), thin (85μ) hairs grow on the specific gland, and short, thick hairs (9 mm and 110μ) are recorded. The withers hairs grow in groups of four to seven hairs, and 3,420 to 7,000 per 1 cm². On the breast, the hairs grow singly, and 11,100 per 1 cm².

The skin of the flying membrane is 0.27 mm thick, epidermis 17μ, and the firm stratum corneum 11μ. The dermal bundles of elastin fibers are up to 220μ in diameter. The sebaceous glands are rather large (73 × 85; 50 × 67μ), with two at every hair. Pigment cells are concentrated near the glands and the body skin. The gland ducts open halfway up the hair root. The hairs grow singly and far apart.

SUBORDO MICROCHIROPTERA

The body skin is thin (60 to 440μ on the withers). The epidermis is thin in places where the unpigmented stratum spinosum consists of one to two rows of cells. The dermal collagen fiber bundles are horizontal. Elastin fibers, which are plentiful, form a loose, horizontal plexus bordering on the subcutaneous muscles. The papillary layer is nearly always thicker than the reticular, their borders being level with the hair bulbs or the bottom of the sweat glands. There are occasional fat cells in the reticular layer.

The sebaceous glands are unilobal, oval or round, with one behind each hair. They lie deep in the dermis or are level with the nongrowing hair bulbs. In the flying membrane they are bigger and there is a group at each hair. Most *Microchiroptera* have saclike, or straight tubular, or twisted apocrine sweat glands in the body and wing skin. Their ducts open into the hair funnels. The sweat glands on the flying membrane are usually simplest in shape. In the muzzle skin they are very long and twisted. The total number of the sweat glands may reach 9,373 (in *Myotis lucifugus;* Sisk, 1957). The number of the sweat glands per 1 cm^2 varies from four on the wings and the intercostal membrane to 673 on the muzzle. The number decreases from front to rear on the trunk. They are at their largest and most active in summer, but continue to function even during hibernation (Quay, 1970). The main component of the sweat is neutral glycoprotein or mucoprotein (*Myotis lucifugus;* Sisk, 1957). Specific glands may occur on the trunk, head, and wings. There are no arrectores pilorum muscles.

In addition to pile and fur hairs there may be guard hairs (*Noctulidae,* Volzhina, 1947). The upper part of the pile has slight thickening and a weak granna. The fur hairs are of uniform thickness, tapering only at the tip. Usually neither has a medulla. In most of the species, the cuticle cells are ring-shaped in all hair categories. The upper shape of the scale varies—it is divided into thirteen types: entire, repand, sinuate, emarginate, equal hastate, unequal hastate, denticulate, dentate, erose, crenate, irregular crenate, broad lobate, and narrow lobate (Benedicts, 1957) (fig. 46). The hairs grow singly, sometimes in tufts but rarely in groups. Some members of the *Familia Molossidae* have peculiar hairs with a bent, spoonlike tip on the upper lip and on the first and last digits. Many *Microchiroptera* have numerous bare skin protuberances on the muzzle which vary in size.

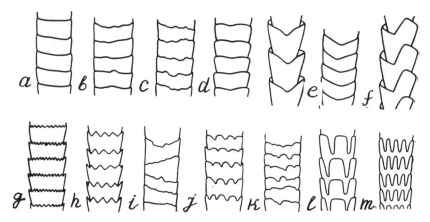

Fig. 46. Diagram of types of cuticle scales of *Microchiroptera* (Benedict, 1957): *a*, entire; *b*, repand; *c*, sinuate; *d*, emarginate; *e*, equal hastate; *f*, unequal hastate; *g*, denticulate; *h*, denate; *i*, erose; *j*, crenate; *k*, irregular crenate; *l*, broad lobate; *m*, narrow lobate.

FAMILIA RHINOPOMATIDAE

Genus rhinopoma

R. microphyllum's hairs are all of one type (Benedicts, 1957). They are short (7 mm long), straight, with a thickening of up to 34μ in the distal part and with a broken medulla. The ring-shaped hair scales are broad lobate (the scale is 20.4μ long, 28μ wide halfway along the hair) except for the distal section, where they are dentate, 18.7 long, 34μ wide. Melanocyte granules, distributed throughout the hair and most numerous in the distal third, form small accumulations at the base of every medullary vacuole. Specific glands formed of sweat and sebaceous glands are on the sides of the muzzle (Magkour, 1961).

FAMILIA EMBALLONURIDAE

All species have only one type of hair (Benedicts, 1957) of uniform thickness. The cuticle is ring-shaped and denticulate, except in *Taphozous,* which has dentate cuticle scales. Most of the species have individual scales with a W-shaped notch. The melanin granules are concentrated in the distal half of the scales, forming girdles along the entire shaft of all genera except *Cornura, Tophozous,* and *Diclidurus,* where the pigment (if present) is dispersed evenly. The medulla is completely absent. The genera fall into three groups for hair structure: (1) *Emballonura* and the closely related *Coleura, Rhynchiscus, Saccopteryx, Cornura, Peropteryx, Peronymus, Centronycteris, Balantiopterix;* (2) *Taphozous;* and (3) *Diclidurus.* The first group's hair length ranges from 7 to 12 mm, thickness 27 to 37μ; the cuticle scale length (halfway up the hair) from 7 to 12 mm, thickness 25 to 37μ; the length of the cuticle scales in the middle of the hair, 12 to 17μ; the breadth of the cuticle scales, 25 to 37μ. In *Taphozous* the length of the hair is 7 to 8 mm, thickness 24 to 84μ; the length of the scales of the cuticle 14 to 25μ. In *Diclidurus virgo,* the length of the hair is 12 mm, thickness 34μ; the length of the scales of the cuticle 29 mm, the breadth 34μ.

Volzhina (1951) gives different data for the pelage of *Tophozous melanopogon* from Benedicts (1957). According to Volzhina, *Tophozous melanopogon* has three hair categories: guard, pile, and fur. The guard hairs are 6.2 mm long and 13μ thick, straight and round, divided into two parts, the lower thin and round, the upper spindle-shaped. The upper edges of the ring-shaped cuticle scales are dentate. The pile hairs, up to 5.6 mm long and 10μ thick, have a spindle-shaped upper section at a 15 to 20° angle to the lower round section. The fur hairs have two small waves, one 5.1 mm long and 7μ thick.

Most members of the species, especially the adult males, have a well-developed specific humeral gland, a sac opening dorsally with a proliferation of epidermal stratum corneum (Schaffer, 1940; Starck, 1958). The sebaceous glands in the sac walls are not well developed. The other specific glands recorded are: labial (in *Saccopteryx*), comprised of sweat glands (Starck, 1958); mandibular (in *Saccopteryx* and some species of *Taphozous*), formed mostly of sweat glands which secrete heavily in males during the mating season; the gland in the nasal region, comprised of sebaceous glands with some sweat glands (Dalquest and Wagner, 1954); and, in certain species of *Taphozous* and *Saccopteryx,* the gular glandular region, or a sac made up of tubular glands (Starck, 1958). In several *Taphozous* species this sac is well developed in both sexes, in others in the male only or absent from both sexes. In one group of species this gland is better developed during the mating season. Some species of *Taphozous* have an intercostal-membrane thoracic gland (*Diclidurus* and *Depanycteris*). No histological analysis has yet been made of most of these glands.

FAMILIA NOCTILIONIDAE

Genus Noctilio

Very short hairs and short, appressed entire or repand cuticle scales are characteristic of the genus (Benedicts, 1957). The hairs categorize into pile and fur, the pile hairs being club-shaped. In the absence of the medulla, melanin granules are distributed throughout the whole hair.

The hair indexes common to *N. leporinus* and *N. labialis* are as follows (Benedicts, 1957): pile hair, 5 mm long in the middle and 26μ thick in the distal part; fur hairs, 5 mm long in the middle and 22μ thick in the distal part. The cuticle scales at the middle of the pile hairs are 14μ long and 22μ wide; the fur hairs are 12μ and 19μ, respectively.

FAMILIA NYCTERIDAE

Genus Nycteris

Unlike the related family *Megadermatidae,* the cuticle scales are ring-shaped, entire or repand, and the medulla is absent (Benedicts, 1957). Melanin granules are for the most part evenly distributed along the shaft. The hairs divide into pile and fur.

Their indexes (common to *N. arge, N. aurita, N. hispida*) are as follows (Benedicts, 1957): pile hairs are 13 to 16 mm long, maximum thickness (distal shaft) 22.1 to 23.8μ, the fur hairs are 13 mm long, maximum thickness (distal shaft) 18.7 to 20.4μ. The cuticle scales halfway up the pile hair are 17μ long and 17μ wide, at maximum thickness 11.9 to 15.3μ long and 22.1 to 23.8μ wide. The cuticle scales halfway up the fur hairs are 15.3 to 17.0μ long and 15.3 to 18.7μ wide, and at maximum thickness are 13.6 to 15.3μ long and 18.7 to 20.4μ wide.

FAMILIA MEGADERMATIDAE

The length of the single category of hairs (genera *Megaderma,* subgenus *Zyroderma*), *Macroderma, Lavia,* and *Cardioderma,* ranges from 11 to 18 mm, the thickness from 17 to 34μ. In genus *Megaderma,* subgenus *Megaderma* are pile hairs (length 20 mm, thickness 24μ) and fur hairs (length 15 mm, thickness 11μ) (Benedicts, 1957). The cuticle scales are broad lobate, 17μ long and 17μ wide, in the pile hairs and 14μ long and 19μ wide halfway up the fur hairs (*Megaderma,* subgenus *Megaderma*) *Macroderma, Lavia, Carioderma* (the last two scales are 10 to 19μ long and 14 to 32μ wide). They may be repand (genus *Macroderma,* subgenus *Lyroderma,* 17μ by 22μ). The medulla is broken. Melanin granules are few and are distributed along the whole hair but mostly in the distal section. At the base of each medulla vacuole is a group of melanin granules.

The hairs are thickest toward base and tip (pile hairs, *Megaderma spasma*), or at the tip only (fur hairs, *M. spasma*), or base only (*M. sinensis, M. gigas*), or base and middle (*L. frons, C. cor*).

Genus *Lavia* males have a specific gland of unknown histological structure on the back.

FAMILIA RHINOLOPHIDAE

Rhinolophus bocharicus

Material. Skin from withers and breast (table 11) of two specimens (body length, 5.2 cm, 5.7 cm), taken at Tachta-Bazar, Murgab river, April 27.

On the withers most of the hairs are fully grown. Fat cells (36 × 63μ or 51 × 54μ) are abundant in the withers but rare in the breast reticular layer.

TABLE 11

RHINOLOPHUS BOCHARICUS: SKIN MEASUREMENTS

Indexes	|	Thickness (μ)						Sebaceous glands (size in μ)
	|	Skin	Epidermis	Stratum corneum	Dermis	Papillary layer	Reticular layer	
Withers	|							
(1)	|	168	17	7	151	95	56	30 × 42, 30 × 48
(2)	|	132	20	8	112	85	27	21 × 48, 22 × 39
Breast	|							
(1)	|	90	11	4	79	67	12	27 × 45, 21 × 39
(2)	|	89	12	5	77	60	17	34 × 39, 28 × 45

Sweat glands are not recorded. The hairs divide into two categories. Pile hairs, 7.8 mm long, 16μ thick, are slightly thicker in the upper third of the shaft. The club has two to three waves. Fur hairs are 5.8 mm long, 13μ thick, with up to four slight waves. Ring-shaped cuticle cells in both categories are 11μ high. There are 23,500 hairs per 1 cm^2 (withers) and 24,400 (breast).

The pelage was studied in *R. akbae, R. affinis, R. ferrumequinum, R. lanosus, R. malayanus* (Benedicts, 1957), *R. hipposideros* and *R. mechelyi* (Volzhina, 1951). There are pile and fur hairs (Volzhina, 1951, also distinguished guard hairs 8.2 to 10.8 mm long and 10 to 14μ thick). Pile hairs are 7.8 to 22 mm long; 9 to 24μ thick; the fur hairs are 5.8 to 10 mm long, 8 to 19μ thick. The pile hairs are distally thickest; the fur hairs are medially and distally thickest. The fur hairs have seven to eight waves (*R. hipposideros*) or nine to ten (*R. mechelyi*), with no medulla in either category; melanin granules are evenly distributed along the whole hair. The cuticle scales in the medial section are long and highly sinuate, alternating with short, hastate scales. Distally, they are short and repand. The scales on the pile hair are 10 to 22.5μ long, 13.6 to 23.8μ wide, on the fur hair 8.5 to 22.1μ and 11.9 to 18.7μ, respectively. In *R. mechelyi,* the hairs grow in groups (Volzhina, 1951). Vibrissae are well developed in *R. ferrumequinum,* growing on the skin projections of the snout, the upper and lower lips at the corner of the mouth, the eyelids and throat (Schneider, 1963). The vibrissae have large blood sinuses and numerous nerve terminals, branches of *N. infraorbitalis.* In *R. hipposideros* vibrissae develop prenatally, earlier than other hairs in ontogenesis (Klima and Gaisler, 1967*a, b*). Newborns are covered with a well developed fur hair. Genus *Rhinolophus* have well developed specific skin glands on the sides of the snout in which sweat glands prevail. In two species of this genus, *R. landeri* and *R. alcyone,* the adult males appear to have specific skin glands in the armpit region, though their histological structure is obscure.

FAMILIA HIPPOSIDERIDAE

The pelage, as in *Rhinolophidae,* divides into pile and fur hairs (except for *Anthops triaenops,* where no categories are distinguishable). There is no medulla, and melanin granules are evenly distributed along the whole hair (except in genus *Coelops,* where they form a strip on every cuticle scale). In most genera, long entire or repand cuticles alternate with short hastate scales in the middle section of the hair. The pile hairs are 8 to 16 mm long, 20 to 34μ thick; the medial cuticle scales are 14 to 26μ long and 9 to 24μ wide. The figures for the fur hairs are 7 to 14 mm, 14 to 20μ; 17 to 24μ and 14 to 17μ; respectively. The hairs are thickest in the distal section, except in *Triaenops,* where they are thickest in the proximal section, and in *Cloeotis,* where the fur hairs are medially thickest. Genera *Hipposideros, Anthops,* and *Coelops* have a frontal glandular sac or a papilla, which is better developed in the male. In some species, *Hipposideros* males have anal specific skin glands. The histological structure of these glands is obscure. During the embryonic period, vibrissae develop earlier than other hairs (*Asellia tridens*) (Klima and Gaisler, 1967b). Unlike *Rhynolophidae, Hipposideridae* newborns have only well developed follicles.

FAMILIA PHYLLOSTOMATIDAE

The many species in this family have closely similar pelage (Benedicts, 1957). *Chilonycteris, Pteronotus, Phyllostomus, Phylloderma, Lionycteris, Sturnira, Brachyphilla, Ardops, Erophylla, Phyllonycteris,* and *Reithronycteris* have pile and fur hairs, as does *Uroderma planirostris* (Volzhina, 1951). The remaining genera have no separate categories. The pile hairs range from 4 mm (in *Chilonycteris*) to 18 mm long (in *Chrotopterus*), and from 17 to 43μ thick; the cuticle scale (middle of the hair) from 14 to 29μ long, and 12 to 29μ wide. The fur hairs are 4 to 26 mm long, 12 to 29μ thick; the cuticle scale (middle of the hair) is 12 to 26μ long, 12 to 29μ wide. For the genera with only a single category, the figures are 5 to 18 mm long, and 20 to 32μ thick; the cuticle scale (middle of the hair) is 12 to 49μ long, 17 to 32μ wide. The medulla is absent in all species. There are three main types of cuticle scale: entire, denticulate, and dentate, the last being rare. A few species have other types. *Brachyphylla,* for instance, have erose scales, *Phylloderma* and *Sturnira,* sinuate and erose. The hair pigmentation pattern is similar in the whole family. Melanin granules are distributed throughout the hair, but more heavily in the distal section. *Ectophylla* differs in that the melanin granules are very large and far apart.

Molt appears to be annual (*Mormops megaplophylla*) (Constantine, 1958).

Members of the family have various specific skin glands: in *Sturnira,* brachial tufts of elongated hairs are described which appear to grow on the specific gland. *Phyllostomus, Artibeus,* and *Trachops* have specific labial glands in which sebaceous glands prevail. The whole area, however, has a higher sensory than secretory specialization (Quay, 1970), and the glands are developed rather poorly (Dalquest and Werner, 1954). A throat gland is recorded in *Chrotopterus, Centurio,* and *Phyllostomus.*

FAMILIA DESMODONTIDAE

The hair divides into fur and pile in *Desmodus* and *Diaemus,* and does not divide into categories in *Diphylla* (Benedicts, 1957). The medulla is absent. Melanin granules are dispersed along the whole hair in *Diaemus* and are arranged in the central part in *Desmodus.* For *Desmodus* and *Diaemus,* a bulb-shaped basal swelling of the hair shaft is characteristic. The pile hairs are 8 to 10 mm long, and 34 to 44μ thick; the length of the scales of the cuticle in the middle of the hair is 9 to 15μ thickness, breadth 29 to 34μ. The respective sizes of the fur hairs are 4 to 8 mm, 17 to 31μ, 9 to 17μ, and 17 to 29μ. In *Diphylla caudata,* the hair length is 12 mm, thickness 22μ; the cuticle scales are 9μ long and 15μ wide. The hairs are thickest in the distal section. No specific skin glands are recorded on the muzzle of *Desmodus rondus* (Dalquest and Werner, 1954).

FAMILIA NATALIDAE

Genus Natalus

The hair divides into pile and fur hairs (Benedicts, 1957). The straight pile hairs are only slightly coarser than wavy fur hairs. There is no medulla. Cuticle scales are long or short, entire or repand. Pigment granules are dispersed. The pile hairs are 6 to 9 mm long and 19 to 29μ thick; the length of the cuticle in the medial part of the hair is 14 to 19μ, thickness 14 to 19μ. The fur hairs are 6 to 9 mm, 14 to 19μ, 10 to 14 mm, and 14 to 15μ, respectively. The hairs are thickest in the distal section. Males have a specific gland on the head, in front of and between the eyes. It is evidently formed by sweat glands, but this requires further study. There are also specific glands on the sides of the muzzle formed by sweat and sebaceous glands (Dalquest and Werner, 1954).

FAMILIA FURIPTERIDAE

No hair categories are distinguished (Benedicts, 1957). In the members of the genus *Furipteridae,* the hairs have bulb-shaped swellings at the base. The medulla is absent. Numerous pigment granules are arranged in strips as beads in the median section of the hair. The cuticle scales are denticulate. The hairs are 9 to 12 cm long, 20 to 26μ thick; the cuticle scales in the middle of the hair are 14 to 17 cm long, 20 to 26μ thick. The hairs attain the greatest thickness in the proximal and medial sections.

FAMILIA THYROPTERIDAE

Genus Thyroptera

The hair of *T. tricolor* divides into pile and fur hairs (Benedicts, 1957). There is no medulla. The pile hairs are 9 mm long, 20μ thick, fur hairs are 9 mm long, 17μ thick. Cuticle scales are emarginate and hastate. In the middle of pile hairs they are 12 mm long and 17μ thick; in the distal section, they are 10 mm long and 20μ thick, respectively. In the middle of fur hairs, they are 19μ long, 17μ thick. The pile hairs are thickest distally, and the fur hairs thickest medially.

The skin suckers on the first digit of the wing and on the foot are characteristic. They are arranged on small stalks. The sucking organ is cup-shaped and consists of fat tissue separated by connective tissue strands fixed on cartilaginous plate at the base of the sucker (Schliemann, 1970).

FAMILIA MYZOPODIDAE

Genus Myzopoda

The hair of *Myzopoda aurita* does not divide into categories (Benedicts, 1957). There is no medulla, but pigment granules are dispersed through the hair, mostly in the distal third. Some hairs are completely without pigment. The hairs are 13 mm long and 17μ thick. The maximum thickness of the hair is recorded in its distal part. The scales of the hair are denticulate, 12μ long and 15μ wide medially, and 11μ long and 17μ wide distally. There are sucking discs without stalks on the first digit of the wing and on the foot, consisting of a fatty tissue separated by connective tissue strands into cylindrical parts arranged longitudinally (Schliemann, 1970).

FAMILIA VESPERTILIONIDAE

The structure of the pelage of the members of numerous genera of the family is fairly homogenous, though some genera show specific features. In *Chalinolobus, Mimetillus, Nycticeius* (subgenus *Scoteins*), *Dasypterus,* and *Miniopterus* no hair categories are recorded (Benedicts, 1957). According to our data *Miniopterus schreibersi* has pile and fur hairs. The pile hairs are 4 to 15 mm long and 17 to 31μ thick, length of cuticle in the middle of hair 9 to 24 and 12 to 26μ. In the fur hairs the figures are as follows: 3 to 17 mm, 12 to 19μ, 7 to 24μ, 9 to 19μ. Pile hairs have clublike thickenings at the distal end. There is no medulla. The pigmentation pattern varies in various genera. Commonly, pigment granules form girdles in the proximal section of the hair shaft and are dispersed in the distal section.

As a rule the cuticle scales are hastate, and they differ in basal thickening. In the distal section, the scales may be hastate, sinuate, or lobate. There are some other differences. In fact, genus *Chalinolobus* has full scales in the proximal third of the shaft, and dentate scales in the rest; in genus *Scotophillus* the scale is lobate; genus *Mimetillus* has dentate scales. In *Miniopterus* the cuticle scale is similar to some species of *Phyllostomatidae,* where long, whole scales alternate with short, hastate scales. In *Harpiocephalus,* the scale is repand or sinuate, and in *Tomopeas,* it is denticulate.

The vibrissae appear first in the embryo, then the occipital and scapular region hairs (*Myotis emarginatus, Plecotus austriacus*) or the sacral and intercostal membrane areas (*Miniopterus schreibersi*) (Klima and Gaisler, 1967). The young are born partially covered with fur.

All members of several genera of this family have various specific skin glands. In fact, *Nycticeus schlieffeni* has a throat gland. A well-developed gland, in which sebaceous glands prevail, is on the sides of the muzzle in *Plecotus, Chalinolobus, Nyctalus, Pipistrellus, Euderma, Vespertilio,* and *Eptesicus.* In *Nyctalus noctula,* for example, it has been demonstrated that these glands are innervated by the fibers of the maxillary branch of the tri-geminal nerve (Porta, 1910). In *Chalinolobus,* increased size has been noted in these glands in both sexes in the period of reproduction (Dwyer, 1966). The secretion of the glands is whitish or yellowish and has a musky smell. Specific labial glands (poorly or moderately developed) are recorded in *Lasiurus* and *Dasypterus* (Werner and Dalquest, 1952). These are mostly formed by sebaceous glands. *Pipistrellus* and *Myotis* have perianal glands represented by sebaceous and sweat glands developed moderately or poorly (Schaffer, 1940).

Myotis mystacinus

Material. Skin from withers (table 12) of four specimens: specimen (1) (body length 4.0 cm) taken in July, and (2) (body length 3.7 cm) in August in Ashkhabad; (3) (body length 5.0 cm) in July in Kamchatka; (4) (body length 4.0 cm) in August in Mongolia.

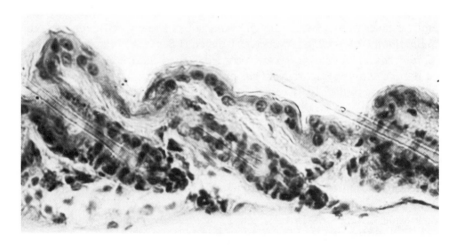

Fig. 47. *Myotis mystacinus.* General structure of withers skin.

The thickness of the skin varies (fig. 47). In *M. mistacinus,* with hair in a state of growth, the skin is considerably thickened. The hair bulbs adhere to the subcutaneous muscles so that no reticular layer is distinguishable. There are rather few fat cells (9×30, $9 \times 42\mu$) in the dermis. The sebaceous glands reach 18×27; $28 \times 64\mu$. The sweat glands have saccular, nontwisted secretion cavities 28 to 50μ in diameter, and their long axis is parallel to the body surface. They are level with the hair bulbs.

TABLE 12

VESPERTILIONIDAE: SKIN MEASUREMENTS (μ)

	Total	Epidermis	Stratum corneum	Dermis	Papillary layer	Reticular layer
Myotis mystacinus						
Withers						
Kamchatka	150	12	3	138	138	
Mongolia	95	10	3	85	85	—
Ashkhabad, (1)	220	17	6	203	203	—
Ashkhabad, (2)	62	10	3	52	46	6
Plecotus auritus						
Withers						
Baikal, (1)	374	28	11	346	326	20
Baikal, (2)	110	17	6	93	81	12
Altai, (1)	132	22	11	110	100	10
Altai, (2)	130	22	12	108	80	28
Sayany	200	17	6	183	173	10
Naryn	60	22	11	38	28	10
Leningrad (winter)	84	12	6	72	62	10
Breast						
Leningrad (winter)	67	12	6	55	45	10
Barbastella barbastella						
Breast	70	12	6	58	46	12
Nyctalus noctula						
Withers						
Caucasus	640	22	6	618	398	220
Kuldja, (1)	240	22	11	218	198	20
Kuldja, (2)	220	17	8	203	148	55
Pipistrellus kuhli						
Withers						
Caucasus, (1)	110	22	12	88	68	20
Caucasus, (2)	85	27	10	58	46	12
Algeria, (1)	83	15	8	68	41	27
Algeria, (2)	75	12	6	63	51	12
Breast						
Caucasus, (1)	72	15	8	57	42	15
Caucasus, (2)	58	12	6	46	38	8
Algeria, (1)	60	12	6	48	36	12
Vespertilio murinus						
Withers						
Belovezh	440	22	16	418	110	308
near Kazan	303	22	16	281	259	22
Kalmykia	275	17	8	258	246	12
near Naryn	106	17	10	89	67	22
Miniopterus schreibersi						
Withers						
(1)	440	22	11	418	390	28
(2)	385	22	11	363	346	17

The hair divides into pile and fur. According to Volzhina (1947), *M. mistacinus* also has guard hairs. They are very sparse, which may explain their failure to get into the skin samples. The straight guard hairs gradually thicken halfway or two-thirds down the length, tapering at the tip. The hair indexes are given in table 13.

TABLE 13

INDEXES OF WITHERS HAIRS OF M. MYSTACINUS

Origin	Length (mm)		Thickness (μ)	
	Pile	Fur	Pile	Fur
Kamchatka	9.3	5.8	18	10
Mongolia	6.0	5.0	11	8
Ashkhabad	5.7	4.3	15	11

Distally, the pile hairs thicken into a weak granna for one-third or one-half of the shaft length. The lower part of the hair is wavy with three to five waves. The height of the cuticle scales is 12 to 13μ. The fur hairs are wavy with seven to nine waves. The length of the scales on the fur cuticle is 11 to 15μ. There are 33,300 hairs (the Kamchatka specimen) per 1 cm^2. Molt is annual, earlier in males (*Myotis velifer*) than females.

Plecotus auritus

Material. Skin from the withers (table 12), specimens taken in summer: in Baikal in July, specimens (1) (body length 4.5 cm) and (2) (body length 4.4 cm); in the Altai in July, specimens (1) (body length 4.6 cm) and (2) (body length 4.0 cm); on the Sayan mountains, September 3 (body length 4.0 cm); on the Barguzin ridge, August 7 (body length 3.6 cm); near Naryn, July 8 (body length 4.3 cm) and in winter; near Leningrad during hibernation, February 5 (body length 4.5 cm).

In summer, skin thickness varies a great deal in the presence or absence of growing hair. In specimens with nongrowing hair, the thickness of the skin is 60 to 110μ. There are fat cells in the dermis of all *P. auritus*. Cell size varies from 12 x 15 to 30 x 33μ. Fat cells are especially abundant in specimens with growing hair. The size of the sebaceous glands is 28 x 45, 30 x 70μ. Sweat glands are saccular, straight, and 40 to 45μ in diameter.

The hairs divide into three categories: pile, intermediate, and fur (according to Benedicts, 1957, only pile and fur) (table 14). The pile hairs have a weak granna for one-third to one-half of the shaft length. Their lower section is wavy (four to six waves). Intermediate hairs are similar to pile,

TABLE 14

PLECOTUS AURITUS: HAIR MEASUREMENTS

Hair category	Origin	Length (mm)	Thickness (μ)	Height of cuticular cell (μ)
Pile	Crimea	12.5	18	15
	Baikal	9.0	15	9
	Barguzin	11.0	16	8
	Naryn	8.5	15	9
	Leningrad	11.0	19	11
Intermediate	Crimea	9.0	19	—
	Barguzin	9.0	12	—
	Leningrad	8.0	14	—
Fur	Crimea	6.0	11	14
	Baikal	7.0	12	17
	Barguzin	8.0	13	14
	Naryn	6.5	12	17
	Leningrad	7.0	13	14

though their lower section is curlier (up to seven to nine waves), with the thickening in the upper part of the shaft being shorter. Fur hairs have no granna; they are twisted the whole length and have up to eleven waves.

There are 23,300 hairs per 1 cm² on the specimen from the Barguzin mountains, and in those from the Altai and Sayan mountains, 17,800 hairs. Hairs grow singly.

In winter, the skin, epidermis, and the stratum corneum are thinner than in most of *P. auritus* taken in summer (table 12). No morphological features are revealed in the skin to distinguish it from summer bats.

Fat cells (15 x 18; 18 x 21μ) are abundant, especially in the lower part of the dermis. The sebaceous glands (17 x 22; 28 x 40μ) are inactive. Large sweat glands, 56μ in diameter, occur less often than in summer *P. auritus*.

The hair divides into three categories. Their shape is similar to other *P. auritus* and they are much the same in size (table 14), but there are fewer hairs—16,700 per 1 cm².

Barbastella barbastella

Material. Skin from the withers (table 12) of *B. barbastella* taken in summer in the Caucasus near Borzomi (body length 5.0 cm).

Fat cells sometimes occur in the reticular layer. Sebaceous glands (17 x 22; 22 x 28μ) are few and inactive. Sweat glands are not recorded. The hair divides into two categories: pile and fur. The pile hairs are 9.8 mm long and

13μ thick, with a weak granna in the upper third. The lower part of the shaft has seven waves. The height of the cuticle scales is 20μ. Fur hairs are 6.9 mm long and 11μ thick. The entire shaft is wavy (up to eight waves). The height of the cuticle scales is 10μ).

Nyctalus noctula

Material. Skin from the withers (table 12) of three specimens taken on July 21 in the Caucasus near Kutaisi (body length 7.3 cm) and two on June 6 in Kuldja, Western China (body lengths 7.0 cm and 7.3 cm).

In the Caucasian specimen the hair is growing and the skin is thicker. Only the upper part of the dermis, about 110μ thick, has a dense plexus of horizontal collagen bundles. The rest is a continuous mass of fat cells (39×56; $34 \times 56\mu$). A thin layer of collagen bundles is found only on the border with the subcutaneous muscles. Under it is a horizontal bundle of elastin fibers. In the Kuldja specimen a reticular layer attains a considerable thickness and has small fat cells (6×11; $14 \times 14\mu$). There are sebaceous glands (12×22; $28 \times 35\mu$) in the upper compact part of the dermis. They are not found at every hair and are not active (fig. 48). Sweat glands, rare in *N. noctula* from the Caucasus, have a saccular, straight secretion section 35 to 56μ in diameter beneath the dense part of the dermis (fig. 48); their long axis is parallel to the hair roots. The specimen from Kuldja has numerous sweat glands, 35 to 40μ in diameter, tubular, twisted, sometimes forming a small glomeruli.

The hair divides into two categories: pile and fur. Pile hairs are 5.6 to 9.7 mm long and 10 to 17μ thick, with a weak granna somewhat less than half the shaft. The lower section of the hair has one to two big, flattened waves. The cuticle scales are 13μ high. The fur hairs are short (3.4 to 6.4 mm) and thin (10 to 11μ) with five to eight waves. The height of the cuticle scales is 1μ. Hair measurements of these specimens are close to the indexes of Benedicts (1957) obtained for *N. noctula* and *N. azoreum*. There are 13,300 hairs per 1 cm² on the specimen from the Caucasus and 31,100 on the Kuldja specimen.

Pipistrellus kuhli

Material. Skin from the withers, breast (table 12), and flying membrane of four specimens: two taken in September in the Caucasus at Aralykch (body lengths 3.6 cm and 3.4 cm), and two taken in July near Biskra in Algiers (Body lengths 3.8 cm and 3.6 cm).

The size of the sebaceous glands is 17×28; $34 \times 56\mu$. Sweat glands on the withers are saccular, elongated, slightly twisted, 22 to 30μ in diameter (specimen from Algiers), or tubular, twisted, 22 to 45μ in diameter (specimens from the Caucasus). On the breast, the sweat glands are saccular and not twisted.

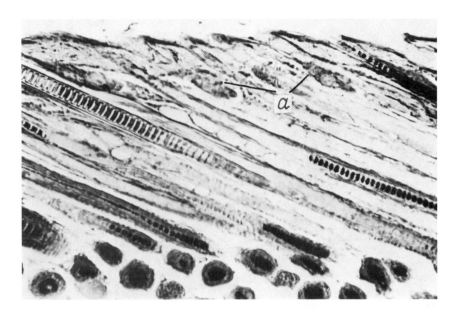

Fig. 48. *Nyctalus noctula.* Withers skin: *a,* sebaceous glands.

The hair is of one category, with or without a slight thickening in the distal section (Benedicts, 1957, distinguishes pile and fur hairs; wavy 4.9 to 5.2 mm long, 14 to 15μ thick), 23,300 hairs per 1 cm². The hairs form no tufts or groups.

The epidermis of the flying membrane is 12μ thick, of which stratum corneum takes up 6μ. Roots of sparsely set hairs have two to four large sebaceous glands (40 x 62μ). Sweat glands are not recorded. In the middle of the flying membrane are thick (to 40μ) bundles of elastin fibers.

Vespertilio murinus

Material. Skin from withers (table 12) of *V. murinus,* taken in Belovezh in June (body length 6.0 cm), near Kazan in July (body length 6.0 cm), in Kalmykia in summer (body length 6.0 cm), and near Naryn in summer (body length 5.3 cm).

The thickness of the skin is highly variable. In the Kazan and Kalmykia specimens the hairs are in a state of growth. The skin of the *V. murinus* from the Belovezh is peculiar. The upper layer, about 80μ thick, is a solid plexus of collagen fiber bundles. The rest of the skin beneath the hair bulbs is filled out by a continuous mass of fat cells. It is not clear whether to classify this part of the skin as a reticular layer or as subcutaneous fat tissue. Collagen bundles are almost completely absent. It seems more likely to be a reticular layer filled with fat cells. In the dermis of most of the specimens, fat cells (15 x 42; 24 x 36μ) are numerous. Sebaceous glands are found, but

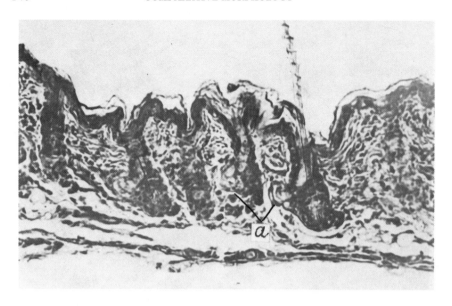

Fig. 49. *Vespertilio murinus.* Withers skin: *a*, sebaceous gland.

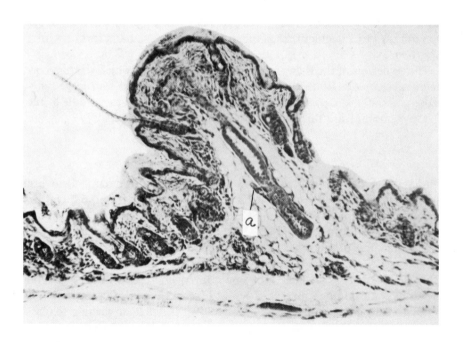

Fig. 50. *Vespertilio murinus.* Withers skin: *a*, sweat gland.

not at each hair (fig. 49). Their size is about the same in all the specimens (17 x 28; 28 x 45μ). The few sweat glands are saccular, elongated, or slightly twisted, 25 to 50μ in diameter (fig. 50). In the Belovezha specimen they are in the layer of fat cells beneath the hair roots.

The hair divides into two categories: pile and fur. The pile hairs are 6.9 to 7.1 mm long, 12 to 13μ thick. There is thickening in the upper third of the hair. The lower part of the hair shaft has two to four waves. The height of the cuticle scales is 9 to 15μ. Fur hairs are 5.3 to 5.5 mm long and 9 to 11μ thick, with seven to nine waves. The height of the cuticle scales is 14μ. In the specimen from Kazan 27,800 hairs are recorded per 1 cm^2 of skin. In the specimen from the Naryn area 18,000 hairs are recorded. The hairs do not form groups.

Miniopterus schreibersi

Material. Skin from withers (table 12) of two specimens taken at the village of Dakovskaya, July 27 (body lengths 5.2 cm and 6.0 cm).

The hair of both specimens is in the process of growth. Fat cells are abundant in the dermis. Their maximum size is 21 x 39; 30 x 36μ. The size of the sebaceous glands is 28 x 56, 30 x 78μ. Saccular, straight sweat glands 25 to 40μ in diameter occur rarely. Their long axis runs parallel to the hair roots.

The hair is divided into two categories: pile and fur (Benedicts, 1957, does not distinguish any categories). The pile hairs are 5.9 mm long, and 13μ thick. The upper fifth of the shaft manifests a slight thickening. The lower part of the hair has up to eight waves. The height of the cuticle scales is 10μ. Fur hairs are 5.1 mm long and 12μ thick, with nine waves. The height of the cuticle scales is 14μ. There are 37,800 hairs per 1 cm^2. The hairs form groups and are subject to seasonal variation.

FAMILIA MYSTACINIDAE

The pelage of *Mystacina tuberculata* does not divide into categories (Benedicts, 1957). There is no medulla. Pigment granules are dispersed along the hair and their concentration varies greatly. The length of the hair reaches 9 mm, thickness 29μ. In the middle of the hair is an elongated swelling. Cuticle scales are whole. In the middle of the hair they are 26μ long and 26μ wide, and in the distal part they are 12 and 29μ, respectively.

FAMILIA MOLOSSIDAE

The hair does not divide into categories and has no medulla. Melanin granules are dispersed or concentrated as longitudinal bands. The hairs are usually short (4 to 7 mm) reaching 9 mm only in three genera. The cuticle scales are dentate with a deep V-shaped notch. The length of the hairs ranges from

4 to 9 mm, the thickness from 17 to 32μ. The length of the scales of the cuticle in the middle of the hair ranges from 7 to 17μ and the thickness from 17 to 32μ. In *Myopterus, Molossops, Cynomops,* and *Tadarida,* the hairs are thickest in the median and distal parts, and in the rest of the family they are even along the whole length. Molt occurs once a year, males molting earlier than females (*Tadarida brasiliensis,* Constantine, 1957). In *Cheiromeles torquatus,* the skin is virtually devoid of pelage. Sparse hairs found on the throat, belly, and sacrum grow to 77μ and should evidently be classified as pile hairs (Benedicts, 1957). Cuticle scales are very short (length 9 and breadth 77μ), whole, repand, or sinuate.

Most of the genera (*Cheiromeles, Tadarida, Platymops, Otomops, Molossus, Promops, Eumops, Austronomus*) possess a specific throat gland (Schaffer, 1940; Harrison and Davies, 1949; Harrison and Fleetwood, 1960; Werner and Lay, 1966; Horst, 1966). In the members of this family, this gland is better developed than in any other *Subordo Microchiroptera* family. Sexual dimorphism in the level of development of this gland may be absent (*Otomops wroughtoni, Platymops barbatogularis, P. macmillani*) (Brosset, 1962; Harrison and Fleetwood, 1960), or manifest. In *Molossus rufus* the gland is equally developed in both sexes in the late embryonic stages, though it does not continue to develop after birth in females (Horst, 1966). Hormonal injections of the females can induce continued development of the gland, and castration of the males results in involution. The gland functions throughout the year, but it is most active during reproduction. The throat gland in *Molossidae* comprises both sebaceous and sweat glands. In addition, on the muzzle of *Tadarida cynocephala,* very large, multilobate sebaceous glands are near the coarse, bristle-like hairs (Werner, Dalquest, and Roberts, 1950). It is interesting to note that male *Tadarida brasiliensis* have a peculiar odor (caused by the secretion of the skin glands) lacking in the females (Herreid, 1960).

ORDO PRIMATES

Compared with other mammalian orders, primates include the greatest number of species (in relation to this order's total species number) whose skin has received systematic and adequately detailed treatment. The skin from different body areas of numerous primates, with samples taken from several members of each species, has been studied by W. Montagna and his collaborators in the Oregon Regional Center. Along with morphological studies, comprehensive histochemical investigations have been performed.

The trunk epidermis is either thin, smooth interiorly, comprised solely of strata malpighii and corneum, or thick, with processes and ridges interiorly formed by the strata malpighii, granulosum, and occasionally by lucidum and corneum. Epidermis pigmentation varies in intensity in different species and in different parts of the body. The inner surface of the sole epidermis is alveolar. On the trunk, the dermal papillae are absent, small, or large. The dermal bundles of collagen fibers are invariably parallel to the skin surface. The blood vessel network may be loose, moderately dense, or dense. There are many or few pigment cells in the dermis. These may be absent in the majority of the body areas. Sebaceous glands of the trunk are unilobal, with ducts opening into the hair bursae or directly onto the skin surface. In some parts of the body, sebaceous glands may be large and multilobal. Specific glands may be comprised of sebaceous glands. Sweat glands commonly divide into apocrine and eccrine, but some primates have transitional sweat glands, combining features of both. Sweat glands may form specific glands.

Arrectores pilorum muscles are usually well developed. The hair may be dense, moderately dense, or sparse. Vibrissae may be absent (as in man). The hairs may grow singly or in tufts. They may or may not form groups. With insufficient evidence available, a comprehensive account of most primate families cannot be made.

Familia Tupaiidae

Skin and epidermis are thin. The trunk stratum malpighii is one to four cells deep; the stratum granulosum may be a broken layer of flattened cells; the stratum lucidum is absent. The trunk epidermis is unpigmented. The trunk dermis has horizontal bundles of collagen fibers forming a dense plexus. Elastin fibers are rare. The sebaceous glands of the trunk are small and unilobal. The trunk skin has both apocrine and eccrine glands. Arrectores pilorum muscles are noted. The hairs grow singly and may form rows.

Tupaia glis

Material. Skin of withers, breast, and hind sole of adult animal.

Our data support those of Montagna and his collaborators (Montagna et al., 1962), differing only in occasional details.

The withers skin (190μ) is thinner than that on the breast (605μ), and epidermis and stratum corneum are thicker (respectively, 31 and 22μ against 18 and 7μ on the breast). No stratum granulosum is recorded in the epidermis, or it may be one cell deep and discontinuous. The stratum malpighii is one to four cells deep. On the hind sole, the epidermis is thick, with peculiar interior and exterior outlines: exteriorly, ridges form alternating processes in cross section. These processes correspond to those of the epidermal inner aspect. The inner process ridges contain numerous pigment cells. In the process region, the epidermis reaches 440μ, and between the processes, 275μ in thickness. The stratum corneum thickness is much the same at the processes and between them, 165μ.

The hair is not in growth, but the papillary layer (159μ on the withers and 217μ on the breast) is thicker than the reticular layer (103μ on the withers

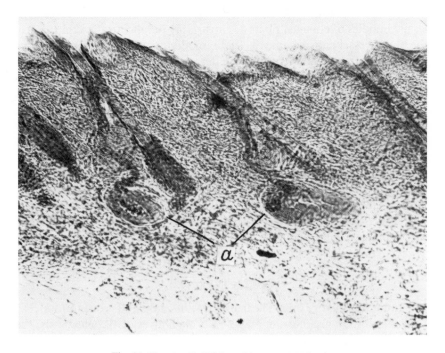

Fig. 51. *Tupaia glis*. Withers skin: *a*, sweat gland.

and 101μ on the breast) (fig. 51). The boundary between the layers occurs at
the sweat glands. In some areas of the reticular layer on the withers are horizontally elongated clusters of fat cells. These measure 17 x 22μ and 17 x
28μ. The papillary layer contains large pigment cells. The sebaceous glands
(56 x 62μ on the withers and 56 x 140μ on the breast) are situated singly,
posterior to each hair, deep in the dermis. Broad excretion ducts open into
the hair canal roughly level with its half length. The apocrine sweat glands
usually have saccular straight or slightly twisted secretion sections, with the
long axis horizontal below the hair bulbs. The excretion ducts open into hair
funnels posterior to the hairs. Each hair has one apocrine sweat gland (fig.
51). In addition, the trunk has eccrine sweat glands with long, narrow, irregularly twisted secretion cavities situated deep in the dermis. The excretion
ducts open on the epidermis surface. A thick layer of collagen bundle fibers,
110 to 140μ thick, adjoins the hind sole epidermis. Under this layer, subcutaneous fat tissue surrounds large glomi of ten to twenty highly tiwsted
eccrine sweat glands with secretion cavity diameters of 28μ (fig. 52). The
ducts open onto the skin surface on the ridges. No arrectores pilorum
muscles are revealed in the individuals under study, and in the skin of those
examined by Montagna et al. (1962) these are very thin.

The hair divides into three categories. The guard hairs are straight, gradually thickening from the base until half to two-thirds of the shaft, becoming

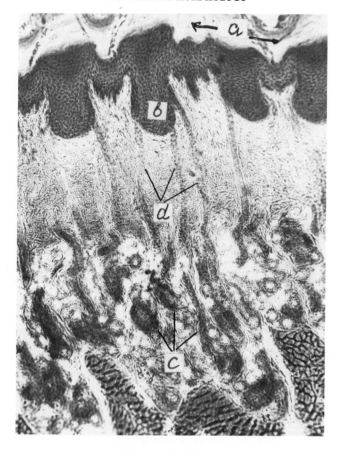

Fig. 52. *Tupaia glis.* Skin of hind sole showing: *a,* corneous
layer and *b,* malpighian layer of epidermis; *c,* secretion cavity;
d, sweat gland ducts.

sharply thinner towards the apex. Their length is 14.5 mm, thickness 78μ.
The medulla is well developed throughout the entire hair and comprises
84.7% of its thickness.

The pile hairs have a rather thin distal third of the shaft which gradually
dilates to maximum thickness in the apical three-fourths to four-fifths. At
the apex, a sharp narrowing of the hair is recorded. The proximal thin portion of the shaft may be somewhat twisted (one to two waves). The length of
these hairs is 12.8 mm, maximum thickness 61μ. The medulla is well developed, comprising 83.7% of the hair thickness.

In the fur hairs, a definite thickening occurs at the base of the first order
hairs. The maximum thickness (21μ) is noted at one-sixth to one-eighth of
the shaft, whereupon the hair gradually tapers toward the apex. The hair
length is 9.5 mm. The medullary layer, passing through the whole length of

the hair, comprises 71.5% of the hair thickness. Fur hairs are occasionally slightly wavy (two to three waves). Second-order fur hairs are much fewer. These differ from the first-order fur hairs only in size: length 4.7 mm, thickness 18μ, 77.8% comprised of the medullary layer. The hairs grow singly, forming no groups. A count of 8,000 hairs per 1 cm^2 of the skin surface is recorded on the withers, and 3,000 on the breast.

FAMILIA LEMURIDAE

On the trunk the epidermal stratum malpighii is one to four cells deep. Occasionally, there occurs a one-cell-deep stratum granulosum. Horizontal collagen fiber bundles predominate in the dermis, frequently perpendicular to one another and forming a dense plexus. On the trunk, the sebaceous glands are commonly large and unilobal. At every hair tuft is an apocrine sweat gland with a saccular, slightly twisted secretion cavity. There are slightly differentiated eccrine glands, somewhat twisted and sunken deep in the fat tissue of soles and finger pads. The hairs grow in tufts, four to twenty-four hairs each.

Lemur catta

Material. Skin from withers and breast of adult individual.

The skin on the withers is thinner than on the breast (660μ against 900μ). The epidermis and stratum corneum are similar in thickness on the withers and breast (respectively, 56μ and 40μ). Strata granulosum and lucidum are absent. The epidermis is strongly pigmented. The stratum malpighii is one to two cells deep (or four rows, after Montagna and Yun, 1962c). The stratum corneum is loose. The outside of the sole epidermis shows ridges corresponding to the underside ridges (Montagna and Yun, 1962). In the withers dermis, the papillary layer (414μ) is considerably thicker than the reticular (190μ). The border between these occurs at the sebaceous gland lower parts.

The reticular layer of the dermis is a nearly compact accumulation of fat cells measuring 34×56, $45 \times 62\mu$. Bordering the subcutaneous muscles runs a zone of adjoining horizontal collagen bundles. The fat cells ascend to the lower parts of the papillary layer between the hair tufts. There is no reticular layer in the dermis of the breast, and the sebaceous glands touch the subcutaneous fat tissue. The papillary layer is 404μ thick. In addition to the predominant horizontal collagen bundles, the upper and medium papillary layers have bundles arranged at various angles to the skin surface. The elastin bundles and nerves are few.

On the breast, the subcutaneous fat tissue, 440μ thick, contains only thin, small bundles of collagen fibers. The fat cells measure 34×50, $34 \times 72\mu$. The peculiar sebaceous glands are irregularly saccular, measuring 100×200, $143 \times 253\mu$, with large vacuoles in secretion cavities. The sebaceous glands

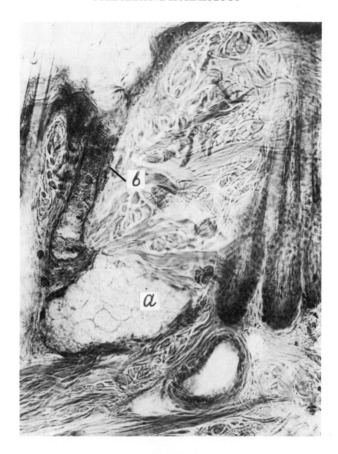

Fig. 53. *Lemur catta.* Withers skin: *a,* sebaceous gland;
b, glandular duct.

are surrounded by pigment cells. There are three glands at every hair tuft in
the withers and one on the breast. Their secretion cavities are deeper than
the hair bulbs, and their ducts run upward to open into the bursa near the
hair funnel (fig. 53). The sebaceous glands are specifically numerous on the
lips, scrotum, and perianal regions (Montagna and Yun, 1962c). At every
brachium, an amygdaloid glandular field at the juncture of the collar bone
and scapula is formed by multilobal sebaceous glands. Their ducts open
into a peculiar sac, which contacts the environment by a small orifice (Montagna and Yun, 1962c); (Montagna and Ellis, 1963).

The apocrine sweat glands are not visible at every hair tuft. They are
arranged below the hair bulbs. The antebrachial gland on the flexor surface
of the antebrachium is comprised of apocrine and eccrine sweat glands. The
ducts reach up the internal epidermal ridges, twisting when passing through
the epidermis to open to the outside. There are sole eccrine glands. No

arrectores pilorum muscles are revealed. The withers hair divides into three categories. The first category hairs are straight, and 24.7 mm long. They thicken progressively toward one-half of the shaft to taper at the apex. Their maximum thickness is 52μ. The medullary layer is 61.6% of the hair thickness and is absent only at the tip and the base.

The second category hairs are uniformly thick through the lower three-fourths of the shaft (with a negligible thickening upward). The upper quarter of the shaft is somewhat thicker. The hairs are slightly wavy with six to nine waves. The medulla is well developed. These hairs divide into two orders. The first-order hairs are 21.5 mm long, 44μ thick, medulla 50.0%. The respective indexes for the second-order hairs are 17.9 mm, 28μ, and 56.9%.

The third-category hairs are fur, uniform in thickness through the entire shaft (22μ), with up to twelve waves. The medulla passes through the entire hair, comprising 72.8% of its thickness. The hair grows in tufts of four to eleven hairs each (or 6 to 24, after Montagna and Yun, 1962c), one hair pertaining to the first category, the rest to the other categories. No groups are noted. The withers has 666 first-category hairs and 8,750 other hairs. The breast has 166 first-category hairs and 10,166 other hairs.

Lemur macaco

Material. Skin from various body areas of adult male (Montagna et al., 1961b).

The epidermis attains the greatest thickness on the muzzle, in the carpal, scrotal, and perianal regions. The stratum spinosum is three or more cells deep. The stratum granulosum, if existent, is a single broken cell layer in a thick epidermis. The relatively thick stratum corneum has compact underparts and loose outer parts. The epidermis is slightly pigmented. A small quantity of pigment cells is found elsewhere between the basal cells. Their numbers are higher in the epidermis of the muzzle, scrotum, perianal region, and front and hind soles. The dermal collagen bundles are horizontal and vertical. The fibroblasts are abundant through the entire dermis. The pigment cells occur only in the head dermis. There are many blood vessels in the dermis. The subcutaneous fat tissue is thickest in the scapular and sternal regions. The subcutaneous muscles are specifically well developed on the back and on the sides.

The sebaceous glands are large in the skin throughout the whole body and are particularly large in the head, scrotum, and perianal regions. There are three types of sebaceous glands. Most of the body surface shows nonspecific, saccular oval or round glands with a duct from the gland's center opening onto the skin surface beside the hair funnel. The terminal part of the duct is twisted, occasionally spiral. Among the sebaceous gland cells are pigment cells containing some quantity of melanin granules. The secretion is of black or brown color. The second-type glands occur in the skin of the face, lips, perianal region, and scrotum. They are large, with a few lobes opening into the common duct to bring the secretion directly to the skin surface. There are no pigment cells in these glands, so the secretion is never colored. In contrast to the unilobal sebaceous glands, the multilobal glands contain glycogen.

The sebaceous glands of vibrissae, the third type of sebaceous glands under study, are similar to those of other mammals. They open into the pilary canal. Dilated, somewhat twisted, unilobal apocrine sweat glands open on the skin surface by very narrow ducts. The hairs grow in tufts, six to fourteen hairs in each. The ratio of pile hairs to fur is 1:3 or 1:4 or 1:2 (on the head). All the hairs have a well-developed medulla. The cortical and medullary layers are strongly pigmented. Most hair bursae have no arrectores pilorum muscles. There is a tuft of vibrissae on the inner side of the antebrachium closer to the ulnar joint. The follicles of these vibrissae have no glands.

Lemur mongoz

Material. Skin from various parts of the body of adult individuals: two males and two females (Montagna and Yun, 1963b). The skin structure is nearly identical to that of *L. macaco* and *L. catta.* Characteristically, male sebaceous glands are much better developed than female and, in contrast to the other lemur species under study, are multilobal. Phosphorylase activity in the apocrine sweat glands is specific.

FAMILIA LORISIDAE

The epidermis on the trunk is thin with a smooth inner surface. The stratum malpighii comprises one to five cell rows. A one-cell-deep stratum granulosum may occur. No pigment cells are recorded in the trunk skin epidermis. The dermis is thick. There are no dermal papillae in the trunk skin. Collagen fiber bundles are parallel or perpendicular to skin surface. The elastin fibers are few except in the deep parts of the dermis. No pigment cells are noted in the trunk dermis. The trunk skin sebaceous glands are unilobal, small, and commonly round. Their ducts open into the hair canals. In the hair-covered trunk areas are one or two apocrine sweat glands. They are tubular and twisted. The secretion cavities are deep in the dermis. On the soles are eccrine sweat glands, specifically abundant in the finger pads. Their secretion cavities are large in size. The hair grows in tufts of three to twenty.

Loris tardigradus

Material. Skin from various body areas of three males kept in captivity (Montagna and Ellis, 1960b).

The skin is very thin. It is thicker on the back and extensors than on the abdomen and flexors. The stratum malpighii is two or three cells deep (it is much thicker only in the scrotum skin). The broken, barely visible stratum granulosum is comprised of flattened cells. The stratum corneum is about ten cells deep. The very thick epidermis of the foot and hand volar surfaces has tall ridges on the undersurface. The apexes of these ridges are reached by the sweat gland ducts. Mitotic activity is confined to the cells at these apexes. At these points, the stratum malpighii is very thick, the stratum granulosum is three to four cells deep, the stratum corneum is very thick, and the stratum granulosum is barely visible.

Pigment cells are found only in the scrotum epidermis and occasionally in the ear and perianal region epidermis. The dermis does not abound in blood vessels except for scrotum and brachial gland skin. There is a small quantity of pigment cells in the scrotum dermis. Mast cells are few except for the dermis of the scrotum, brachial gland, lips, and front and hind soles, where they are arranged next to the blood vessels.

On the head and scrotum, the sebaceous glands are larger, with a shorter duct. With a high esterase content and no glycogen and phosphorylase, the sebaceous glands appear to be functioning weakly. The larger sebaceous glands form part of the eyelid Meibomian glands. As in the majority of mammals, these glands are surrounded by a large number of thin, specific, cholinesterase-containing nerves. The apocrine type sweat glands on the head, muzzle, and ventral body regions are the least both in size and number. They are large and numerous on the antebrachium and thigh medial surfaces, being disproportionately large on the brachial medial surface.

Although the glands are arranged next to the follicles, their ducts open onto the skin surface independently by funnel-shaped dilations near the pilary canal orifices. All the glands are in different functional stages, unconnected with the functional states of the nearby follicles. The excretion duct is about 20μ in diameter. It is formed by two layers of very low cuboidal cells. The glandular epithelium may be cubical or tall columnar with cell apexes jutting into the glandular lumen.

In contrast to the apocrine glands of other mammals, those of *L. tardigradus* have a high phosphorylase content. Equally developed in both sexes, the brachial specific gland is situated on the medial side of the shoulder region (Hill, 1956; Montagna and Ellis, 1960*b*). Some watery, yellowish, malodorous liquid is exuded in this region in live animals. In structure, the glands comprising this organ are indistinguishable from the trunk apocrine glands except for very tall glandular cells and larger myoepithelial cells. Short, narrow ducts open onto the skin surface irrespective of hair follicles.

Brachial organ glands differ from other glands in their rich content of alkaline phosphatase and in their surrounding nerves containing cholinesterase. Straight, narrow excretion ducts of the apocrine glands open along the epidermal ridges. The intraepidermal ducts are straight or slightly twisted.

The glandular cells are all similar, differing from both the higher primate eccrine gland dark and light cells. Their cytoplasm is filled with colorless globules interspersed with occasional small basophilic granules. They have no glycogen. The hair follicles are arranged in groups of four to twenty or more. Every hair's pilary canal opens onto the skin surface independent of other canals. Arrectores pilorum muscles are present only at the largest hairs. The hair growth is haphazard, with growing hair to be seen in any body area. The growing hair bulbs penetrate into the subcutaneous fat tissue. The active hair follicles grow parallel to the skin surface where the skin is thin. Resting hair follicles are short with bulbs halfway down the dermis depth. Both active and resting follicles contain glycogen and phosphorylase.

Nycticebus coucang

Material. Skin from various body areas from two adult males and three females (Montagna et al., 1961a).

The trunk epidermis has a stratum granulosum containing one row of flattened cells with a small amount of keratohyalin granules. The stratum malpighii is two or three cells deep. The stratum corneum is of nearly uniform thickness through the entire body. A small number of pigment cells is found in the face and scrotum epidermis. The sole epidermis is thick, with underside ridges reached by excretion ducts at certain intervals. Despite a considerable thickness of the epidermis, the strata granulosum and lucidum are thin in these sites with the stratum corneum being thick. A small number of pigment cells is revealed at the base of the epidermal ridges around the sites reached by the sweat gland ducts. Small, hardly visible, acetylcholinesterase-positive nerves ascend from the dermis to the epidermis and appear to penetrate between the epidermal cells. Dermal pigment cells are rare even in the scrotum. The blood vessels of the dermis are arranged mostly at the hair tufts. Occasionally, capillary loops are visible under the epidermis. In the soles, large nerves ascend from the subcutaneous fat tissue to the upper dermis to break into inverted, arch-shaped branches beneath the epidermis. Many of these nerves terminate by small, round Meissner's corpuscles.

The active hair sebaceous glands are larger, with short excretion ducts. The sebaceous glands of lips, muzzle, brachial gland, eyelids, and genitals are larger than those of other parts of the body. Long, very narrow excretion ducts of the apocrine sweat glands run parallel to the hair bursae, dilating conically to open onto the skin surface independently. A layer of very large myoepithelial cells surrounds the glandular epithelium. The secretory cell cytoplasm contains rare, fine basophilic granules and small yellow pigment granules as well as some quantity of positive granules not containing glycogen. The nerves make up a lax network around the sweat gland secretion cavities. They contain cholinesterase which is Nycticebus specific.

Thin, long ducts opening into the hair bursae run from large sweat glands in the brachial gland. Front and hind limb sole and digit sweat glands resemble both apocrine and eccrine glands. The thick, freely twisted secretion cavities turn sharply into a very thin and straight excretion duct, which twists terminally within the epidermis. The apocrine sweat glands commonly contain phosphorylase. In other primates, with the exception of *Loris,* phosphorylase is found only in the eccrine glands.

The hair of the brachial region is sticky with a yellow, malodorous substance. The entire skin of *N. coucang* is odorous, but the brachial skin has a particularly pungent smell. The hair includes soft fur and, rarely, coarse pile. The vibrissae, small in size, are abundant on the upper lip, cheeks, brows, and eyelids. The number of follicles is six to twenty. They are arranged in islets surrounded with smooth skin. The dorsal hair follicles are larger than the ventral. The follicles are close to one another, opening onto the skin surface individually. The follicles may have different states of activity within a single hair group, some resting, others growing. The terminal nerve forms a small ring around a follicle directly beneath the sebaceous gland excretion aperture. One or several nerves run to the sweat gland.

Arctocebus calabarensis

Material. Skin from various body areas from two adult and one young female (Montagna, Machida, and Perkins, 1966*b*).

The skin structure resembles that of other African lorises. The dermis has a poor network of blood vessels. There are archlike nerves with sausage-shaped endings under the lip epidermis. The sebaceous gland excretion ducts open into the deepest part of the pilary canal. The largest sebaceous glands occur in the skin of the lips, muzzle, perianal region, and scrotum. The narrow ducts of apocrine sweat glands open into the large hair funnels. The hairs grow in tufts of seven to thirteen. There are no arrectores pilorum muscles.

ORDO EDENTATA

Extant *Edentata* are highly specialized mammals with very few common characteristics regarding skin structure. No stratum granulosum and stratum lucidum occur in the epidermis, which is pigmented lightly or not at all. Ossifications may occur in the skin. The epidermal covering of these ossifications has turned into corneous scales. A peculiar pattern of collagen bundles in the dermis (perpendicular but parallel to the skin surface) is characteristic of *Edentata.* Sebaceous glands are small. Sweat glands, if not absent, are tubular or slightly twisted into glomeruli. Arrectores pilorum muscles may or may not be absent. The pelage may be well developed or sparse.

Familia Myrmecophagidae

Not enough work has been done on the skin of anteaters. A peculiar structure of the plexus of the bundles of collagen fibers is reported in the dermis (Braun and Lebedev, 1947) as well as some data on the structure of the skin of *Tamandua* (Machida et al., 1966). We studied the skin of *Tamandua tetradactyla*. The available material is inadequate for a full description of this family. There are no ossifications in the skin. The sweat glands are tubular and slightly twisted. The hairs are developed. There are four hair categories in addition to vibrissae, and all hairs grow singly.

Galago crasicaudatus

Material. Skin from various body areas of adult male (Montagna and Yun, 1962*a*).

The skin is similar in structure to that of *Galago senegalensis*. The stratum malpighii is three to five cells deep. The stratum granulosum is broken and barely visible. Pigment cells are found in the epidermis of the ears, scrotum, soles, and eyelids. In the embryonal epidermis, pigment cells are abundant throughout the entire body. There are characteristic horny spines on the soles. Sebaceous glands are large and multilobal on the face, eyelids, lips, and scrotum. Sweat glands are few in the whole body skin with a single gland per three to five hair groups. These small glands resemble apocrine glands but combine the histochemical features of both apocrine and eccrine glands. Their function appears to be not thermoregulatory but odorous. There are no eccrine glands in the sole regions with horny spines.

Galago senegalensis

Material. Skin from various body areas from one adult male and two females (Yasuda, Aoki, and Montagna, 1961).

The epidermis is thin. The stratum malpighii has three to four cells. The stratum granulosum is one cell deep and broken. The stratum corneum is eight to ten cells deep, being similar to the stratum malpighii in thickness. Arborescent pigment cells are noted on the epidermal ridge apexes of foot soles, scrotum, and vulva.

The dermis pigment cells are found only in the perianal region, scrotum, and ears. The fibroblasts are few. The network of blood vessels is lax. Capillaries, few in number, are located under the epidermis and around the hair follicles. The superficial dermis layers contain a network of nerve fibers, rich in cholinesterase. Larger and more complex sebaceous glands are found in the scrotum, vulva, perianal region, and lips. These glands have a poor blood supply. The narrow, straight duct of the apocrine gland with a terminal, ampule-shaped dilation opens straight onto the skin surface near the hair funnel. The secretion cavity of the gland is deep in the dermis. The trunk glands appear inactive. The largest glands are noted in the lips, eyelids, and perianal regions. The sweat glands are few on the feet and hands and numerous on the pads. Their secretion cavities are thick and slightly twisted, and the excretion ducts are narrow and long, opening onto the skin surface. The glandular cells are hard to differentiate into dark and light. Hair groups include three to six hairs. The hairs have a developed medullary layer. Arrectores pilorum muscles are noted only in the hair follicles of the tail.

Galago demidovi

Material. Skin of various body areas from an adult male and adult female (Machida et al., 1966).

The skin structure is similar to that of other African lorises. The stratum malpighii has one to two layers. Very few pigment cells are found in the deep parts of the stratum malpighii except for the breast, back, and perianal regions. The sebaceous glands are occasionally bilobal with ducts opening into the pilary canal halfway down its length. There are large, multilobal glands in the lips, eyelids, and genitals. In the head, the glands have long ducts. In contrast with other galagoes, the glands are alkaline phosphatase positive.

The apocrine sweat glands are large in the head, eyelids, lips, nose, cheeks, and anogenital regions. The ducts open into the inside of the hair funnel. The eccrine sweat glands are found only in the front and hind soles. The glandular epithelium divides into dark and light cells. The tufts contain three to four hairs. Arrectores pilorum muscles are found only in the tail skin. The epithelial sheath of the resting hair root has no glycogen, resembling that of *Tupaia.*

FAMILIA TARSIIDAE

The epidermis is thin. The stratum malpighii has one to two rows. No stratum granulosum and stratum lucidum are recorded in the trunk epidermis. The trunk epidermis is not pigmented. There are no dermal papillae. In the dermis, collagen fiber horizontal bundles prevail, forming a dense plexus. The elastin fibers are few. In the trunk skin, the sebaceous glands are single-lobed and small. The trunk skin has apocrine sweat glands; the sole skin has eccrine glands. There are specific skin glands: circumanal, labial, and some others. The hairs grow in tufts of five to nine.

Tarsius spectrum

Material. Skin from the withers and hind sole of an adult.

The outer withers skin surface is wavy. The skin is thin (319μ). The hind sole epidermis is very thick with a wavy inner surface corresponding to the outer ridges. The stratum granulosum is pronounced. In the dermis (297μ thick) the reticular layer is absent, as the sweat glands reach up to the subcutaneous muscles. The sebaceous glands, measuring 34×84, $28 \times 84\mu$, are set one at every hair at the apical one-third or half of the root's length. A large circumanal gland is formed by huge sebaceous and apocrine sweat glands (Schaffer, 1940; Hill et al., 1962; Montagna and Machida, 1966). The sweat glands have elongated, sacciform, sausage-shaped, or nearly straight secretion cavities, 50μ in diameter. They are set below the bulbs, with the longer axis parallel to the skin surface.

In the hind sole, the eccrine sweat glands are few, widely-spaced, and nearly straight, with sausage-shaped secretion cavities and a long excretion duct. There are no arrectores pilorum muscles. The hair divides into two categories. The first-category hairs are few in number, straight or slightly wavy, and 21.7 mm long. They thicken very gradually from the base upward to a maximum of 21μ halfway or two-thirds up the shaft to taper at the tip. The medullary layer is not recorded at the base and tip. It accounts for 81% of the hair thickness.

The second-category hairs are slightly wavy, with up to nine waves. The shaft is slightly dilated at the base and remains uniform in thickness nearly to the tip, before which it is somewhat thicker. The medulla is developed through the whole hair except for the tip and base. The second-category hairs divide into two orders: the first-order hairs are 16.9 mm long, 23μ thick, medulla 47.8%. The second-order hairs are 16.3 mm long, 21μ thick, medulla 71.5%. The hair grows in lax tufts of five to nine with no groups. There are 1,083 hairs per 1 cm² of skin surface. The breast has fewer hairs, with only 666 per 1 cm² of the skin surface.

T. syrichta

Material. Skin from various body areas of adults: two females (Montagna and Machida, 1966) and from one female and two males (Arao and Perkins, 1969).

The skin structure is similar to that of *T. spectrum*. The outer ear, nose, tail, hand, and foot, have few pigment cells. Only the eyelid and lip epidermis is well pigmented. Pigment cells are also found in the eyelid, outer ear, and tail dermis. There are numerous mast cells around hair roots. The sebaceous glands are small in the trunk and huge in the lateral parts of the upper lip, where they form a labial specific skin gland. Large sebaceous glands along with apocrine sweat glands are concentrated in the circumanal region (the circumanal gland). Very large sebaceous glands are also recorded in the eyelid skin. Every hair tuft has an apocrine sweat gland. The glands open onto the skin surface near the hair funnel. The eccrine sweat glands are found only in the soles. Their secretion cavities are sacciform, resembling apocrine sweat glands. The glandular cells are similar to the light type cells.

There is a specific skin gland in the male belly epigastric region (Hill, 1951) formed by very large sebaceous glands as well as by a few large apocrine sweat glands. The glandular secretion is orange in color. The tufts contain six to nine hairs. De Meijere (1894) reported three hairs.

FAMILIA CEBIDAE

The stratum malpighii of the trunk epidermis is one to seven cells deep. The stratum granulosum is either absent or weak. The epidermis underside is smooth. The epidermis pigmentation level varies widely. There are no dermal papillae in the trunk dermis. Horizontal collagen fiber bundles prevail, set at right angles to one another. The pigment cells in the dermis are either rare or absent. The sebaceous glands of the trunk skin are small or medium sized, uni- or bilobal with short ducts opening into the pilary canal. Apocrine glands are found throughout the entire hair-covered skin. Every hair tuft has a single gland. The secretion ducts open onto the hair funnels or onto the skin surface near them. The eccrine sweat glands are invariably located in the soles. The glandular epithelium contains both light and dark cells. The glandular secretion cavities are in the deep parts of the dermis or subcutaneous fat tissue. The hair tufts contain two to thirteen hairs.

Aotes trivirgatus

Material. Skin from various body areas of adults: two females and two males (Hanson and Montagna, 1962).

Most of the body epidermis contains a weak stratum granulosum and a relatively thick stratum corneum. There are few fibroblasts in the dermis. The mast cells are found largely in the perivascular tissue. The pigment cells are situated near the blood vessels, hair follicles, and sebaceous glands. The dermis blood-vessel network is loose. The sebaceous glands vary in shape and size. The resting hair glands are smaller than those of the growing hairs. Larger glands open into the pilary canal or straight onto the skin surface.

The eccrine/apocrine gland excretion ducts open onto the skin surface beside the hair funnel. Some glands resembling apocrine sweat glands form the pectoral specific gland. Their ducts open into the pilary canal not far from the skin surface. There is a tail specific gland on the ventral surface of the tail base (Hill et al., 1958-1959) which is comprised of twisted apocrine glands with ducts opening onto the skin surface beside the hair funnels. In the sole skin are weakly differentiated eccrine glands resembling apocrine glands. The hair tufts contain three to ten or more hairs. The arrectores pilorum muscles are well developed.

Cacajao rubicundus

Material. Skin from various body areas of one adult female and two young males (Perkins et al., 1968).

The stratum malpighii of the bulk of the trunk has one to three cell layers. The stratum granulosum is found only in the soles. The adult female epidermis pigment cells occasionally occur in the eyelids, nose, soles, in the perianal region, genitals, and tail. The dermal papillae are well developed in the circumanal region, genitals, and soles. The elastin fibers are few. The pigment cells are rather numerous in the dermis of the young males and are rare in the adult female dermis. There are many blood vessels in the dermis. The sebaceous glands are small. Larger glands are found in the muzzle and anogenital region. The gland ducts open into the pilary canal. No specific glands comprised of sebaceous glands have been revealed. The apocrine glands are most abundant in the breast, where their number equals that of the hair follicles. The glandular secretion cavities are in the deep parts of the dermis, with their excretion ducts opening into the pilary canals. The eccrine sweat glandular secretion cells are weakly differentiated. The hair tufts contain two to five hairs. The hairs are arranged in lines.

Alouatta caraya

Material. Skin from various body areas of adults: one female and two males (Machida and Giacometti, 1968).

The trunk stratum malpighii is four to seven cells deep. There is a stratum granulosum and (in the soles) a stratum lucidum. The epidermis is well pigmented. Melanin granules occur even in the stratum corneum. Elastin fibers are abundant, specifically around the hair follicles and arrectores pilorum muscles. Sebaceous glands of the trunk are small. They are multilobal in the eyelids, lips, and perianal region. The glands open into the pilary canals by short ducts. The eyelid Meibomian glands have long, straight ducts which open onto the skin surface. The glands are surrounded by pigment cells. The apocrine sweat glands with twisted secretion cavities open into the hair funnel. Larger apocrine glands are located in the eyelids, lips, and perianal region. The eccrine sweat glands occur throughout the entire body surface except eyelids, lips, ears, and perianal region. They are fewer in number than the apocrine sweat glands in the trunk. The eccrine gland ducts open onto the skin surface. The hair tufts contain three to five hairs. Every hair follicle has a sebaceous and an apocrine sweat gland.

Cebus capucinus

Material. Skin from withers and breast of an adult.

The skin on the withers (605μ) is thicker than on the breast (440μ). Its outer surface is very irregular. The withers epidermis is 72μ thick, breast 56μ, stratum corneum 55 and 45μ. The stratum corneum has many layers and is loose. The stratum granulosum and stratum lucidum are absent. The epidermis is not pigmented. The dermis visibly divides into papillary and reticular layers. The boundary between them passes on the level of the hair bulbs.

In the papillary layer, the collagen fiber bundle is laxer, with bundles set in various directions (horizontal bundles predominate). In the reticular layer their plexus is more compact. The elastin fibers are rather abundant. In the papillary layer they are located at various angles to the skin surface, and in the reticular layer they are mostly parallel to the skin surface. In some areas of the withers dermis papillary layer there occur large pigment cells.

The sebaceous glands have a single lobe. They are round-shaped and small in size: 62×73, $84 \times 84 \mu$. They are located singly posterior to every hair. The excretion duct is perpendicular to the hair root. The apocrine sweat glands have long (up to 0.66 mm in the withers skin and up to 1 mm in the breast), sausage-shaped, slightly twisted secretion cavities 34μ in diameter running parallel to the skin surface posterior to the hairs. The arrectores pilorum muscles are 22μ in diameter.

The withers hairs divide into two categories. The first-category hairs are straight or slightly wavy. The hair is somewhat thickened at the base, but remains uniformly thick through the remaining hair length. The hair divides into two size orders. The first-order hairs (length 34.2 mm, thickness 51μ) have a relatively well developed medullary layer (41.1%), absent only in the base and tip. The second-order hairs (length 28.0 mm, thickness 44μ), have a poorly developed medullary layer (36.3%), which is sometimes broken.

The second-category hairs have up to four waves. Their length is 17.9 mm. Thinner hairs (28μ) have no medulla. Thicker hairs (39μ) have a medulla of 33% of their thickness, this layer being broken at long spaces. The hairs grow in loose tufts of thirty-three. There are 1,833 hairs per 1 cm².

Cebus albifrons

Material. Skin from adults: nine males and two females, samples taken from thirty-four body areas (Perkins and Ford, 1969).

The trunk stratum malpighii contains one to three cell layers. In the soles, the stratum malpighii is ten to twenty cells deep. The stratum granulosum is well developed, and the stratum corneum is compact. The pigment cells are abundant in the soles and genitals; they also occur in moderate numbers in the eyelids, lips, ears, nose, and circumanal region. They are rare in other body areas. The elastin fibers and mast cells are few. Spindle-shaped pigment cells occur in the upper dermis of the head, brows, nose, cheeks, abdomen, and genitals. They are rare in the other body areas. The sebaceous glands are small and unilobal. They occur singly or in twos at every hair follicle. In the nose, ears, and anogenital region the glands are multilobal and much larger in size. Usually, every hair tuft has a single apocrine gland, but in the breast there is one gland per hair. The excretion duct opens in the immediate vicinity of the hair funnel. The hair tufts contain two to six hairs. Both resting and growing hairs normally occur in the same tuft. Arrectores pilorum muscles are well developed in the back and tail.

Saimiri sciurea

Material. Skin from various body areas of two adult females (Machida et al., 1967).

The stratum malpighii is two to three cells deep. Only the sole epidermis has a well developed stratum corneum and stratum granulosum. There are no pigment cells. The dermis is relatively thin. The elastin bundles are few. In the upper portion of the dermis are small elastin fiber bundles perpendicular to the skin surface. Abundant pigment cells are commonly located in the dermis in small groups. Sebaceous glands are small and single-lobed. Multilobed glands are found only in the head, muzzle, and anogenital region. The apocrine sweat glands are larger in the eyelids, lips, and perianal region. The secretion cavities are deep in the dermis with ducts opening into the pilary canal or funnel. On the breast, apocrine glands occur in small groups. The eccrine sweat gland ducts open onto the skin surface to reach the inner epidermal ridge apexes. The hair tufts contain seven to thirteen hairs on the head and muzzle, and three to seven hairs on the other parts of the body. Hair roots on the tail have well developed arrectores pilorum muscles.

Ateles geoffroyi

Material. Skin from various body areas of an adult male (Perkins and Machida, 1967).

In the hair-covered areas, the stratum malpighii is four to seven cells deep. The stratum granulosum, one to two cells deep, is noted only in the scrotum and tail. The epidermis is heavily pigmented. The dermal papillae are well developed. In the dermis, collagen bundles run parallel to the skin surface, perpendicular to one another. There are many elastin fibers. There are no pigment cells in the dermis. The sebaceous glands are medium in size. Large, multilobed glands are located in the eyelids, lips, ears, and anogenital region. Pigment cells are peripheral to many of the glands.

Apocrine sweat glands are more abundant than eccrine. They have large, twisted secretion cavities deep in the dermis. The ducts open onto the skin surface or into the funnels. The eccrine and apocrine sweat glands are located in the soles, armpits, groin, back, scrotum, and the hair-covered portion of the raptorial tail. The glands lie deep in the dermis or subcutaneous fat tissue. Where eccrine glands are found together with apocrine, eccrine gland ducts open onto the skin surface near those of the apocrine glands. They are very alkaline-phosphatase rich. The hair tufts contain three to eight hairs. There are one to three apocrine glands connected with every tuft, each hair having a sebaceous gland.

Lagothrix lagotricha

Material. Skin from various body areas of two adult males (Machida and Perkins, 1966).

In a number of features, the *L. lagotricha* skin is intermediate between the lower narrow-nosed monkeys of the Old World and the broad-nosed monkeys of America. The stratum malpighii is three to seven cells deep. The epidermis is pigmented and is very thick in the hair-covered areas of the hand, foot, and tail. The dermal papillae are not pronounced. The elastin fibers are noted only in the pads. The mast cells are abundant in the dermis. The capillary network is rather well developed. Large multilobed glands are recorded only in the lip and eyelid skin. Nearly all the glands have peripheral pigment cells. The glands are surrounded by alkaline- and phosphatase-positive capillaries and have no glycogen. The apocrine glands are small. They are at their largest in the eyelids, lips, breast, and back. Glomerulus-twisted secretion cavities lie deep in the proximal dermis parts or in the subcutaneous fat tissue. Their ducts open either into the funnel or straight onto the skin surface beside it. Both apocrine and eccrine glands are recorded in the genitals and the hair-covered tail area, their ratio being 1:1. Each tuft contains four to fifteen hairs.

There is a sebaceous and an apocrine gland at every hair tuft. Arrectores pilorum muscles are specifically well developed in the tail.

Familia Callithricidae

The stratum malpighii of the trunk is formed by one to three cell layers. The trunk epidermis has no stratum granulosum. The epidermis is either not pigmented or it has a large number of pigment cells. Mast cells are common in the dermis. The pigment cells in the trunk dermis are present or absent. Horizontal collagen bundles prevail in the deep portions of the dermis. The sebaceous glands are small or medium sized, single- or bilobed, one or two at a hair follicle. Small- or medium-sized apocrine glands are arranged singly at a hair tuft with ducts opening into the funnels or beside them. The eccrine sweat glands are noted only in the soles. Their glandular cells divide into two types: light and dark. The arrectores pilorum muscles are well developed. The tufts contain two to six hairs.

Callimico goeldii

Material. Skin from adults: two males and one female (Perkins, 1969).

The epidermal stratum malpighii is formed by one to two cell layers. It is at its thickest in the soles (10 to 25 cell layers). The epidermis is heavily pigmented. Pigment cells are common throughout the entire body. Blood vessels are numerous in the dermis. Larger, multilobed sebaceous glands are in the lips, nose, and cheeks. Very large, multilobed sebaceous glands form specific glands: circumanal and pectoral. Glands of a similar size are arranged at the hair roots of the manubrial tuft at about 6 to 8 mm posterior to the sternum and in the pubic-genital region. Their ducts open beside the hair funnels. The apocrine glands of the manubrial hair tuft, pubic, circumanal, and genital regions are much larger than in the other areas. The eccrine sweat gland secretion cavity diameter is three times the diameter of the ducts. One-third of the eccrine gland length comprises the excretion duct and two-thirds comprises the secretion cavity. The hair tufts contain three to five hairs on the trunk. The tufts are arranged in lines. There are five to six vibrissae on the ventral side of the antebrachium.

Callithrix jacchus

Material. Skin from withers of an adult.

The withers skin (308μ) has an irregular outer surface. In the epidermis (34μ) the stratum malpighii is two to three cells deep. The epidermis is not pigmented. In the dermis the papillary layer (162μ) is thicker than the reticular layer (112μ). In the medial and upper portions of the papillary layer, collagen bundles run in various directions (fig. 54) forming rather a dense plexus, which becomes looser with depth. There are no pigment cells in the

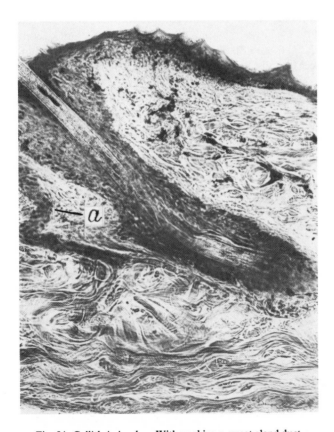

Fig. 54. *Callithrix jacchus.* Withers skin: *a,* sweat gland duct.

dermis. The sebaceous glands, single-lobed and small (17×45, $28 \times 39\mu$), are not found at some hair roots. Short and large sebaceous gland ducts are well defined at all the roots. The apocrine glands, with sausage-shaped, slightly twisted secretion cavities 23μ in diameter, lie level with the hair bulbs horizontal to the body surface. The secretion cavities attain a length of 150μ. The ducts open near the hair funnel.

Large sebaceous glands are found in the skin of the lips (Miraglia and Santos, 1971). They open into a hair channel, and occasionally, directly to the skin surface. The fatty material in these glands is a mixture of neutral lipids (triglycerides and cholesterol esters) and acidic lipids (fatty acids and a small amount of phospholipids). The acini of the glands are rich in cytochrome oxidase and succinic dehydrogenase while the ducts present only a weak reaction. Nonspecific esterase exists in fairly large amounts in the differentiated cells, in small quantities in mature cells, and not at all in the cells of the ducts. Glycogen alkaline phosphatase, acid phosphatase, and cholinesterases are not found in the sebaceous glands of the lips.

The growing hair is of two categories. The first-category hairs are straight or slightly curly and of uniform thickness throughout their length (somewhat thinner at the base). This category divides into four size orders, 29.2, 22.7, 15.6, and 14.7 mm long, and 48, 37, 24, and 19μ thick, with a medulla of 62.5, 69.2, 50.0, and 36.8%. The third- and fourth-order hair medullary layers are occasionally broken. The second category includes wavy hair (up to four waves), of uniform thickness (19μ) throughout the length (8.8 mm) with a weak, broken medullary layer (36.9%). The hairs grow in tufts of two to four. The most common number of hairs per tuft is three. Large hairs occur singly or as part of a tuft. In section parallel to the body surface, hair categories and orders are barely distinguishable. There are 18,330 hairs per 1 cm² of the withers.

Callithrix argentata

Material. Skin from various body areas of two adult males (Perkins, 1969).

The skin structure is similar to that of *C. jacchus*. Mast cells occur throughout the dermis of all the body areas under study. Pigment cells occur in the upper lip and the eyelid edge. The sebaceous glands are large and multilobed in the lips, ears, nose, and anogenital regions. The apocrine sweat glands open into the hair funnels through their ducts or beside them.

In the scalp skin, the apocrine glands are small. In the face, they are medium sized, with one to three hair tufts. The thighs, belly, throat, tail, and back have a small gland at every hair tuft. Large apocrine glands are located two to three at every hair tuft in the breast, and one to two in the circumanal region. The eccrine sweat glands are abundant in the pads. Hair tufts include three to five hairs. The roots are surrounded by peculiar neurovascular plexuses.

Callithrix pygmaea

Material. Skin samples from various body areas of nine adult males (Perkins, 1968).

The skin structure is similar to that of other *Callithricidae*. The pigment cells occur mostly in the stratum basale and stratum spinosum of the head, the dorsal part of the trunk, and the dorsal surfaces of the limbs. Elastin fibers in the dermis are numerous. The sebaceous glands are not large. The apocrine gland ducts open beside the hair funnels, or independent of them in the pubic region. The hair tufts contain three hairs each.

Saguinus oedipus

Material. Skin from various body areas of two adults (Perkins, 1969*b*).

The stratum malpighii of the trunk is two to three cells deep. It is thirty cells deep in the soles. The epidermis is heavily pigmented except for breast, belly, armpits, and soles. There are no dermal papillae in the trunk. The mast cells occur throughout. Spindle-shaped pigment cells occur in the upper dermis of all parts of the body except breast, belly, armpits, and soles. The sebaceous glands are single-lobed, small, two at each hair follicle of the head, thighs, and armpits and one in the back, breast, belly, and tail, medium-sized with multilobed glands (one to two at each hair) in the muzzle. Large, multilobed muscles are present in the lips, concha auriculae, and anogenital regions. Sebaceous gland ducts open into the pilary canals. In the breast, medium-sized apocrine sweat glands occur, one or two at a tuft. In the anogenital regions. armpits, and suprapubic regions are large apocrine sweat glands. The ducts open onto the skin surface beside the hair funnels. There are three to five hairs in a tuft.

Saguinus midas

Material. Skin from withers, breast, and hind sole of an adult.

The hair is growing and the skin is thickened (770μ on the withers and 990μ on the breast). The stratum corneum has not been preserved. The epidermis is lightly pigmented. Hair growth has resulted in a greater thickness of the papillary layer (580μ on the withers and 770μ on the breast) compared with the reticular layer. In the dermis, horizontal collagen bundles prevail, but in the papillary layer (particularly in its upper and middle portions) ascending bundles are rather numerous. The collagen-bundle plexus of the reticular layer is looser than in the papillary. The sebaceous glands are single-lobed, oval, small in size (28×39, $28 \times 56\mu$). Located singly, posterior to the hair, the sebaceous glands are more pronounced in nongrowing hairs. No sweat glands are recorded in the withers. In the breast, the apocrine sweat glands are large, sausage-shaped, and slightly twisted, with relatively short ducts running along the hair roots. The secretion cavities are

0.40 mm long and 56μ in diameter, disposed posterior to the hair roots. In the sole skin, numerous eccrine sweat gland glomeruli lie deep below the dermis (at a distance of 220-450μ from the outer skin) surrounded by fat cells. The glandular secretion cavities are 30μ in diameter. At the growing hair, the arrectores pilorum muscles start roughly at half the root. The withers hairs divide into two categories. The first-category hairs are straight with a small narrowing at the base, above which they do not change their thickness. They taper only at the tip. This category divides into two size orders, 30.9 and 28.3 mm long, 48 and 37μ thick, medulla 64.6 and 52.1%, respectively. The second-category hairs have few differences from those of the first category. They are straight or slightly wavy, tapering toward the tip from two-thirds of the proximal. The medulla is well developed. Two size orders are distinguished, 15.8 mm and 13.1 mm long, 50μ and 37μ thick, medulla 66.0% and 64.9%. The hairs grow singly, forming groups of three set in a row. Occasionally these groups stretch into rather a long line.

Saguinus nigricollis

Material. Skin from various body areas of adults: two males and one female (Perkins, 1966).

In the hair-covered areas, the stratum malpighii contains one to two cell layers; in bare areas, two to fifteen layers. The pigment cells and granules are dispersed in the epidermis basal layer throughout the body. The dermal papillae are weak in the trunk. In the dermis, moderately large collagen bundles form a dense plexus. Single-lobed sebaceous glands of the trunk are small. In the skin of the eyelids, lips, and genitals, the sebaceous glands are large. In the posterior part of the belly a ventral gland, 10 to 12 mm in diameter, is comprised of huge, multilobal sebaceous glands. Their ducts open directly onto the skin surface.

The apocrine sweat glands are abundant throughout the whole body. There is one gland per three hairs, except for the lips and the ulnar gland, with one gland per vibrissa, and the eyebrows and peri-inguinal regions, where some glands open onto the skin surface irrespective of the hair follicles. The ulnar gland is a small elevation at the palm with vibrissae 2 mm in diameter. It contains apocrine glands surrounded by pigment cells. The sternal gland is 4 mm in diameter and is comprised of apocrine sweat glands over the sternum. The glandular secretion ducts open into the hair bursae. The hair tufts contain two to six (commonly three to four) hairs.

FAMILIA CERCOPITHECIDAE

The body epidermis, varying in thickness, may or may not have a stratum granulosum. The epidermis pigmentation is slight or nonexistent. The dermal papillae of the trunk may be developed well, poorly, or not at all. The collagen bundles are mostly parallel to the skin surface. The elastin fibers are numerous. Pigment cells may occur. The sebaceous glands are small and unilobal, occasionally bilobal. The apocrine and eccrine sweat glands occur at all hair-covered areas. The hairs are arranged in tufts with up to six hairs in each.

Macaca nemestrina

Material. Skin from various body areas of adults: one male and one female (Perkins et al., 1968).

The body stratum malpighii is one to three cells deep, on the ears and scrotum three to six cells, on the foot soles up to twenty cells. Stratum granulosum is found only in the soles. The epidermis is pigmented. There are only occasional pigment cells. Large, multilobal sebaceous glands occur only in the lips and anogenital regions. Apocrine sweat glands are found in all the hairy areas. The lower portions of the glands are frequently in the subcutaneous fat tissue. The excretion ducts open near the hair funnels. There is one gland per three to five hair tufts. The eccrine sweat glands are found throughout the entire body except for the eyelids, lips, ears, armpits, and ischial callosities. On the tail, thighs, and genitals, their ratio to the apocrine sweat glands is 1:1; in other areas, 1:2 or 1:3. The glandular epithelium contains both dark and light cells. The hair tufts have two to five hairs.

Macaca mulatta

Material. Skin from various body areas of adult specimens: one male and one female (Montagna et al., 1964).

The epidermis of the underside of the trunk is smooth. The stratum malpighii is three or more cells deep. There is a stratum granulosum. The stratum corneum of the ischial callosities is particularly well developed. Melanocytes occur only in the foot sole and face epidermis. The collagen bundles of the dermal reticular layer are mostly parallel to the skin surface. The dermis contains melanocytes. The trunk sebaceous glands are small in size. Larger, multilobal glands are found on the face and anogenital regions. The sweat glands are similar to those of *M. nemestrina*. The hair tufts have two to seven hairs (commonly three to four).

Macaca speciosa

Material. Skin from various body areas of adult specimens: two males and one female (Montagna, Machida, and Perkins, 1966*a*).

The epidermis of the body skin is smooth and thin, except on the forehead, the hairy part of the head, face, and lips. It is unpigmented. There are dermal papillae on the head, lips, ischial callosities, and foot soles. The blood vessels are abundant throughout the dermis. Individual cells or cell clusters contain pigment granules. The sebaceous glands are much larger on the face and lips. In males, sebaceous glands are larger than in females. On the forehead, top of the head, and some other sites, sebaceous glands are surrounded by small pigment cells. The eccrine and apocrine sweat glands occur throughout the entire body. On the breast and neck they are more abundant than on other sites. The eccrine sweat ducts open directly onto the skin surface; those of apocrine glands open near the hair funnels. On the body, the hair tufts have two to three hairs. Arrectores pilorum muscles are well developed.

Macaca cyclopis

The skin is similar to that of *M. speciosa* (Hayasaki, 1959; Setoguti and Sakuma, 1959). The thinnest epidermis is on the belly. The epidermis is pigmented rather intensively throughout the entire body, specifically where hair is sparse or absent. The pigment cells are widespread in the dermis, particularly on the limbs. The trunk sweat glands are scarce. They are well developed in the perianal and pubic areas. Apocrine glands are more abundant than eccrine glands in the armpit and perianal regions. The sweat glands are deep in the dermis, the apocrine glands being deeper than the eccrine. The eccrine glands are largest on the fingertips. Hair is densest on the scalp and thinnest on the breast and belly. The hairs grow singly or in tufts, up to nine hairs in a tuft. Typically, the tufts have two to three hairs.

Cerocebus torquatus (= C. atys)

Material. Skin from a female and a male (Machida et al., 1965).

In the hairy skin areas the epidermis is thin. In the skin of the feet and ischial callosities, it is thick and has a stratum granulosum. There are no pigment cells in the trunk epidermis. The dermal mast cells are few. Large clusters of pigment cells are found in the dermis of the scalp, breast, belly, back, thighs, and tail, as are uni- or bilobal sebaceous glands. Large, multilobal glands occur in the skin of the eyelids, lips, and genitals. The apocrine sweat glands are largest on the eyelids, lips, and genitals. The gland ducts open into the hair canal. There are two to three apocrine glands per hair tuft. The number of eccrine sweat glands in hairy areas in relation to that of apocrine glands is about 1:2 to 1:3. The hair is arranged in tufts with four to six hairs in each.

Papio anubis

Material. Skin of a young male from various body areas (Montagna and Yun, 1962*b*).

The epidermis is medium in thickness with a thin, discontinuous stratum granulosum. It is highly pigmented in the skin of the face, brows, lips, armpits, buttocks, and scrotum. Only the basal layer is pigmented in the head, breast, and belly epidermis. In the limbs and sole epidermis, the pigment is concentrated at the epidermal ridge tops. There is very little pigment in the other body areas. In the dermis are well-developed dermal papillae. The distribution of large and small polymorphous pigment cells shows regional differences. These never ascend to the papillae. There are varying numbers of branched pigment cells on the lobe periphery of nearly all the sebaceous glands. In contrast with the majority of the Old World monkeys, *P. anubis* sebaceous glands contain alkaline phosphatase. Except for the skin of the lips and the ischial callosities regions, the eccrine glands are plentiful elsewhere. They are largest in number in the upper portion of the breast, near the alae nasi, and on the soles.

The secretion cavity is a glomus deep in the dermis. Short, rather thick nerves containing specific cholinesterase surround the secretion cavity of the gland. Large apocrine-type glands are scattered sparsely throughout the whole of the body. These are more numerous in the anterior portion of the breast and belly and in the perianal region, being fewer than the eccrine glands elsewhere. The glands have a thick secretion cavity twisted to an unusual extent and a thin, straight duct opening into the hair canal or onto the skin surface near it. The anterior part of the breast has a concentration of sweat glands resembling the armpit organ of man and the anthropoids. The hairs grow in tufts in threes (1 to 6). Each tuft contains a large hair with two smaller ones. Hair growth follows a mosaic pattern, with the same tuft containing both growing and nongrowing hairs.

Pygathrix cristatus (= *Presbytis pyrrus*)

Material. Skin from two adult females and one male (Machida and Montagna, 1964).

On the trunk, the epidermis has a stratum malpighii of four to five layers and a discontinuous stratum granulosum. Except on the back, the epidermis is highly pigmented by virtue of melanocytes and pigment granules in the stratum basale cells. No elastin fibers are found in the papillary layer, but small numbers of them occur in the deep parts of the reticular layer. The dermis has melanocytes only on the back. The trunk skin is weakly vascularized. Relatively small, multilobal sebaceous glands open into hair sheaths via very short ducts. There are melanocytes near sebaceous glands. A number of melanin granules are found in the fully differentiated sebaceous glands and their secretions. There is no alkaline phosphatase in the glands. The numbers of apocrine and eccrine glands are nearly equal on the trunk. The apocrine glands have thickened, nearly straight secretion cavities. The absence of specific glands (except on the soles) is noteworthy. Hair tufts have three to five hairs.

FAMILIA PONGIDAE

On the trunk, the epidermis is thin or rather thick, with stratum granulosum and occasionally with stratum lucidum. The epidermal undersurface contains ridges or alveoli. There are normally dermal papillae. Except in the dermal papillae, the dermis contains collagen bundles mostly parallel to the skin surface. Normally there are no pigment cells in the trunk dermis. The trunk sebaceous glands are small and unilobal. There are eccrine and apocrine glands on the trunk, the former being larger in number and better developed. The presence of axillary specific glands is characteristic. Arrectores pilorum muscles are well developed. The hairs are arranged in tufts of two to five or grow singly.

Hylobates hoolock

Material. Skin from various body areas of two immature females and one adult (Parakkal et al., 1962).

On the trunk, the epidermis is thin. A network of ridges occurs on the foot sole epidermis underside. There are few pigment cells among the stratum basale cells. The stratum granulosum is present only on the most mechanically disturbed sites. The papillary layer has thin collagen bundles, a small number of fibroblasts, and numerous mast cells. There are no pigment cells. The nerve endings contain alkaline phosphatase and cholinesterase (Winkelmann, 1962). Larger sebaceous glands are found on the lips, eyelids, the anal region, and the genitals. Most glands open into the hair bursae; some open directly onto the skin surface near the hair funnel. Numerous arborescent pigment cells occur among the secretion duct cells and on the glandular lobe surface. The majority of glandular cells have pigment granules. On the body surface, the eccrine sweat glands are less numerous than the apocrine. The apocrine sweat glands have larger and less tightly twisted glomeruli than the eccrine. Ducts of the majority of these glands open into the hair bursae, some opening directly onto the skin surface near the hair funnel. The hairs grow in tufts of three.

Pan troglodytes

Material. Skin from various body areas of one immature female and three mature males (Montagna and Yun, 1963a).

In the hairy skin areas, the epidermis is thickest on the face, scalp, and buttocks. The epidermis is pigmented, with the greatest melanin concentration being found at the epidermal ridge apexes. Pigment granules are well developed in the stratum granulosum (1 to 2 cells deep), and occasionally in the stratum corneum. The stratum lucidum is one to two cells deep. It is specifically well developed in the scalp skin. There are few fat cells in the subcutaneous fat tissue. Numerous fibrocytes and mast cells occur in the dermis. Pigment cells are found in the dermis of the scalp, eyelids, lips, nose, and genitals. Elastin fibers, scarce elsewhere, are concentrated in the dermal papillae and deep parts of the reticular layer. The blood vessels are numerous.

In the immature female, the sebaceous glands are recorded only on the face. In males, the sebaceous glands are large and multilobal on the face. The eccrine glands are distributed throughout the body (except the nose, lips, female external genitals, and male preputium). These are most numerous on the foot soles, face, and head and most scarce on the back. The glands are large with a compact glomus in the deep parts of the dermis and a long, straight secretion duct. The glandular epithelium contains dark and light cells. All the features of the chimpanzee eccrine glands are similar to those of humans. In the immature female, the apocrine sweat glands are much smaller in size than in the males. These are much fewer in number than the eccrine glands and occur throughout the body, attaining the largest numbers on the face. They are also plentiful on the meatus acousticus externus, axillary, and pubic regions. They form a specific gland in the axillary region. On the throat and armpits, there are two to three follicles per gland or one hair group per gland. Every duct of the apocrine gland opens directly onto the skin surface near the pilary canal. The duct is very thin as compared with the thick secretion cavity. The glandular cells of the axillary gland contain glycogen and ionic iron. The glands surrounded by nerves containing cholinesterase show a moderate phosphorylase activity. The distribution of eccrine glands is similar to that in man. The hair is rather dense in young animals and sparse in adults. In the sexually immature female, two to three (mostly three) hair follicles form a tuft. In the adult chimpanzee, these tufts are not well defined. However, they are visible throughout the body. Some hair bulbs are embedded in the subcutaneous fat tissue. Arrectores pilorum muscles are well developed. The hair structure is similar to that of the gorilla. The articular pads of the second finger phalanges of the front limbs have a similar specialization to that of the gorilla.

Gorilla gorilla

Material. Skin from various body areas of a young, sexually mature (Strauss, 1950) male of about three years (Ellis and Montagna, 1962).

The skin on the back is thicker than on the belly. The entire skin surface is black. In some places the epidermis pigmentation is greater than in others. The epidermis measurements are presented in table 15.

The stratum granulosum comprises one to two cell rows (Strauss, 1950, notes that stratum granulosum is found only on the foot soles). The stratum lucidum can be seen in the epidermis of all areas after staining. Dermal papillae are found throughout the body, being better developed in some

TABLE 15

Skin values of three-year-old gorilla male (Ellis and Montagna, 1962)

Body area	Skin layers (μ)			
	Stratum malpighii	Stratum lucidum	Stratum corneum	Total thickness
Thigh (lateral portion)	55	—*	90	145
Breast	55	—*	35	90
Back under scapulae	35-55	—*	35	70-90
Belly	70-90	—*	55	125-145
Head	70-90	—*	90	160-180
Foot sole	125-215	35	450-540	610-790
Palm of hand	110-270	50	575-720	735-1040
Armpit	55	15	55-70	125-140

*Included in stratum corneum.

places. In the dermis, the collagen bundles are parallel or perpendicular to the body surface. The collagen bundles are most solid on the foot and palm digits. There are no pigment cells in the dermis. Mast cells are plentiful.

The hair follicles are surrounded by sebaceous glands. These are largest in size on the cheeks, upper lip, and eyebrows. The eccrine sweat glands are more plentiful and larger in size in hairless skin areas. The lowest portions of the gland nearly adhere to the subcutaneous fat tissue. The twisted part of the glands in the hairless skin regions is nearly twice as large as in the haired.

The glandular cells divide into two types. The first type is characterized by a pyramidal shape, a moderately basophilic cytoplasm stained with toluidine blue with pH 5.0, and little or no glycogen. The second type cells are larger and less regular in shape with a cytoplasm nonstained by basic dyes and with no glycogen. These two cell types are comparable to the dark and light cells of human eccrine sweat glands.

Small apocrine glands are sparsely distributed throughout the bulk of the haired skin. These are largest and most twisted on the breast, cheeks, around the nipple, and in the perianal region. The apocrine glands are largest in number and in size in the axillary region, where they combine with eccrine glands to form the glandular organ. The apocrine gland ducts open into the hair bursae between the sebaceous gland orifice and the hair funnel. The apocrine glands are invariably connected with the hairs. On the thighs (lateral surface) every hair group has one to two glands. In the armpit, the apocrine glands reach the canal of nearly every hair. The ratio of apocrine to eccrine glands varies depending on the body area, the average being 1:3.5. If foot soles and palms are not taken into account, this ratio is 6:7.

The hair is sparse. Large and small hair follicles make up tufts of two to five. The growing hair follicles are relatively thin and short. Arrectores pilorum are large with several portions. Most of the hair follicles are surrounded by a nerve ring situated immediately over the sebaceous gland secretion orifice. The innervation of haired skin regions in the gorilla resembles that of furbearing animals (Winkelmann, 1961).

The outer parts of the joints of the front limb fingers (front limbs being used by gorilla for support when walking), have well-developed pads. The epidermal stratum corneum is well developed. Plentiful eccrine sweat glands and specific nerve endings similar to those on the foot sole and palm are also situated there. In the chimpanzee, which uses finger joints when walking, joint pads are less developed. Similar modifications are found in the prehensile portion of the tail in spider and woolly monkeys (*Brachyteles* and *Lagothrix*).

The development of numerous sensitive nerve endings in the skin subject to strong mechanical effects is a remarkable feature of evolution. These endings are numerous in the joint pads of the gorilla, rare in the joint pads of the chimpanzee, and numerous in the ventral pad of the *Ateles* grasping tail. In the terminal nerve organs of gorilla skin (and also in chimpanzee and orangutang) cholinesterase is recorded (Winkelmann, 1961, 1962).

The skin varies topographically. On the breast, the thickness of the epidermis varies from 0.05 to 0.1 mm, dermis 1.5 to 2.1 mm. There are a few low, randomly scattered dermal papillae. Hair cover is sparse: approximately one hair per 2.9 cm of skin. Hairs group in tufts of two to three. Arrectores pilorum muscles are poorly developed and occur rarely. Sebaceous glands may be absent. Sweat glands divide into two categories: apocrine and eccrine.

On the belly, the epidermis is thin (on the average 0.05 mm); the dermis is relatively thick (2.5 to 3.0 mm). The dermal papillae are absent. The hairs are rather numerous: 11.7 per 1 cm of the skin. Hair tufts are formed by three, rarely by two, hairs. Occasionally hairs grow singly. Arrectores pilorum muscles are well developed. Sweat glands are nearly as numerous as in the breast skin. The apocrine glands are larger in number than the eccrine. There are a few intermediate type glands.

In the middle of the back, the skin is very thick, largely due to the dermis (medium thickness, 3.8 to 4.5 mm). The epidermis is thin (on the average 0.08 mm). The dermal papillae are absent. The hairs are numerous (on the average 21.9 per 1 cm of the section). Hair tufts are formed by two to three hairs (occasionally by four to five hairs). Arrectores pilorum muscles are poorly developed or absent. The sweat glands (of eccrine type only) are as numerous as in the breast and belly skin. The axillary specific gland is formed by apocrine-type sweat glands. These are highly developed and grouped by sections separated from one another by connective tissue interlayers. The axillary specific gland in the male is better developed than in the female.

In the gorilla, in the parietal region, there is a skin "pad" up to 5 cm thick, 7.5 cm long, and 5 cm wide (Straus, 1942). The epidermis reaches a thickness of 0.25 mm. The dermis is formed by a dense plexus of large collagen bundles passing in different directions. The epidermis is particularly thick in the skin of front and hind soles (0.56 to 1.22 mm). The sweat glands (eccrine type only) are more abundant than in other parts of the body (see table 14).

Tamandua tetradactyla

Material. Skin from withers and breast of *T. tetradactyla* (body length 45 cm) taken in Brazil.

The skin is the same thickness on the breast and withers (3.08 mm), but the breast epidermis is thicker (39μ) than on the withers (28μ). The stratum malpighii is two to four cells deep. The stratum corneum (11μ thick) is solid on withers and breast. The epidermis has no pigment. According to Machida et al. (1966), the epidermis is five to ten cells deep. The stratum granulosum is very pronounced. Melanocytes are absent in the epidermis of the back, buttocks, anus, genital region, toes, and proximal part of the tail. However, they are found in the epidermis of the head, eyelids, lips, nose, ears, cheeks, breast, and tail tip.

The dermis (3.05 mm on the withers, 3.04 mm on the breast) has a thick papillary layer (2.05 mm on the withers, 2.04 mm on the breast) and a thinner reticular layer. Only the outer 60 to 80μ section of the papillary layer has thin collagen bundles, which run in all directions. In the remainder of the papillary layer, thick horizontal bundles of collagen fibers are mutually perpendicular (Braun and Lebedeva, 1947, no species indicated), with a few vertical bundles. In the reticular layer are considerably thinner horizontal bundles (fig. 55). Elastin fibers, mostly oblique, are numerous only in the outer papillary layer (Machida et al., 1966, found no elastin fibers).

The sebaceous glands are small (34×67; $22 \times 56\mu$), unilobal, and pressed up against the hair bursae. Posterior to every hair is a sebaceous gland. Machida et al. (1966) found sebaceous glands only at the hair follicles of the eyelids.

The secretion cavities of nonfunctioning apocrine sweat glands are slightly twisted and are 34μ in diameter on the withers and 28μ on the breast. They adjoin the hair bursae. The excretion ducts open into the hair funnels posterior to the hairs. In some cases, two excretion ducts open into the same hair funnel, anteriorly and posteriorly. In the skin of the outer genital organs, huge glomi or tubular glands are found near the follicles of the bristle-like hairs (Schaffer, 1940). In the skin of the lips, muzzle, and toes eccrine sweat glands are recorded (Machida et al., 1966). Arrectores pilorum muscles are large.

The hairs of the withers divide into four categories. First-category hairs

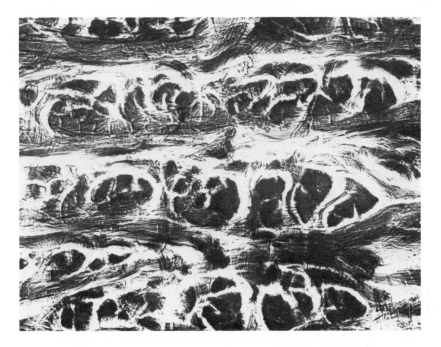

Fig. 55. *Tamandua tetradactyla*. Network of collagen fiber bundles in reticular layer.

gradually thicken from the base to the middle of the shaft and then gradually taper into a long, thin tip. There is no medulla, and the average length is 6.4 mm, maximum thickness being 114μ. Second-category hairs are similar to the first, but have shorter tips and are considerably thinner and shorter, 9.1 mm long and 81μ thick. Third-category hairs have a short club with a thickening over it. These hairs are thickest in the proximal section, from which they gradually begin to taper. Their length is 7.8 mm, thickness 126μ, and there is no medulla.

Fur hairs grow to 8.2 mm long and 16μ thick and belong to the fourth category. They are of uniform thickness the whole length. The fur hair is slightly wavy (up to four waves) with a medulla through the entire length. This is the least abundant category. These hairs grow singly and do not form groups. The first and second categories average 166 hairs, and the third and fourth categories average 1,166 hairs per 1 cm². On the breast the hairs are considerably sparser, averaging 416 hairs per 1 cm².

FAMILIA BRADYPODIDAE

A brief account is available of the structure of the skin, sweat glands, and pelage (Wislocki, 1928), as well as data on the specific dorsal gland (Schaffer, 1940). We studied the skin of *Bradypus tridactylus*. The skin has no ossifications. Skin, epidermis, and stratum corneum are rather thicker on the breast than on the withers. The epidermis is thick owing to a highly developed stratum corneum. The pelage is dense. On the breast the hair is longer, though there are fewer hairs than on the back. There are three hair categories, and the hairs grow in tufts.

Bradypus tridactylus

Material. Skin from withers (1.54 mm) and breast (1.60 mm) of one specimen (body length 32 cm).

The skin is 3 mm thick on the throat; 3 mm on the dorsal surface of the neck; 2.5 mm on the back; 2 mm on the belly; 2 mm on the top of the head; 2 to 2.5 mm on the legs (Wislocki, 1928). The thickness of the epidermis (withers) is 118μ, (breast) 123μ. The stratum corneum is multilayered and rather loose, 106μ thick on the withers and 112μ on the breast. The stratum malpighii is one to two cells deep. There is no pigment in the epidermis. The papillary layer of the dermis is thicker than the reticular layer (1.5 mm and 0.27 mm on the withers, and 1.25 mm and 0.22 mm on the breast). The border between them is level with the hair bulbs. The upper part (40 to 50μ) of the papillary layer consists of small bundles of collagen fibers running in various directions and forming a compact plexus.

In the remaining part of the papillary layer, thick collagen bundles run horizontally, perpendicular to one another. They are crossed by a few bundles that rise vertically upward. The collagen fibers in the reticular layer are horizontal and run parallel to one another. Elastin fibers are few in number in the dermis, and fat cells are absent. The skin adheres tightly to the subcutaneous muscles (Wislocki, 1928). The sebaceous glands are unilobal and small, 34×67; $22 \times 56\mu$, and are present only anterior and posterior to hair tufts. We find no sweat glands. This may be because they are very small and sparse (Wislocki, 1928). Large sweat glands are recorded on the bare part of the muzzle in *Choloepus hoffmanni* (Wislocki, 1928), though sweat glands are absent in the naked pads of the front and hind feet of that species. In *Bradypus* such pads are not recorded at all (Wislocki, 1928). There appears to be a skin specific gland between the shoulder blades in males (Schaffer, 1940).

The hair of the withers (part of it growing) divides into three categories. First-category straight hairs gradually thicken from a short club to half or two-thirds of the shaft and then taper sharply toward the apex. Their average length is 21.9 mm, their average thickness 217μ. The medulla runs the length of the hairs. Second-category straight hairs have a long club, which gradually turns into a long, thick granna. Before the granna, the hair is bent. The club has no medulla, but in the granna the medulla is developed (missing only in the tip). The second category hairs subdivide into two size orders. The first order is 32.3 mm long and 209μ thick. The thickness of the club is 54μ. The length of the granna is 22.4 mm; the medulla in the granna is up to 83%. The second order is shorter (31.8 mm), but thicker (218μ). The club at the base is 35μ thick. The length of the granna is 20.9 mm.

The third category is fur with no medulla. There are two size orders, both of the same thickness the whole length. The first-order hairs are slightly wavy, 24.7 mm long and 29μ thick. The second-order hairs are wavier (four waves), shorter (12.8 mm), and thinner (22μ). The hairs grow in tufts of one thicker (of the first or second category) and three to four thinner hairs, though there are tufts of four to six thin hairs only. There are 666 thick hairs and 5,333 thin hairs per 1 cm^2.

The breast hair categories are the same, the only difference being size. First category, first order is 27.3 mm long and 207μ thick; the second order is 38.8 mm long and 214μ thick, with the granna taking up 26.0 mm. The thickness of the club is 63μ. The second-order, second-category hairs are shorter than those in the preceding category (31.4 mm), and thinner (158μ). The length of the granna is 13.1 mm, and the thickness of the club is 62μ. The first-order fur hairs are 26.5 mm long and 27μ thick; the second-order are 16.5 mm long and 25μ. The hairs grow in tufts of one big and four to six small hairs, tufts with no big hair being more sparse here than on the withers (they contain four to five hairs). There are 666 large and 4,666 small hairs per 1 cm^2.

Familia Dasypodidae

There are some data available on the skin structure of nine-banded armadillos (Cooper, 1930) and of specific skin glands of armadillos (Schaffer, 1940). We studied the skin of *Euphractus sexcintus*.

The dorsal and lateral parts of the body are protected by an osseous test. On the abdominal side, the epidermis is thick. It has a solid corneous layer. The subcutaneous fat tissue is well developed. The sweat gland secretion cavities are twisted into large glomi. The pelage is sparse. The hair divides into three categories. The hairs grow singly.

Euphractus sexcinctus

Material. Skin from the breast (5.3 to 6.6 mm thick).

The outer surface of the skin is irregular. The epidermis is 112μ thick. The stratum malpighii is five to six cells deep; the stratum corneum is 67μ thick and very solid. The epidermis is not pigmented. The dermis is 1.3 to 2.6 mm thick, and the boundary between the pipillary and reticular layers is not pronounced, lying about 1.8 mm from the surface of the skin, level with the hair bulbs. The outer surface of the papillary layer, 150 to 200μ thick, has a dense plexus of small collagen bundles running in various directions.

The remainder of the dermis has large horizontal collagen bundles perpendicular to each other (except in the inner reticular layer).

In the inner reticular layer, horizontal bundles of mostly parallel collagen fibers form a loose plexus. Fat cells occur between them. The subcutaneous fat tissue is a thick (4.0 mm) accumulation of fat cells crossed by sparse bundles of collagen fibers running horizontal or at an acute angle to the skin surface. There are two sebaceous glands (100 x 123μ) at every hair. Their ducts are long and perpendicular to the hair roots. Sweat glands occur only at the hairs, with the secretion cavities (up to 110μ in diameter) forming large, very twisted glomeruli (0.44 x 1.00 mm). In the skin of the front sole are large glomi (0.6 x 1.0 mm) of highly twisted eccrine sweat glands. The secretion cavities of these glands are 28μ in diameter. There are four, or sometimes five ducts of the dorsal specific gland (Schaffer, 1940). This is formed by a combination of large sebaceous glands and tubular gland glomi opening into a round "cistern" that opens out through the ducts. There are anal glands.

Three hair categories are recorded. The first category has bristle-like, slightly flattened straight hairs varying in thickness from 13.7 to 24.6 mm (average 18.3 mm). They are thickest (217μ) at the bottom and middle and gradually thin down at the tip. There is no medulla. The second-category hairs are short, straight, and bristle-like. They thicken from the base to the dorsal third and then gradually thin down again. There is no medulla. The hairs in this category are 6.4 mm long, and 123μ thick. The third-category fur hairs are wavy and of uniform thickness (18μ) all down the shaft. Their length ranges from 7.5 to 15.0 mm (average 12.2 mm). The medulla is well developed through the whole length except base and tip. They grow singly, and there are 250 hairs per 1 cm^2.

ORDO PHOLIDOTA

There are published data on the scales and hair of members of the *Ordo Pholidota* (Rahm, 1956; Mohr, 1961) and their specific skin glands (Pocock, 1924; 1926*b*; Weber, 1928). We studied the skin of *Manis pentadactyla*.

Unlike other mammals, their bodies are covered with mobile corneous scales of epidermal origin. The scales are fixed to the skin by their anterior edge, and their sharp posterior edges overlap in a tile pattern. The scales are variously triangular, ellipsoid, or truncated. The scales are absent on the lateral parts of the muzzle, chin, throat, belly, the inner surface of the legs, and, in some species, on the outer sides of the legs. The number of the scales remains constant all through the animal's life. They grow only in size (Rahm, 1956). At the base of the scales of the Asian species are short, bristle-like hairs (one to four hairs at every scale) (Mohr, 1961). The skin areas which are not covered with scales are black and have sparse hair. The hair of the head, front legs, and breast is black or dark brown; the hair of the belly and hind legs is reddish brown (Rahm, 1956).

Manis pentadactyla

Material. Skin from the sacrum and breast (body length 42 cm), taken May 16, southern China. The skin is in poor condition.

The structure of the skin differs considerably on the dorsal and the ventral sides of the body (the structure of the scales has not been studied). The sacrum skin is up to 1 mm thick, and the breast skin 0.85 to 2.2 mm. The stratum malpighii of the epidermis (breast) is thin, while the stratum corneum, highly developed, reaches a thickness of 110μ. Large bundles of collagen fibers are just under the bursa scales. They twist, following the outline of the bursa. Thick layers of large bundles of collagen fibers are arranged deep down. In the inner layers of the dermis, between the bursae of the scales, collagen bundles are horizontal. In the other dermal layers they run at various angles to the skin. Elastin fibers are absent. The breast dermis has an intricate plexus of collagen bundles (especially in the upper layers of the 120 to 150μ thick dermis), which are oblique to the skin. Elastin fibers are present only in the inner layers where, rather numerous, they usually lie parallel to the skin surface. Hair bursae sometimes occur. Arrector pili muscles are absent.

No sebaceous or sweat glands are found in the anal region. There are sebaceous glands which are very enlarged (Weber, 1928), or large, paired glandular formations (Pocock, 1924). On the neck, slightly behind the line between the ears, is a rectangular glandular field formed by swollen sebaceous glands (Pocock, 1926*b*). There are no arrector pili muscles in the breast skin.

There are two categories of hair. The first category includes bristle-like, straight hairs which are slightly bow-shaped with broken tips and no medulla. An average of 5.5 mm long, their thickness is uniform the whole length (tapering gradually toward the apex) at 101μ. The second category includes wavy fur hairs uniformly thick down their whole length. Some have no medulla (length 7.4 mm, thickness 16μ). Others (length 6.5 mm, thickness 17μ) have a medulla (52.9%). The hairs grow singly, 166 hairs per 1 cm^2.

ORDO LAGOMORPHA

The skin of *Ochotonidae* has not been studied, but the skin of *Leporidae* has been rather thoroughly investigated. There are several works on the skin structure, especially the pelage (Zharkov, 1931; Fetisov, 1931; Zubin, 1938; Pavlova, 1951; Kuznetsov, 1952; Gerasimova, 1955; Marvin and Shuma-

Fig. 56. *Lepus timidus.* Horizontal network of collagen fiber bundles in reticular layer.

kova, 1958; Tserevitinov, 1958; Kogteva, 1963). The skin specific glands of *Leporidae* have been studied, especially the wild rabbits (Mykytowicz, 1966*a, b;* Schaffer, 1940).

We studied the skin of *Ochotona rufescens* and *Lepustidus l. europaeus,* and the pelage of *L. tolai.* The skin is not very thick. On the body the epidermis is thin. Stratum granulosum and stratum lucidum are absent. Stratum malpighii is one to two cells deep. The epidermis is not pigmented. In the dermis, the bundles of collagen fibers are mainly horizontal (fig. 56). Sebaceous glands are small with single lobes, one at every hair. Sweat glands are absent. Specific glands are numerous, such as sole, anal, chin, circumoral. There are arrectores pilorum muslces (fig. 57). The hair divides into guard, pile, and fur.

FAMILIA OCHOTONIDAE

Ochotona rufescens

Material. Skin of withers of two specimens (body lengths 185 and 190 mm) taken in April, near Kzyl-Arvat, Turkmenistan.

The hairs are not growing. The skin is slightly wrinkled, 365 and 220μ

Fig. 57. *Lepus tolai.* Arrector pili muscle in withers skin: *a*, sebaceous gland; *b*, fur hairs; *c*, pile hair; *d*, muscle.

thick. The epidermis is 22 and 22μ, and the stratum corneum is 11 and 11μ. The stratum malpighii is two cells deep.

The papillary layer (200 and 121μ) is only slightly thicker than the reticular layer (143 and 77μ). The plexus of bundles of collagen fibers in the papillary layer and in the upper parts of the reticular layer is solid (fig. 58). Elastin fibers are numerous in the papillary layer. There are groups of fat cells (39 x 50; 33 x 50μ) in the reticular layer. Sebaceous cells are 28 x 39; 22 x 56μ. In *Ochotona alpinus* there is an anal and a preputial gland (Schaffer, 1940).

Guard hairs (length 16.2 mm, thickness 61μ, medulla 91%) are straight, gradually thickening from the base to the distal two-thirds of the shaft. The hair has a thin tip into which a narrow strip of the medulla projects deeply. The pile hairs, which have a weak granna, divide into two orders. The first

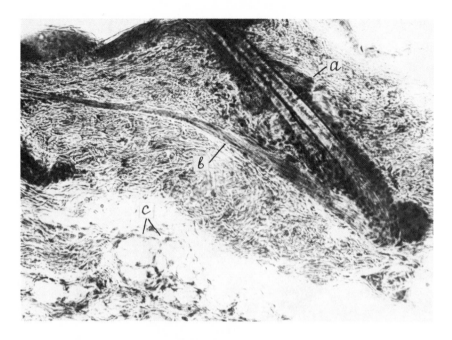

Fig. 58. *Ochotona rufescens.* Withers skin: *a,* sebaceous gland; *b,* arrector pili muscles; *c,* fat cells.

order is 14.2 mm long; thickness 38μ, medulla 90%. In the second order (12.5 mm long, 25μ thick, medulla 80%), the shaft is slightly wavy. The fur hairs, wavy and uniformly thick down their length, are 9.5 mm long, 14μ thick, medulla 79%. The hairs grow in tufts of two to three with two to three tufts in a group. There are 11,333 hairs per 1 cm^2.

Familia Leporidae

Oryctolagus cuniculus

Wild rabbits have anal glands consisting of a compact mass situated at the sides of the rectum, two bodies 15 to 25 mm long, 3 to 10 mm broad, 6 mm thick, weight 110 mg. (Schaffer, 1940; Mykytowicz, 1966*a, b*). With the onset of sex maturity they become larger in the male. The greatest thickness and activity of the anal gland coincides with the reproductive season. The anal gland is formed of highly twisted tubular glands. The castration of young rabbits depresses the development of the anal gland in males but has almost no effect in females.

Inguinal glands are well developed in rabbits. They are pouch-shaped, formed by sebaceous and deep lying sweat glands (Schaffer, 1940; Mykytowicz, 1966*a*, *b*). In males these glands are better developed. In the reproduction period, inguinal glands are at their largest and most active. The chin gland is formed by an accumulation of tubular glands. These are apocrine glands (developed from the hair follicle, some of their cells losing part of the protoplasm during secretion) (Lyne et al., 1964). At the same time, in the chin gland are accumulations of glands or individual glands with cells following the eccrine secretion pattern. There are also glands having both type cells, with apocrine and eccrine secretion patterns. The effect of castration on the development of the chin gland is similar to the effect on the anal gland. There are other glands as follows: preputial (formed by sebaceous glands), circumoral (of tubular glands), planum nasolabiale (of tubular glands), and pigment (in the middle of planum nasolabiale, formed by sebaceous glands) (Schaffer, 1940).

Lepus timidus

Material. Skin from withers of a specimen with winter pelage (body length 55 cm) taken in November, on the northern Dvina River.

The thickness of the skin is 476μ, the epidermis 22μ, and the stratum corneum 14μ. There are four zones of varied thickness on the body (Kogteva, 1963). The thickest skin (0.43 to 0.55 mm) is on the sacrum; it is thinner (0.26 to 0.32 mm), lower down the spine, still thinner (0.19 to 0.25 mm) over much of the body, such as the sides, middle of the back, shoulder blades, and neck, and thinnest (0.11 to 0.18 mm) on the belly. The stratum malpighii of the withers epidermis is one to two cells deep. In the papillary layer of the dermis (thickness 174μ), bundles of collagen fibers form a loose plexus, and in the reticular layer (280μ) a denser one. Elastin fibers are numerous (Zubin, 1938). No fat cells are recorded in the subcutaneous fat tissue. Sebaceous glands are round, $22 \times 28\mu$.

Hairs with long roots lie in the papillary layer of the dermis almost parallel to the skin. The straight guard hairs gradually thicken toward the upper two-thirds of the shaft. They are 16.2 mm long, 61μ thick, medulla 91%. In the pile hairs, the granna is more marked. There are two size orders, 35.1 and 33.1 mm long, thickness 102 and 48μ, medulla 93 and 78%. The fur hairs are wavy and are of uniform thickness (17μ) down the whole length (30 mm), medulla 65%.

Kogteva (1963) has obtained different values for the hairs (table 16). Sometimes intermediate hairs are distinguished (Zharkov, 1931). The hairs grow in groups of three to four tufts near guard or large pile hair with some twenty fur hairs to a tuft. In winter, 16,500 to 18,800 hairs are recorded per 1 cm² of skin on the withers, 17,200 to 20,500 on the back, 24,800 to 29,000 on the sacrum, and 9,100 to 10,300 on the belly (Fetisov, 1931; Kogteva, 1963). In summer, the number of hairs on the back decreases to as little as 9,300. For length of pelage on the body of *L. timidus* there are four distinguishable zones; the inguinal area, with the longest hair; the ventral side of the body, with somewhat shorter pelage; the sides, shoulder blades, and sacrum area, with medium length hair; and the back, with the shortest pelage (Kogteva, 1963).

Lepus tolai

Material. Skin from the withers of *L. tolai* (body length 40 cm) taken in January in Repetek, Turkmenia.

The skin is 440μ thick, epidermis 28μ, stratum corneum 8μ, papillary layer 258μ, and reticular layer 154μ. In structure, the skin is similar to *L. timidus.* Arrectores pilorum muscles are well developed. The size of the sebaceous glands is 28×39; $34 \times 56\mu$. The categories and orders of hair are similar to *L. timidus.* The guard hairs are 42.6 mm long, 108μ thick, medulla 90%. The pile hairs, first order are 26.1 mm long, 103μ thick, medulla 90%. The pile hairs, second order are 22.4 mm long, 68μ thick, and medulla 89%. The fur hairs are 19.8 mm long, 17μ thick, and medulla 83%. In *L. tolai,* five zones are distinguished on the body for length of pile hairs. The ventral side of the body has the longest pelage (Tserevitinov, 1958).

Lepus europaeus

Material. Skin from withers of an adult specimen, taken in summer in the Crimea.

Specific skin glands are similar to those of the rabbit, except that the chin and anal glands are smaller and the inguinal gland is bigger (Mykytowycz, 1966a, b). In *L. europaeus* the differences in the size of the anal gland and the variability of the inguinal gland vary less with the season than they do in the rabbit.

The hair structure if similar to that of *L. timidus.* The guard hairs are 26.8 mm long, 123μ thick, medulla 94%. The pile hairs, first order, are 22.6 mm, 48μ, and 79%, respectively. Second order is 18.0 mm, 22μ and 50%. Fur hairs are 18.0 mm, 22μ, and 50%. Some researchers also distinguish intermediate hairs (Gerasimova, 1955). In the winter season the hairs of the *L. europaeus* become longer and thinner (table 17) and more numerous (table 18). Geographic variability affects hair density.

ORDO RODENTIA

Little work has been done on rodent skins. There are few detailed skin studies of the enormous number of species in the order. Those that do exist are mainly of commercial species: *Sciurus vulgaris* (Kuznetsov, 1927, 1928;

TABLE 16

LEPUS TIMIDUS FROM LENINGRAD AND NOVGOROD REGIONS (KOGTEVA, 1963): HAIR MEASUREMENTS

Category	Summer hair					n	Winter hair				
	Length (mm)		Thickness (μ)				Length (mm)		Thickness (μ)		
	M±m	±	M±m	±			M±m	±	M±m	±	
Withers											
Guard											
Pile	30.7±0.35	1.73	96.40±0.31	1.5		36	43.8±0.22	1.32	92.8±0.09	0.54	
1st order	26.5±0.17	0.83	105.60±0.17	0.85		50	41.4±0.16	1.12	96.4±0.1	0.73	
2d order	24.6±0.26	1.3	82.00±0.49	2.45		45	39.0±0.3	2.0	78.8±0.2	1.34	
3d order	23.1±0.27	1.36	68.00±0.21	1.05		45	37.2±0.12	0.8	62.5±0.11	0.74	
4th order	20.9±0.26	1.33	34.50±0.09	0.45		60	34.8±0.17	1.32	31.5±0.16	1.23	
Fur	16.4±0.01	0.03	15.20±0.32	1.6		75	25.7±0.14	1.21	12.0±0.15	1.31	
Sacrum											
Guard											
Pile	31.2±0.25	1.27	108.80±0.23	1.17		75	44.5±0.37	3.18	100.1±0.21	1.82	
1st order	25.8±0.14	0.72	115.20±0.31	1.53		75	40.3±0.09	0.84	108.8±0.18	1.56	
2d order	24.5±0.24	1.22	82.00±0.24	2.21		75	39.0±0.17	1.47	80.0±0.15	1.3	
3d order	23.2±0.14	0.73	65.16±0.53	2.65		75	37.7±0.13	1.13	63.0±0.31	2.69	
4th order	22.0±0.1	0.5	33.30±0.26	1.31		75	36.5±0.11	0.95	32.9±0.22	1.91	
Fur	18.2±0.11	0.56	15.80±0.25	1.28		75	27.3±0.15	1.3	12.2±0.13	1.13	
Belly											
Guard											
Pile	39.5±0.23	1.15	72.50±0.27	1.35		25	49.5±0.25	1.26	64.5±0.34	1.72	
1st order	33.7±0.22	1.1	79.20±0.21	1.05		25	45.3±0.23	1.15	77.0±0.28	1.4	
2d order	27.9±0.21	1.06	38.90±0.16	0.8		25	43.8±0.27	1.34	37.8±0.23	1.15	
3d order	25.3±0.25	1.28	26.30±0.23	1.15		25	40.5±0.24	1.22	25.6±0.22	2.1	
4th order	23.6±0.24	1.23	20.50±0.24	1.21		25	37.5±0.34	1.71	19.8±0.13	0.65	
Fur	18.5±0.27	1.15	13.30±0.25	1.25		25	28.2±0.28	1.4	11.5±0.23	1.16	

*Measurements were obtained for 25 summer specimens.

TABLE 17

Seasonal variability of sacral region hair of Lepus europaeus from the Ukraine (mean values) (Gerasimova, 1955)

| Season | Pile | | | | Intermediate | | Fur | |
| | 1st order | | 2d order | | | | | |
	Length (mm)	Thickness (μ)	Length (mm)	Thickness (μ)	Length (mm)	Thickness (μ)	Length (mm)	Thickness (μ)
Summer	33.5	113	28.1	85	26.2	49	20.2	19
Winter	35.9	103	34.0	67	29.5	34	23.5	16

TABLE 18

Geographical and seasonal variations in hair density of Lepus europaeus per 1 cm² of skin (Gerasimova, 1955)

Mountain ridge (origin)	Axial	Inter-mediate	Wool	Mountain ridge (origin)	Pile	Inter-mediate	Fur
Menzelinsky	584*	590	8,876	South-	382	504	8,156
	1,382	1,988	25,812	eastern	968	1,260	18,012
Central	388	564	6,188	Trans-	412	528	6,320
	982	960	22,038	caucasian	748	788	16,848
Ukraine	373	612	6,840				
	814	640	20,398				

*Upper figure: summer measurements. Lower figure: winter measurements.

Tserevitinov, 1958), *Spermophilopsis leptodactylus* (Lavrov and Naumov, 1934; Sokolov, 1964), *Castor fiber* (Tserevitinov, 1951, 1958; Belkovich, 1962), *Ondatra zibethica* (Lavrov, 1944; Marvin, 1974; Gudkova-Aksenova, 1951; Tserevitinov, 1951, 1958; Kuznetsov, 1952; Pavlova, 1955), *Chinchilla lanigera* (Pavlova, 1968, 1969). There are several descriptions of other rodents (Tserevitinov, 1958). Specific skin glands have been investigated in a number of rodent species, but still in only a minority of the *Rodentia* order (Schaffer, 1940; Ortman, 1960; Quay, 1954; Vorontsov and Gurto-voi, 1959; Sokolov and Skurat, 1966; Skurat, 1969, 1972 and others).

We have studied the skin of *Aplodon rufa, Sciurus vulgaris, S. persicus, Funambulus palmarum, Tomeutes lacroides, Spermophilopsis leptodactylus, Marmota bobac, M. caudata, Citellus undulatus stramineus, Citellus undulatus leucosticus, Pteromys volans, Geomys bulbivora, G. floridanus, Pterognathus bailey, Dipodomus merriami, D. deserti, Liomys pictus, Heteromys desmarestianus, Castor fiber, Cricettulus migratorius, Peromyscus maniculatus, Gerbillus nanus, Meriones unguiculatus, Meriones meridianus, M. m. penicilliger, M. m. meridianus, Meriones nogarium, Rhombomys opimus, Ondatra zibethica, Ellobius talpinus, Prometheomys schaposhnikovi, Lemmus lemmus, Lemmus obensis, Myopus schisticolor, Dicrostonyx torquatus, Alticola argentatus, Lagurus lagurus, Arvicola terrestris, Microtus gregalis, M. nivalis, M. oeconomus, Myospalax terrestris, M. aspalax, Spalax microphthalmus, Nesokia indica, Rattus norvegicus, R. rattus, Mus musculus, Apodemus agrarius, A. sylvaticus, Micromys minutus, Glis glis, Dyomys nitedulla, Ellomys quercinus, Muscardinus avellanarius, Selevenia betpakdalensis, Sicista subtilis, Zapus setchuanus, Euchoreutes naso, Allactaga jaculus, Allactagulus acontion, Pygerethmus platyurus, Scirtopoda telum, Paradipus ctenodactylus, Eremdipus lichtensteini, Hystric leucura, Myocastor coypus, Chinchilla lanigera,* and *Heterocephalus glober.*

The differences in the structure of skin proper is slight in rodents. The skin is thin or medium. In most of the species the epidermis is thin. Stratum spinosum has two to three cell layers, while stratum granulosum and stratum lucidum are usually absent. The epidermis is pigmented only in a few rodents (mostly in *Soricidae*). In the dermis, the boundary between the reticular and papillary layers is level with the hair bulbs. The papillary layer is thicker than the reticular. The bundles of the collagen fibers in the dermis of most rodents are horizontal; the network of elastin fibers is well developed. Some rodents have pigment cells in the dermis of the body skin. The sebaceous glands are well developed but always fairly small, with single lobes. Sweat glands are absent. Specific glands are numerous: sole, anal, in the corners of the mouth, abdominal, lateral. Arrectores pilorum muscles are present in the skin.

The hair usually divides into vibrissae, guard, pile, and fur. The medulla is well developed in all hair categories except vibrissae in most rodents. In many species the guard hairs have a long, medulla-free tip. The hairs grow in tufts or groups. Hair characteristics usually determine the nature of the skin in individual rodent families. In some cases, hair may not correspond to the above categories or may turn into spines.

FAMILIA APLODONTIIDAE

Aplodon rufa

Material. Skin from the withers and breast of an adult specimen.

The hairs in the skin are not growing. The withers skin is 1.12 mm thick, the breast skin is 1.59 mm, epidermis is 28 and 56μ, stratum corneum is 16

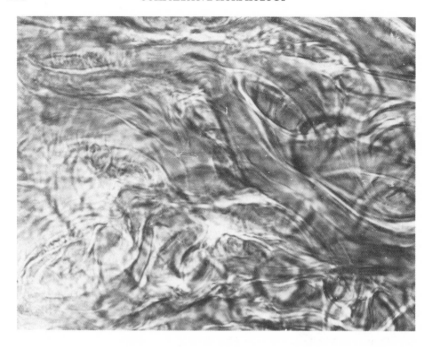

Fig. 59. *Aplodon rufa*. Network of collagen fiber bundles in reticular layer of withers skin.

and 34μ, dermis is 1.10 and 1.54 mm, papillary layer is 0.33 and 0.44 mm, reticular layer is 0.77 and 1.10 mm. The reticular layer of the dermis is much thicker than the papillary layer. In the dermis, horizontal bundles of collagen fibers dominate, although there are quite a few rising bundles (fig. 59). The collagen fibers form a dense plexus. Elastin fibers occur in small numbers in the papillary layer, are absent in the reticular layer, and form a dense plexus on the border with the subcutaneous muscles. No sebaceous glands are recorded in the skin of the withers, and in the breast skin they are small (11 x 56μ), oval, and elongated, with one at every hair tuft. An anal gland opens into the anus (Schaffer, 1940). The mouth-corner gland is formed of highly increased sebaceous glands; tubular glands are absent (Quay, 1965). Arrectores pilorum muscles are absent.

The hairs on the withers divide into four categories: guard, pile, intermediate, and fur. The guard hairs have a long, thin (about 22μ), medulla-free club which gradually expands into a shaft, thickest at three-quarters of its length. The hairs of this category subdivide into three size orders; length 17.9; 17.0 and 16 mm; thickness 99, 78 and 57μ; medulla 49.5, 48.9, and 57.9%, granna length 10.1, 6.5, and 5.4 mm. In the pile hairs the length is 14 mm, thickness 44μ, medulla 56.9%. The club is curved, and in some hairs the medulla thins before the granna (which is 4.1 mm long). Intermediate hairs are 12.3 mm long, 24μ thick, medulla 41.6%. They have a very small granna (length 2.4 mm). They are wavy (up to seven waves). The fur hairs are the same thickness (19μ) along their whole length (10.3 mm). Their shaft is wavy (up to seven waves) with medulla 47.3%. The hairs grow in tufts of seven to thirteen, some with and some without a guard hair. There are 333 guard and 7,666 other hairs per 1 cm².

FAMILIA SCIURIDAE

Sciurus vulgaris

Material. Skin from withers (table 19) of specimen (1) taken in June on the Sudzukchinski reservation (Soviet Far East), and specimens (2) and (3) taken in summer in the Moscow region.

In (1) and (3) the hair is growing, so the skin is thick. In (2) the hair is nongrowing. There are three skin thickness zones (Tserevitinov, 1958). The thickest skin is in the sacral area and on the head and neck. The second zone includes the whole back and borders the first zone on the side. The third zone covers the entire belly and part of the sides. The stratum malpighii of the epidermis is two to three cells deep. The epidermis is not pigmented. Due to the hair growth, the papillary layer of the dermis in specimens (1) and (3) is thicker than the reticular, while in (2) it is thinner. The collagen fiber bundles in the dermis are predominantly horizontal and form a compact plexus

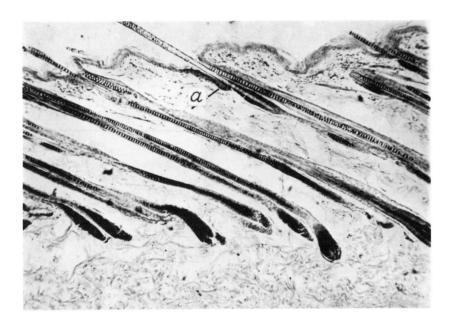

Fig. 60. *Sciurus vulgaris.* Withers skin: *a,* sebaceous gland.

TABLE 19

SCIURIDAE: SKIN MEASUREMENTS (μ)

	Skin	Epidermis	Stratum corneum	Dermis	Papillary layer	Reticular layer
Sciurus vulgaris						
Withers						
(1)	952	17	6	935	605	330
(2)	556	39	17	517	165	352
(3)	678	18	6	660	495	165
Breast						
(1)	792	22	6	770	220	550
(2)	550	22	6	528	418	110
(3)	902	22	10	880	605	275
Sciurus persicus						
Withers						
(1)	592	19	6	473	143	330
(2)	1,183	28	11	1,155	220	935
(3)	999	22	6	977	187	790
Breast						
(1)	666	28	11	638	528	110
(2)	853	28	14	825	165	660
Funambulus palmarum						
Withers						
(1)	650	45	24	605	110	495
(2)	727	45	20	682	132	550
Breast						
(1)	518	34	23	484	440	44
(2)	694	39	22	655	165	490
Tomeutes lacroides						
Withers	892	67	28	825	330	495
Breast	655	50	18	605	220	385
Atlantoxerus getulus						
Withers	968	—*	—*	968	88	880
Breast	898	28	14	870	100	770
Spermophilopsis leptodactylus						
Withers						
(1)	850	50	5	800	200	600
(2)	860	25	8	835	260	575
Marmota bobac						
Withers						
(1)	1,010	55	22	955	715	240
(2)	1,810	50	17	1,760	1,760	absent
(3)	2,160	50	17	2,110	1,450	660
Breast						
(1)	1,145	45	17	1,100	880	220
(2)	1,348	28	6	1,320	770	550
(3)	1,385	45	17	1,340	990	350

TABLE 19—*continued*

	Skin	Epidermis	Stratum corneum	Dermis	Papillary layer	Reticular layer
Marmota caudata						
(1)	1,630	90	23	880	550	330
(2)	1,466	36	*	1,430	770	660
Breast						
(1)*	832	62	28	550	440	110
Citellus undulatus stramineus						
Withers						
(1)	672	67	28	605	275	330
(2)	473	45	18	428	308	120
Breast						
(2)	658	28	8	630	330	300
C. u. leucostictus						
Withers						
(1)	457	39	17	418	187	231
(2)	540	45	17	495	275	220
Breast						
(2)	645	40	18	605	275	330

*The thickness of the subcutaneous fat tissue is indicated.

(fig. 60). In the papillary layer many bundles run in different directions. Elastin fibers occur in the papillary layer and in the deeper section of the reticular layer.

Sebaceous glands are small (22 x 56; 28 x 112μ) and elongated along the hair bursae. There are specific skin glands at the corners of the mouth consisting of multilobal (up to seven lobes) sebaceous glands 333 x 464μ. Beneath these glands are tubular glands whose ducts open into the skin surface near the lips (Skurat, 1972). Around the anal opening is an anal gland arranged as a ring 0.22 mm broad, formed by sebaceous glands 224 x 560, 448 x 504μ (Schaffer, 1940; Skurat, 1972). The ducts of the glands open into hair bursae found in the entrance of the anal opening. In the skin of the front and hind foot soles, sole glands are well developed. They consist of numerous glomeruli of tubular glands attaining the size 0.7 x 1.1; 1.6 x 4.4 mm (Schaffer, 1940; Skurat, 1972). In the preputial area are expanded sebaceous glands at the hair bursae (Schaffer, 1940). The thickness of arrectores pilorum muscles is 10 to 15μ.

Six hair length zones can be distinguished in the pelage of the squirrel (Tserevitinov, 1958). The longest hairs (over 25 mm) are recorded in the sacral area. The second zone (22 to 25 mm) takes in the buttocks and the middle of the back. The third (19 to 22 mm) zone is a narrow strip surrounding the second zone. The fourth zone (16 to 19 mm) is on the withers,

TABLE 20

SEASONAL CHANGES IN THE HAIR OF SCIURUS VULGARIS (MEAN VALUES) (KUZNETSOV, 1928)

	Length (mm)				Thickness (μ)			
	Guard	Pile	Inter-mediate	Fur	Guard	Pile	Inter-mediate	Fur
Back								
Summer	16.3	11.1	8.8	7.0	88.4	79.1	27.6	18.0
Winter	28.0	19.8	17.3	12.4	89.1	77.9	27.8	17.6
Side								
Summer	21.9	16.3	12.8	9.4	82.5	70.4	28.3	17.2
Winter	36.0	27.4	22.1	16.7	80.1	68.4	29.0	16.4
Belly								
Summer	—	16.4	—	10.1	—	78.7	—	17.3
Winter	—	20.8	—	15.7	—	77.0	—	16.5

sides, and the lower part of the belly. The fifth zone (12 to 16 mm) covers the front of the belly and approaches the head, and the sixth zone (12 mm) is the head. The hair is divided into guard, pile (three orders), and fur (two orders) (Tserevitinov, 1958). In the skin of the withers, the length and thickness of the hairs are as follows: guard, 29.1 mm and 71μ; pile, first order, 25.6 mm and 88μ; pile, second order, 21.2 mm and 80μ; pile, third order, 17.6 mm and 45μ; fur, first order, 14.7 mm and 34μ; fur, second order, 12.5 mm and 16μ. The hairs grow in two types of tufts (Kuznetsov, 1927, 1928); pile tufts, a single pile hair with four to five and occasionally six fur hairs around it; guard tufts, one guard hair with an intermediate hair on either side, two to three fur hairs between the guard and the intermediate hairs, and four fur hairs outside each of the latter (a total of 15 to 17 hairs).

The pelage of the squirrel varies sharply with the season. In winter the hairs are considerably longer, and the number is greater, but the thickness of the hair remains virtually unchanged (Kuznetsov, 1928) (table 20). In winter there are 8,000 to 10,000 hairs per 1 cm² of skin and 3,000 on the belly (Kuznetsov, 1952). Geographic variability of the fur is also very well expressed. Squirrels living in colder climates have longer, denser pelage (table 21). Vibrissae in squirrels grow in groups of three pairs and one single hair in various parts of the head, at the base of the forepaw, and on the belly (one to two pairs) (Kuznetsov, 1927, 1928).

Sciurus persicus

Material. Skin from withers, breast (table 19), and hind sole of three specimens taken in the Caucasus on May 22, in autumn, and on September 2.

TABLE 21

COMPARISON OF WINTER HAIR: MOSCOW REGION (S. v. VARIUS) AND EAST-SIBERIAN (S. v. CALOTUS) SQUIRRELS (MEAN VALUES) (KUZNETSOV, 1928)

	Hair length (mm)				Hair thickness (μ)				Number of hair groups per 10 mm² of skin
	Guard	Pile	Intermediate	Fur	Guard	Pile	Intermediate	Fur	
Back									
S. v. varius	28.0	19.8	17.3	12.4	89.1	77.9	27.8	17.6	89
S. v. calotus	31.8	25.6	21.5	16.4	87.2	77.8	29.5	18.3	104
Side									
S. v. varius	36.0	27.4	22.0	16.7	80.1	68.4	29.0	16.4	79
S. v. calotus	37.3	28.9	24.0	18.0	80.7	67.2	30.1	17.2	92
Belly									
S. v. varius	—	20.8	—	15.7	—	77.0	—	16.5	66
S. v. calotus	—	26.4	—	18.1	—	76.3	—	16.8	66

The hair is not growing on any of the three specimens. The stratum malpighii of the withers epidermis is two cells deep. The epidermis of specimens (1) and (3) is not pigmented; specimen (2) is lightly pigmented. The epidermis of the hind sole is thick, 605μ, and has a solid corneous layer, 330μ. The alveoli on the inner surface of the epidermis are 187μ high; there is a two- to four-row stratum granulosum, but no stratum lucidum. The epidermis is not pigmented. In the dermis on the withers, collagen fibers bundles are predominantly horizontal and their plexus is rather dense. There are few elastin fibers. The sebaceous glands are small (22×34; $28 \times 56\mu$), one at every hair. On the hind soles, tubular glands form large glomi. The diameter of the secretion section is 34μ, and the excretion ducts open onto the skin surface. Arrectores pilorum muscles are thin (10 to 17μ).

The hair divides into three categories: guard, pile, and fur. The guard hairs are 13.2 mm long, 89μ thick, and straight. They thicken gradually from the lower third, tapering only in the last quarter. The medulla is absent only from base and tip. The three orders of pile hairs have a weak granna in the upper third. The first order, 11.1 mm long and 75μ thick (length of the granna 6.5 mm), is straight. The upper part of the second and third orders (length 9.5 and 7.6 mm, thickness 44 and 27μ) is wavy and the granna is 4.9 and 3.0 mm long. The medulla is well developed along the whole shaft. The fur hairs (6.6 mm long and 18μ thick) are also wavy with up to five waves. The medulla is missing only at the base and tip. The hairs grow in tufts, usually of one big and four to six smaller hairs, though there are tufts of only three or four small hairs. The tufts do not form groups. There are 666 big and 7,000 small hairs per 1 cm².

Funambulus palmarum

Material. Skin from withers and breast (table 19) of two females, taken in Sri Lanka, June 28.

The hairs are not growing (though there are individual growing hairs). The stratum malpighii of the epidermis is two to three cells deep. The epidermis is not pigmented. In the papillary layer, bundles of collagen fibers run in several directions. In the reticular layer they are predominantly horizontal, though many bundles rise to the surface at various angles. The plexus of collagen bundles in the dermis is dense; the elastin fibers are few. In specimen (1) there is a thin layer of fat cells (one to three rows) 45×62 and $50 \times 84\mu$ in the under part of the dermis almost where the subcutaneous muscles begin. The sebaceous glands are small (22×34; $28 \times 50\mu$), and round, anterior to and one behind each hair (fig. 61). No arrectores pilorum muscles are recorded. All the withers hairs are alike in shape. They gradually narrow from the base to the upper half or third of the shaft and then thin toward the tip. The hairs are straight, and it is impossible to distinguish between them in terms of common categories, though they do divide into five orders for size (table 22).

Fig. 61. *Funambulus palmarum*. Withers skin: *a,* sebaceous glands.

The hairs grow in tufts of three, a big one in the middle with two smaller at the sides, though there are tufts of only two small hairs. There are 1,333 big and 6,333 small hairs per 1 cm².

Tomeutes lacroides

Material. Skin from withers and breast (table 19) of one specimen, taken in India, February 28.

The hairs are not growing, but the skin is thick. The epidermis is not pigmented. The reticular layer is thicker than the papillary. Bundles of collagen fibers in the greatest part of the dermis run toward the surface at various angles and, in the deeper part of the reticular layer, only parallel to it. The plexus of bundles is looser in the reticular layer. Sebaceous glands are not recorded. Large accumulations of cells occur near the hair roots, which could be reduced sebaceous glands.

TABLE 22

FUNAMBULUS PALMARUM: HAIR MEASUREMENTS

Orders	Length (mm)	Thickness (μ)	Medulla (% of hair thickness)
1st	12.5	59	89.9
2d	9.6	63	84.2
3d	8.8	61	90.2
4th	7.7	53	88.9
5th	4.8	262	73.1

The withers hair divides into three categories. The guard hairs, which are straight, thicken gradually upward from the base to a maximum at two-thirds up the shaft. They are 16.2 mm long, 73μ thick, medulla 85%. There are few hairs of this category. The pile hair is similar except it bends in the lower shaft (which is weakest in the first order and strongest in the third). The length of pile hairs (first order) is 15.1 mm, thickness 67μ, medulla 83.6%; (second order) 11.7 mm, 45μ, 84.5%; and (third order) 11.6 mm, 36μ, 72.3%. The thickening of the shaft is more pronounced. The fur hairs (length 9.7 mm, thickness 29μ, medulla 75.9%) are wavy down the whole length, with up to seven waves. The upper part of the shaft is slightly thickened and straight. The hairs grow in tufts of mostly three hairs. There are 9,666 hairs per 1 cm^2 on the withers and 5,000 on the breast.

Atlantoxerus getulus

Material. Skin from withers and breast (table 19) of a female specimen taken in Morocco.

The skin is thick, with large folds on the outer surface and a thick epidermis. The stratum malpighii is four to seven cells deep, and the whole epidermis is densely pigmented (especially its basal layer). On the hind sole around the pad of the foot, the epidermis becomes very thick (400μ). The stratum corneum is very compact (231μ). On the inner side, the epidermis has high (up to 132μ) alveoli; the stratum granulosum is pronounced; the stratum basale of the epidermis is pigmented.

In the dermis of the body skin the reticular layer is much thicker than the papillary. The bundles of collagen fibers along the whole dermis (except deep in the reticular layer) rise to the surface at various angles. In the lower reticular layer the collagen bundles are horizontal, while elastin fibers occur in small numbers only in the dermal papillary layer. Adjoining the subcutaneous muscles is an almost continuous layer (five to eight rows) of fat cells, 34 x 45; 45 x 50μ. It is interesting that pigment cells are numerous in the upper and the middle papillary layers. The sebaceous glands are round

(39 x 56; 34 x 73μ) and fairly numerous, with two at every hair, one posterior, one anterior. In the hind sole dermis are small glomeruli of tubular glands surrounded by fat cells, the diameter of the secretion cavity being 28μ. The ducts open into the skin surface. Arrectores pilorum muscles are not very large and are 11μ thick.

Spermophilopsis leptodactylus

Material. Skin from withers (table 19) of two females (body length 23 and 22 cm) taken in Repetek, Turkmenistan, in May.

The epidermis is pigmented equally on back, belly, and sides. There is little pigment in the hind sole epidermis, and the scrotum skin is most heavily pigmented (Sokolov, 1964). The glands in the corners of the mouth of *S. leptodactylus* cannot be seen on the outside. As in *Sciurus vulgaris*, this gland is formed by superficial sebaceous glands (302 x 406μ) and deep-lying glomeruli of tubular glands (glomeruli size 627 x 884μ) (Skurat, 1972). The anal gland is a thin girdle around the orifice of the anus, formed by sebaceous glands of up to 403 x 569μ (Skurat, 1972). The sole glands formed by glomeruli 190 x 460μ (Skurat, 1972) of eccrine sweat glands are in the foot callosities. The number of the glomeruli in the callosities varies from fourteen to twenty.

No categories are found in the summer withers pelage, which is in three size orders (table 23). All are highly flattened, with a short, nonmedullary club thickening sharply toward the upper half of the shaft (in the third order, toward the lower third).

TABLE 23

SPERMOPHILOPSIS LEPTODACTYLUS: HAIR MEASUREMENTS

Orders	Length (mm)	Thickness (μ)	Medulla (% of hair thickness)
1st	7.6	123	92
2d	4.8	109	90
3d	2.5	43	84

A different subdivision of *S. leptodactylus* summer fur has been proposed (table 24). Lavrov and Naumov (1934) have distinguished guard, pile, intermediate first order, and fur hairs.

In winter, the hairs divide into three categories: guard, pile, and fur (two orders). Guard hairs are straight and gradually thicken toward the upper two-thirds of the shaft (17.8 mm long, 94μ thick, medulla 88%); they have a long (1.5 mm), medulla-free tip.

The pile hairs are similar to the guard hairs, but have a more pronounced granna. The fur hairs have no granna. The pile hairs are 13.1 mm long, 104μ thick, 90% medulla; the fur hairs (no medulla) (first order) 10 mm and 21μ; (second order) 9.4 mm and 16μ.

The hairs grow in tufts and groups (Lavrov and Naumov, 1934). In winter on the sacrum are groups of one central pile and ten fur hairs, in two lots of five (sometimes three or six) on either side of the pile hair. On the belly, the number of hairs in a group ranges from six to twelve. In summer the largest tufts on the sacrum have nine hairs; most have seven. The largest tufts on the belly have five, more often two or three.

In winter there are 4,436 fur and 528 other hairs per 1 cm² on the middle of the back, and 3,548 fur and 4,300 others on the middle of the belly. The figures for summer are 1,032, 868, 344, and 276.

S. leptodactylus has much hair on the front soles, even more on the hind soles, and a hair lining on the edge of the front and hind toes and the outer edge of the foot.

Marmota bobac

Material. Skin from withers and breast (table 19) from three specimens taken in the Starobelsk Steppes, Ukraine, June.

The hair is growing. The stratum malpighii of the body epidermis is three to four cells deep. There are large melanocytes in the basal layer of the epidermis. The hind sole epidermis attains great thickness, 750 to 880μ; the stratum granulosum, which is two to three cells deep, is well expressed; the stratum corneum and the stratum lucidum are 455 to 605μ. The alveoli on the inner side are rather far apart, 210 to 275μ deep; the stratum basale is pigmented.

Owing to the pelage growth on the body, the reticular layer of specimen (2) is absent (the hair bulbs adhere directly to the subcutaneous fat tissue); the reticular layer of specimen (1) is thinner than the papillary. Only specimen (3) has equal dermal layers. The dermal collagen fiber bundles are mostly horizontal and form a dense plexus.

Elastin fibers are few. Under the subcutaneous fat tissue, next to the subcutaneous muscles, is a horizontal bundle of elastin fibers. In the reticular layer, specimens (1) and (3), groups of fat cells extend down from the hair bulbs to the subcutaneous fat tissue. The subcutaneous fat tissue, well developed in the withers skin, is 0.88, 0.66, and 0.66 mm thick, respectively, in specimens (1), (2), and (3), with only rare bundles of collagen fibers. The fat cells in the subcutaneous tissue are 55 x 110; 77 x 121 in (1), and 34 x 56; 28 x 56 in (3). The sebaceous glands are oval or round, and small (28 x 56; 56 x 84), one behind each hair.

In the corners of the mouth, a specific gland 2.2 mm long, 0.4 mm thick, is formed of sebaceous glands reaching 243 x 373μ (Skurat, 1972). The

TABLE 24

SUMMER HAIR OF SPERMOPHILOPSIS LEPTODACTYLUS (LAVROV AND NAUMOV, 1934)

	Guard		Pile		Intermediate 1st order		Intermediate 2d order		Fur	
	Length (mm)	Thickness (μ)	Length (mm)	Thickness (μ)	Length (mm)	Thickness (μ)	Length (mm)	Thickness (μ)	Length (mm)	Thickness (μ)
Middle of back	8.5	150	5	103	2	36	1.4	21	1	10
Middle of belly	7.5	90	6.5	78	3.5	58	0.9	18	Absent	Absent

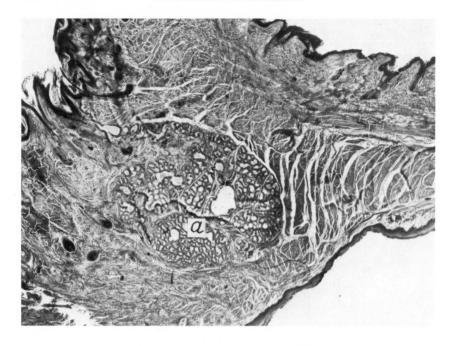

Fig. 62. *Marmota bobac.* Complex of sweat glands: *a,* corner of mouth.

ducts are connected with the hair bursae, opening into the mouth cavity. There is a complex of tubular glands under the hair bulbs (fig. 62), with individual segments of 0.33 x 0.36, 0.26 x 0.84 in size. The anal glands form three pouches on the dorsal and lateral sides of the orifice (Schaffer, 1940; Ortman, 1960; Skurat, 1972). The duct of the central pouch leads to the gland cavity, which is 10 mm long and 10 mm at the basal cross-section. The lateral pouches are smaller. Their walls are made up of glandular tissue 0.67 mm thick formed by alveolar glands. Under the sebaceous gland layer are a few large tubular glands with glomi reaching 0.67 x 1.79 mm. A similar structure is found in the anal glands of *Marmota marmota* (Kratochvil, 1968). In the hind sole dermis, large glomi of eccrine sweat glands (119 x 309, 224 x 717μ) lie surrounded by fat cells. The diameter of the glandular secretion cavities ranges from 45 to 65μ. Well-developed arrectores pilorum muscles are 33 to 55μ thick.

Marmota caudata

Material. Skin from withers, breast (table 19), and hind sole of two-year-old male (1) taken in Western Tien-Shan, August 19, and adult female (2) taken in Central Pamir, June 18.

The hair of neither animal is growing, but the skin is rather thick. The stratum corneum of the epidermis is compact, the well-pronounced stratum granulosum is one to two cells deep. The female's epidermis is pigmented, and the hind sole epidermis is thick (0.60 to 1.10 mm). The stratum granulosum is two to three cells deep; the stratum corneum is well developed (440μ); the inner surface of the epidermis is alveolar. The alveoli are from 110 to 440μ high; the epidermis is lightly pigmented. In the dermis of the body skin horizontal bundles of collagen fibers predominate. Their plexus is rather loose. The elastin fibers, few in number, lie horizontally in the upper and middle papillary layers. In specimen (2) are small groups of fat cells (28 x 56; 39 x 56μ) in the lower reticular layer and near the hair bulbs. Only specimen (1) has fat cells in the subcutaneous tissue, which is 660μ on the withers, and 220μ on the breast. The sebaceous glands are small (28 x 78; 39 x 50μ), one at each hair. Sole glands are made up of large eccrine sweat gland glomi (diameter of the secretion section, 45 to 50μ). Arrector pili muscles are thin (22 to 44μ). The withers hairs divide into guard, pile (three orders), and fur. The guard hairs are straight, thicken somewhat at the base, and remain the same thickness for most of the length, tapering only toward the tip. The pile hairs gradually thicken from the base to the upper two-thirds of the shaft and then taper toward the tip. The pile hairs (first order) are slightly wavy, the second and third orders are curly (seven to eight waves). The fur hairs are of uniform thickness the whole length. Guard hairs are 44.2 mm long, 137μ thick, medulla 87.3%; pile hairs (first order) 38.6 mm, 80μ, and 73.5%; (second order) 33.2 mm, 52μ, and 52.0%; and (third order) 30.1 mm, 46μ, and 58.7%; fur hairs are 11.8 mm, 18μ and 55.6%. The hairs grow in tufts, one guard hair in the middle with eleven to eighteen other hairs at the sides. There are 250 guard and 3,083 other hairs per 1 cm^2.

Citellus undulatus stramineus

Material. Skin from withers, breast (table 20), and hind sole of two specimens taken in Jungar Ala-Tau, June.

The hairs are in the earliest stage of growth. The stratum malpighii is two to four cells deep. The epidermis is strongly pigmented in the withers and lightly in the skin of the breast. The hind sole has a thick (330μ) epidermis; the stratum granulosum is two to three cell rows deep; the stratum lucidum is well expressed; the stratum lucidum and stratum corneum are 132μ thick; the inner surface of the epidermis is alveolar, the alveoli being 110μ high; there is no pigment in the epidermis. In the dermis of the body skin, horizontal bundles of collagen fibers predominate and form rather a compact plexus. In the upper papillary layer they run in several directions. In the middle of the withers reticular layer are one to two layers of fat cells (56 x 60; 50 x 67μ). The sebaceous glands are small (22 x 56; 34 x 56μ), ball-shaped or oval, one at each tuft, two at each guard hair. The specific glands

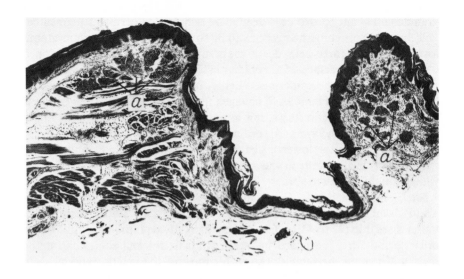

Fig. 63. *Citellus undulatus stramineus.* Skin of hind foot sole: *a,* sweat gland glomi.

in the corners of the mouth cannot be seen on the surface. As in the other members of the family they are formed of large, superficial sebaceous glands (up to 347 x 437μ) lying deeper than the sweat glands (the size of their glomi is up to 0.78 x 1.34 mm, Skurat, 1972). There are three pockets of anal glands (the largest on the dorsal and the two others on the lateral side of the orifice, Skurat, 1972). In the walls of the pockets is a layer of sebaceous glands 1.0 x 1.3 mm with tubular glands underneath. Similar anal glands are found in *Citellus richardsoni* (Schaffer, 1940).

The hind foot dermis has large eccrine sweat gland glomi (fig. 63) with a secretion cavity diameter of 34 to 39μ. Arrectores pilorum muscles are thin (15 to 28μ). The withers hair divides into guard, pile (four orders), and fur. The guard hairs (there are not many) gradually thicken toward the upper two-thirds of the shaft and then taper to the tip. The pile hairs have a slight granna. The pile hairs (first, second, third, and fourth orders, all of which are thickest about two-thirds up the shaft) are 10.6; 8.1; 7.7; 5.9 mm long. The fur hairs are curly (up to five waves). The guard hairs are 1.68 mm long, 94μ thick, medulla 87.3%; the pile hairs (first order) 11.3 mm, 117μ, and 90.6%, (second order) 12.5 mm, 110μ, and 86.4%, (third order) 12.1 mm, 116μ, and 83.7%, (fourth order) 10.0 mm, 80μ, and 86.3%; fur hairs 8.2 mm and 14μ. The hairs grow in tufts of one guard and ten to fifteen other hairs. Sometimes the pile and fur hairs are arranged in two groups on either side of the guard hair. There are 666 guard and 8,000 pile and fur hairs per 1 cm^2.

Citellus undulatus leucostictus

Material. Skin from withers and breast (table 19) of two specimens taken on the Anadyr ridge.

The hair on the withers is not growing, and on the breast it is in the initial stage of growth, so the skin is thinner on the withers than the breast. The stratum malpighii is three cells deep. The epidermis is lightly pigmented (less than in *C. u. stramineus*). In the dermis, horizontal bundles of collagen fibers predominate and form a dense plexus. There are few elastin fibers. Fat cells occur in the reticular layer.

The sebaceous glands are small (17 x 45; 28 x 75μ). There are large glomi of eccrine sweat glands in the hind sole dermis. The diameter of their secretion cavity is 34 to 39μ. The arrectores pilorum muscles are 10 to 12μ thick.

FAMILIA PTEROMYIDAE

Pteromys volans

Material. Skin from withers, breast, and flying membrane of one specimen.

The withers and breast skin are 763μ and 317μ thick, respectively, the epidermis 28 and 22μ, stratum corneum 14 and 10μ, dermis 735 and 295μ, papillary layer 715 and 185μ, reticular layer 20 and 110μ. Hair growth has caused the withers skin to thicken. The stratum malpighii is two to three cells deep. The outer dermal layer, which is about 330μ thick on the withers, is formed by bundles of collagen fibers lying in several directions to form a dense plexus. The remaining part of the papillary layer is filled with fat cells 28 x 62; 34 x 73μ. In the dermis of the breast most of the collagen bundles are parallel to the surface and there are no fat cells. The sebaceous glands (22 x 62; 28 x 75μ) are 100 to 120μ from the outer surface and singly posterior to each hair. The arrectores pilorum muscles, 10 to 12μ thick, lie in the upper half of the papillary layer.

All the hairs belong to the same category, but four size orders are distinguished. The hairs are straight, thicken from the base, remain the same thickness most of the way, and then taper very gradually toward the tip. The first and second orders have very long, very thin nonmedulla tips, which are much shorter in the third and fourth orders. The first, second, third, and fourth orders are 16.8, 15.5, 13.5, and 11.3 mm long, respectively, 50, 60, 25 and 20μ thick; medulla 76.0, 93.4, 76.0, and 70.0%. There are many growing hairs of varying length in the skin. The hairs grow in tufts of five to seven. There are 22,000 hairs per 1 cm² on the withers and 7,333 hairs per 1 cm² on the breast. The skin of the flying membrane, which has many characteristic outer folds, is 220 to 600μ thick. There are bundles of cross-striated muscles in the middle of the flying membrane.

Familia Geomyidae

Geomys bulbivora

Material. Skin from withers and breast (table 25).

The hair is growing in the withers but not on the breast. Despite this, the breast skin is considerably thicker than the withers skin. Long roots of growing hairs with a well-developed medulla pierce the whole dermis of the withers skin. Their large bulbs adhere to the subcutaneous muscles, resulting in the absence of the reticular layer in the withers skin. In the dermis, a compact horizontal plexus of collagen fibers is in the upper 275μ. The whole thickness of the dermis below this level is filled with an almost continuous mass of fat cells, up to 39×62; $34 \times 73\mu$. The dermis of the breast has a regular, dense plexus of collagen fibers through almost its whole thickness. Only in the inner part of the reticular layer is the plexus loose. The collagen fiber bundles are mostly horizontal. No sebaceous glands are recorded, although ducts are visible in many hair tufts. In members of *Geomyidae* family (*Geomys bursarius* and *Thomomys umbrinus*) the gland in the corner of the mouth is formed of poorly developed sebaceous glands and large mucous glands (Quay, 1965). Arrectores pilorum muscles are absent.

The hairs divide into three unusual categories. The hairs that begin to thicken from the base to a maximum three-fourths of the way up belong to the first category. They have a short tip and a well-developed medulla that is absent only from the base and tip. There are two size orders in this category, length 10.1 and 9.6, thickness 57 and 52μ, medulla 89.5 and 88.5%. The second category (length 8.4, thickness 57 and 52μ, medulla 89.5 and 82.9%) thickens only very slightly, and the third category (6.7 mm long, 22μ thick, medulla 72.9%) is wavy, thickens toward the middle of the shaft, and has a long tip without a medulla. The hairs grow in tufts of ten to twelve. There are 8,166 hairs per 1 cm². The hairs do not form groups.

The breast has similar categories to the withers. The first category (first order) is 9.1 mm long, 71μ thick, medulla 91.6%; (second order) 8.7 mm, 57μ, 89.5%; the second category is 7.9 mm, 35μ, 82.9%, and the third category is 6.8 mm, 22μ, 86.4%. Long bases without medulla are characteristic of all the hairs. They grow in tufts of ten to fourteen. A count of 4,000 hairs is recorded per 1 cm².

G. floridanus

Material. Skin from withers and breast (table 25).

The hair is not growing. The outer surface of the skin has large folds, and the breast skin is much thicker than the withers skin. The withers papillary

TABLE 25

GEOMYIDAE AND HETEROMYIDAE: SKIN MEASUREMENTS (μ)

	Skin	Epidermis	Stratum corneum	Dermis	Papillary layer	Reticular layer
Geomys bulbivora						
Withers	600	—*	—*	600	600	—
Breast	2,230	30	17	2,200	330	1,870
Geomys floridanus						
Withers	476	56	34	420	330	90
Breast	2,230	30	17	2,200	220	1,980
Perognathus bailey						
Withers						
(1)	484	22	16	462	440	22
(2)	237	17	11	220	110	110
Breast						
(1)	194	17	8	177	100	77
Dipodomys merriami						
Withers	198	—*	—*	198	121	77
Dipodomys deserti						
Withers						
(1)	413	28	17	385	220	165
(2)	737	22	11	715	715	—
Breast						
(1)	275	22	11	253	143	110
(2)	374	22	11	352	220	132
Liomys pictus						
Withers						
(1)	446	28	17	418	165	253
(2)	1,128	28	12	1,100	1,100	—
Breast						
(1)	468	28	17	440	385	55
(2)	578	28	14	550	550	—
Heteromys desmarestianus desmarestianus						
Withers	1,122	22	11	1,100	1,100	—

*Badly preserved.

layer of the dermis is considerably thicker than the reticular layer, but this is reversed in the breast skin. In the dermis, horizontal bundles of collagen fibers predominate and form a loose plexus. There are few elastin fibers except for a thick (28μ) bundle parallel to the surface at the border with the subcutaneous muscles. There is an almost continuous layer of fat cells (28 × 45; 28 × 32μ) under the hair bulbs. The sebaceous glands are somewhat oval and large (fig. 67) (28 × 84; 34 × 67μ), at every hair tuft. Arrectores pilorum muscles are not recorded. The withers hairs divide into three categories, which do not correspond to the common categories. The first (length 8.7 mm, thickness 112μ, medulla 94.7%) has a thin, long (0.55 mm) club, thickening rather sharply to the shaft, then gradually, for three-quarters of the length, tapering to the tip. The tip has no medulla and is short. In the second category the club is shorter (1.6 mm) and thicker, with no sharp thickening. These subdivide into two orders: length 7.4 and 7.0 mm, thickness 86 and 81.0μ, medulla 87.3 and 81.0%. The third category (length 6.4 mm, thickness 39μ, medulla 77.0%) is slightly wavy. The thickening in the shaft is slight, and in the fourth category it is virtually absent. The fourth category is 4.7 mm long, 25μ thick, medulla 88.0%. The hairs grow in tufts of four to six. There are 7,666 hairs per 1 cm² on the withers.

FAMILIA HETEROMYIDAE

Perognathus bailey

Material. Skin from withers (table 25) of two adult specimens.

In specimen (1) the withers pelage is growing, so the skin is thicker than in specimen (2), where it is not. The stratum malpighii is one cell deep. In the dermis, horizontal bundles predominate and form a rather solid plexus. Specimen (1) has small groups of fat cells (17 × 28, 22 × 28μ) in the inner papillary layer near the hair bulbs. Specimen (2) has not. The elastin fibers are small, a horizontal bundle lying on the border with the subcutaneous muscles.

The sebaceous glands are small (17 × 45; 39 × 45μ), heart- or oval-shaped, one or two at each hair, most of them nonfunctioning.

In the genus *Perognathus* a caudal specific gland, comprised of numerous, multilobed sebaceous glands (Quay, 1965a, b), is one-fourth to halfway down the ventral surface of the tail. The mouth-corner gland (in *P. ampulus, P. parvus, P. hispidus, P. artus, P. californicus*) is formed of large sebaceous glands, largest in *P. ampulus* and smallest in *P. californicus* (Quay, 1965). Arrectores pilorum muscles are 10μ thick. The hairs divide into three unpronounced categories (it might be more accurate to treat them as orders). The first-category hairs (length 11.8 mm, medulla 96.2%) have a short (about 0.3 mm) club above which the shaft gradually thickens to a maximum of 70μ in the bottom third. They gradually thin, ending in a very long (up to 4 mm), thin, medulla-free tip. The second category differs from the first only in size (length 8.3 mm, thickness 60μ, medulla 95%), and in a shorter tip (2 to 3 mm long). In the third category (length 7.4 mm, thickness 57μ, medulla 91.3%) the tip is very short. The hairs grow in tufts of four. A count of 7,666 hairs is recorded per 1 cm^2.

Dipodomys merriami

Material. Skin from the withers (table 25) of adult taken in New Mexico, October 11.

The stratum malpighii is two to three cells deep. The plexus of the collagen fiber in the dermis is dense and the bundles tend to be horizontal. The sebaceous glands are small (17 $\times$ 22μ). The mouth corner gland is formed of large sebaceous and mucous glands (Quay, 1965). The arrectores pilorum muscles are thin (6μ). The hair, not growing, divides into two categories. The first category, 21.2 mm long, with 88.3% medulla, has a short (about 150μ) club above which it gradually thickens to reach a maximum of 51μ about three-fourths up. The hair has a very long (up to 5.5 mm) medulla-free tip. The second category of hair is rather different. It thickens from the base, reaching its maximum a fifth of the way up the shaft, then gradually tapers toward the tip. There are two size orders, 15.0 and 14.5 mm long, both 42μ thick, medulla 85.8 and 92.9%. The tip is short; 0.8 to 0.9 mm (first order) and 0.6 to 0.6 mm (second order). The hair grows in tufts of six to seven. There are 16,670 hairs per 1 cm^2.

D. deserti

Material. Skin from withers, breast, and hind soles (table 25) of two specimens taken in June.

In specimen (1) withers, the hair is not growing and the skin is thinner than in specimen (2), with growing hair. The stratum malpighii of the epidermis is two to three cells deep. The hind sole sample is fur-covered; the epidermis is rather thin; there are no alveoli on its inner aspect. The stratum corneum is thick (56μ), but not very firm. The stratum granulosum and stratum lucidum are not recorded; the epidermis is not pigmented. The bulbs of growing hairs on the body skin of (2) adhere to the subcutaneous muscles, for there is no reticular layer. The collagen fiber bundles in the dermis are compact and predominantly horizontal. There are few elastin fibers. Specimen (1) has no fat cells in the dermis, but in specimen (2) fat cells (50 $\times$ 62; 45 $\times$ 67μ) fill out the inner layers of the skin.

The sebaceous glands are small (28 $\times$ 50; 39 $\times$ 56μ), oval or bean-shaped; in the cross section two glands are seen, one anterior, one posterior to each tuft of hair. An oval glandular formation (diameter 4.4 mm) is found in the withers skin of both specimens. Near the center of the gland the skin thickens to 1.2 mm. Six to eight sebaceous glands are visible in a cross section of this glandular formation (specimen 1) 0.56 $\times$ 0.71 to 0.49 $\times$ 0.83 mm, and (specimen 2) 0.77 x 1.2 to 0.66 x 0.80 mm. Each of the glands has twenty to twenty-four spherical lobes. There is no intermediate between these and the common sebaceous glands, which have smaller glandular cells and very broad ducts that do not open into a hair bursa but onto the skin surface. The duct diameters are 550 to 660μ. The data obtained are consistent with the description by Quay (1954).

On the hind sole are two sebaceous glands at every tuft of hair. The sebaceous glands are larger (35 x 70; 44 x 65μ) than on the body. The tubular glands are not recorded in the samples. The mouth corner gland in *D. deserti, D. ordii, D. microps, D. panamintinus, D. elephantinus, D. heermanni,* and *D. nitratoides* is formed of medium mucous glands and large sebaceous glands, the largest being in *D. microps* (Quay, 1968). Arrectores pilorum muscles are thin (5 to 12μ). The withers hairs divide into two categories: first, 22.0 mm (medulla 94.6%) with a short club that gradually thickens to 55μ a third from the proximal, then tapers towards the distal into a long (3.5 to 4.4 mm), thin, medulla-free tip. Both orders of the second category reach maximum thickness one-third up from the proximal and taper toward the tip, but their tip is considerably shorter (first order 0.8 mm, second order 0.5 mm). Their overall length is 19.2 and 15.2 mm, thickness 49 and 38μ, medulla 93.9 and 92.2%, respectively. The hairs grow in tufts of four to seven, 8,000 hairs per 1 cm².

Liomys pictus

Material. Skin from withers and breast (table 25) of two specimens taken June 24.

The hair of specimen (1) is not growing, so the skin is much thinner than in specimen (2) with growing hair. The outer skin surface is plicate. The stratum malpighii is two to three cells deep. No reticular layer is recorded for (2). Hair bulbs adhere to the subcutaneous muscles, and large hair roots occupy the whole dermis. Only the upper 0.44 mm has a rather compact plexus of collagen bundles. Deeper, the skin is filled out with fat cells (39 x 56; 39 x 62μ). The papillary layer is thinner than the reticular in specimen (1). In the dermis, horizontal bundles of collagen fibers predominate. A network of large elastin fibers, mostly horizontal in arrangement, covers the whole papillary layer almost evenly. Deep in the reticular layer are two to three layers of fat cells (17 x 34; 22 x 28μ).

The sebaceous glands (22 x 56; 39 x 112μ) are oval, one at each hair, with thin arrectores pilorum muscles (10 to 22μ). The withers hair (three categories) is rather odd. The first category comprises straight, flattened, bristle hairs (13.8 mm long), with a short, medulla-free club (300 to 400μ long) that thickens sharply to the hair shaft. The maximum thickness (346μ) is recorded in the proximal third, after which there is a gradual tapering toward a long, whiplike tip which is not pointed. In cross-section it is convex-concave, convex side forward. The hair bulges slightly at both ends, with a well-developed medulla, while the middle section is a thin strand. The second-category hairs, which are short-tipped, thicken gradually to a maximum two-thirds to three-quarters from the proximal. The two size orders are 11.8 and 9.2 mm long, 63 and 36μ thick, medulla 82.6 and 88.9%, respectively. Third-category hairs, 7.3 mm long, medulla 84.9%, are thickest in the proximal section and have a short tip. The hairs grow in groups; a first-category hair in the middle with a tuft with five to seven other hairs on either side. There are 583 first-category and 5,500 other hairs per 1 cm².

Heteromys desmarestianus

Material. Skin from withers, one specimen (table 25).

The hairs are growing and the skin is rather thick, with a two-cell layer stratum malpighii. Extremely broad bursae of growing hairs (up to 300μ) run through the whole dermis. There is no reticular layer, and the bulbs of the hairs adhere to the subcutaneous muscles. Collagen bundles in the dermis are mainly arranged along the hair roots. Small groups of fat cells (17 x 28; 17 x 22μ) occur in the middle and inner dermis. There is a small sebaceous gland (22 x 67μ), nonfunctioning at every hair. The 17μ arrectores pilorum muscles are found only in the upper dermis. The hairs divide into three categories. The first category has flattened bristles with a short (660μ), thick (130μ), medulla-free club, with an immediate, sharp thickening that gradually thins into a long (2.2 mm), thin, medulla-free tip. In cross section these hairs are similar to those of *Liomys pictus,* with a convex anterior and concave posterior. The anterior septum of the hair is thicker than the posterior. The hair bulges on the edges, the medulla being thicker there than in the middle.

The first category divides into two size orders: length 11.6 and 12.4 mm, and thickness 348 and 190μ. The second-category hairs have an even thickness with a small bulge, two-thirds toward the distal (first order) and three-quarters (second order), and a short tip, both orders being 39μ thick and 11.4 and 9.9 mm long, respectively, medulla 84.7 and 74.4%. The third-category hairs are thickest in the proximal third of the shaft, 8.2 mm long, 29μ thick. The 79.4% medulla is not continuous in many of these hairs. First-category hairs grow singly, the other hairs grow in tufts of six to nine. There are 1,000 first-category and 4,666 other hairs per 1 cm².

FAMILIA CASTORIDAE

Castor fiber

Material. Skin from withers and breast (table 26) of two specimens (body length 75 cm and 65 cm) taken in the Voronezh Reservation, November 28 and December 18.

The hairs are growing. The outer skin has many tiny folds. A loose

TABLE 26

CASTOR FIBER: SKIN MEASUREMENTS

	Skin (mm)	Epidermis (μ)	Stratum corneum (μ)	Dermis (mm)	Papillary layer (mm)	Reticular layer (mm)
Withers						
(1)	8.76	78	67	8.69	1.65	7.04
(2)	6.18	90	77	6.09	1.87	4.22
Breast						
(1)	5.02	72	61	4.95	1.76	3.19
(2)	4.36	78	66	4.29	2.09	2.20

stratum corneum occupies most of the epidermis. The stratum malpighii is only two cells deep, and the reticular layer is thick. There is a compact plexus of collagen fiber bundles in the dermis arranged in various directions. Small groups of fat cells (28×33, $33 \times 39\mu$) occur in the inner reticular layer near the blood vessels.

The ventral dermis has a peculiar arrangement of smooth muscle bundles, unconnected with hairs (a peculiarity not previously noted). They run parallel to the surface throughout the papillary layer, 50 to 112μ from the inner aspect of the epidermis. The diameter of the muscle bundles reaches 30μ.

The large (45×112; $84 \times 140\mu$) one- to three-lobed sebaceous glands are active (fig. 64), one anterior, one posterior to every hair tufts. The 1 to 3 mm thick septa of the anal glands (paired sacs, 6.6 to 8 cm long and 2.1 to 3 cm wide) contain multilobed, large sebaceous glands (Schaffer, 1940). Sacciform preputial glands are well developed, measuring 8.26 to $12 \times 2.75 \times 0.85$ to 1.48 cm (Schaffer, 1940). Kattsnelson and Orlova (1956) demonstrated that the so-called preputial gland is actually a nonglandular skin sac lined with a typical skin epithelium. Due to cornification of the epithelium, numerous strata corneum are formed, the sac stretching with accumulation of the layers. Acting as capillaries, the slits between the strata corneum absorb the urine from the urogenital sinus into which the preputial sac opens. The urine gradually penetrates the inner areas of the gland differentiating the corneous plates, loosening and macerating them. The salts precipitating from the urine form accumulations of small granules. An analysis of the preputial sac content shows that it contains about 35% of Ca salts. There are no arrector pili muscles.

The back hair divides into guard, pile (three orders), and fur hairs, with fur and four orders of pile on the ventral side (Tserevitinov, 1951). On the back the guard hairs are 66 mm long, 194μ thick; pile hairs (first order) 58.2 mm and 194μ (second order) 40.7 mm and 185μ (third order) 31.5 mm and 101μ; fur hairs, 23.7 mm and 16μ. On the ventral side, pile hairs (first

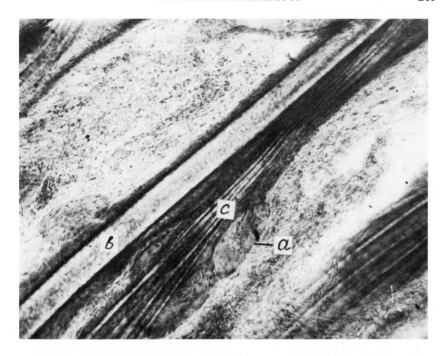

Fig. 64. *Castor fiber.* Outer layer of withers skin: *a,* sebaceous gland; *b,* pile hair; *c,* fur hairs.

order) 34.6 mm, 121μ; (second order) 20.0 mm and 129μ; (third order) 22.9 mm and 96μ; (fourth order) 20.4 mm and 65μ; fur hairs, 16.4 mm and 11μ. the Belly is covered with a denser fur, 34,180 hairs per 1 cm^2, and the back 30,980 (Tserevitinov, 1958).

A comparison of the hair of ten adult beavers and six underyearlings taken in December-February in Byelorussia indicates that in the latter the hairs are invariably shorter and thinner (Zamakhaeva, 1972*a, b*). (table 27).

Hair density of adult beavers is much higher than that of an underyearling (Zamakhaeva, 1972) (table 28).

Familia Cricetidae

Cricetulus migratorius

Material. Skin from withers (table 29) and hind sole of three specimens: (1) (body length 9 cm) taken in Tashauz region, Turkmen SSR, April 27; (2) (body length 10 cm) taken in the Caucasus, May; (3) (body length 10 cm) taken in Azerbaijan, February 17.

The hair of the spring specimens is not growing. The outer surface of their skin is wrinkled. The hairless epidermis from the hind sole is up to 140μ thick, with a clear two- to three-cell layer stratum granulosum, and a

TABLE 27

HAIR GROUPS AND CATEGORIES (M ± m)

Age	Hair groups and categories (M ± m)					
	Guard	Pile 1st order	Pile 2d order	Pile 3d order	Inter-mediate	Fur
	Hair length on middle of back (mm)					
Adults	56.1±0.9	43.6±1.4	36.4±0.56	28.5±0.02	25.4±0.51	19.7±0.29
Under-yearlings	55.6±1.2	46.8±0.5	38.5±0.77	29.6±0.5	27.0±0.16	20.5±0.67
	Hair thickness on middle of back (μ)					
Adults	199	176±1	137±1	85±2	24	15
Under-yearlings	145±4	153±3	116±1	58±4	22±1	12
	Hair length on belly (mm)					
Adults	31.3±0.16	26.5±0.14	24.0±0.23	20.9±0.25	18.9±0.11	16.8±0.11
Under-yearlings	29.6±0.76	23.3±0.16	22.8±0.36	21.0±0.21	19.7±0.17	16.8±0.19
	Hair thickness on belly (μ)					
Adults	142±1	123±1	95	60±2	18	12
Under-yearlings	92	92±1	72±1	47±1	19	10

thick (73μ), compact stratum corneum. Alveoli are present on the inner aspect of the epidermis, 39μ in height. In the dermis, collagen bundles, mainly horizontally arranged, form a compact plexus, with elastin fibers abundant in the papillary layer. The sebaceous glands are large (34×56; $56 \times 84\mu$), unilobed, oval, one at every tuft of hair, their lower extremity almost reaching the hair bulbs. The distal aperture of the ducts is round. Multilobal sebaceous glands form a specific gland on the middle of the belly (male and female) (Vorontzov and Gurtovoi, 1959) which is also found on other members of the *Cricetulus* genus—*C. triton, C. eversmanni, C. barabensis, C. longicaudatus, C. kamensis.* The eccrine sweat glands in the sole form large glomi (0.13×0.25; 0.23×0.42 mm) with secretion cavities 17 to 22μ in diameter oddly intertwined with elastin bundles.

A 0.7 mm wide glandular ring forms the anal gland around the anal ori-

TABLE 28

AGE VARIABILITY OF HAIR DENSITY IN BEAVER (ZAMAKHAEVA, 1972*a, b*)

Age	Body area	Number of hairs per 1 cm^2 (thous.)			Overhair to fur hair ratio
		Total	Overhairs (guard + pile + intermediate)	Fur	
Adults	back	26.8	123	26.7	1 : 227
	belly	30.2	221	30.0	1 : 135
Underyearlings	back	18.9	132	18.8	1 : 142
	belly	22.8	334	22.6	1 : 96.5

fice (Skurat, 1972). Some sebaceous glands measure 0.67 x 0.78; 0.62 x 1.70 mm. In *C. barabensis,* the mouth corner gland is formed of large sebaceous glands and very large mucous glands (Quay, 1965c). Thin arrectores pilorum muscles (6μ) are alike in spring and winter and are arranged in the compact upper zone of the papillary layer. The hair divides into two categories, which can be treated as guard and fur hairs. The guard hairs, with a slight thickening in the distal third, divide into three size orders: 12.2, 10.1, and 10.0 mm long; 35, 37 and 36μ thick; medulla 68.6, 91.9 and 91.7%. The first order has a thin, medulla-free tip 2.2 to 2.4 mm long. The second and third orders do not. The three orders also differ in the length of the apical thickening (first order 5.1 mm; second order 3.6 mm; and third order 2.9 mm). The fur hairs have two ligatures but no thickening, and the medulla structure is different in the distal part. They measure 8.5 mm long, 17μ thick, medulla 76.5%, and they have short tips. The hairs grow in tufts of four to six, 10,000 per 1 cm^2.

The hair of the winter specimens is growing and the skin is thick (table 29). The upper papillary layer has a compact plexus of collagen bundles, 100 to 120μ thick; in the rest of the dermis it is loose. The arrangement is mainly horizontal, and fat cells (45 x 67, 40 x 84μ) are apparently inactive.

The hair differs from that of the spring animals, dividing into guard, pile, intermediate, and fur. The guard hairs (first order) with the long whip tip are missing, though second and third orders are very similar, 10.5 and 8.9 mm long, 28 and 27μ thick, medulla 78.6 and 88.9%. In the pile hairs (9.3 mm long, 24μ thick, medulla 79.2%) the 3 mm granna is very pronounced, divided lower down by a neck with a narrowed medulla. The intermediate hairs (8.4 mm, 18μ), with neither granna nor thickening, have one ligature 1.9 to 2.0 mm from the tip. Sometimes there is a slight secondary ligature. The medulla in the distal section (83.4%) differs from the proximal. The fur hairs (6.9 mm, 17μ) have no ligatures. The medulla (82.4%) is the same along the whole length.

COMPARATIVE MORPHOLOGY

TABLE 29

CRICETIDAE: MEASUREMENTS OF WITHERS SKIN (μ)

	Skin	Epidermis	Stratum corneum	Dermis	Papillary layer	Reticular layer
Cricetulus migratorius						
(1)	133	12	6	121	110	11
(2)	220	22	16	198	143	55
(3),	496	28	14	468	440	28
Peromyscus maniculatus						
(1)	118	22	11	96	84	12
(2)	103	22	11	81	73	8
Gerbillus nanus						
(1)	265	17	7	149	132	17
(2)	156	22	11	134	112	22
Meriones meridianus penicilliger						
(1)	484	22	8	462	440	22
(2)	179	17	8	162	134	28
Meriones meridianus meridianus						
n=11, June	253	30	15	223	184	39
n=9, November	196	26	13	170	146	24
M. m. nogaiorum						
n=7, June	190	29	14	161	136	25
n=11, November	174	26	13	148	114	34
Ondatra zibethica Withers						
(1)	1,139	28	17	1,111	1,100	11
(2)	1,263	33	22	1,230	1,100	130
(3)	1,272	28	17	1,144	1,100	44
Breast						
(2)	1,238	28	17	1,210	990	220
(3)	1,244	34	17	1,210	660	550
Ellobius talpinus Summer						
(1)	242	22	17	220	187	33
(2)	253	33	22	220	176	44
Winter						
(3)	237	28	20	209	176	33
(4)	162	22	11	140	123	17
Prometheomys schaposchnikovi						
(1)	292	17	7	275	275	
(2)	252	28	14	224	140	84

TABLE 29—*continued*

	Skin	Epidermis	Stratum corneum	Dermis	Papillary layer	Reticular layer
Lemmus lemmus						
(1)	418	22	11	396	396	
(2)	165	22	15	143	110	33
Lemmus obensis	Badly preserved					
(1)	152			152	132	20
(2)*	235	28	17	84	84	
Myopus schisticolor	112	17	10	95	84	11
Dicrostonyx torquatus						
Summer						
(1)	178	22	15	156	123	33
(2)	162	28	17	134	112	22
Winter						
(3)	246	22	11	224	224	—
Alticola argentatus	379	17	11	362	352	10
Lagurus lagurus	131	22	11	109	99	10
*Arvicola terrestris**	1,584	22	11	462	242	220
Microtus gregalis						
Summer						
(1)	287	12	6	275	275	
(2)	364	17	9	347	330	17
Winter						
(3)	150	17	11	133	123	10
(4)	294	14	7	280	280	
Microtus arvalis						
(1)	162	22	11	140	110	30
(2)	418	22	7	396	363	33
Microtus oeconomus						
Summer						
(1)	182	17	10	165	110	55
(2)	627	22	11	605	605	
Winter						
(3)	132	22	16	110	110	
(4)	95	17	8	78	67	11
Myospalax myospalax	1,021	11**	*	1,010	460	550
M. aspalax						
(1)	792	22	11	770	770	
(2)	264	11**	*	253	231	22

*badly preserved.
**without stratum corneum.

Peromyscus maniculatus

Material. Skin from withers of two adults taken in June (table 29).

The hair is not in growth. The outer surface of the skin is wrinkled. The stratum malpighii is two to three cells. Collagen bundles, mainly horizontal, form an intricate plexus in the dermis. Elastin fibers are few. There are sebaceous glands (39 x 56; 56 x 84μ) but not at all tufts, and a large proportion are inactive. The gland in the corner of the mouth is formed of medium sebaceous and mucous glands and small tubular glands (Quay, 1965c). There is a midventral gland in almost all the following *Peromyscus maniculatus* subspecies: *Peromyscus maniculatus artemisiae, P. m. bairdii, P. m. gracilis* (in six of the twelve males under study), *P. m. labecula, P. m. nebrascensis* (Doty and Kart, 1972). This gland is not present in all the female individuals: in *P. m. artemisiae,* it is revealed in three females of the six under study; in *P. m. bairdii* in twenty-one of the thirty-one; in *P. m. gracilis* in one of the ten; in *P. m. labecula* in three of the fourteen; in *P. m. nebrascensis* in none of the three females under study. The gland size is correlated with the weight of the testicles. Arrectores pilorum muscles (4μ in diameter) are noted only in specimen (1).

There are two orders of guard and one each of pile and fur hairs. The straight guard hairs that thicken gradually from proximal to distal are first order. They are 12.1 mm long, 39μ thick, medulla 84.7%, with a long (1.8 mm), thin tip without medulla. The second order are much the same, but with a shorter tip. Their shaft may be wavy. They are 10.1 mm long, 35μ thick, medulla 82.9%.

The pile hairs have a pronounced granna divided by a proximal neck and a short tip. There are two sixe orders: 9.8 mm, 29μ, medulla 79.4%; and 9.1 mm, 19μ, medulla 84.3%. In the second order the neck is less pronounced. In the first order the granna is 3.1 mm; in the second order, it is 2.3 mm. The fur hairs (9.0 mm, 16μ, medulla 81.3%) are slightly twisted and have no granna. About 1.0 to 1.2 mm from the tip, the medulla changes structure and thins 2.2 to 2.6 mm at 1.2 to 1.4 mm from the tip and again 1.0 to 1.2 mm up.

Gerbillus nanus

Material. Skin from the withers of two adult animals: one, date unknown, from southeastern Iran; one taken December 31, in eastern Iran (table 29).

The hair is not in growth on either animal. The stratum malpighii is one to two cells deep. The horizontal collagen fiber plexus in the dermis is compact and there are few elastin fibers. Specimen (2) has small groups of fat cells (22 x 28; 22 x 34μ) in the reticular layer, while fat deposits are found in the subcutaneous fat tissue (99μ thick) in specimen (1). These fat cells are

22 x 34; 28 x 45μ. The sebaceous glands are rather large (22 x 39; 28 x 50μ), nearly oval, one behind each hair tuft, the lower extremity reaching the hair bulbs. The mouth corner gland in *G. gerbillus* is formed of medium sebaceous and large mucous glands (Quay, 1965c).

The hair is somewhat peculiar. It divides into categories treatable as guard and pile hairs. The guard hairs (two size orders) have a small, rather sharp proximal thickening, above which they very gradually become still thicker, reaching the maximum at five-sixths from the proximal. The tip is short, and the medulla is missing from base and tip. The two orders are 10.9 and 7.8 mm long, 39 and 23μ thick, with medulla 92.4 and 87.0%, respectively. The pile hairs (length 8.4 mm, thickness 20μ, medulla 85.0%) thicken at the base and then stay uniform the whole length, with a neck 2 to 2.5 mm from the tip, but no granna. Above the neck the structure of the medulla is different and the tip is very long (1.6 to 1.7 mm). The hairs grow in tufts of two or three, usually rather close together, although individual tufts do occur. A count of 31,000 hairs is recorded per 1 cm^2.

Meriones unguiculatus

There is an abdominal specific gland on the middle of the belly. Castration results in the involution of the gland; testosterone injection restores its function (Mitchell, 1965).

In ontogenesis, hair on the abdominal gland develops earlier than on the belly (Feldman and Mitchell, 1968). In the abdominal gland, hair tufts are formed by three proliferating follicles which pass through the same hair follicle and sebaceous gland canal. The nonspecific sebaceous glands reach full size at twenty days old, whereas sebaceous glands of the abdominal gland continue to grow until the animal reaches maturity. The absence of fat in the subcutaneous fat tissue underlying the ventral gland and the absence of subcutaneous muscles is characteristic. The mouth corner gland in *M. tristrami* is formed of small sebaceous and mucous glands (Quay, 1965c).

Meriones meridianus

Material. Skin from withers (table 29) of two *M. m. penicilliger* taken in July in Anau; twenty-two *M. m. meridianus* (thirteen taken in June and nine in November); and nineteen, *M. m. nogaiorum* (eight taken in July and eleven in November).

M. m. penicilliger

In specimen (1), the hair is in growth and the skin is rather thick. The skin surface has large wrinkles. The stratum malpighii is two cells deep. In specimen (1) only the outer dermis (200 to 220μ thick) is a compact plexus of col-

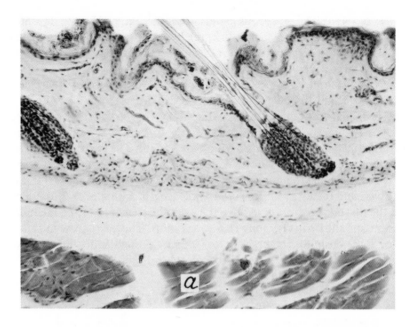

Fig. 65. *Meriones meridianus.* Withers skin: *a,* subcutaneous muscles.

lagen bundles. In the remainder, the collagen bundles are loose and fat cells (22 x 45; 22 x 39μ) occur. In specimen (2), the collagen fiber bundle plexus in the whole dermis is compact and there are no fat cells (fig. 65). Elastin fibers are numerous in both. The sebaceous glands are large (28 x 50; 34 x 56μ), one at every tuft of hair. In the middle of the belly in males and females, an abdominal gland is formed by large sebaceous glands (Sokolov and Skurat, 1966*a, b*). This is larger in males (average for seven—13 x 5 mm) than in females (average for eight—7 x 2.7 mm). Arrectores pilorum muscles are thin (6 to 17μ).

Although the hair does not divide into the ordinary categories, we will distinguish guard, pile, and fur hairs. The guard hairs, 12.2 mm long (medulla 87.0%), gradually thicken from the base to a maximum 46μ, two-thirds proximal with a long, medulla-free tip (2.0 to 2.2 mm). The two orders of pile hairs are thickest between six-sevenths and seven-eighths proximal, both with a short tip. Their length is 11.1 and 10.5 mm, thickness 48 and 32μ, medulla 91.8 and 90.7%, respectively. The fur hairs are narrow in two places in the shaft (sometimes only two narrowings of the medulla) and have a long (1.0 to 1.1 mm), thin tip without medulla. Their length is 6.7 mm, thickness 19μ, medulla 84.3%. The hairs grow in tufts of five to eight, forming no groups. There are 24,000 hairs per 1 cm^2.

M. m. meridianus

In summer, hair of eleven of the animals is not in growth. Only two are in growth (these two animals are not included in table 29). Their skin is thick (495 and 506μ). The stratum malpighii is one to two cells deep. The collagen fiber plexus in the dermis is compact; the bundles are mostly arranged horizontally. There is a sebaceous gland (22 x 39; 17 x 45μ) anterior and posterior to every tuft of hair. The arrectores pilorum muscles are thin (11μ). The hairs fall into three categories: guard hairs (two orders), pile, and fur. The hairs, which gradually thicken from the proximal to reach the maximum in the upper two-thirds of the shaft, are treated as guard hairs. Their tip is long (1.0 to 1.2 mm), thin, medulla-free. The guard hairs (second order) are similar except that they lack the long tip. Their measurements are 14.2 and 11.9 mm, 49 and 47μ, medulla 83.7 and 93.7%. The granna (4.4 mm), separated by a neck at the bottom, is well developed in the pile hairs. Pile-hair length is 11.5 mm, thickness 40μ, medulla 92.5%. The fur hairs (8.6 mm, 20μ, medulla 85.0%) have no granna. In some there is a narrowing. A characteristic (1.5 to 1.7 mm) long, thin, medulla-free tip is recorded. The hairs are arranged in tufts of four to nine, 21,966 hairs per 1 cm^2.

In winter the hair is not in growth, and the skin is thinner than in summer. The thickness of the epidermis and stratum corneum is about the same as in the summer animals, as is the structure of the epidermis. The papillary layer is considerably thicker than the reticular. Horizontal bundles of collagen fibers form a compact plexus. No fat cells are found in the dermis and subcutaneous tissue and the sebaceous glands are small. The arrectores pilorum muscles are thin (10 to 11μ). The hairs grow in tufts, 30,874 per 1 cm^2.

M. m. nogaiorum

In summer, in most of the specimens the hair is not in growth and the skin is not very thick. One animal has growing hair and its skin is thick (440μ) (this specimen is not included in table 29). The structure of the epidermis and dermis is similar to that of other gerbils. The sebaceous glands and arrectores pilorum muscles are similar to M. m. meridianus, and the hair categories to M. m. meridianus. The first order of guard hairs is 14.8 mm long, 53μ thick, medulla 81.2%; the second order 11.6 mm, 48μ, 91.7%; the pile hairs, 10.9 mm, 41μ, 92.7%; the fur hairs, 8.9 mm, 21μ, 85.7%. The pile hair granna is 4.3 mm. The hairs grow in tufts, 27,999 per 1 cm^2.

In winter, the hair is not in growth, and skin, epidermis, and stratum corneum are somewhat thinner than in summer animals. The structure of epidermis and dermis is very similar except that the papillary layer is much thinner and the reticular layer is thicker. An average of 35,916 hairs are recorded per 1 cm^2.

Rhombomys opimus

On the median line of the body, approximately in the middle of the belly, there is an abdominal specific gland, a small skin swelling two to three times as thick as the adjoining skin, with 40 to 67 ducts (Sokolov and Skurat, 1966*a, b*). The glandular field is an irregular oval, elongated lengthwise. The gland is longer in males (average for seven, 22 x 7 mm) than in females (average for six, 15 x 4 mm). Finger pressure on the gland expresses a whitish secretion with a slight odor of burnt horn. The abdominal gland is a complex of densely-adhering, multilobal sebaceous glands measuring 0.38 x 1.10; 0.27 x 1.21; 0.88 x 3.41; 0.71 x 1.87 mm. This gland is considerably larger than the single-lobed sebaceous glands at the hair bursae in general skin areas (17 x 28; 46 x 52μ). The glandular lobes open into a common duct widening at the base into a peculiar reservoir. The upper part, which is considerably narrower, opens into a small surface aperture. Every duct is beside a hair bursa.

In ontogenesis the abdominal gland first becomes evident in ten-day-old females (Sokolov and Skurat, 1969), though it does not yet function at that age. In males it is 12 mm long the day after birth. The gland at the corner of the mouth is well developed, 2 to 4 mm long (Skurat, 1972). The multilobal sebaceous glands forming it are 0.70 x 1.59 mm. The anal gland is a small, narrow ring surrounding the anal orifice (Skurat, 1972). The thickness of the glandular girdle is 0.56 to 0.73 mm, and the size of the individual sebaceous glands is 0.37 x 0.62; 0.41 to 1.06 mm. Sole glands are present only in the sole callosities (Skurat, 1972). The glomi of the tubular glands forming this specific gland are 137 x 146μ.

Ondatra zibethica

Material. Skin from withers and breast (table 30) of three adults, specimen (1) female (body length 28 cm); specimen (2) male (body length 28 cm); specimen (3) female (body length 25 cm) taken in summer, Kuban Plavni.

The thickness of epidermis and stratum corneum is about the same on withers and breast. The stratum malpighii is two to three cells deep. The hairs are in growth; their very long roots form an acute angle with the surface. Only the upper papillary layer has a compact plexus of collagen bundles; the papillary layer is considerably thicker than the reticular layer. Fat cells (56 x 73μ) occur in the dermis. The few elastin fibers are mostly in the upper dermis. Numerous (especially in the breast skin) bundles of smooth muscles not connected with hairs are arranged in the upper 0.2 to 0.3 mm of the dermis.

The sebaceous glands are small (22 x 56; 28 x 100μ) (fig. 66). There are specific skin glands (5 x 6 mm in the female and 3 x 10 mm in the male) in the mouth corners (Skurat, 1972), formed by a complex of hypertrophied

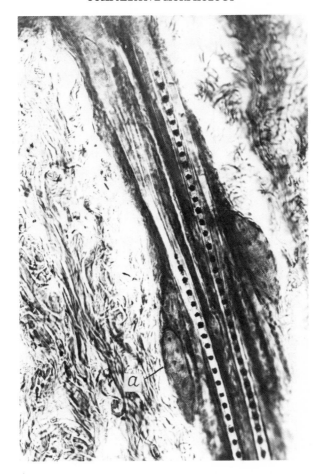

Fig. 66. *Ondatra zibethica*. Withers skin: *a*, sebaceous glands.

sebaceous glands 0.5 x 2.3; 0.7 x 2.6 mm in size, situated over the glomi of tubular glands. There are also sebaceous glands at the labial hairs (0.78 x 1.06 mm) and the cheek hairs (0.50 x 1.42 mm). The anal gland is a girdle of large sebaceous glands measuring 48 x 134μ, 56 x 162μ (Skurat, 1972) around the anus. Large sebaceous glands are also found here (Quay, 1965c). O. zibethica has paired prepucial glands whose secretion contains cyclopentadecanol, cyclopentadecanone, cycloheptadecanol, and cycloheptadecanone. The distilled secretion contains 58.0% of cycloheptadecanol, 40% of cyclopentadecanol, and 2% of respective ketones (Stevens and Erickson, 1942). Two other macrocyclic alcohols, cyclotridecanol and cyclonandecanol have also been reported (Stevens, 1945). No specific glands are found on the sole.

The muskrat skin divides into five zones for pile hair length (Tserevitinov,

TABLE 30

ONDATRA ZIBETHICA: HAIR MEASUREMENTS (PAVLOVA, 1955)

	Guard		1st order		2d order		3d order		4th order		Fur	
	Length (mm)	Thickness (μ)	Length (mm)	Thickness (μ)	Length (mm)	Thickness (μ)	Length (mm)	Thickness (μ)	Length (mm)	Thickness (μ)	Length (mm)	Thickness (μ)
Summer, Leningrad region (mean for 4 pelts)												
Middle of back	28.0	103	25.8	109	23.6	110	20.2	75	19.4	41	13.5	16
Belly	23.3	109	21.4	99	19.9	93	17.9	50	—	—	13.4	15
Winter, Archangelsk region (mean for 3 pelts)												
Middle of back	36.4	100	31.5	109	27.3	110	23.5	70	—	—	17.4	15
Belly	25.6	95	23.1	96	21.3	88	20.0	52	—	—	15.3	15

1958). The first zone, with the longest hairs (over 35 mm) is on the sacrum. The second zone (30 to 35 mm) is along the backbone. The third zone (26 to 30 mm) is beside and behind it, taking in the shoulder blades and sides. The fourth zone (21 to 26 mm) is on the belly and withers; the fifth zone (less than 21μ) covers the front legs, head, and throat.

The longest fur hairs (over 19 mm) grow in two areas halfway down the sides of the body; the second zone (16 to 19 mm) is along the backbone and around the first zone approaching the belly; the third zone (13 to 16 mm) covers the upper neck and extends down the belly; the fourth zone (up to 13 mm) is on the lower section of the belly, hind legs, head, throat, and around the front legs.

The hair divides into guard, pile and fur (table 31). The numbers of hairs per 1 cm^2 of the skin ranges from 10,440 on the sacrum and 10,661 on the scapula to 12,303 on the belly (Tserevitinov, 1951). Not only topographical but also seasonal variability of hair density is observed (E. A. Pavlova, 1955) (table 31).

Geographic differences in the winter pelage of animals from Buryat-Mongolia (Kuban and Barguzin regions), Kurgan area, Kazakhstan (Balkhash and Syr Darya region), and the Arkhangelsk area are slight (Tserevitinov, 1951). The winter male from Kurgan has the highest count of fur hairs on the sacrum (69 per pile hair). The Balkhash muskrat is next with 63, the Arkhangelsk with 59, the Kuban with 57, and the Barguzin with 51. The longest guard and pile hairs are found on the Kurgan muskrat, then the Syr Darya, the Barguzin, and the Kuban. The Barguzin, Kuban, and Leningrad muskrats have the finest hair; the Balkhash and Syr Darya muskrats have rather thick hair; the Kurgan muskrat has the thickest.

Somewhat different data on muskrat skin and hair measurements, with animals taken in March in the Leningrad region, have been obtained by Mkhitaryantz (1974). The skin on the back is 1.6 mm thick, on the belly 0.74 to 0.81 mm thick; the epidermis on the back is 14 to 22 mm thick; on the belly 26 to 29 mm thick; the size of the sebaceous gland is $140 \times 35\mu$; the heat conductivity of the pelt on the back is 0.094 [mlcal/cm^2]/sec.

Ellobius talpinus

Material. Skin from withers (table 29) of specimens (1) and (2) (body lengths 8 cm and 9 cm) taken May 30, Karaganda region, and specimens (3) and (4) (body lengths 9 cm and 10 cm) taken March 10 (end of winter) near Murgab. Breast pelage of the summer animals was also studied.

In summer, specimen (1) has most hairs in the initial growth stage, with a minority actively growing. The measurements are from areas with hair in the initial growth stage. Specimen (2) hair is also in the initial stage. The stratum malpighii is one to two cells deep. The collagen fiber plexus in the dermis is rather dense and predominantly horizontal. The middle and inner parts of the papillary layer have many thick horizontal elastin fibers.

TABLE 31

DENSITY OF THE HAIR AND RATIO OF HAIR CATEGORIES OF SUMMER AND WINTER HAIR OF
ONDATRA ZIBETHICA FROM SYR-DARYA (PAVLOVA, 1955)

	Number of hairs per 1 cm^2 of skin surface					
	Guard	Pile			Fur	Total hairs per cm^2
		1st order	2d order	3d order		
March						
Females						
Middle of back	8	15	47	112	7,057	7,239
Belly	4	13	36	85	8,372	8,510
Males						
Middle of back	5	14	47	94	7,232	7,392
Belly	4	13	35	60	7,820	7,932
Mean for 10 pelts						
Middle of back	6	14	47	103	7,144	7,315
Belly	4	13	36	72	8,096	8,221
February						
Females						
Middle of back	9	17	71	123	11,027	11,227
Belly	7	14	51	133	12,612	12,817
Males						
Middle of back	7	12	61	133	11,848	12,061
Belly	6	10	35	142	12,252	12,445
Mean for 10 pelts						
Middle of back	8	14	56	128	11,438	11,644
Belly	6	12	43	138	12,432	12,631

Groups of fat cells (28 × 33; 22 × 33μ) occur in the reticular layer of specimen (1). The sebaceous glands are large (45 × 84; 56 × 140μ), oval or round (fig. 67). A section shows that they are arranged singly posterior to the hair tufts. The lower parts of the sebaceous glands characteristically penetrate almost down to the hair bulb level. The arrectores pilorum muscles are thin (6 to 11μ).

The hair divides into three categories: guard, pile, and fur. The two orders of guard hairs (10.6 and 8.8 mm long, 47 and 43μ thick, medulla 80.9 and 93.3%) are straight, gradually thickening from the base to halfway up the shaft, then tapering toward the tip, which is not very long. The pile hairs (9.2 mm, 37μ, medulla 89.5%) have a pronounced granna (2.8 mm

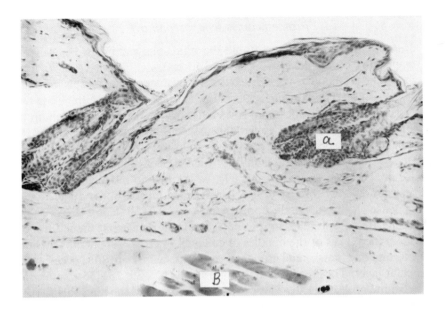

Fig. 67. *Ellobius talpinus*. Skin of withers: *a,* sebaceous gland; *b,* subcutaneous muscles.

long) separated from the lower shaft by a neck, where the medulla thins. The tip is also short. The fur hair (length 7.7 mm, thickness 17μ), with no granna, has two slight ligatures where the hair is wavy, and a short tip. The 82.4% medulla has a peculiar structure over the upper ligature. The hairs grow in tufts of five to eight, 14,333 per 1 cm².

In late winter the hairs are not growing and the skin is thinner than in the summer animals, though the structure of epidermis and dermis is the same. In specimen (3) the lower papillary layer contains round fat cells (39 x 39; 33 x 56μ). The sebaceous glands (28 x 56; 56 x 112μ) and arrectores pilorum muscles (6μ) are the same as in the summer animals. The hairs have broken tips, so that all the hair lengths fall short. The hair categories are the same as for summer animals. The guard hairs (first and second orders) are 7.1 and 6.2 mm long, 28 and 25μ thick, medulla 60.8 and 88.0%; the pile hairs are 6.3 mm, 18μ, and 83.4%; and the fur hairs are 5.6 mm, 16μ, medulla 81.3%. The hairs grow in tufts of six to eleven, 26,666 per 1 cm².

The summer breast hair categories are similar to those of the withers, with one order of guard hairs and two of pile hairs, with a granna length of 3.2 and 2.1 mm, respectively. The fur hairs have up to five waves. The guard hairs are 9.5 mm long, 48μ thick, medulla 85.5%; the pile hairs (first order) 8.2 mm, 41μ and 92.7%; (second order) 7.3 mm, 24μ, and 83.4%; the fur hairs are 6.4 mm, 18μ, and 83.4%.

Prometheomys schaposchnikovi

Material. Skin from withers (table 29), hind sole, and breast pelage from two specimens (body lengths 14 cm and 13 cm).

Specimen (1) has growing hair, so the skin is thicker than on specimen (2), which does not. The skin surface is irregular. The stratum malpighii is two cells deep. The rather thick hind sole epidermis (165μ; 44μ) is taken up by stratum corneum; the inner epidermal surface is alveolar; the height of the alveoli is 66μ; the pronounced stratum granulosum is two to three cells deep.

Due to the hair growth, the bulbs of the lengthened hair roots adjoin the subcutaneous muscles in specimen (1), so there is no reticular layer in the dermis. Bundles of collagen fibers are arranged mostly horizontally, forming a dense plexus only in the upper dermis, which is 56 to 90μ thick. Below this zone are fat cells (28×56; $28 \times 45\mu$) and loose connective tissue. In specimen (2) the plexus of the bundles of collagen fibers is compact throughout the dermal thickness; small groups of round fat cells (28 to 30μ) occur in the reticular layer. The sebaceous glands are small (17×39; $28 \times 56\mu$) and oval, one behind each hair tuft. There are no skin glands in the hind sole though there are bulges ("pads") in the skin containing accumulations of fat cells. The arrectores pilorum muscles are thin (5 to 6μ).

The withers hair divides into three categories: guard, pile, and fur. The guard hairs (length 13.9 mm, thickness 39μ, medulla 69.0%) are straight, thickening gradually from the proximal, with a weak granna one-third up the shaft. The pile hairs, which subdivide into three orders, have a more pronounced granna (4.0; 2.9; and 2.2 mm long) separated by a thinning medulla and a wavy club. Their length is 14.2, 13.3, and 13.0 mm; thickness 36, 29, and 19μ; medulla 66.7, 65.6, and 73.7%. The fur hairs have no granna, medulla, ligatures, or shaft. In some hairs the cells in the upper medulla are different from the lower. They are 11.5 mm long, 17μ thick, medulla 82.4%. The hairs grow in tufts of five (which do not form groups), with 24,666 hairs per 1 cm^2.

In the breast pelage the hair categories are much the same. The granna is not pronounced in the guard hairs, which are 10.7 mm long, 40μ thick, medulla 72.5%. The pile hairs are similar to pile hairs on the withers, but there are only two orders (length 10.3 and 10.1 mm; thickness 32 and 25μ, medulla 65.7 and 68.0%, granna 3.3 and 2.2 mm). Unlike the withers, the fur hairs (length 8.0 mm, thickness 21μ, medulla 66.7%) have a weak granna (2.4 mm), though some hairs do not.

Lemmus lemmus

Material. Skin from withers (table 29) of two specimens taken in summer (body lengths, 11 and 10 cm).

Specimen (1) has growing hair and, therefore, a thicker skin than specimen (2), where hair growth is coming to an end. The stratum malpighii is one to two cells deep. The bulbs of the growing hairs of specimen (1) adjoin the subcutaneous muscles, so the reticular layer is absent. In specimen (1) there is a plexus of collagen fibers in the upper layer of the dermis (50 to 80μ) only, the remainder being filled out by fat cells (39×56; $33 \times 67\mu$). The sebaceous glands are oval and larger in specimen (1) (34×73; $34 \times 56\mu$) as they are active. In specimen (2), most of the glands are inactive. The glands in the corners of the mouth are 3 to 4 mm long and are comprised of hypertrophied sebaceous glands (0.83×1.18 mm) (Skurat, 1972). The gland ducts open into the hair bursae. In *L. trimucronatus,* poorly developed tubular and mucous glands have been found in the corner of the mouth gland (Quay, 1965c).

The anal gland is a group of large sebaceous glands surrounding the anal orifice and extending along the rectum beyond the intestinal crypts (Skurat, 1972). The gland ducts open into the anal orifice. Sebaceous hair glands near the anal region are very well developed ($240 \times 520\mu$). The buccal gland is a group of sebaceous glands extending from eye to ear. Their ducts open into the hair bursae. The sacral gland is clearly visible on the surface, as the pelage is sparse and the surface porous and uneven (Skurat, 1969). In some males it takes up the whole sacral region right to the base of the tail, reaching an average of 27×31 mm. The largest sacral glands are recorded in the males with the most developed testicles. The sacral gland is made up of hypertrophied multilobal sebaceous glands up to $124 \times 274\mu$. The ducts open into hair bursae. Wallin (1967) believes it is the highly thickened epidermis that has the secretory function in the sacral gland. Preputial glands are recorded (Schaffer, 1940). Arrectores pilorum muscles are thin (5 to 6μ).

The hairs divide into three categories: guard, pile, and fur. The guard hairs (length 16.1 mm, thickness 65μ, medulla 80.0%) are straight, with a short club that evolves into a granna 7.8 mm long without neck or narrowing of the medulla. The hair tips are not long. There are two size orders of pile hairs, length 13.5 and 12.2 mm, thickness 51 and 46μ, medulla 78.5 and 84.8%. There is a narrowing of the medulla (more pronounced in the second order) just before the granna. The granna is 4.3 and 2.8 mm long. Fur hairs have no granna. In the first order (length 10.2 mm, thickness 22μ, medulla 77.3%) there are two narrowings of medulla and hair. In the second order (length 8.7 mm, thickness 15μ, medulla 80.0%), the medulla narrows only once. In both orders, the medulla structure in the distal part is different from the proximal, and the second order hairs are slightly wavy. The hairs grow in tufts, one big hair with four to ten smaller hairs around it, though there are tufts of only five to seven smaller hairs. The tufts form groups. There are 3,000 big and 32,000 smaller hairs per 1 cm^2.

Lemmus obensis

Material. Skin from withers (table 29) of two specimens (body length, 10 cm and 12 cm) taken on Taimyr, June 8, and Novaya Zemlya, October 18.

In summer the skin is not very well preserved. The hair is not in growth, and the papillary layer is rather thicker than the reticular. There are no fat cells.

In early winter the hair is not growing, but the skin is thicker than in summer owing to a thick, subcutaneous fat tissue. The stratum malpighii is one to two cells deep. The reticular layer is absent, so the hair bulbs touch the underlying fat tissue. The dermal plexus of collagen fiber bundles is compact, the bundles are mainly horizontal. The subcutaneous fat tissue is a continuous layer of two to four fat cells (39×62; $39 \times 56\mu$) beneath the hair bulbs. The sebaceous glands are nonfunctioning and small ($17 \times 34\mu$). The sacral gland is similar to *L. lemmus* (Skurat, 1969), the individual sebaceous glands that form it being $280 \times 294\mu$. Well developed arrectores pilorum muscles, 11μ thick, can be seen at every hair tuft.

The hair of specimen (2) divides into three categories: guard, pile, and fur. The guard hairs (length 19.6 mm, thickness 47μ, medulla 46.8%) are straight, with a long, thin club that evolves, without a neck, into a 7.3 mm granna. There is a long, medulla-free tip.

All three size orders of pile hairs are similar in shape but have a short tip (length 17.2, 16.2, and 15.0 mm; thickness 59, 64, and 38μ; medulla 62.8, 82.9, 47.3%). There is a small narrowing in the medulla of the second order just before a granna that is still more pronounced in the third order. The granna length for the three orders is 6.9, 5.9, and 3.5 mm. The fur hairs have no granna, are slightly wavy, and usually have two narrowings, with the structure of the medulla being slightly different above the upper one. The fur hairs also subdivide into three orders (length 13.4, 11.7, and 5.2 mm; thickness 25, 19, and 15μ; medulla 56.0, 47.1, and 53.3%). The hairs grow in tufts of three to five, these being grouped in twos and threes. The June animal has 26,666 hairs per 1 cm².

Myopus schisticolor

Material. Skin from withers (table 29) (body length 10 cm) taken on August 26.

The hair is not growing. The malpighian layer is one cell deep and the network of horizontal bundles of collagen fibers is compact in the dermis. The sebaceous glands are small (22×45; $22 \times 56\mu$), with one behind each hair tuft. The small area of the gland in the corners of the mouth (2×3 mm) is covered with coarse, light-colored hair (Skurat, 1971). Individual sebaceous glands are $418 \times 526\mu$. The anal gland is a glandular girdle, 6.7 mm wide and 3 mm thick, surrounding the anal orifice. The sebaceous glands forming it are 0.5×1.1 mm. The arrectores pilorum muscles are thin (7μ).

The hairs divide into two categories: pile and fur. No guard hairs are found. The pile hairs have three size orders (length 14.0, 12.3, and 12.3 mm; thickness 62, 53, and 41μ; medulla 82.3, 88.7, and 87.9%). They are straight, with short tips, and their transition into granna is gradual without a neck. The granna length is 7.1, 5.6, 3.4 mm. The fur hairs have three size orders (11.4, 9.5, and 8.5 mm; 19, 15, and 15μ; medulla 63.1, 73.4, and 80.0%). The fur hairs have no granna. First-order hairs have two narrowings where the shaft is twisted; the second order has only one slight narrowing. The shaft is twisted in three to four waves. The third-order hairs have no ligature. However, as in the other fur hairs, the distal medulla differs in structure from the proximal.

Dicrostonyx torquatus

Material. Skin from withers (table 29) of two summer and one winter lemming (body lengths 11 cm, 12 cm, 13 cm), specimens (1) and (2) taken in June and July. Specimen (3), date unknown, has winter fur.

In summer, the hair is not in growth. The outer surface of the skin is irregular and there is a loose network of horizontal bundles of collagen fibers. The sebaceous glands are small (28×62; $50 \times 67\mu$), and are visible in longitudinal section, one at every tuft (fig. 68). In the breast are two large glands at each hair tuft (fig. 73). The mouth corner gland in *D. groenlandicus* is formed of small sebaceous and sweat glands and large mucous glands (Quay, 1965c).

The hairs divide into three categories: guard, pile, and fur. The long, straight guard hairs thicken gradually from the proximal to halfway up the shaft. They subdivide into two orders, lengths 14.7 and 12.3 mm; thickness 56 and 60μ, medulla 78.6 and 83.4%. The long, medulla-free tip is longer in the first order. In the pile hairs, dividing into two orders, the granna (4.3 and 3.8 mm long) is pronounced. The second-order pile hairs bend at a thinning of the medulla below the granna. The pile hairs are straight (length 11.8 and 12.3 mm; thickness 60 and 55μ, medulla 80.0 and 83.9%) and their tips are short.

The fur hairs have no granna. The first-order fur hairs have two to three waves. The medulla structure differs in the distal section, below which there is a medulla thinning. In the second order, the medulla is uniform the length of the shaft. There are no narrowings. The hairs are slightly wavy. The fur hairs are 10.5 and 8.6 mm long, 22 and 20μ thick, medulla 77.3 and 75%. The hairs grow in tufts of two to four except the guard hairs, which grow singly, surrounded by groups of other hairs. There are 250 guard and 6,333 other hairs per 1 cm².

In winter the hair is growing and the skin is thicker. The roots of the growing hair, which are very long, run almost parallel to the skin. Their

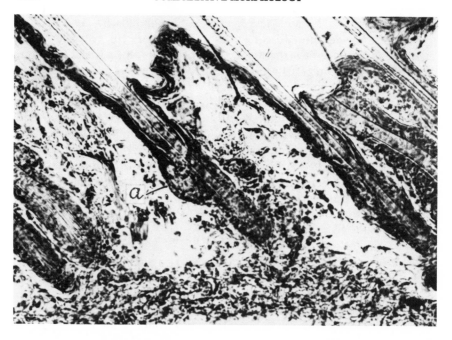

Fig. 68. *Dicrostonyx torquatus.* Withers skin: *a,* sebaceous gland.

bulbs reach the subcutaneous muscles, so that the reticular layer is absent. Only the upper 40μ of the dermis has a compact plexus of collagen bundles, the remainder being filled out with fat cells. The sebaceous glands are smaller (22 x 45μ) than in summer, one singly behind the hair tufts.

There are three categories of hair: guard, pile, and fur. The guard hairs are straight, thickening from the proximal half or two-thirds up the shaft, 24.4 mm long, 46μ thick, medulla 82.7%. They have a very long, thin, medulla-free tip. The two orders of pile hairs that do not have a pronounced granna are thickest in the distal third or half of the shaft. The second order has a slight thinning in the medulla just before the distal thickening. This is not recorded in the first order. The pile hair tips, still rather long, are shorter than in the guard hairs, and their club is slightly wavy. They are 20.7 and 18.2 mm long, 46 and 32μ thick, medulla 91.4 and 87.5%.

The fur hairs have no dilations. The tip, though long, is shorter than in the guard hairs, and there is no thickening in the shaft. In the first order the medulla differs in the distal and proximal shaft after a slight narrowing, which is missing in the second order, where the medulla is unchanged in the whole shaft. The length of the fur hairs is 15.6 and 13.9 mm, thickness 16 and 17μ, medulla 81.3 and 82.4%. The hairs grow in tufts of four to seven, 11,083 to the 1 cm².

Alticola argentatus

Material. Skin from withers (table 29) and hind sole (body length 12 cm) taken in summer, Central Pamir.

The hair is growing, so the skin is rather thick. The stratum malpighii is only one cell deep. The pads of the hind sole have a thickened epidermis (257μ), a solid stratum corneum (140μ), and a pronounced two- to three-layer stratum granulosum; on the inner surface of the epidermis are alveoli 56μ high.

The dermis of the body skin is entirely filled with broad roots of growing hairs. Only in the upper papillary layer (about 60μ), horizontal bundles of collagen fibers form a compact plexus. Despite hair growth, there are no fat cells in the dermis.

The sebaceous glands are small (17×40; $17 \times 44\mu$), singly posterior to hairs. The gland in the corners of the mouth (2.2 to 3.5 mm) is distinguished by a light yellow skin area covered with sparse, coarse, light, short hairs (Skurat, 1972). The large sebaceous glands forming it are 0.34×0.78; 0.43×0.78 mm. The excretion ducts open into hair bursae. The anal gland is made up of sebaceous glands (0.41×1.62; 0.6×1.12 mm) surrounding the orifice of the anus, with their excretion ducts opening into it (Skurat, 1972).

There are large glomi of eccrine sweat glands (0.34×1.00 mm) in the hind foot dermis. The secretion cavity diameter is 22μ. Arrectores pilorum muscles are not recorded (evidently because the hair is growing).

The hair divides into four categories: guard, pile, intermediate, and fur. The guard hairs thicken at the base, remain uniform for some distance, then thicken again in the distal section where they have a weak granna. There are two distinct size orders, 17.0 and 15.4 mm long, 5.9 and 5.2 mm thick, medulla 83.7 and 93.9%; length of granna 5.9 and 5.2 mm. The first-order hairs have rather a long tip (1.3 to 1.6 mm). The pile hair granna is much more pronounced, starting from a thinning in the medulla. There are two orders of pile hairs, 15.1 and 14.1 mm long; 45 and 27μ thick; medulla 88.9 and 77.8%; length of granna 4.3 and 4.2 mm. Their tips are short. In the intermediate hairs (length 13.0 mm, medulla 76.5%) the distal shaft is separated by a medullary thinning. Beyond this point the medulla structure changes, though the thickness (17μ) remains the same. The fur hairs (11.5 mm long, 15μ thick, medulla 63.4%) have no thinning of shaft or medulla, though the medulla is different in structure in the distal section.

Lagurus lagurus

Material. Skin from withers (table 29) (body length 10 cm) taken April 27, Karaganda region.

The hair is not growing. The outer surface of the skin is plicate. The stratum corneum takes up half the epidermis; the stratum malpighii is two cells deep. The dermis has a compact, mainly horizontal network of collagen fiber bundles, while elastin fibers are few. The sebaceous glands are small (17 x 39; 17 x 34μ) and heart-shaped, singly posterior to the hair tufts. The lateral specific glands in *L. curtatus* are developed poorly and only in adult males (Bailey, 1897, 1900, 1936). In *L. przewalskii* both sexes have well-pronounced lateral glands (Quay, 1968). The gland in the corner of the mouth in *L. curtatus* is comprised of medium sebaceous and mucous glands and small tubular glands (Quay, 1965c).

The arrectores pilorum muscles are thin (6μ), and the hair categories are guard, pile, intermediate, and fur. The guard hairs thicken gradually from the base, become uniform for some length, then thicken again near the distal end. They have no granna. The two orders are 14.2 and 12.2 mm long, 51 and 50μ thick, medulla 84.4 and 92.0%. The first order has tips 1.8 to 2.2 mm long; the second order has very short tips, the distal thickening is more pronounced and forms a granna. In the two orders of pile hairs (length 11.1 and 10.7 mm, thickness 56 and 34μ, medulla 92.9 and 88.2%, length of granna 3.9 and 2.5 mm). The second order of pile hairs has a thinning in the medulla below the granna. The tip of the pile hairs is short. Intermediate hairs (8.6 mm long, 18μ thick, medulla 66.7%) are granna-free. However, their distal end is separated by a medullar thinning, and the medulla structure in the distal section of the shaft differs from that of the proximal section. In the fur hairs (7.0 mm long, 18μ thick, medulla 77.0%) the medulla structure is homogenous and there is no thinning. The hairs grow in tufts of four to seven, the guard hairs growing singly. There are 23,333 hairs per 1 cm².

Arvicola terrestris

Material. Skin from withers (table 29) of one specimen (body length 22 cm) taken June 30, Karaganda region.

The hair is not growing. The papillary layer is almost equal to the reticular layer in thickness. In the papillary layer, the collagen fiber bundles are mainly horizontal (though some run at various angles to the surface). In the reticular layer nearly all are horizontal. There are very few elastin fibers. The subcutaneous fat tissue, which is clearly separated from the dermis, is thick (1.1 mm). The fat cells in the outer part (150 to 200μ) are large and round (56 x 73; 56 x 78μ), but are smaller (34 x 50; 34 x 56μ) in the inner part. There are few bundles of collagen fibers in the subcutaneous fat tissue. The sebaceous glands are small (fig. 69). The glands in the corners of the mouth, 3.5 mm long, 1.3 mm wide, are covered with bristle-like hair (Skurat, 1972). The constituent multilobal sebaceous glands are 0.53 x 1.20 mm. There are large multilobal sebaceous glands (0.33 x 1.12 mm) elongated along the hair roots on the lower lip. The anal gland, surrounding the

Fig. 69. *Arvicola terrestris.* Withers skin: *a,* sebaceous glands; *b,* subcutaneous muscles.

anal orifice, is formed by 0.39 x 1.34 mm sebaceous glands (Skurat, 1972). Poorly developed sole glands (Skurat, 1972) are found only in some of the callosities. The tubular gland glomi are 112 x 202; 106 x 129μ. The clearly defined side glands on the inner surface of the pelt, are formed by sebaceous glands (470 x 490 ; 420 x 488μ). They are oval along the body, their individual size being (male) 5.7 mm (female) 11.2 mm (Skurat, 1972), though their size and activity varies with the season (Stoddart, 1972); according to Vries (1930) the same holds true of *Arvicola scherman.* At the beginning of the breeding season, the lateral gland is flattened and haired. Before long the gland thickens, secreting actively through the hair follicles, with the hair occasionally coming out. By the end of the breeding season, the gland surface is almost hairless. There are also preputial glands (Schaffer, 1940).

The pelage divides into five zones for hair length (Tserevitinov, 1958). The longest hairs are on the lower back. Hairs divide into guard, pile (two orders), and fur. The guard hairs (16.7 mm long, 72μ thick, medulla 42%) thicken very gradually from the proximal to the distal seven-ninths of the shaft. The tip (0.8 to 12.8 mm) is rather long. Both orders of pile hairs (15.5 and 12.8 mm long, 61 and 45μ thick, medulla 58 and 60%, respectively) have a granna. The fur hairs (10.2 mm long, 18μ thick, medulla 62%) do not. They curl, forming up to eight waves. The hairs grow in tufts of five to nine, with 666 guard and 31,333 other hairs per 1 cm^2. Kozhukhovskaya

(1969) gives different hair measurements of a water vole taken in the Leningrad region in March.

Somewhat different data concerning skin and hair measurements of water voles taken in March in the Leningrad region are obtained by Mkhitaryantz (1974). The skin on the back is 620 to 650μ thick; that on the belly is 460 to 500μ; the epidermis on the back is 14 to 18μ thick; that on the belly is 20 to 22μ; the sebaceous glands are 118 to 120μ long, 4 to 6μ wide; the hair is not growing; the length of guard hairs on the back is 23 to 27 mm; 16 to 21 mm on the belly; 16,532 per 1 cm^2 on the back, and 14,644 on the belly. Heat conductivity of the back pelt is 1,015 [mlcal/cm^2]/sec, and that on the belly is 1,002 [mlcal/cm^2]/sec.

Microtus gregalis

Material. Skin from withers (table 29) of four specimens (body lengths, 9 cm, 10 cm, 8 cm, and 9 cm); specimens (1) and (2) taken in June near Chita; (3) and (4) taken on December 1, in Kirghizia.

In summer, the hair on both animals is growing and the skin is thick, with an irregular surface. The stratum malpighii is one cell deep.

Only the upper 80 or 90μ of the dermis forms a rather compact plexus of mainly horizontal collagen fiber bundles. The inner layers of the dermis are filled by large growing hair roots (most of them parallel to the surface), loose connective tissue, and a few fat cells (22 x 34; 17 to 28μ). There are a few elastin fibers in the dermis. In specimen (1) the roots reach the subcutaneous muscles. Specimen (2) has a fine reticular layer in between.

The sebaceous glands are rather large (28 x 56; 45 x 95μ) and oval, one at every hair tuft. The gland at the corners of the mouth is rather poorly developed (Skurat, 1972; Sokolov and Skurat, 1975) smaller than in others of the genus (2.1 to 2.5 mm long, 0.22 mm wide, 1.49 mm thick), comprised of sebaceous glands of up to 0.42 x 0.89 mm. The well-developed anal gland, comprised of sebaceous glands, forms a solid ring around the anal orifice and projects at the side into the rectum beyond the intestinal crypts (Skurat, 1972; Sokolov and Skurat, 1975). The skin thickens somewhat where large sebaceous glands (225 x 403μ) form side glands (fig. 70). The large sole glands are more highly developed than in other common voles (Skurat, 1972; Sokolov and Skurat, 1975). The glomeruli of tubular glands reach 89 x 168μ. The arrectores pilorum muscles are thin (4 to 5μ) and are found only in the compact zone of the dermis.

There are three categories of hair. (Marvin, 1974, in his description of the hair of *Microtus gregalis, M. arvalis, M. oeconomus, M. agrestis, M. middendorfi, Clethrionomys glareolus, C. rutilus, C. rufocanus, Rattus norvegicus, Mus musculus, Apodemus agrarius, A. sylvaticus,* and *Micromys minutus,* for some reason distinguishes only pile and fur hairs, without size orders; due to this confusion we do not quote Marvin's hair measurements).

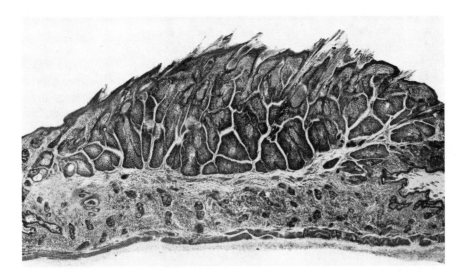

Fig. 70. *Microtus gregalis*. Side gland.

Straight guard hairs (13.2 mm long, 58μ thick, medulla 81.1%), thicken from proximal to distal, where they form a 6.3 mm granna. The tip is short, only 0.77 to 0.90 mm. The pile hairs (3 size orders) have a very pronounced granna above a medulla thinning (10.7, 9.9, and 8.6 mm long; 54, 36, and 20μ thick; medulla 94.5, 83.4, 60.0%; granna 3.6 mm, 2.7, and 1.5 mm). The first order is straight; the second has a wavy lower shaft (proximal to the granna). The third order is still more wavy. The intermediate hairs (7.7 mm long, 15μ thick, medulla 66.7%) have a thinning of the medulla toward the distal; the medulla structure differs in the upper part of the shaft, where it is also somewhat thicker. Fur hairs are absent. There are 22,000 hairs per 1 cm^2,

Winter specimen (3), with nongrowing hair, has considerably thinner skin than (4), with growing hair. The stratum malpighii is one to two cells deep. The growing hair of (4) has long roots, the bulbs reaching the subcutaneous muscles. Specimen (3) has a thin reticular layer between roots and muscles. In (4) only the upper dermis (50 to 60μ thick) network of horizontal collagen bundles is fairly compact; in the remainder of the dermis it is very loose with many large fat cells (22×56; $28 \times 50\mu$) interspersed. Specimen (3) has small oval fat cells (diameter 11 to 17μ) in the lower papillary layer. There are single sebaceous glands (17×56; $28 \times 73\mu$), which are much the same size as in summer, at each hair tuft. Most of the glands are not functioning in (3) and are active in (4). The arrectores pilorum muscles are thin (6μ).

The hair divides into four categories: guard, pile, intermediate, and fur.

The guard hairs (15.8 mm long, 58μ thick, medulla 81.1%) gradually thicken from the base (the thickening is more pronounced in the distal section) forming a weak 7.6 mm granna. They have a long tip (1.0 to 1.3 mm). The three orders of pile hairs (13.1, 12.8, 12.4 mm long; 63, 65, and 59μ thick; medulla 92.1, 95.4 and 95.0%) have a more pronounced granna (5.9, 4.9, and 3.9 mm long) in the third order, separated by a medulla narrowing and short tips. The slightly wavy intermediate hairs have a distal medulla contraction; the distal medulla differs in structure from the proximal. There are two orders, 10.2 and 9.6 mm long, 20 and 13μ thick, medulla 60.0 and 77.0%. The fur hairs (7.1 mm, 15μ, medulla 80.0%) have no medullary narrowings and the medulla structure is constant.

Microtus arvalis

Material. Skin from the withers (table 29) of two specimens (body length 10 cm and 11 cm) taken June 19, near Bobruisk.

The hair of specimen (1) is not growing, and the skin is thinner than in (2), whose hair is in growth. The stratum malpighii is one to three cells deep. Specimen (1) has a compact network of mainly horizontal collagen bundles in the dermis, and a reticular layer containing small groups of fat cells (22 × 28; 17 × 28μ). In specimen (2) this compact network is only in the outer papillary layer (100 to 150μ thick), the remainder of the dermis being filled with large roots of growing hairs, loose connective tissue, and fat cells (28 × 45; 28 × 50μ).

The reticular layer is loose and there are few elastin fibers in the dermis of either. The sebaceous glands are rather large, round or oval (22 × 58; 45 × 56μ), one behind each hair. In specimen (2), few are inactive. The mouth corner gland is similar to that of other voles (Skurat, 1972; Sokolov and Skurat, 1975) (fig. 71), and does not normally exceed 3 mm (2 to 5 mm) in length, with the longest of its sebaceous glands no more than 0.28 × 0.44; 0.39 × 1.00 mm. The anal gland is large, reaching 10 mm (Schaffer, 1940; Skurat, 1972; Sokolov and Skurat, 1975), bigger in males than in females, its size apparently varying seasonally. Its sebaceous glands are 0.34 × 1.03; 0.86 × 2.96 mm. The sole glands are formed by sweat-gland glomeruli, which are largest in summer (Skurat, 1972; Sokolov and Skurat, 1975). The sacral gland (lateral pair) is found in males from Central Europe (Lehman, 1967). It is comprised of increased sebaceous glands. The sebaceous glands are well developed in the preputium skin (Schaffer, 1940). The arrectores pilorum muscles are pronounced in both animals (8 to 10μ thick), but are found only in the dense zone of the papillary layer in specimen (2).

There are four categories of hair: guard, pile, intermediate, and fur. The two orders of guard hairs bulge a little at the base, and again in the distal section, but do not form a granna. Their length is 14.4 and 12.2 mm, thickness 5 and 64μ, medulla 87.1 and 84.4%, the first order being longer (1.5 to 2 mm) in the tip, as compared with the second order (1.0 to 1.5 mm). The

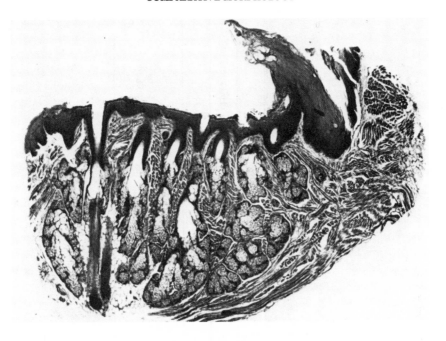

Fig. 71. *Microtus arvalis*. Gland in corner of mouth.

two orders of pile hairs (10.9 and 9.6 mm long, 57 and 33μ thick, medulla 82.5 and 78.8%) have a pronounced granna (4.7 mm and 3.3 mm long) separated from the proximal section by a medullar narrowing. Their tips are short (0.30 to 0.35 mm). The intermediate hairs (9.7 mm long, 23μ thick, medulla 52.2%) have no granna, but a medullar narrowing divides the distal section, where the structure of the medulla is different. This narrowing is also present in the fur hairs (8.6 mm, 17μ, medulla 82.4%) though the medulla structure remains unchanged. The hairs grow in tufts, 17,000 per 1 cm².

In *M. arvalis,* taken in March (Leningrad region), Kozhukhovskaya (1969) distinguishes three hair categories: guard (one order), pile (two orders), fur hairs (one order). Their measurements are close to those obtained by us (table 32).

Somewhat different measurements of skin and hair of common voles (March, Leningrad region) are obtained by Mkhitaryantz (1974): skin on the back 416 to 468μ thick, epidermis on the back 17 to 18μ, epidermis on the belly 15 to 16μ, sebaceous glands 90 to 120μ long, 3 to 5 mm wide; the hair is not growing; pile hairs on the back are 12 to 13.5 mm long; on the belly 9.5 to 11.5 mm; fur hairs on the back 10 to 11.5 mm long, on the belly 6 to 10 mm. Two granna size orders are distinguished: a broader granna 50 to 64μ, and a narrower granna 20 to 40μ. Hair conductivity is 1.003 [milcal/cm³]/sec.

TABLE 32

M. ARVALIS HAIR MEASUREMENTS (MARCH, LENINGRAD REGION) LENGTH AND THICKNESS (KOZHUKHOVSKAYA, 1969)

Hair	n	Midback				Belly			
		Length (mm)		Thickness (μ)		Length (mm)		Thickness (μ)	
		M±	±	M±	±	M±	±	M±	±
Guard	50	13.6±0.21	0.74	64.3±0.90	3.15	11.5±0.04	—	61.3±0.79	3.07
Pile (1st)	50	12.5±0.34	0.98	57.4±0.73	3.28	10.6±0.10	—	56.0±0.64	2.41
Pile (2d)	50	11.4±0.30	0.94	32.5±0.71	2.84	9.2±0.26	—	49.6±0.83	2.39
Fur	50	6.7±0.08	0.69	15.8±0.86	2.16	6.9±0.09	—	15.2±0.71	2.01

Microtus oeconomus

Material. Skin from withers (table 29) of specimens (1) and (2) (body length 11 cm and 12 cm), taken July 29; and specimens (3) and (4) (body length 11 cm and 10 cm), taken February 24, all near Irkutsk.

In summer, the skin of specimen (1) is considerably thinner, as its hair is not growing, while specimen (2) has hair in growth. The stratum malpighii is one to two cells deep. In specimen (2) only the upper part of the dermis, 140 to 150μ thick, is formed by a rather compact collagen fiber plexus. The remainder of the dermis is filled by large hair roots, loose collagen bundles, and numerous fat cells (39 × 67; 34 × 67μ). The hair bulbs nearly reach the subcutaneous muscles, so there is no reticular layer. In specimen (1) the plexus of horizontal collagen fibers is not very dense in the papillary layer and is very loose in the reticular layer. There are not many elastin fibers in the dermis, and fat cells (17 × 33; 22 × 33μ) are only in the reticular layer. There are single sebaceous glands, small (22 × 56; 39 × 50μ), oval or round behind the hair tufts, active in (1), inactive in (2). The glands at the corners of the mouth are well developed, maximum length 5.5 mm and breadth 2.2 mm (Skurat, 1972; Sokolov and Skurat, 1975). Large sebaceous glands (0.63 × 1.23 mm) are elongated along the hair bursae. The large anal gland is largest in males (6.5 × 13; 7 × 13 mm) (Skurat, 1972; Sokolov and Skurat, 1975). The sole glands are formed by large eccrine sweat-gland glomi, 127 × 230 (Skurat, 1972; Sokolov and Skurat, 1975). The twin sacral gland (Lehman, 1966, refers to it as a lateral gland) is better developed in males (5 × 12 mm) than females (5 × 8 mm) (Skurat, 1969; Sokolov and Skurat, 1975). Its size varies directly with the size of the testicles. The size of the sebaceous glands forming the sacral gland is 0.38 × 0.44 mm. Thin arrectores pilorum muscles (6 to 8μ) are found in a compact zone of the dermis in specimen (2).

The hair divides into three categories: guard, pile, and fur. The guard hairs (length 14.3 mm, thickness 87.8μ) thicken somewhat at the base, then stay unchanged until they thicken sharply in the distal section. The difference in thickness is considerable: 11μ at the base and 49μ at the thickest point. The length of the distal dilation is 6.5 mm. There is no neck or medulla narrowing.

The tip is long (0.8 to 1.3 mm) and medulla-free. The pile hairs are fairly similar but without the long tip and with a medulla narrowing before the granna. There are two size orders, 12.5 mm long, 53 and 47μ thick, medulla 86.8 and 93.7%, length of the granna 5.4 and 3.6 mm. The fur hairs are of even thickness with a shaft ligature about 2.0 mm from the tip and a medullar narrowing. The second ligature is very slight. The distal section of the medulla differs from the proximal. There are two orders of fur hairs (10.8 and 9.5 mm long, 23 and 19μ thick, medulla 47.8 and 52.6%). The hairs grow in tufts of four to seven, 21,000 per 1 cm².

The hair of the winter animals is not growing, and their skin is thin. The

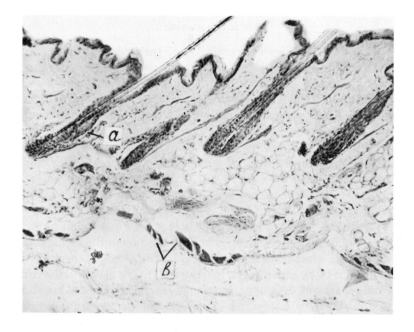

Fig. 72. *Microtus oeconomus*. Withers skin, winter: *a*, sebaceous gland;
b, subcutaneous muscle.

epidermis and stratum corneum are nearly the same as those of the summer animals; the stratum malpighii is one to two cells deep. Although the winter hair is not growing, the bulbs are close to the subcutaneous muscles. In specimen (3) the reticular layer is absent altogether, and in (4) it is very thin (fig. 72). The dermis collagen fiber bundles are mostly horizontal. The sebaceous glands are small (15 × 33; 28 × 56μ) and mostly inactive.

As for the four categories of hair, the guard hairs (15.7 mm long, 59μ thick, medulla 83.1%) are similar to the summer hairs. The distal thickening (a weak granna) is 5.6 mm long. The two orders of pile hairs (17.1 and 15.0 mm, 57 and 50μ, medulla 87.8 and 86.0%) have a medullar narrowing before the granna (6.3 and 4.2 mm). The first order of fur hairs is similar to the summer fur, but the second order does not have ligatures or medullar narrowings, and their medulla structure remains unchanged down the shaft. The fur hairs are 15.0 and 10.3 mm long, 20 and 16μ thick, medulla 60.0 and 81.3%.

Genus Clethrionomys

Specific glands at the corners of the mouth, anus, and sole are well developed (*C. rufocanus, C. glareolus, C. rutilus*) (Skurat, 1972). The length of the mouth corner glands is 2.5 to 4.5 mm, breadth 1.5 to 3 mm; their sebaceous glands (0.68 × 1.12; 0.37 × 0.67 mm) are arranged along the hair roots. The anal glands range from 2 to 4 mm in length and from 2 to 5 mm in breadth, their size being subject to sex and seasonal dimorphism. The glands, bigger in the male, enlarge in summer. Their multilobal sebaceous glands vary from 0.67 to 1.40 mm.

The sole glands on the hind feet are comprised of eccrine sweat gland glomeruli (83 to 128 × 109 to 224μ). There are no sex differences, and the gland appears to shrink in winter. The side glands are often developed in the American *Clethrionomys* species (Bailey, 1954). Lehman (1962) notes that in *C. glareolus* of Central Europe, small, multilobal side glands occur in reproducing males. However, secretion manifests itself only rarely. Side glands are comprised of hypertrophied sebaceous glands.

Side glands are also found in *C. rufocanus* from the Central and Polar Urals (Semenov, 1975) (the author calls them sacral). These are paired oval glands at the sides of the sacral region. They are better developed in males (average size 23 × 11 mm) than in females (18 × 8 mm). The development of lateral glands coincides with the reproductive period. They disappear upon its completion. These glands are absent in *C. rufocanus* in winter (from November through March) and are not found in the majority of young animals. However, in young animals which have been involved in reproduction, oval pigmented patches are noted in the sacral region (average size 17 × 18 mm) (recorded in 10 animals of the 20 examined).

The gland in the corner of the mouth in *C. glareolus, C. gapperi* and *C. occidentalis* is comprised of large sebaceous glands as well as medium mucous and tubular glands (Quay, 1965c).

Myospalax myospalax

Material. Skin of withers (table 29), hind sole, and the tip of the nose of *M. myospalax* (body length, 22 cm) taken in summer.

In the skin of the withers, the hairs are in the final stage of growth or have finished growing, but the skin thickness is still considerable. The hind sole epidermis attains considerable thickness; the stratum lucidum is well developed. On the nose, the epidermis is 605μ; from the inner side it has alveoli 121μ high. The stratum granulosum is two cells deep. The stratum corneum is rather thick (374μ) and compact. The withers reticular layer is well developed and thicker than the papillary. In the upper dermis the bundles of collagen fibers are predominantly oblique; in the middle and lower dermis they are horizontal. Small groups of fat cells extend downward from the bulbs that cross the reticular layer. The sebaceous glands are small (28 × 45; 34 × 39μ), one at every hair, with hypertrophied, multilobal sebaceous glands (0.46 × 1.98; 0.44 × 1.90 mm) adjoining the tufts; their duct opens near, not into, the hair follicle. The duct is 308μ in diameter. No such glands are recorded on breast or sacrum. There are no glands in the skin of the hind soles or nose. The arrectores pilorum muscles are 22μ in diameter.

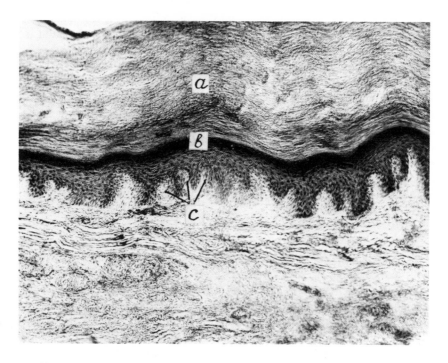

Fig. 73. *Myospalax aspalax*. Skin of hind foot sole: *a,* stratum corneum of epidermis; *b,* stratum granulosum of the epidermis; *c,* dermal papillae.

Myospalax aspalax

Material. Skin from withers, breast (table 29) of the front and hind sole from two specimens (body length 20 cm, 21 cm) taken in summer.

The hair is growing on the withers of specimen (1) and the skin is considerably thicker than in specimen (2), where the hair has almost completed growth. The breast skin is thinner. The stratum corneum comprises about half of the epidermal thickness. The stratum malpighii is two to three cells deep. In the skin of the front foot sole the epidermis is thick, 396μ, and the stratum corneum is very compact, 176μ thick; the alveoli on the inner surface of the epidermis reach a height of 88μ. The stratum granulosum is three to four layers deep; the stratum lucidum is absent. In the skin of the hind foot sole, the thickness of the epidermis reaches 418μ; its alveoli are 88μ high (fig. 73). The stratum corneum is thick (286μ); there is no stratum lucidum; the stratum granulosum is two to three cells deep.

In specimen (1) the hair is growing, so there is no reticular layer on the withers. Long hair roots run at acute angles to the surface across the whole dermis, the hair bulbs nearly reaching the subcutaneous muscles. In specimen (2) the withers reticular layer is considerably thinner than the papillary.

Horizontal collagen bundles predominate in the dermis, except in the outer dermis where many bundles are arranged in various directions. The collagen fibers form a compact plexus. In the middle and inner dermis layers, specimen (1) has groups of fat cells (22×39, $28 \times 39\mu$) near the hair sheaths. In specimen (2) only a few fat cells (11×28, $11 \times 22\mu$) are at the hair roots, while elastin fibers are numerous in the upper dermis. The sebaceous glands are small (22×50; $34 \times 62\mu$) and mostly inactive, one at each hair (fig. 74a). Along with common sebaceous glands, huge sebaceous glands are found near the hair tufts. Their secretion cavities are under the hair follicles, and the broad ducts open onto the surface of the skin together with the hair funnels (fig. 74b). The glands on the hind and front sole are absent (fig. 73), although skin callosities are pronounced. The arrectores pilorum muscles (17 to 22μ thick) are well developed.

The withers hairs divide into three categories: guard, pile and fur. The two orders of guard hairs (18.9 and 16.9 mm, 59μ and 46μ, medulla 72.9 and 76.1%) which are straight, gradually thicken into a granna (10.5 and 6.8 mm) in the distal section. The tips are short, and the medulla is absent only from base and tip. The pile hairs differ only in size and in a medulla narrowing just before the granna. They also subdivide into two orders (length 16.7 and 16.0 mm, thickness 44 and 33μ, medulla 79.6 and 66%, length of granna 6.2 and 4.1 mm). The fur hairs subdivide into two orders

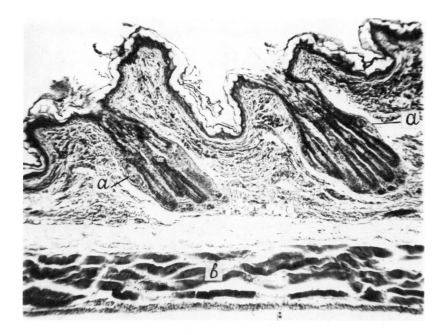

Fig. 74a. *Myospalax aspalax.* Withers skin: *a,* sebaceous glands; *b,* subcutaneous muscles.

Fig. 74b. *Myospalax aspalax.* Withers skin: *a,* huge sebaceous glands;
b, their ducts; *c,* subcutaneous muscles.

(14.6 and 11.3 mm long, 2.0 and 19μ thick, medulla 65.0 and 68.5%). They
are slightly wavy, have no granna, and their medulla has a somewhat different structure in the distal section. The hairs grow in tufts of six to eight,
three or four tufts forming a group. There are 6,750 hairs per 1 cm².

Familia Spalacidae

Spalax microphthalmus

Material. Skin from withers, breast (table 33), and hind sole of two specimens (body length 22 cm and 23 cm) taken May 22, Odessa region, and
January 15, near Surchany village.

The skin on the withers is thicker than on the breast but, characteristically, the thickness does not vary much in different parts of the body (Tserevitinov, 1958). There are only two thick skin zones. The thicker skin is on
the sacrum, on half of the back and neck, and on the anterior part of the
back, sides, and belly. Its surface is irregular. The hair is not in growth. The
stratum malpighii is two to three cells deep. The epidermis is not pigmented.
In the skin of the hind sole the thickness of the epidermis reaches 605μ. Its
inner surface is alveolar, the height of the alveoli being 187μ. The stratum

TABLE 33

SPALAX MICROPHTHALMUS: SKIN MEASUREMENTS (μ)

	Skin	Epidermis	Stratum corneum	Dermis	Papillary layer	Reticular layer
Withers						
Summer	557	45	33	512	292	220
Winter	661	56	39	605	330	275
Breast						
Summer	397	56	42	341	275	66
Winter	292	72	60	220	210	10

corneum (363μ thick) is very compact. There is a one to two layer stratum granulosum. The epidermis is pigmented. In winter, the epidermis of the hind sole is 520μ thick. The stratum corneum is 308μ. The height of the alveoli on the inner surface of the epidermis is 165μ. The stratum granulosum is three cells deep. The stratum lucidum is absent. The epidermis is not pigmented.

The papillary layer is thicker than the reticular on the body, on which it borders at the hair bulbs. The collagen fiber bundles are largely horizontal in the dermis (especially in the reticular layer), the plexus being very compact in the papillary layer and much looser in the reticular. It is interesting to note that elastin fibers are most numerous in the inner papillary and outer reticular layer. There is also a rather compact network of elastin fibers in the inner reticular layer, with predominantly horizontal bundles everywhere. In the breast of the winter specimen, single bundles of cross-striated muscles rise in the dermis above the hair bulb level.

The summer specimen has large accumulations of fat cells (45 x 56; 45 x 50μ) in the reticular layer. The sebaceous glands are small and oval (34 x 78; 22 x 56μ), with one anterior and one posterior to every tuft of hair (fig. 75). Skin glands are absent from the front and hind feet (fig. 76). The specific glands at the mouth corners are not outwardly visible. They are formed by large (0.54 x 0.74; 0.37 x 1.34 mm), multilobal sebaceous glands (Skurat, 1971). An anal gland comprised of large, multilobal sebaceous glands in the end of the rectum reaches a length of 7 mm and a breadth of 9 mm (Skurat, 1972). Large sebaceous glands (0.27 x 0.48 mm) are in the perianal region near the hair bursae. Arrectores pilorum muscles, 17 to 28μ thick, are clearly visible at every hair tuft on the withers, but are not on the breast.

There are four body zones for pelage length (Tserevitinov, 1958). The longest hairs are in three areas, two on the sides of the body, one on the lower third of the back. No guard hairs are found in the sample under study; only pile, intermediate, and fur. Both orders of pile hairs (11.1 and

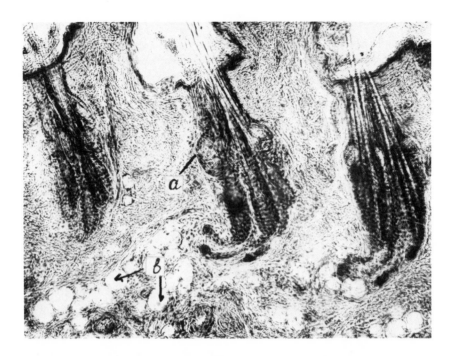

Fig. 75. *Spalax microphthalmus*. Withers skin, summer: *a*, sebaceous glands; *b*, fat cells.

11.5 mm long, 50 and 42μ thick, medulla 76.0 and 66.7%) have a granna
(4.2 and 3.5 mm long), which is more pronounced in the second order. The
first order hairs have a neck with a medulla narrowing. The tip is short. The
intermediate hairs (10.9 mm long, 23μ thick, medulla 56.6%) have a sinuate
club and a small granna (2.4 mm), and a neck beneath the granna has one or
(less often) two slight ligatures. The fur hairs (9.8 mm, 21μ, medulla 33.3%)
have two or three ligatures (the medulla is broken at the upper ligature).
Part of the shaft above the upper ligature is somewhat thinner than below,
and the medulla is different.

The hairs grow in tufts of seven to eleven or in groups of two tufts with
one big hair between them, with 330 big hairs and 18,670 others per 1 cm^2.

In winter the hair is not growing and the skin, the epidermis, and the stratum corneum are thicker than in summer. The outer surface of the skin is
wrinkled. The stratum malpighii is two cells deep; the stratum corneum is
loose. The dermis is similar in structure to the summer specimen, except
that the fat cells (56 x 56; 39 x 56μ) in the reticular layer are few. Individual
bundles of cross-striated muscles rise rather high (above the level of the hair
bulbs) in the breast skin. The sebaceous glands are somewhat larger (45 x 84,
34 x 62μ) than in the summer specimen.

The categories of hairs are similar. The few guard hairs (16.5 mm long,

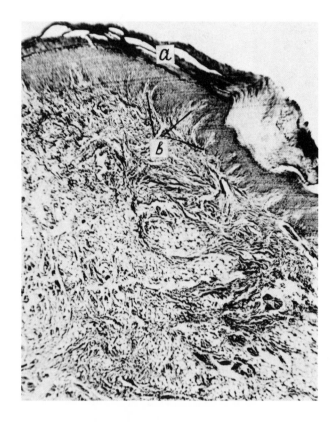

Fig. 76. *Spalax microphthalmus*. Skin of sole: *a*, stratum corneum;
b, dermal papillae.

56μ thick, medulla 50.0%) are straight, thicken somewhat in the distal two-thirds, and have a long, medulla-free tip with no medulla at the base. The pile hairs divide into two orders (13.9 mm and 14.3 mm long, 5.8 and 5.1μ thick, medulla 83.7 and 79.1%, granna 5.8 mm). In the second order the club is wavy (up to six waves) and the granna is more pronounced (5.8 and 5.1 mm long). The medulla thins in a neck below the granna. In the intermediate hairs (13.5 mm long, 22μ thick, medulla 63.7%) the granna (2.8 mm) is less pronounced. In addition to the neck there is an extra shaft contraction with a medulla narrowing. The fur hairs are similar to the summer hairs (11.5 mm long, 22μ thick) and are distinguished by a long, medulla-free tip. The medulla (63.7%) is much better developed than in summer. Guard hairs grow singly, others in tufts of two to seven, although tufts do occur with up to eighteen hairs. One guard hair and three to four tufts form a group. There are 1,333 guard and 32,000 other hairs per 1 cm^2.

Familia Muridae

Nesokia indica

Material. Skin from withers (table 34) and hind sole from specimen (1) (body length, 19 cm), April 28, Tashauz, specimen (2) (body length, 18 cm), December 27, near Bairam Ali, and specimen (3) February 27, near Samarkand.

In summer the hair is at a relatively early stage of growth, but the skin has thickened. The stratum malpighii is two to three cells deep. The epidermis of the hind sole is very thick (table 34); the stratum corneum is thick and compact; the inner surface of the epidermis is alveolar; the stratum granulosum is well developed; the epidermis is not pigmented.

TABLE 34

EPIDERMIS OF NESOKIA INDICA'S HIND FOOT (μ)

Thickness	Specimen (1)	Specimen (2)
Epidermis	495	165
Stratum corneum	220	66
Epidermal alveoli	165	77

The withers dermis has a thicker reticular layer than papillary layer. The mainly horizontal (especially in the reticular layer) collagen bundles form a compact plexus with elastin fibers (numerous in the papillary layer). There are large groups of fat cells (28×67, $45 \times 70\mu$) in the middle and inner reticular layers. The hind sole dermis has numerous large pigment cells. The sebaceous glands are large (22×62; $22 \times 84\mu$), deep set, one at every hair. The gland in the corner of the mouth is 3 to 4 mm long (Skurat, 1972), covered outwardly with sparse hairs. Its sebaceous glands reach 0.50×1.00 mm. Their duct diameter is 0.39 mm.

The anal gland forms a ring around the anal orifice (Skurat, 1972). The ring is 0.28 mm thick and 1 mm wide. The ducts of the individual, multilobal sebaceous glands of the anal gland open into the anal orifice. There are preputial glands (Schaffer, 1940). In the skin bulges of the hind foot callosities, tubular gland glomi are larger in specimen (1). The diameter of the secretion cavities is 50μ in specimen (1) and 28μ in specimen (2). In the winter specimen (2) the gland glomi are surrounded by fat cells, the glands being less active than in summer specimen (1). Arrectores pilorum muscles are rather well developed, 12 to 28μ in diameter.

There are four categories of hairs. Guard hairs (22.8 mm long, 103μ thick, medulla 80.6%) have a long club that is thickest three-fourths to four-fifths up the shaft. The three orders of pile hairs (19.6, 17.9, and 14.3

mm long; 106, 107, and 49μ thick, medulla 89.7, 86.0, and 87.2%) have a pronounced granna (13.2, 10.8, and 6.7 mm) but no neck or pregranna narrowing.

Some very specific hairs (14.6 mm long, 114μ thick, medulla 90.4%) belong to a third category. These have a short, medulla-free club that thickens sharply into a shaft, then gradually tapers toward the tip. The fourth category is fur hairs. These are slightly wavy and have no ligatures or medulla narrowings. Their thickness (24μ) is constant along their whole length (11.4 mm). The medulla is 54.2%. The hairs grow singly or in pairs, forming dense groups. Eight to twelve hairs of different types are grouped around a guard hair. Sometimes four to nine hairs of the second, third, and fourth categories form a group without a guard hair. There are 330 guard hairs and 7,330 hairs of other types (3,640 to 5,460 on the back and 2,300 to 2,600 on the belly per 1 cm² (Kao Van Shung, 1970).

In winter, specimen (2) hairs are not growing and the skin is thin. Specimen (3) hairs are at a later stage of growth than the summer specimen, and the skin is rather thick. Both winter specimens have a thicker epidermis and stratum corneum than the summer animal. The stratum malpighii is two to three cells deep. Owing to hair growth, the reticular layer is absent in specimen (3). In the dermis, horizontal collagen bundles predominate. Groups of fat cells (44 × 56μ) directed to the hair bulbs traverse the specimen (2) reticular layer. In specimen (3), fat cells (28 × 56μ) occur in the papillary layer. Both have subcutaneous fat tissue, 330μ (2) and 110μ (3) on the withers, and 165μ (2) and 110μ (3) on the breast, with rare bundles of collagen fibers in all of them. The sebaceous glands, about the same size as in the summer specimen (22 × 62; 45 × 84μ) are deep in the skin, one at each hair. The hairs divide into four categories, but no guard hairs are recorded on the small area of skin under study (the presence of guard hairs is confirmed by Kao Van Shung, 1970). The pile hairs (17.3 and 16.6 mm long, 81 and 67μ thick, medulla 86.5 and 88.1%, granna length 9.3 and 7.9 mm) differ in shape from the summer hairs. The third category of hairs (13.6 mm long, medulla 88.6%) has a very small thickening (35μ). The fur hairs have none, their length being 13.3 mm, thickness 23μ, medulla 78.3%. They are slightly wavy with two or three slight ligatures. The hairs grow in tufts of eight to ten, with 12,000 hairs per 1 cm² (3,420 to 3,800 on the back and 1,990 to 3,310 on the belly, according to Kao Van Shung, 1970).

Rattus norvegicus

Material. Skin from withers (table 35) and soles of two specimens (body length 21 cm, 19 cm) taken in Grodno region.

The hairs of both are growing, though specimen (1) is more advanced and its skin is thicker. The stratum malpighii is two cells deep. In the skin of the hind sole the epidermis is 0.66 mm thick, of which 330μ is the stratum corneum. The inner surface of the epidermis is alveolar. The height of the alve-

COMPARATIVE MORPHOLOGY

TABLE 35

MURIDAE: SKIN MEASUREMENTS (μ)

	Skin	Epidermis	Stratum corneum	Dermis	Papillary layer	Reticular layer
Nesokia indica						
Withers						
Summer						
(1)	812	22	9	790	350	440
Winter						
(2)	818	28	15	460	350	110
(3)*	1,018	28	13	880	880	—
Breast						
Summer						
(1)*	578	28	17	550	275	275
Winter						
(2)*	385	22	11	198	176	22
(3)*	358	28	14	220	220	—
Rattus norvegicus						
Withers						
(1)	1,418	28	15	1,390	1,115	275
Withers						
(2)	869	22	7	847	605	242
Breast						
(1)	892	22	7	770	330	440
Rattus rattus						
Withers						
Summer						
(1)	1,232	22	8	1,210	1,100	110
(2)	578	28	12	550	462	88
Winter						
(4)	457	17	6	440	297	143
Breast						
Summer						
(1)	682	22	11	660	594	66
(2)	627	22	11	605	550	55
Winter						
(3)	253	22	11	231	187	44
Mus musculus						
Withers						
(1)	281	28	17	253	165	88
Withers						
(2)	220	22	11	198	121	77
Apodemus agrarius						
Withers						
Summer						
(1)	155	15	5	140	123	17
(2)	498	22	11	476	454	22
(3)	129	22	12	107	90	17

	Skin	Epidermis	Stratum corneum	Dermis	Papillary layer	Reticular layer
(4)*	190	22	11	123	123	—
(5)	213	28	17	185	157	28
(6)	212	22	14	190	140	50
Breast						
Summer						
(1)	715	22	11	693	627	66
(3)	158	18	9	140	112	28
(4)	297	17	6	280	168	112
Winter						
(5)	139	10	4	129	95	34
(6)	174	17	10	157	123	34
Apodemus sylvaticus						
Withers						
(1)	140	17	6	123	95	28
(2)	151	17	6	134	100	34
Breast						
(1)	160	17	6	143	110	33
(2)	193	17	6	176	121	55
Micromys minutus						
Withers						
Summer						
(1)	107	17	11	90	73	17
(2)	242	22	11	220	220	—
(3)	117	22	11	95	84	11
(4)	366	28	11	338	338	—
(5)	385	22	7	363	363	—
Winter						
(6)	133	17	11	116	101	15
(7)	101	17	11	84	67	17
Breast						
Summer						
(1)	123	11	4	112	84	28
(2)	101	17	7	84	67	17
Winter						
(6)	164	17	10	147	113	34
(7)	154	14	7	140	123	17

*Thickness of the subcutaneous fat tissue is given in the text.

oli is 220μ. The border between the stratum malpighii and the stratum corneum is irregular. In the withers dermis, the reticular layer is thinner than
the papillary, though this is well developed. Collagen fiber bundles are
arranged in various directions in the papillary layer and are largely horizontal in the reticular, the network being compact in the whole dermis.
Specimen (1) has large groups of fat cells (28 × 50; 39 × 50μ) in the reticular
layer under the hair bulbs. Sebaceous glands are rather big (34 × 129; 39 ×
134μ) and oval, singly behind the hairs. The sebaceous glands are well
developed on the tail (Schaffer, 1940). The specific glands are similar to
Nesokia indica (Skurat, 1972).

The mouth corner glands are 2.5 mm long, 2.5 mm wide and 1.3 mm
thick. The sebaceous glands are 0.3 to 0.80 × 0.67 to 1.23 mm. The multilobal sebaceous glands of the anal gland range from 0.50 × 0.56 mm to
1.36 × 1.06 mm, opening either independently or into hair bursae (Schaffer,
1940). Preputial glands are 16 to 17 × 6.55 to 7 × 1 to 3.17 mm, consisting
of alveolar glands (Schaffer, 1940). On the hind soles, tubular glands form
small glomi (112 to 190 × 218 to 391μ). The diameter of their secretion
cavity is 39μ. The ducts open into the skin surface. Some glomi are deep in
the subcutaneous muscles. The arrectores pilorum muscles are thin (10 to
17μ).

In albino rats, foot sole glands in hind feet are concentrated in eleven skin
plantar tubercules: five apical pads at the tips of the digits, four interdigi-
tals, and two metatarsals; in the front foot there are only ten such pads
(Ring and Randall, 1974). Sole sweat glands of albino rats have only one
type of secretion cell (Munger and Brusilov, 1971). The secretion of these
glands shows very high levels of potassium.

There are three hair categories: guard hairs (16.1 mm, 85μ, medulla
85.9%) have a thin club that expands into a granna (length 8.2 mm) in the
distal section. There is neither neck nor medulla narrowing before the
granna. The three orders of pile hairs (13.9, 13.5, and 12.2 mm; 79, 62, and
47μ; medulla 92.5, 80.7, and 74.5%) have a pronounced granna (6.0, 6.9,
and 3.6 mm) with a neck and a medulla narrowing beneath. The hairs are
straight. The two orders of fur hairs (10.9 and 6.9 mm, 24 and 20μ, medulla
66.7 and 70.0%) have two to three ligatures (first order) or three ligatures
(second order), and no granna. The hairs are bent at the ligatures. The hairs
are arranged singly or in pairs, six to eight forming a group, 7,666 per cm².
(According to Kao Van Shung, 1970, in winter there are 5,350 to 5,710 hairs
per cm² on the back, and 6,800 to 8,610 on the belly; in summer there are
4,460 to 5,600 hairs and 3,880 to 5,950, respectively).

Rattus rattus

Material. Skin from withers and breast (table 35) from three specimens,
taken in summer (body length 17 and 16 cm) July 14, in Grodno region and

June 21, in Novgorod region; and in winter (body length 16 cm) taken December 15, Grodno region.

In summer the withers hair of specimen (1) is growing and the skin is thicker than in specimen (2), with hair not growing. The stratum malpighii is two cells deep. On the withers, the papillary layer is considerably thicker than the reticular layer. In specimen (1) only the upper zone of the papillary layer, about 440μ thick, is a compact plexus of collagen bundles arranged in various directions, the remainder being filled by large hair roots with interspersed fat cells (39 x 62μ). The reticular layer has a lax plexus of collagen bundles and major accumulations of fat cells.

In specimen (2) the collagen plexus in the dermis is compact. The bundles run in various directions in the papillary layer, but are mostly horizontal in the reticular. There are no fat cells in the dermis of this specimen. Elastin fibers are few in both. The structure of the breast skin is similar to that of the withers. The sebaceous glands (34 x 112, 68 x 73μ) are rather large, ovally twisted, pointing downward, one behind each hair. In a Hawaiian population of *Rattus exulans,* an abdominal skin specific gland, formed by sebaceous glands, is biggest in adult males (average 43 mm long and 4.7 mm wide) (Quay and Tomich, 1963). Thin arrectores pilorum muscles (8 to 14μ) are recorded only in specimen (2).

There are three categories of hairs. The guard hairs (23.2 mm long, 71μ thick, medulla 53.5%) are straight, thickening gradually from the proximal up three-fourths of the shaft. The two orders of second category hairs (14.9 and 14.2 mm long, 143 and 61μ thick, medulla 93.1 and 90.2%) can be classified for convenience as pile hairs. They expand rather sharply at the base, remaining almost the same thickness until they begin to taper at three-fourths of the shaft. They are considerably shorter than the first category, and their medulla is much better developed. The fur hairs (three orders) are of uniform thickness their entire length. The first order has a single ligature, the second category has two, the third category has three. The medulla is broken at the thinner ligatures where the hair is bent at an obtuse angle. The length of the fur hairs is 12.9, 9.1, and 8.9 mm; thickness 45, 26, and 22μ; medulla 84.5, 57.7, and 54.6%. The hairs grow singly, in unpronounced groups of three to five, with 500 first category hairs and 11,166 others per 1 cm².

In winter the withers hairs are not growing, and the skin, the epidermis, and the stratum corneum are thinner than in summer. The stratum malpighii is two cells deep. The reticular layer is thicker on the withers than in the summer animals. Horizontal collagen bundles predominate in the dermis, forming a compact plexus. There are small groups of fat cells (28 x 39, 26 to 44μ) in the reticular layer. The sebaceous glands are somewhat larger than in summer (45 x 112; 50 x 112μ). A horizontal section shows bilobal sebaceous glands beside first category hairs. The arrectores pilorum muscles are thin (12μ). The hair categories are the same as for the summer specimens. The guard hairs measure in length 19.3 mm, thickness 116μ, medulla 89.7%. The pile hairs (first order) 15.2 mm, 191μ, and 96.9%. The pile hairs (second order) measure 14.4 mm, 102μ, and 92.2%. The fur hairs (first order) measure 13.3 mm, 61μ, and 89.8%. The fur hairs (second order) 11.2 mm, 46μ, and 32.4%. The fur hairs (third order) 8.7 mm, 24μ, and 66.7%. In the fur hairs, first order, part of the shaft below the ligature is thicker than the distal part. The second order has only one ligature. The third order has two ligatures. Most of the hairs grow singly (some in pairs). Up to twenty other hairs are grouped around each guard hair. There are 1,000 guard hairs and 16,333 others per cm^2.

Mus musculus

Material. Skin from withers (table 35) of two specimens (body lengths 9 cm and 8 cm) taken August 24, Kara-Kul, Uzbek SSR.

The hair is not growing in either animal. The stratum malpighii is two cells deep. The reticular layer of the dermis is rather thick, but is thinner than the papillary layer. The plexus of collagen fiber bundles in the dermis is compact. The majority of bundles are horizontal, but bundles run in various directions. There are few elastin fibers and there are no fat cells in the dermis. The sebaceous glands are small (22 x 34; 28 x 67μ), oval, and behind the hairs at the upper half of their roots. The anal gland is formed by sebaceous glands (Schaffer, 1940). There are also preputial, upper lip (tubular), and sole (tubular) glands (Schaffer, 1940). Thin (6μ) arrectores pilorum muscles are recorded only in specimen (2).

Apodemus agrarius

Material. Skin from withers (table 35) and hind sole of four summer specimens (body lengths 9, 10, 8, and 9 cm) taken June 14, Krasnodar region; July 4, Bobruisk region; August 20, near Vladivostok; and from two winter animals (specimens 5 and 6) (body lengths 10 and 9 cm) taken January 15 and February 29, Krasnodar region.

In summer no geographic variability of the skin is found. The hair of specimens (1), (3) and (4) is not growing and their skin is thinner than that of specimen (2), with growing hair. The stratum malpighii is one to two cells deep. In the hind sole the epidermis is 129μ thick and the stratum corneum 45μ. The inner surface of the epidermis is alveolar, the height of the alveoli being 39μ. The stratum granulosum is well pronounced. No reticular layer is recorded in specimen (4), where the hair bulbs touch the subcutaneous fat tissue. In two other specimens the reticular layer is thin.

The collagen bundles in the dermis are predominantly horizontal. In specimens (1) and (4), groups of fat cells (28 x 56, 22 x 39μ) occur at the hair bulb level. Specimen (2) has few fat cells and specimen (3) has none. There are large pigment cells in the dermis of the hind soles. Only specimen (4) has fat cells in the subcutaneous fat tissue, two to three layers of cells (thickness 45μ). The sebaceous glands are small (22 x 56; 28 x 84μ), one behind each hair, appearing crescent-shaped in a longitudinal section of the skin with their flat side pressed against the hair bursa. A vertical section shows that each gland envelops the hair root front and back. There are specific skin glands: mouth corner, anal, and on the soles. The mouth corner glands are 1.8 to 2.0 mm long (Skurat, 1972) formed by sebaceous glands 0.20 x 0.31 mm. The width of the anal gland is 2 mm, its sebaceous glands being 0.46 x 0.38 mm (Skurat, 1972). There are rather large eccrine sweat gland glomi (154 x 308μ) in the callosities of the hind soles. The diameter of their secretion sections is 28μ. Arrectores pilorum muscles are well developed, but they are small in diameter (6 to 10μ).

There are three hair categories. Guard hairs (15.8 mm, 119μ, medulla 94.2%) are straight, thicken at the base, and then remain uniform the whole length. There is a pre-apical narrowing three-fourths to four-fifths up the shaft. The pile hairs (13.9 mm long, 45μ thick, medulla 75.5%) have a very pronounced granna, defined by a medulla narrowing but no neck. The granna is 6.2 mm long. The fur hairs resemble those of shrews. In most of the hairs there are two small ligatures with a thinner medulla. The length of the fur hairs is 11.9 mm, thickness 24μ, medulla 58.4%. The hairs grow singly. There are some not very well defined rows of a few hairs. There are 1,333 guard hairs and 8,666 hairs of the second and third category per cm². The vibrissae on the muzzle are shorter and fewer in number than on other members of *Apodemus* (Kratochvill, 1968) and there is no sex dimorphism in the size or number of vibrissae.

In winter the hairs are not growing, but the skin is thick. The stratum malpighii is one to two cells deep. In the dermis, horizontal bundles of collagen fibers predominate and form a dense plexus. In specimen (6) some large (22 x 45; 34 x 50μ) fat cells occur in the reticular layer. The sebaceous glands are much the same size as in summer (22 x 62, 39 x 84μ). The arrectores pilorum muscles are well developed but thin (5 to 10μ).

The hair categories resemble those of summer. There are two size orders each of guard and fur hairs. The first order guard hairs are 17.3 mm long, 95μ thick, medulla 92.7%. The second and third orders are 15.8 mm, 101μ, 94.1%. The pile hairs are 13.9 mm, 45μ, 75.6%. The fur hairs (first order) are 13.8 mm, 22μ, and 50.0%; (second order) 12.4 mm, 20μ, and 60.0%. Most of the hairs grow singly, some in pairs, with 20,333 hairs per cm².

Haitlinger (1968*a, b*) in *Apodemus agrarius* taken in Wroclaw, Poland, in summer and in winter, likewise distinguishes three hair categories: guard, pile, and fur. According to measurements and coloration, each category is divided into several "types" (statistical treatment of measurements is not given, so the significance of the above indexes is open to question) (see tables 36 and 37).

TABLE 36

AVERAGE MEASUREMENTS OF HAIRS FROM DIFFERENT REGIONS OF BODY IN A. AGRARIUS (IN MM.)
(HALFTINGER, 1968*a, b*)

Region	Hair type	Hair length	Granna length	Thickness of granna (μ)
Winter: Middle of back	Guard hairs: Tipped	11.46	4.54	60
	Black-tipped brown	10.11	4.75	75
Posterior part of back	Tipped	12.88	4.67	57
	Black-tipped brown	10.54	4.97	69
Middle of belly	Tipped	7.61	3.06	50
	True	6.44	3.34	69
Summer: Middle of back	Tipped	10.09	7.04	80
	Black	8.85	6.96	112
	Black-tipped brown	7.98	6.46	112
	Brown	7.71	6.62	108
	Brown bent	7.65	6.85	74
Posterior part of back	Tipped	11.15	6.98	71
	Black	9.10	6.80	101
	Black-tipped brown	8.26	6.48	107
	Brown	7.86	6.06	93
Middle of belly	Tipped	6.50	3.95	55
	True	4.47	3.38	89
Winter:	Guard hairs: Guard	9.42	4.25	72
	Acuminate	8.42	5.73	60
	Medium	7.15	3.34	49
	Short	6.64	2.61	32
Posterior part of back	Guard	9.94	3.95	61
	Acuminate	9.37	3.68	53
	Medium	8.10	3.22	40
	Short	7.53	2.72	31
Belly	Chief	5.18	1.22	—
	Intermediate	4.90	1.42	—
	Acuminate	4.71	1.69	
Summer: Middle of back	Chief	7.09	1.22	—
	Intermediate	6.42	1.66	—
	Acuminate	5.88	1.52	—

TABLE 36—*continued*

Region	Hair type	Hair length	Granna length	Thickness of granna (μ)
Posterior part of back	Chief	7.96	1.18	—
	Intermediate	7.45	1.43	—
	Acuminate	7.21	1.89	—
Belly	Chief	3.38	1.07	—
	Intermediate	3.08	1.11	—
	Acuminate	3.05	1.21	—

TABLE 37

HAIR NUMBERS PER 1 CM² OF SKIN SURFACE (HALFTINGER, 1968*a*, *b*)

Region	Summer				Winter			
	Guard	Pile	Fur	Total	Guard	Pile	Fur	Total
Posterior part of back	350	375	5,675	6,400	350	900	9,875	10,825
Middle of back	550	375	4,700	5,625	325	625	8,175	9,125
Middle of belly	525	125	4,400	5,050	375	400	6,590	7,725

Apodemus sylvaticus

Material. Skin from withers and breast (table 37a) of two specimens (body lengths 7 cm and 8 cm) taken August 11, Transcarpathian area.

In specimen (2) the hairs are in the initial stage of growth, so the skin is relatively thick. The stratum malpighii is two cells deep. The papillary layer of the dermis is considerably thicker than the reticular. The plexus of collagen fiber bundles is compact and predominantly horizontal in the dermis. Elastin fibers are few. Specimen (2) has two to three layers of fat cells (17×22; $13 \times 22\mu$) in the lower reticular layer. The sebaceous glands are small (17×56; $28 \times 67\mu$), oval, and set singly behind the hairs. The specific glands at the corners of the mouth, anus, and sole are the same as in other mice (Skurat, 1972). The multilobal sebaceous glands forming the mouth corner gland (0.18 to 0.29 $\times$ 0.45 to 0.62 mm) are elongated along the hair roots. These glands are larger in summer than in winter. The anal gland's multilobal sebaceous glands reach 0.86 $\times$ 2.18 mm. The entire anal gland ranges from 1.28 to 1.64 mm in width, 0.32 to 0.91 mm in thickness, with seasonal

TABLE 37a

AVERAGE MEASUREMENTS OF HAIRS FROM DIFFERENT REGIONS OF BODY IN A. SYLVATICUS (IN MM)
(HAITLINGER, 1968a)

Region	Hair type	Hair length	Granna length	Granna thickness (μ)
Guard hairs				
Winter				
Middle of back	Tipped	12.70	5.10	51
	Black	10.24	4.36	55
Posterior part	Tipped	13.05	5.04	50
of back	Black	10.96	4.48	46
Belly	Tipped	9.38	4.34	48
	True	7.07	3.45	54
Summer				
Middle of back	Tipped	11.08	4.77	48
	Black	8.98	4.59	49
Posterior part	Tipped	12.51	4.81	38
of back	Black	9.58	3.94	43
Belly	Tipped	7.72	3.78	38
	True	5.97	3.43	51
Guard hairs				
Winter				
Middle of back	Guard	9.19	3.51	45
	Acuminate	8.12	3.04	42
	Medium	7.60	2.71	37
	Short	7.27	2.15	30
Posterior part	Guard	10.37	3.47	41
of back	Medium	9.14	2.49	33
	Short	8.63	2.04	27
Belly	Guard	6.66	3.07	48
	Acuminate	6.35	2.71	42
	Medium	5.62	2.49	36
	Short	5.27	2.20	27
Posterior part	Guard	9.49	3.54	37
of back	Acuminate	8.82	2.88	32
	Medium	8.05	2.43	28
Belly	Guard	5.48	2.61	—
	Acuminate	5.03	2.44	—
Fur hairs				
Winter				
Middle of back	Chief	7.43	1.18	—
	Intermediate	6.64	1.64	—

TABLE 37a—*continued*

Region	Hair type	Hair length	Granna length	Granna thickness (μ)
	Acuminate	5.74	1.69	—
	Giant	6.84	1.65	—
	Chief	8.70	1.27	—
	Intermediate	7.99	1.60	—
	Acuminate	7.66	2.06	—
Belly	Chief	5.18	1.22	—
	Intermediate	4.90	1.42	—
	Acuminate	4.71	1.69	—
Summer Middle of back	Chief	7.09	1.22	—
	Intermediate	6.42	1.66	—
	Acuminate	5.88	1.52	—
Posterior part of back	Chief	7.96	1.18	—
	Intermediate	4.45	1.43	—
	Acuminate	7.21	1.89	—
Belly	Chief	4.58	1.11	—
	Intermediate	4.19	1.25	—
	Acuminate	3.95	1.33	—

variations in size; the glands are smaller in winter. Eccrine sweat gland glomeruli, forming the sole gland, are 79 to 123×162 to 217μ, (57 to 102×99 to 179μ in winter).

Among the members of the family under investigation, only in *A. sylvaticus* and in *A. flavicollis* is a caudal gland recorded (Skurat, 1972) on the median line of the lower surface of the tail. It looks like a skin bulge. The size of this gland varies considerably from one individual to another. The caudal gland is formed by multilobal sebaceous glands (0.09 to 0.47×0.21 to 0.70 mm) (fig. 77). Sex and seasonal dimorphism in size is noted in the caudal gland, which is larger in males than in females and is larger in summer than in winter. The arrectores pilorum muscles are thin (5 to 6μ).

Hair divides into three categories. The guard hairs (9.8 mm long, 57μ thick) are rather peculiar, having a long, gradually thickening club with neither neck nor medulla narrowing proximally. The length of this thickening is 6.6 mm. The pile hairs (8.8 mm long, 33μ thick, medulla 81.9%) have a long, straight club and a granna (length 3.4 mm) with a neck before it and a thin medulla. There are three orders of fur hairs (8.3, 7.5, and 6.9 mm long; 22, 17, and 13μ thick; medulla 50, 58.9, and 53.9%). The first order has two to three ligatures; the second order has three ligatures; and the third order

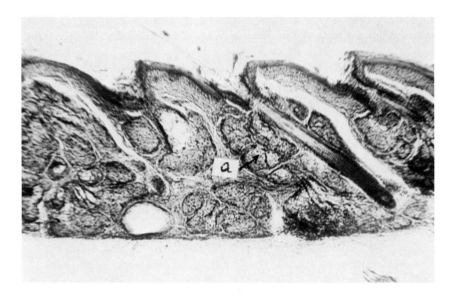

Fig. 77. *Eliomys quercinus.* Withers skin: *a,* sebaceous gland.

has two ligatures where the hair is bent at an obtuse angle. The hairs grow singly, but form compact groups of three or four, with 15,000 hairs per 1 cm². Haitlinger (1968*a*) likewise distinguishes three hair categories in *Apo-demus sylvaticus* taken in Wroclaw, Poland, in summer and in winter (table 37a). The vibrissae on the muzzle are longer and more numerous than in *A. agrarius* (Kratochvil, 1968).

Micromys minutus

Material. Skin of the withers, breast (table 35) and hind sole; specimens (1) to (5) (body length 5 cm) June 4 (body length 6 cm) August 14, Amur region (body length 4.5 cm) June, Krasnodar region (body length 5.5 cm) July, Mogilev region. Specimens (6) and (7) (body lengths 5 and 5.5 cm) January 7, Kurgan region.

In summer no geographic variability is found in the skin. The hair of specimens (4) and (5) is growing and their skin is particularly thick. The hair of specimen (2) is in the initial stage of growth and the skin is somewhat thickened. The hair of specimens (1) and (3) is not growing and the skin is thin. The stratum malpighii in the specimens with growing hair is two to three cells deep, in the others, one to two cells. The epidermis is not pigmented. The hind sole epidermis is rather thick (134 to 185μ); the stratum corneum is highly developed (28 to 45μ); the alveoli of the inner surface of the epidermis are 56 to 95μ high.

The withers dermis in specimens (1) and (3) has a thin reticular layer with a compact plexus of collagen fiber bundles, mainly horizontal. Elastin fibers are few, fat cells absent. In the specimens with growing hair, the reticular layer is absent, as the hair bulbs adjoin the subcutaneous muscles. Only the upper dermis has a compact collagen plexus, which occupies half the dermis thickness in specimen (2), is 78μ in specimen (4), and is 112μ in specimen (5). Here most of the collagen bundles are horizontal. Elastin fibers are rare. The rest of the dermis is made up of large hair roots with fat cells (22×34; $28 \times 39\mu$) along them. There are pigment cells in the hind sole dermis. The sebaceous glands in the withers skin are of uniform size (34×56; $40 \times 73\mu$) in all the specimens, and are oval or elongated, singly placed behind the upper part of the hair roots. In specimens (4) and (5) most of the glands are inactive. Where the skin of the hind sole is thickened, there are large eccrine sweat gland glomi. These glands are of uniform size in summer and winter animals. The diameter of the secretion section of all is 22μ. The arrectores pilorum muscles are thin.

The hair divides into three categories. Guard hairs (two orders) have a long, thin club that comprises a third to half the hair (first order). There is a thickening, which is not a granna. In the second order the club is shorter. The guard hairs are 7.2 and 5.3 mm long, 43 and 50μ thick, medulla 81.6 and 64.3%. The pile hair (5.3 mm long, 28μ thick, medulla 64.3%) granna is well pronounced and 2.0 mm long. Beneath is a neck and a medulla narrowing. The fur hairs (length 5.0 mm, thickness 17μ, medulla 65.8%) have two ligatures in which the medulla is thinner and the hair is bent obtusely. The hairs grow singly and form no groups. There are 21,666 hairs per cm^2.

Winter specimen (6) has growing and nongrowing hairs. The hair of specimen (7) is not growing. In the area with growing hairs, the skin and epidermis are thickened. The epidermal structure is similar to that of summer. In the dermis, horizontal collagen fiber bundles predominate. The bulbs of the growing hairs of specimen (6) reach the subcutaneous muscles and, in areas where the reticular layer is absent, there are fat cells in rows between the hair bursae. In specimen (7) the reticular layer is filled with fat cells (22×45, $34 \times 56\mu$). The sebaceous glands are about the same size as in summer speimens (17×39; $28 \times 84\mu$).

The hair categories are the same as in summer. The guard hairs (first order) are 8.0 mm long, 40μ thick, medulla 72.5%; (second order) 7.0 mm, 39μ, and 77.0%. The pile hairs are 6.8 mm, 27μ, and 59.5%. The fur hairs are 5.9 mm, 20μ, and 85.0%. The pile hair granna length is 2.0 mm. The fur hairs have two to three ligatures. The hairs grow singly. There are 27,666 hairs per 1 cm^2.

FAMILIA MYOXIDAE

Glis glis

Material. Skin from withers (table 38) and hind sole from two specimens

TABLE 38

Myoxidae and Seleviniidae: measurements of withers skin (μ)

	Skin	Epidermis	Stratum corneum	Dermis	Papillary layer	Reticular layer
Glis glis						
Withers						
Summer	391	17	5	374	154	220
Winter	330	28	17	302	112	190
Breast						
Summer	364	28	17	336	112	224
Winter	173	17	11	156	84	72
Dyromys nitedula						
Summer						
(1)*	897	17	6	660	660	—
(2)*	470	22	11	196	112	84
Eliomys quercinus						
Autumn						
(1)	1,018	28	17	990	990	—
(2)	798	28	14	770	715	55
(3)*	583	22	7	264	132	132
Muscardinus avellanarius						
Summer	118	28	17	90	73	17
Selevinia betpakdalensis						
Withers	313	28	11	285	123	162
Breast	264	17	7	247	118	129

*Thickness of the subcutaneous fat tissue is given in the text.

(body lengths 16 and 15 cm) taken August 2, in Transcarpathia, and February, in the Caucasus.

In summer there are three zones for thickness of skin (Tserevitinov, 1958). The thickest skin is in the middle of the back and on the head. The second borders the first on the sides, sacrum, and withers. The third is on the legs. The stratum malpighii is two or three cells thick.

The epidermis and the stratum corneum are thick on the hind sole (224 and 140μ). The height of the alveoli at its inner surface is 45μ. The stratum lucidum and stratum granulosum are pronounced. The withers hairs are not growing, and the dermis has a characteristic high development of the reticular layer which is thicker than the papillary layer. Horizontal, very wavy bundles predominate in a dense plexus of collagen fibers. In the papillary layer horizontal elastin fibers are abundant. The sebaceous glands are round or oval (17 x 39; 22 x 45μ), one behind each hair. The glands at the mouth corners are not outwardly visible because they are not fur-covered, as are those of most rodents (Skurat, 1972). The multilobal sebaceous glands comprising this gland are 0.36 x 0.55 mm (Schaffer, 1940, reports tubular glands as well in the upper lip).

The anal gland is 7.5 to 11 mm wide (Skurat, 1972). According to Schaffer (1940) a twin anal gland is egg-shaped, 6 mm long. In addition to its multilobal sebaceous glands, there are also large sebaceous glands (0.17 x 0.36; 0.39 x 0.48 mm) on the periphery of the anus. According to Hrabě (1971) the glands comprising the anal gland belong to the tubo-alveolar type.

There are twin preputial glands formed by sebaceous glands (Schaffer, 1940). In the callosities of the hind sole skin are large, eccrine sweat gland glomi (Schaffer, 1940). Their secretion cavities are 28μ in diameter. Arrectores pilorum muscles 5 to 10μ thick are seen beside only some of the hairs.

There are two hair categories. The guard hairs (14.7 mm, 45μ) are straight, with a sharp dilation of the upper part forming an 8 mm granna. Most hairs have a medulla thinning just below the granna. In a few of the hairs the medulla is broken. There is a long, thin medulla-free tip. The medulla is not very well developed (40%). All the other hairs belong to the second category (which, for convenience, we will call pile). Their granna is well pronounced although there is no neck below it, and in some hairs there is no medulla narrowing at all. The club has six to nine waves; the tip is short. There are four orders: length 14.2, 13.1, 10.4, and 10.3 mm; thickness 48, 42, 26, and 23μ; medulla 45.8, 52.3, 57.6, and 65.2%; granna length 6.7, 4.8, 2.7, 2.9 mm. The hairs grow singly or in pairs and form no groups. There are 22,300 hairs per 1 cm². The pelt divides into three zones for length (Tserevitinov, 1958). The longest hair is on the sacrum; the shortest is on the ventral side of the body.

In winter the skin is thinner than in summer. The epidermis and stratum corneum are thicker. The structure of epidermis and dermis is similar in the summer animals. The reticular layer is thicker than the papillary. The hind sole epidermis is 224μ thick. The height of the epidermal alveoli is 101μ; the stratum corneum is very thin at 56μ; the stratum granulosum and stratum lucidum are present in the epidermis. The shrunken sebaceous glands (12 x 34; 12 x 28μ), found only at some hair roots in the body skin, are inactive. Mostly only the ducts remain. In the hind sole callosities, the eccrine sweat gland glomi are as large as in summer (diameter of their secretion sections is 28μ), with small gland glomeruli, with excretion ducts opening directly on the smooth surface beyond the skin, although the hair is growing.

Dyromys nitedula

Material. Skin from withers (table 38) and hind sole of two specimens (body lengths 9.5 cm and 8 cm) taken August 24 and September 1, Belovezhskaya Puscha.

In specimen (1) the withers hairs are growing and the skin is thick. The hair of specimen (2) is not growing and its skin is much thinner. The stratum malpighii is two cells deep. The hind sole epidermis 280μ thick; the alveoli on its inner surface are 112μ high; a thin stratum granulosum and stratum lucidum are clearly defined; a compact stratum corneum is 112μ thick. In specimen (1), the growing hair bulbs on the withers adjoin the subcutaneous fat tissue, so there is no reticular layer. The upper dermis (290μ) is a compact collagen fiber plexus. The remainder of the dermis is filled with fat cells.

Specimen (2) has a marked reticular layer, not much thinner than the papillary. The collagen fiber bundles in the dermis of both specimens are mostly horizontal, although there are numerous oblique bundles. The few elastin fibers are mainly in the upper papillary layer. The subcutaneous fat tissue is thick, 220μ in specimen (1) and 250μ in specimen (2), probably because autumn fat is accumulated against hibernation. The fat cells in the subcutaneous fat tissue are 56 x 84; 40 x 73μ. The small, bean-shaped sebaceous glands (22 x 50; 34 x 62μ) are set behind the hairs. There are very large eccrine sweat glands glomi in the hind sole callosities (diameter of secretion section 28μ). The anal gland is similar to that in *Glis glis,* minus the dorsal part and smaller (Hrabě, 1971). Arrectores pilorum muscles (5μ thick) are found only in specimen (1).

There are four categories of hair: guard, pile, intermediate, and fur. The guard hairs (13.5 mm long, 35μ thick, medulla 54.2%) are straight, thickening gradually from the proximal to the upper four-fifths to five-sixths of the shaft, then tapering into a long, very thin medulla-free tip. The pile hairs (11.4 mm, 27μ, medulla 74.0%) have a well-defined granna (2.9 mm) with no neck. Their club is straight; their tip is short. Intermediate hairs (9.5 mm, 22μ, medulla 72.9%) have a small granna (1.8 mm) and a wavy club (six to seven waves) with small shaft narrowings in some, the medulla being thinner. The tip is short. The fur hairs (5.9 mm long, 17μ thick, medulla 64.7%) have no granna. They have up to five waves. The hairs grow singly and do not form groups. There are 23,300 hairs per 1 cm^2.

Eliomys quercinus

Material. Skin from withers (table 38) and hind sole of specimens (1) to (3) (body lengths 11.5 cm, 12 cm, and 12 cm) taken August 14, August 20, September 20 in northern European USSR and near Tbilisi.

In specimens (1) and (2) the withers hairs are growing and the skin has thickened. The skin of specimen (3) is thinner, as most of the hairs have completed growth (fig. 78). The stratum malpighii is two to three cells deep. The hind sole epidermis is thick (220μ) (fig. 79); the alveoli measure 95μ on the inner surface; the stratum granulosum is two to three cells deep; the stratum lucidum is not pronounced; the stratum corneum (84μ) is very compact and resembles the stratum lucidum. Specimen (1) has no reticular layer on the withers, where growing hair bulbs touch the subcutaneous muscles. In specimen (2) this layer is thin, and in specimen (3) it equals the papillary

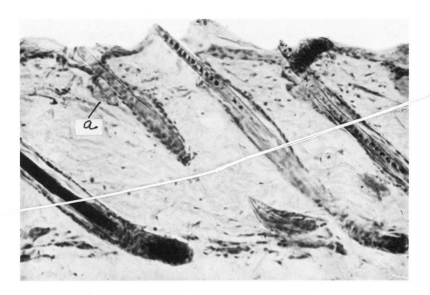

Fig. 78. *Eliomys quercinus: a,* sebaceous glands.

layer. Specimens (1) and (2) have a compact collagen bundle plexus only in the upper papillary layer, 550μ thick in specimen (1), 385μ in specimen (2). In the remainder of the dermis, fat cells (45×56; $45 \times 50\mu$) and loose connective tissue lie between the thick roots. In specimen (3) the dermal collagen plexus is compact (fig. 81). Horizontal bundles predominate, though some oblique ones rise to the skin surface at various angles. Numerous elastin fibers, especially prevalent in the outer papillary layer, are also horizontal. Fat deposits in the subcutaneous fat tissue (thickness 297μ) are found only in specimen (3) (fig. 81), which, being taken later (September 20) had been accumulating fat for hibernation (size of fat cells: 56×90; $56 \times 84\mu$).

The sebaceous glands are bean-shaped and small (17×56; $22 \times 50\mu$) (fig. 81). In a transverse section, one gland is seen behind each hair, and two occur on some. There are huge eccrine sweat gland glomi in the hind sole callosities surrounded by fat cells (fig. 82). The glomi stretch dorso-ventrally. Glomi also occur in other parts of the sole, but they are small and parallel to the skin surface. The diameter of the secretion sections of the glands is 28 to 30μ. Arrectores pilorum muscles, not found in specimens (1) and (2), are 6μ thick in specimen (3).

There are four categories of hairs. The guard hairs (13.4 mm, 47μ, medulla 48.9%) are typical in shape. They gradually thicken for one-fifth of the shaft, then taper into a very long, thin medulla-free tip. The pile hairs (two orders, 12.2 and 10.2 mm long, 37 and 22μ thick, medulla 67.6% and 72.8%) have a pronounced granna (5.7 mm and 2.1 mm), not separated by

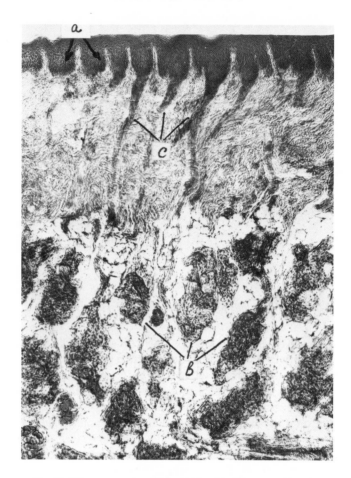

Fig. 79. *Eliomys quercinus*. Sole gland on hind foot: *a,* epidermis;
b, glomi of sweat glands; *c,* gland ducts.

a neck. The long club has up to seven waves; the tip is short. The intermediate hairs (8.2 mm long, 26μ thick, medulla 69.3%) have a weak granna. The fur hairs (8.3 mm long, 19μ thick, medulla 79%), which are also wavy, have none. The distal medulla differs in structure from the proximal. The hairs grow singly, sometimes in pairs. There are 17,660 hairs per 1 cm².

Muscardinus avellanarius

Material. Skin from withers (table 38) (body length 8 cm).

Some of the hairs are in growth, so the papillary layer is thicker than the reticular. The sebaceous glands were not measured because the skin has not been adequately preserved. The sebaceous glands are well developed in the meatus acoustics (Schaffer, 1940). The preputial glands are absent (Schaffer, 1940). The anal gland is similar to that in *Dyromys nitedula* (Hrabě, 1971).

There are three categories of hair. The guard hairs (10.9 mm, 36μ, medulla 66.7%) are straight, thickening gradually for four-fifths of the shaft from the base. They have long medulla-free tips. The pile hairs are similar in shape, but without the long tip. They are shorter (8.9 mm), and thinner (24μ), but the medulla (75.0%) is better developed. The fur hairs, which do not thicken distally, have two (rarely three) ligatures with the shaft bent at an angle. There are two size orders: 5.9 and 7.2 mm long, 21 and 20μ thick, medulla 81.0 and 85.0%.

FAMILIA SELEVINIIDAE

Selevinia betpakdalensis

Material. Skin from withers and breast (table 38) (body length 7.5 mm) taken June 25, Pet-Pak-Dale.

The hair is in an early stage of growth or is not growing. The stratum malpighii is mostly two cells deep, and the reticular layer is thicker than the papillary. Sparse bundles of collagen fibers form a lax plexus in the papillary layer, and the number of horizontal elastin fibers is small. In the reticular layer the collagen fiber plexus is considerably more compact, the bundles running at various angles to the surface. The sebaceous glands (17 x 50; 50 x 95μ) are oval and squeezed up against the hair bursae (fig. 80). The arrectores pilorum muscles are thin (5μ).

There are three hair categories in the withers. The guard hairs are long (14.7 mm), straight, and of nearly uniform thickness down the whole length, with a small thickening (38μ) about four-fifths to five-sixths up the hair. The medulla (73.7%) extends through the whole hair except base and tip. The second category hairs (12.9 mm long, 30μ thick, medulla 80%) (for convenience termed pile) are not very different. The distal thickening is more pronounced, but there is no neck or contraction in the medulla. The medulla thickening extends about one-fifth of the hair. The third category (for convenience termed fur) are of even thickness (23μ) down the whole length (12.1 mm). The shaft has three (sometimes four) ligatures, where the medulla is contracted and the hair bent. The hairs grow singly or, more rarely, in pairs forming compact groups (three to four hairs in a row). There are 24,660 hairs per 1 cm².

FAMILIA DIPODIDAE

Sicista subtilis

Material. Skin from withers and breast (table 39) of two specimens (body lengths 6 and 5.5 cm) taken August 17, the Urals.

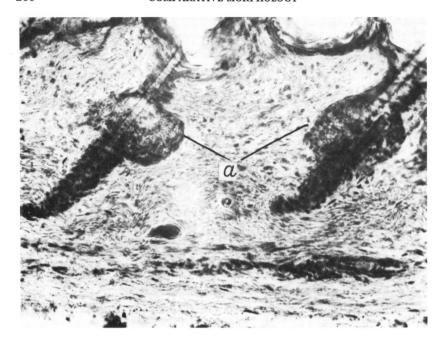

Fig. 80. *Selevinia betpakdalensis.* Withers skin: *a,* sebaceous gland.

In specimen (1) the hair is growing and the skin is thicker than in specimen (2), with hair not growing. The stratum malpighii is one cell deep. Specimen (2) has a compact collagen fiber plexus in the dermis. In specimen (1) this is found only in the outer papillary layer (90 to 100μ thick), the remaining part of the papillary layer and the reticular layer being a very loose network of thin, horizontal collagen bundles. There are few elastin fibers.

Fat cells (28 x 39; 22 x 34μ) occur mostly in the papillary layer in specimen (1). In specimen (2), droplike fat cells are found in the reticular layer, ranging from very small to rather large (11 x 17; 17 x 17μ). The large sebaceous glands (22 x 56; 28 x 78μ) are arranged singly behind the hairs. The glands at the mouth corners, not outwardly visible, are a complex of multilobal sebaceous glands (1.17 to 2.80 mm long and 0.78 to 0.84 mm thick) (Skurat, 1972). The sebaceous glands (0.34 x 0.90; 0.56 x 1.12 mm) stretch along the hair roots. The anal gland, which reaches 0.90 mm in length, surrounds the anal orifice in a girdle of multilobal sebaceous glands (Skurat, 1972). The sole glands are formed by tubular glands twisted into glomi (52 x 134μ) (Skurat, 1972). The arrectores pilorum muscles (7μ thick) are found only in specimen (2).

There are two hair categories differing from the *Gerbellinae.* Straight pile hairs (three size orders: 11.1, 10.1 and 9.7 mm; 48, 37 and 32μ; medulla 83.4, 83.8 and 83.4%) with a pronounced granna have a neck with a medulla narrowing and a granna length of 5.0, 3.5 and 2.5 mm. The second

TABLE 39

DIPODIDAE: SKIN MEASUREMENTS (μ)

Indexes	Skin	Epidermis	Stratum corneum	Dermis	Papillary layer	Reticular layer
Sicista subtilis						
Withers						
(1)	258	17	7	241	196	45
(2)	140	17	10	123	95	28
Breast						
(1)	168	Badly preserved		168	112	56
(2)	157	17	7	140	112	28
Zapus setschuanus						
Withers						
(1)	112	Badly preserved		112	84	28
(2)	101	Badly preserved		101	73	28
Euchoreutes naso						
Withers						
(1)	110	22	11	88	60	28
(2)	167	22	12	145	100	45
Allactaga jaculus						
Withers						
(1)	479	28	17	451	418	33
(2)	627	22	12	605	550	55
Breast						
(1)	218	22	11	196	140	56
(2)	206	22	11	184	134	50
Allactagulus acontion						
Withers						
(1)	163	28	20	135	90	45
(2)	196	28	20	168	112	56
Breast						
(1)	156	22	14	134	112	22
(2)	134	28	17	106	84	22
Pygerethmus platyurus						
Withers						
(1)	235	39	22	196	140	56
(2)	196	28	14	168	112	56
Scirtopoda telum						
Withers						
(1)	165	22	12	143	110	33
(2)	258	28	17	230	196	34
Paradipus ctenodactylus						
Withers	157	17	11	140	123	17
Eremodipus lichtensteini						
Withers						
(1)	172	22	11	150	140	10
(2)*	196	17	11	95	95	—

*The thickness of the subcutaneous fat tissue is given in the text.

category (fur) have no granna. There are two orders (8.8 and 5.4 mm, 21 and 19μ, medulla 81.0 and 84.3%). They have several ligatures (two or three in the first order and one or two in the second order) where the medulla is thinner and the hair is bent. The hairs form loose tufts of three to five, 23,000 hairs per 1 cm².

Zapus setschuanus

Material. Skin from withers (table 39) of two specimens taken in Sun-Non, time unknown.

The outer surface of the skin is slightly wrinkled. The papillary layer is rather thicker than the reticular. The bundles of collagen fibers in the dermis are horizontal, forming a compact plexus in the papillary layer and a loose plexus in the reticular. In specimen (1) large accumulations of fat cells (11 x 22, 17 x 22μ) occur in the reticular layer. The sebaceous glands are small (11 x 34; 17 x 39μ). A section shows one gland anterior or posterior to most hair tufts. The mouth corner gland in *Z. priceps* is formed of very large and medium mucous glands (Quay, 1965c).

There are four categories of hair. The guard hairs (10.9 mm long, 74μ thick, medulla 93.3%) differ from *Gerbellinae* first category hairs. They have a short club with no medulla in the base. This thickens almost immediately into a shaft that maintains its thickness three-fourths to four-fifths of the length. It then thickens slightly and tapers sharply into a medulla-free tip, which is shorter than in other members of this family. The pile hairs (9.7 mm, 28μ, medulla 87.9%) have a long but not very thin club (there is no medulla in the base) and a granna 3.3 mm long. However, they have no neck and no narrowing of the medulla. The tip is short. In the two orders of fur hairs (8.3 and 6.8 mm long, 28 and 17μ thick, medulla 82.2 and 82.4%), the thickening in the distal shaft is barely visible or is absent altogether. The tip is short; the shaft is slightly wavy. The second order has two narrowings where the medulla is decreasing and the hairs are obtusely bent.

Ten jerboas (*Jaculus jaculus*) taken in the neighborhood of Khartoum were studied (season and sex not indicated) (Ghobrial, 1970).

The stratum malpighii is one or two cells thick. The reticular layer of the dermis is compact. The sebaceous glands are found at every hair follicle. They are comprised of four to eight large cells. Hair grows in tufts of up to five hairs on the limbs and up to twenty on the neck. There are arrectores pilorum muscles.

Euchoreutes naso

Material. Skin from withers of two specimens (table 39).

The skin is much thicker in specimen (2) than in specimen (1). The stratum malpighii is one to two cells deep. In the dermis, horizontal collagen fibers bundles predominate forming a compact plexus. The reticular layer

TABLE 40

JACULUS JACULUS: SKIN MEASUREMENTS (GHOBRIAL, 1970)

Skin measurements	Epidermis M (lim)	Corneous layer (lim)	Dermis M (lim)
Neck at top	20 (15-30)	5 (4-7)	—
Neck at bottom	25 (21-33)	6 (4-8)	154 (140-170)
Withers	19 (15-27)	4 (1-5)	370 (200-740)
Sacrum	16 (12-26)	6 (3-9)	270 (200-370)
Breast	19 (12-26)	5 (3-8)	520 (370-800)
Belly	21 (10-30)	5 (2-8)	370 (230-600)
Proximal part of forelimbs			
Exterior	19 (14-23)	6 (5-9)	370 (270-470)
Inside	18 (10-23)	5 (3-8)	—
Proximal part of hindlimbs			
Exterior	23 (12-45)	4 (1-7)	200 (180-230)
Inside	16 (11-23)	4 (3-6)	360 (200-540)

of specimen (1) is a continuous deposit of fat cells (22 x 28; 17 x 28μ). The sebaceous glands are small (11 x 34; 22 x 45μ). A vertical section shows sebaceous glands, one anterior, one posterior to every hair tuft (fig. 81). A horizontal section also shows that every hair tuft has a small sebaceous gland on its side (mostly posterior). The arrectores pilorum muscles are thin (6 to 8μ).

No categories are distinguished in the hair, treatable as guard hairs. There are two size orders. The long hairs (17.5 mm) are of almost even thickness (22μ) the whole length. They taper very gradually toward the tip, which is long (though shorter than, for instance, in *Allactaga jaculus*). Their medulla takes up 86.4%. The second category hairs differ only in size and in their short tip. Their length is 15.0 mm, thickness 19μ, medulla 84.3%. The hair grows in loose tufts of seven to eleven, 27,333 per 1 cm².

Allactaga jaculus

Material. Skin from withers and breast (table 39) of two specimens (body lengths 22 and 24 cm) taken July 11, near Simferopol.

The withers hair is growing and the thick skin is much thinner on the breast. It is noteworthy that the roots of the growing hairs lengthen not vertically to the skin but at an acute angle. The roots are very long—up to 1.8 mm. The papillary layer is considerably thicker than the reticular, its upper area (132 to 154μ thick) being a compact plexus of collagen fiber bundles. The remainder is filled with connective tissue and accumulations of fat cells (28 x 56; 28 x 45μ). Specimen (2) has few fat cells. Horizontal collagen bundles predominate in the dermis. The reticular plexus is very loose and there are fat cells. Horizontal elastin fibers are numerous in the compact zone of

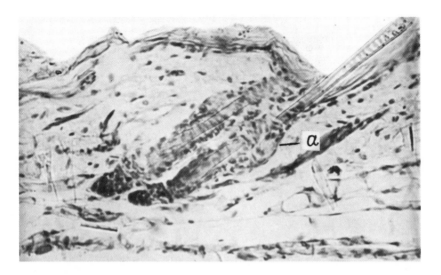

Fig. 81. *Euchoreutes naso.* General structure of withers skin: *a,* sebaceous glands.

the papillary layer. The sebaceous glands (28 × 45; 34 × 84μ) surround the hair tufts, one at every tuft. The gland at mouth corners (length 2.1 to 2.7 mm, width 1.8 mm) (Skurat, 1972) is easily seen due to its sparse, bristle-like hairs. Its multilobal sebaceous glands are 0.27 × 0.70; 0.29 × 0.81 mm. There are two glandular pockets at the sides of the anus, that is, the anal gland (Skurat, 1972). The septa and the bottoms of the pockets are formed by multilobal sebaceous glands (0.34 × 0.61; 0.47 × 0.72 mm). Each pocket has a duct into the anal orifice. There are also large sebaceous glands (0.25 × 0.64 mm) in the skin lining the anal orifice, forming a 2.2 mm glandular net-work. The sole glands are formed by eccrine sweat gland glomeruli (0.11 × 0.38 mm). Thin arrectores pilorum muscles (6 to 8μ) are found in the com-pact zone of the papillary layer.

There are five hair-length zones (Tserevitinov, 1958). The longest hairs are in the central part of the back; the shortest hairs are on the belly. There are two hair categories. Both orders of guard hairs (15.5 and 11.9 mm long, 28μ thick, medulla 60.8 and 67.9%) are straight, thin with a small bulge in the upper three-fourths to four-fifths of the shaft, and a characteristic long, thin, medulla-free tip. The second order may be growing, as there is a medulla at the base. The pile hairs differ in size (12.5 mm long, 22μ thick, medulla 54.6%) and have a short tip and a slightly wavy shaft. The hairs grow in tufts of six to eight.

Allactagulus acontion

Material. Skin from withers, breast (table 39), and hind foot soles of two males (body lengths 9 and 10 cm) taken in July, Kazakhstan.

The hair is in the earliest stage of growth and the skin is not thickened (fig. 82). The stratum corneum is very well developed; the stratum malpighii is two cells deep. In the sole, the epidermis is thick (220μ); the stratum corneum measures 121μ; the stratum granulosum and stratum lucidum are marked; and there are 56μ high alveoli on the inner surface of the epidermis, which is not pigmented. Collagen fiber bundles, largely horizontal in the withers dermis, form a fairly compact plexus in the papillary layer and a loose plexus in the reticular layer, which is filled by fat cells (28 x 56; 22 x 34μ). Elastin fibers are few. The hind sole dermis contains many pigment cells. The sebaceous glands are small (17 x 28; 28 x 39μ) and bean-shaped. There are eccrine sweat glands in the callosities with long ducts 28μ in diameter (fig. 83). The arrectores pilorum muscles are thin (6 to 11μ).

There are two categories of hair on the withers. Guard hairs are long (14.7 mm) and of even thickness (23μ) (medulla 87.0%), except for a very long, thin, medulla-free tip. The tip is short in both orders of pile hairs, which are 11.4 and 10.7 mm long, 19 and 15μ thick, medulla 84.3 and 80.0%. The second order is wavy.

Pygerethmus platyurus

Material. Skin from withers (table 39) and tail of two specimens (body lengths 9 and 8 cm) taken in summer, Aral region.

The outer surface of the withers skin is irregular. Horizontal collagen bundles predominate in the compact dermal plexus. Elastin fibers, also horizontal, are numerous in the middle section of the papillary layer. There are

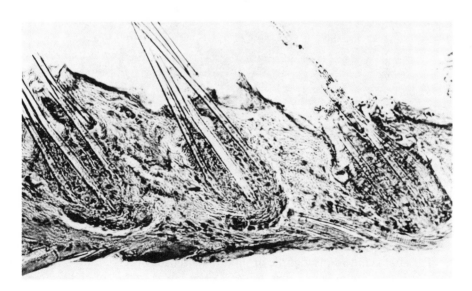

Fig. 82. *Allactagulus acontion.* Withers skin.

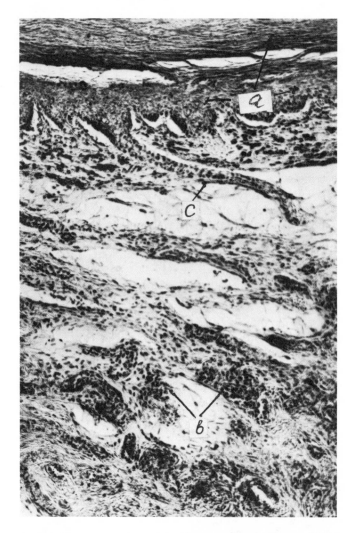

Fig. 83. *Allactagulus acontion.* General structure of sole skin: *a,* epidermis;
b, glomi of sweat glands; *c,* sweat gland duct.

no fat cells in the dermis, but large pigment cells are especially numerous in the reticular layer. A small (28×45; $28 \times 56\mu$) oval sebaceous gland is visible anterior and posterior to each hair in a longitudinal section. Beside or on them is a concentration of pigment cells. The inconspicuous mouth corner glands (3.36 mm long) comprise a complex of sebaceous glands (Skurat, 1972).

The anal gland forms a girdle of multilobal sebaceous glands around the anal orifice 3.4 mm wide (Skurat, 1972). The largest glomi of the eccrine sweat glands comprising the sole gland measure 0.14×0.32 mm.

There is only one category of hair—guard hairs of two orders. They are straight, thin, with no thickenings, 13.2 and 12.9 mm long, 20 and 17μ thick, medulla 85.0 and 82.4%. In the second order the medulla varies in the upper and lower parts. The hairs grow in tufts of five to nine, but not in groups, 33,333 per 1 cm^2.

The hairs grow on the side of the tail, and the unpigmented epidermis is thin with neither stratum corneum nor stratum lucidum. The dermis, about 170μ thick, is a plexus of collagen fibers with a large number of pigment cells in all strata. In the withers skin pigmentation is heavier in specimen (1) than in specimen (2). There is a continuous layer of fat cells below the dermis.

Scirtopoda telum

Material. Skin from withers (table 39) of two males (body lengths 95 and 97 mm) taken May 11, Guriev region.

The outer surface of the skin is irregular. The hair of specimen (2) is not growing, so the skin is thinner than in specimen (1), with growing hair. The stratum malpighii is one to two cells deep. The mainly horizontal dermal collagen bundles form a loose plexus. The sebaceous glands are small (17 x 28; 28 x 34μ), and can be seen in a section anterior and posterior to each hair tuft (fig. 84). The mouth corner glands, made up of multilobal seba-

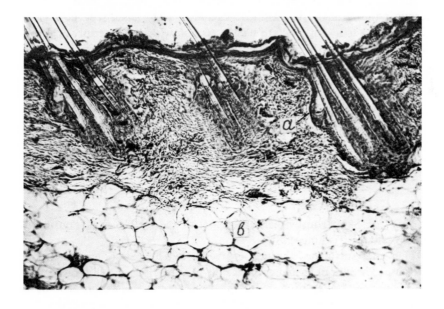

Fig. 84. *Scirtopoda telum*. Withers skin: *a,* sebaceous glands; *b,* fat cells.

ceous glands on the edge of the lower lip, do not show on the outside. They are 1.9 to 3.6 mm long (Skurat, 1972). The ducts of the multilobal sebaceous glands (0.28 × 0.91; 0.43 × 0.67 mm) open into the hair bursae in the anal orifice. The sole glands are formed by eccrine gland glomi (87 × 252μ) (Skurat, 1972). The arrectores pilorum muscles are thin (8 to 11μ). The peculiar hairs divide into three categories (for convenience termed guard, pile, and fur). The guard hairs thicken sharply at the base and retain their thickness to the top, where they taper into a very long, thin, medulla-free tip. They are 18.1 mm long, 37μ thick, medulla 91.9%. The two orders of pile hairs differ by the absence of the apical thickening and long tip. They are 13.7 and 12.7 mm long, 31 and 25μ thick, medulla 90.4 and 85.0%, respectively. The very sparse fur hairs thicken somewhat just above a thin, medulla-free base, then taper gradually up to a long fine tip. They are 10.3 mm long, 23μ thick, medulla 86.6%. The medulla extends well into the tip. The hairs grow in tufts of seven to ten, numbering 33,330 hairs per 1 cm^2.

Paradipus ctenodactylus

Material. Skin from withers (table 39) of one specimen (body length 15 cm) taken spring, Repetek.

The hair is not growing. The outer surface of the skin is irregular. The stratum corneum forms the thickest part of the epidermis, and the stratum malpighii is one to two cells deep. The papillary layer is only slightly thicker than the reticular layer. Horizontal collagen fibers bundles form a compact plexus in the dermis. The elastin fibers are few. The small sebaceous glands (22 × 73; 28 × 67μ) (fig. 84) are seen in section with one anterior, one posterior to every hair tuft. The arrectores pilorum muscles are thin (6μ).

The hair, which does not divide into categories, can be treated as two orders of guard hairs. The first order has long hair (19.2 mm, medulla 71.8%) with short clubs and without medullae. The hair club thickens somewhat into the shaft, which is 39μ thick down the whole length, then tapers into a very long medulla-free tip. The second order (15.5 mm, 11μ, medulla 81.9%) differs only in size and in absence of a long tip. The hairs grow in tufts of six to nine, with 13,666 hairs per 1 cm^2.

Eremodipus lichtensteini

Material. Skin from withers (table 41) of two specimens taken in June near Tashauz, Turkmen SSR.

The outer surface of the skin is irregular. The hair of specimen (1) is growing. The hair of specimen (2) is not. Nevertheless, the skin of specimen (1) is thinner. The stratum malpighii is one to two cells deep. The dermis differs in the two specimens. In specimen (1) the hair bulbs leave a thin reticular layer, yet come close to the subcutaneous muscles. Specimen (2) has a continuous layer of fat cells immediately beneath the hair bulbs, with almost no collagen fibers (fig. 85). This zone should be regarded as subcutaneous fat tissue rather than as the dermal reticular layer.

In the dermis of both specimens, horizontal collagen fiber bundles form a dense plexus with numerous, deep, horizontal elastin fibers. Only specimen (2) has fat cells (56 x 78; 45 x 67μ) in its 84μ thick subcutaneous tissue. The sebaceous glands are small (17 x 45; 28 x 39μ), and, in section, two glands show at every hair, anterior and posterior. The arrectores pilorum muscles are thin (8 to 12μ).

The hair divides into two categories. The guard hair shaft thickens over the base to 42μ and remains unchanged for almost its whole 20.7 mm length, thinning gradually at the beginning of a very long, thin medulla-free tip (medulla in the shaft 76.2%). The two orders of pile hairs (15.5 and 15.6 mm long, 33 and 25μ thick, medulla 84.9 and 88.0%) are slightly wavy and have short tips. The hairs grow in tufts of seven to eleven, with 37,333 hairs per 1 cm^2.

FAMILIA HYSTRICIDAE

Hystrix leucura

Material. Skin from withers and breast (table 41) of a male (body length 75 cm) taken in January, near Repetek, Kara Kum, Turkmen SSR.

This skin, very thick on the withers, thinner on the sacrum, and still thinner on the breast, has a peculiar, complex structure. Large groups of fat cells alternate with bundles of smooth muscles in the thick masses of collagen fibers. The outer surface of the skin is in large folds. The very thick epidermis has a highly pronounced stratum corneum with scales loosely arranged to form air cavities that may act as insulation. The stratum malpighii is only three or four cells deep. The stratum granulosum (one or two cells) and a thin stratum lucidum are clearly distinguishable. The epidermal stratum basale is poorly pigmented in spots, and sometimes not at all. At times the pigmentation is of medium intensity. The pigment cells are large. The reticular layer is thicker than the papillary layer on both breast and withers, but they are equal in the sacrum. The bundles of collagen fibers forming a compact plexus in both layers are mainly horizontal.

In some skin areas fat cells (88 x 121; 100 x 154μ) rise almost to the epidermis, where there is a narrow (100μ) zone of collagen fiber bundles. In the papillary layer very large arrectores pilorum muscles are numerous. In the inner part of the reticular layer and the outer subcutaneous fat tissue are smooth muscles not apparently connected with hairs or spines; they form a loose plexus 8 to 10 mm below the surface. These muscle bundles are separated by fat cells or collagen fiber bundles. Elastin fibers are plentiful, lying in various directions in the papillary layer and horizontally in the rest of the

TABLE 41

Hystrix leucura and Myocastor coypus: skin measurements

	Skin (mm)	Epidermis (μ)	Stratum corneum (μ)	Dermis (mm)	Papillary layer (mm)	Reticular layer (mm)
Hystrix leucura						
Withers	15.73	330	296	11.40	4.20	7.20
Sacrum	10.18	185	157	10.00	5.00	5.00
Breast	2.52	220	190	2.30	1.10	1.20
Myocastor coypus						
Withers						
Summer						
(1)	2.80	56	45	2.75	1.65	1.10
(2)	1.80	56	40	1.75	1.20	0.55
Winter						
(3)	1.87	67	50	1.74	1.32	0.42
(4)	1.65	67	45	1.59	1.26	0.33
Breast						
Summer						
(1)	1.92	56	40	1.87	1.32	0.55
(2)	2.27	67	50	2.21	1.21	1.00
Winter						
(3)	1.95	84	67	1.87	1.54	0.33
(4)	2.28	84	67	2.20	1.54	0.66

dermis, with a thick, loose plexus of elastin fibers just above the subcutaneous fat tissue. In the withers the subcutaneous fat tissue is 4 mm thick. It consists of accumulations of fat cells (100×132; $90 \times 143\mu$) sometimes crossed by layers of collagen fibers with numerous elastin fibers. On the border with the subcutaneous muscles are thin (about 20μ), horizontal bundles of elastin fibers.

There are small sebaceous glands ($121 \times 253\mu$) near hairs and spines, and larger, multilobal sebaceous glands (0.25×0.71; 1.93×3.30 mm) at the roots of large spines. Every spine has one or two glands with a common duct. The mouth corner glands are in the folds of the skin on the inner surface of the cheeks, and at the juncture of upper and lower lips, where there is a small hollow covered with sparse bristles (Skurat, 1972). The gland is 2 cm long, formed by a large number of multilobal sebaceous glands (0.93×2.35 mm) with ducts opening into the hair bursae. There are large sebaceous glands (0.95×2.34; 1.00×2.91 mm) on the lips. Three pouches (two, according to Schaffer, 1940) around the anus form the anal gland (Skurat, 1972). Each measures 15×13 mm and has a cavity lined with a thick, oily secretion. The septa of the pockets are formed by multilobal sebaceous glands (3 × 3 mm). There are also common but very large sebaceous glands (0.78 × 1.18 mm) in the anal orifice. The sole glands, formed by eccrine sweat glands, are found on front and hind feet. The glomeruli (0.54 × 1.14; 0.68 × 1.32 mm) of these glands are most numerous in the skin of the toes. Exteriorly, the porcupine spines are round in cross section. The cortical layer is relatively poorly developed (fig. 85a). It has longitudinal processes. They are tall, thin, and compact, most reaching the spine's center. The spine's medulla is a structure of small alveoli filling the entire cavity (fig. 85b). The sparse hairs fall into three categories, with sizes tabulated below (table 42).

TABLE 42

HAIR VALUES FOR H. LEUCURA

Withers hairs	Category 1	Category 2	Category 3
Length (mm)	55.6	28.9	26.9
Thickness (μ)	214	98	55.3
Medulla (%)	73.0	36.9	absent

FAMILIA MYOCASTORIDAE

Myocastor coypus

Material. Skin from withers, breast (table 41), front and hind sole of specimens (1) and (2), summer (body lengths 45 cm and 30 cm) and specimens (3) and (4) winter males (57 cm and 52 cm) bred on Narofominsk farm.

In summer the hairs are growing and have long roots. The outer surface of the skin is covered with small folds. A loose stratum corneum comprises the major portion of the epidermis. The stratum malpighii is one to three cells deep. In spots, a one-cell-deep stratum granulosum is noted. There is no stratum lucidum in the papillary, and in the upper part of the reticular layer, oblique bundles of collagen fibers predominate. Elastin fibers are numerous in the papillary layer (especially in the upper half) and sparse in the reticular. Groups of fat cells (55 × 110, 56 × 84μ) occur around the hair bulbs and in the reticular layer.

The sebaceous glands, which are rather large (39 × 95; 67 × 224μ), lie singly posterior to every tuft of hair. The mouth corner gland is large, 10 × 10 mm (Skurat, 1972) about 7 mm thick, and covered with sparse bristle hairs. It is formed of large (up to 1.00 × 3.47 mm), multilobal sebaceous glands. The anal gland is a glandular body 25 × 42 mm in size in the terminal section of the ventral rectum (Schaffer, 1940; Ortmann, 1960; Skurat,

(a)

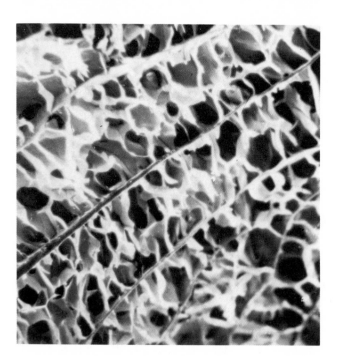

(b)

Fig. 85. *Hystrix leucura*. Spine structure: (a), cross-section;
(b), photomicrograph.

1972). Its 4 mm duct opens into the ventral wall of the anus 12 mm from its edge. Into the duct orifice other smaller ducts open (diameter from 0.2 to 0.7 mm). The anal gland is formed of sebaceous glands 1.2 x 1.3 mm in size. A few tubular glands are found in the hind sole callosities of one specimen only. These glands form small glomeruli (Sokolov, 1964b; Skurat, 1972) (secretion cavity diameter up to 67μ). The lower parts of the arrectores pilorum muscles are 12 to 28μ thick about halfway down the hair roots.

There are three categories of hair: guard, pile, and fur (table 43). The guard hairs are lancet-shaped and slightly wavy. The pile hair has a very pronounced granna with a sharp narrowing where it begins bending at 40 to 45°. The fur hair has up to twelve waves.

TABLE 43

MYOCASTOR COYPUS: HAIR MEASUREMENTS

Hair types	Length (mm)		Thickness (μ)		Medulla (%)	
	Summer	Winter	Summer	Winter	Summer	Winter
Guard	62.2	44.9	195	207	76	81
Pile						
1st order	40.4	32.4	206	201	80	82
2d order	29.5	24.7	190	156	80	80
3d order	21.9	21.0	120	62	67	66
4th order	19.2	19.4	44	43	46	70
Fur						
1st order	17.1	19.3	22	20	23	50
2d order	16.8	15.5	17	13	36	50

In winter the skin structure is like that of the summer skin, with some differences in the layer measurements. The hair categories and orders are also the same (table 43). Fadeev (1955) and Mamed-Zade (1973) obtained hair indexes only slightly different from ours, although Mamed-Zade (1973) does not distinguish orders in pile and fur hairs. In winter the body of M. coypus divides into four zones for the length of pile and fur hairs (Fadeev, 1955). The zone with the longest hairs (over 58 mm) covers the center and middle of the back, and the zone with the longest fur hairs (over 16 mm long) is near the withers. The second zone of pile and fur hairs covers the neck, withers, sides, and sacrum (pile hairs 44 to 58 mm, fur hairs 13 to 16 mm). The third zone for pile hairs (length 30 to 44 mm) is on the sides, neck, breast, and the belly; for fur hairs (10 to 13 mm) it is below on the sacrum. The fourth zone for pile and fur hairs is on the forehead, the end of the muzzle, chin, throat, groin, and thighs (here the pile hair is less than 30 mm long, the fur less than 10 mm). The guard hairs and big first to third order pile hairs grow singly, and the small (fourth to fifth order) pile hairs grow in tufts with fur hairs (Fadeev, 1955). The hairs are arranged in groups of 40 to 150.

The guard hair groups, the most complex, consist of four tufts of fur with 6 to 34 hairs in each, two pile hairs and one guard. The second type consists of two fur hair tufts on either side of a pile hair, with 13 to 30 fur hairs per group. The third type, fur hair groups, has 20 to 26 fur hairs parallel to the groups of guard hairs. The guard hair groups are characteristic of the middle and back of the sides and belly. The pile groups are most common on belly and sides, and less so on the middle of the back. Fur hair groups are common on the belly and less so on the sides.

FAMILIA CHINCHILLIDAE

Chinchilla laniger

Material. Skin from midback and body side of three adult males and two females; killed early January, breeding farm in Kirov.

The skin surface on the middle of the back and the side of the body is slightly wavy (N. Pavlova, 1969). The thickness of the skin on the middle of the back is 370 to 400μ, and it decreases in thickness toward the belly to 300μ or 250μ. The epidermis is the thickest on the middle of the back (12μ); thinner on the side (9μ) and belly (8μ). The stratum malpighii is two or three cells deep on the middle of the back and one or two cells deep on the side and belly. The thickness of the papillary and reticular layers is about equal. The bundles of collagen fibers run in different directions, though they are horizontal in the reticular layer. Deep in the reticular layer, groups of fat cells are especially numerous on the belly. In the skin of the animal examined by Seele (1968), the subcutaneous layer of fat tissue is well developed with deeply buried roots of hairs in growth. There are up to six sebaceous glands ($14 \times 36\mu$ to $28 \times 30\mu$) at every hair tuft, two or three small glands being found at the hair bursae of pile and guard hairs. In the anterior septum of the anus a nonpaired anal gland opens into a deep hollow (Schaffer, 1940). Every hair tuft has a muscle of its own (Wilcox, 1950). The arrectores pilorum muscles are thin (8 to 10μ).

The hairs are only slightly differentiated (N. Pavlova, 1968, 1970, 1971), except that guard hairs resemble the fur and are of equal length. Guard, pile, intermediate, and fur hairs (first and second orders) can be distinguished (table 44). The granna is shifted upward and is comparatively small, averaging 26% of the total guard hair shaft, 22.5% of pile, and 20% of intermediate hairs. The medulla in the guard hair granna is 50μ wide, in pile hair 42μ, in intermediate hair 30μ, in fur hair (first order) 15μ, and fur hair (second order) 10μ. The hairs grow in tufts (N. Pavlova, 1969), the simplest of

TABLE 44

CHINCHILLA LANIGER: HAIR MEASUREMENTS (MEAN DATA) (N. P. PAVLOVA, 1968)

| | Guard | | Pile | | Intermediate | | Fur | | | |
| | | | | | | | 1st order | | 2d order | |
	Length (mm)	Thickness (μ)	Length (mm)	Thickness (μ)	Length (mm)	Thickness (μ)	Length (mm)	Thickness (μ)	Length (mm)	Thickness (μ)
Withers	28.3	54	27.5	42	26.6	26	25.2	18	24.6	13
Sacrum	38.4	59	31.8	45	30.4	28	27.8	19	26.0	17
Belly	26.6	50	25.0	42	22.9	21	21.3	17	20.7	15

which includes 15 to 30 fur and intermediate hairs from a common opening.

This opening leads to a large hollow at the bottom of which individual hair follicles can be seen. The lower parts of the shafts found in the skin hollow, are intricately interwoven. This is different from all other mammals. Two group patterns can be distinguished. Groups of pile hairs in which one pile hair grows beside the fur hairs belong to the first type. Two such groups can be united, the two tufts of fur hairs pressed closely together, with two pile hairs at the sides.

There are also groups in which two pile hairs out of a common hair bursa grow between two fur tufts. This group can be still more complex when, beside one or both tufts of fur hairs, a pile hair grows in a follicle of its own. It is not usually well developed, and three or four weakened pile hairs may be found between two hair tufts randomly in the hair bursa. The second type comprises groups of guard hairs in which two tufts of hair are united by a guard hair in the middle. The guard hair, which can be found in an individual fur hair tuft, sometimes unites a tuft of fur and a tuft of pile hairs or two tufts of pile hairs. Here the pile hairs are less developed than usual. On the sacrum and withers there are up to 25,000 hairs per 1 cm², and on the belly up to 14,000 (Pavlova, 1968).

In the embryo the hairs grow singly, each with a single sebaceous gland and arrectores pilorum muscle (Wilcox, 1950). In the newborn cub, the hairs are arranged in tufts of 8 to 20.

FAMILIA BATHYERGIDAE

Heterocephalus glaber

Material. Skin from withers and breast of a specimen taken in Kenya in January.

The outer surface is irregular, covered with large folds. The hairs are arranged singly very far apart. The thickness of the withers skin ranges from 308μ on the folds to 168μ in the hollows between them. The breast skin is thicker than the withers skin. The epidermis is thick (45μ on the breast and 56μ on the withers, of which 17 to 34μ is stratum corneum). The stratum malpighii is three to five cells deep. There is no stratum granulosum or stratum lucidum. The epidermis is not pigmented. The absence of the alveoli or processes on the inner surface of the epidermis despite its considerable thickness is characteristic.

No layers can be distinguished in the dermis due to very sparse hair. The dermis on the folds of the withers is 252μ thick. Between folds it is 112μ. The plexus of the predominantly horizontal collagen fiber bundles is rather compact. Elastin fibers, also horizontal, are numerous. The lower parts of the dermis on the folds have rather large accumulations of fat cells (28 x 39; 28 x 50μ). Interestingly, there are numerous large pigment cells in the upper withers dermis. There are no fat deposits in the withers subcutaneous fat tissue, which forms a continuous layer under the dermis. Its thickness is much less between folds (down to 110μ) than at the folds (up to 660μ). The fat cells measure 34 × 67; 28 × 73μ. The roots of the hairs are surrounded by compact connective tissue. The hairs do not have the blood lacunae characteristic of the vibrissae. There is a large three-lobal sebaceous gland (or three smaller sebaceous glands) beside each hair (lobe size 56 × 129; 56 × 84; 45 × 84μ). The gland in the corner of the mouth is formed of medium sebaceous and mucous glands (Quay, 1965c).

The hairs on the withers are short and arched backward. They taper very gradually from base to tip. The length of a hair (only one hair has been measured) is 4.7 mm, the maximal thickness is 28μ. There is no medulla.

FAMILIA CTENODACTYLIDAE

Ctenodactylus gundi

Material. Skin from withers, breast, and hind sole of one specimen.

The hair is growing. The external surface of the skin is slightly undulating. The withers skin is 928μ thick, the breast is 429μ, the epidermis is 28 and 17μ, respectively, the stratum corneum is 16 and 10μ, the dermis is 900 and 412μ, the papillary layer is 550 and 352μ, the reticular layer is 350 and 60μ. The stratum malpighii is two cells deep. The epidermis in the skin of the hind sole is 450μ thick; the stratum corneum is compact and highly developed (220μ); the stratum granulosum is pronounced; the stratum lucidum is not recorded; the alveoli on the inner surface of the epidermis are 176μ high; the epidermis is pigmented. Bundles of collagen fibers form an intricate plexus in the withers dermis. In the papillary layer both horizontal and oblique rising bundles occur in large numbers, with horizontal bundles predominating in the reticular layer. In the upper papillary layer, elastin fibers are numerous and horizontal. There are small groups of fat cells (34 × 34; 22 × 34μ) at some hair bulbs in the withers, and a few large pigment cells are found in the upper dermis. The sebaceous glands with active cells are few (they are found only at nongrowing hairs), small (22 × 78; 22 × 84μ), flat and oval, growing singly posterior to hair tufts.

Accumulations of oval cells are found posterior to the growing hair. These will develop into sebaceous glands. There are small glands at the sides of the anus. In the skin of the withers and the sacrum a scattering of sweat glands occurs. Its lower end reaches the hair bulb level. The secretion cavities, which are tubular and very twisted, do not form glomeruli. Unfortunately, we have failed to identify where their ducts open. The diameter of the secretion cavities reaches 57μ. Unlike *C. gundi*'s common specific glands, tubular skin glands do not form compact accumulations. There are large eccrine sweat gland glomi in the hind sole dermis, surrounded by fat cells (secretion cavity diameter 30μ). The arrectores pilorum muscles (6μ thick) are found only on the breast.

There are three categories of hair on the withers. The two orders of guard hairs thicken gradually from the base to reach the maximum two-thirds up, though the difference is not great (22μ at the base and 44μ at maximum). The hairs are straight (17.9 and 15.0 mm) and there is no medulla in the base (up to 0.88 mm in the first order, and 0.33 mm in the second order) or the long tip (especially long in the first order). In the middle of the shaft, the medulla comprises 81.9% of the thickness. The pile hairs differ from guard hairs in size (length 13.8 mm, thickness 29μ, medulla 79.4%) and in their short tip. The fur hairs are uniformly thick their whole length (length 9.1 mm, thickness 17μ, medulla 64.8%).

ORDO CETACEA

Numerous works deal with the skin of *Cetacea*. Some are of historical interest (e.g., Scoresby, 1813; Mayer, 1855). Others give a more or less detailed description of the skin of various species of *Cetacea: Globicephala melaena* (Fjelstrup, 1888), *Odontoceti* (Kukenthal, 1889, 1909; Belkovich, 1956-1962), *Odontoceti* and *Mysticeti* (Sokolov, 1953-1971; Ling, 1974). Other papers describe various parts of the skin (Tomilin, 1947; Scholander and Schevill, 1955; Felts, 1966; Palmer and Weddel, 1964; Harrison and Thurley, 1974). We have studied the skin of *Stenella frontalis, S. longirostris, Delphinus delphus, Tursiops truncatus, Lagenorhynchus obliquidens, Globicephala melaena, Pseudorca crassidens, Orcinus orca, Phocaena phocaena, Neophocaena phocaenoides, Delphinapterus leucas, Kogia breviceps, Physter catodon, Berardius bairdii, Balaenoptera acutorostrata, B. borealis, B. physalus,* and *Eubalaena glacialis.*

In all *Cetacea* under investigation, the skin surface is relatively smooth. The skin may reach a greater absolute thickness than in other mammals, but the ratio between skin thickness and body length is only 0.3 to 1.5%.

Though *Cetacea*'s epidermis is very thick, it has only three layers: stratum corneum, stratum spinosum, and stratum basale. Numerous alveoli are found on the inner epidermal surface (fig. 86). They interdigitate with dermal protrusions. The alveoli are oval (occasionally round) in longitudinal section (fig. 86), with the longer axis of the oval always set in the same direction. The epidermal septa limiting the alveoli from the longer side and passing parallel to the long axis of the alveoli are termed longitudinal septa. The septa limiting the alveoli at the shorter side are termed the transverse septa.

At the upper dermal papilla level, the thickness of transverse and longitudinal septa is equal. Both septa are thin near the base of the papillae, with the transverse becoming thinner than the longitudinal. At one point, the transverse septa disappear altogether. The longitudinal epidermal septa

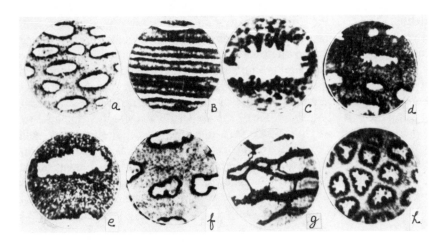

Fig. 86. Shape of epidermal alveoli at sections parallel to surface: *a, Balaenoptera physalus* (middle layers of epidermis); *b, Balaenoptera physalus* (deep part of epidermis showing longitudinal epidermal septa); *c, Lagenorhynchus obliquidens; d, Balaenoptera acutorostrata; e, Orcinus orca; f, Berardius bairdii; g, Physeter catodon* (deep part of epidermis); *h, Physeter catodon* (medial portion of epidermis).

penetrate into the dermis to form parallel ridges with a sinuate lower edge. The ridges extend, continually branching off dichotomously (fig. 86). Individual epidermal cells or processes project into the tissue of the dermal papillae (fig. 86c). These processes stretch along the dermal papilla (fig. 86h), differing in shape and size in various cetaceans. The stratum malpighii cells vary in shape according to their position. The epidermal cells forming the septa between the dermal papillae are spindle-shaped and flattened on both sides. The nearest row of epidermal cells show the dermal papilla their narrow side, while the cells in the middle of the septa turn their broad side to the papillae.

The cells under the dermal papillae apexes are polyhedral. Closer to the outer surface, the epidermal cells are horizontally flattened. The epidermal cell nuclei are ellipsoid, their long axis lined up with the cell's longitudinal axis. The nuclei become more flattened as they approach the stratum corneum.

The epidermis is very thick, yet strata granulosum and lucidum are absent. The upper layers of the epidermis are not usually fully cornified. The majority of all nuclei of stratum corneum are destroyed, though some nuclei in the uppermost parts of the stratum corneum are only deformed. The presence of protein-bound phospholipid, cystein, and cystine, the absence of RNA, and the retention of pyknotic nuclei in the cornified cells resemble pathological parakeratosis in man and the normal parakeratosis of the pouch epidermis in kangaroos (Spearman, 1972).

The cells immediately adjoining the dermal papillae are most highly pigmented. The pigment granules surrounding the cellular nucleus above and laterally are cap-shaped. Pigmentation goes right through into the stratum corneum, though epidermal cells above the papilla apexes are pigmented slightly or not at all. A similar unpigmented band extends upward above the apex of the dermal papillae, continuing into the stratum corneum, though in some cases all epidermal cells above the dermal-papillae apex level are regularly pigmented. On the border of the epidermis and dermis, melanocytes sometimes occur with long processes projecting between the epidermal cells.

In the epidermis of highly reflexogenic zones (such as peripapillary and perianal) there are many nerve terminals, some of them encapsulated (Palmer and Weddel, 1964; Giacometti, 1967). There are fewer nerve endings in the trunk epidermis, and there is a wider range of nerve endings in the dermis. These are free endings, encapsulated glomeruli, and cones (Agarkov and Khadzhinsky, 1970).

Cetacea's dermis consists of a thin layer of densely interwoven fibers and papillae rising through to the epidermal alveoli (fig. 87). Bundles of collagen fibers are arranged in the subpapillary layers at various angles. In the papillae, bundles of collagen and elastin fibers run close together along the long axis of the papilla. Fat cells are not normally found in the papillae. Commonly, the papillae have both incoming and outgoing capillaries uniting at the papilla apex, and they have up to five anastomoses (Palmer and Weddel, 1964). Elastin fibers in the subpapillary layer mostly coincide with collagen fibers in direction. The elastin network is denser in the outer layers of the dermis, though the fibers themselves are thinner than in the deeper layers. The thickest elastin fibers are parallel with the longitudinal epidermal septa. There are fat cells directly beneath the epidermis or some distance below, with a few small ones in the upper dermal layers. Farther from the epidermis, the number increases. As the dermis gradually changes into subcutaneous fat tissue, no firm line can be drawn between them. Surkina (1968) believes that the skin muscle in dolphins clearly separates the dermis from the subcutaneous fat tissue. Blood vessels are numerous in the dermis.

In the subcutaneous fat tissue, the bundles of collagen fibers are far apart (fig. 87), the space between them filled with massive accumulations of very large fat cells, more than in any other mammal. As the compactness of the collagen bundle plexus decreases inward, their thickness increases greatly. *Cetacea* have the thickest collagen bundles of all mammals. The compactness of the collagen bundle plexus begins to increase again near the lower border of the skin, and the bundles become still thicker. In the upper and middle layers of the subcutaneous fat tissue, collagen fiber bundles are usually arranged at various angles. As they approach the subcutaneous muscles, their direction begins to approximate horizontal. Near the subcutaneous muscles the bundles are arranged longitudinally. Elastin fibers are mostly arranged along the bundles of collagen fibers in their immediate

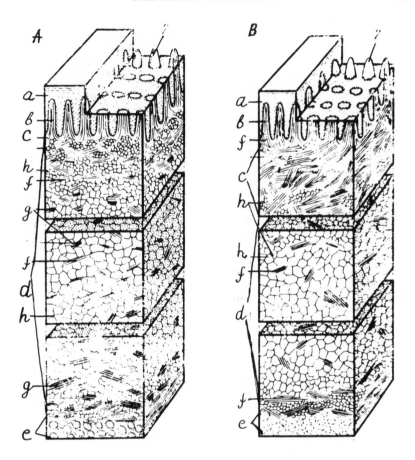

Fig. 87. Structural diagram of *Mysticeti* (*A*) and *Odontoceti* (*B*): *a,* epidermis; *b,* dermal papilla; *c,* subcutaneous layer of the dermis; *d,* subcutaneous fat tissue; *e,* subcutaneous muscles; *f,* collagen fiber bundles; *g,* elastin fiber bundles; *h,* fat cells.

neighborhood. Some cetaceans have thick bundles of elastin fibers (fig. 87). The number and the size of fat cells increase with depth and decrease only near the subcutaneous muscles. Large blood vessels are rare in the subcutaneous fat tissue. The capillary network is extremely well developed, with the capillaries running between the septa of fat cells.

Consideration will be given only to those skin layers in the flippers which are also found in the body (epidermis, dermis, and subcutaneous fat tissue). The underlying layer of tendon fibers parallel to the surface of flippers, dorsal fin, and caudal flukes are not included in this study since they are not a part of the common integument.

In addition to the common integument, the tail flukes consist of three

Fig. 88. *Tursiops truncatus.* Network of collagen fiber bundles in middle layer of tail flukes.

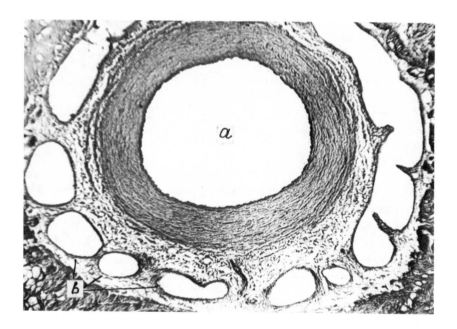

Fig. 89. *Tursiops truncatus.* Coronary blood vessels in tail flukes: *a,* artery; *b,* veins.

layers—one median and two lateral (Roux, 1883). The median layer is made up of numerous, closely interwoven bundles of collagen fibers (fig. 88). The outer layers consist of numerous tendons anterior to the caudal peduncle, which arch back in a peculiar way toward the edge of the fluke. On the border of the tendon layer and the medulla of the tail flukes are numerous blood vessels. They are coronary in type, with a large artery in the center surrounded by veins (Tomilin, 1947; Scholander and Schevill, 1955; Felts, 1966) (fig. 89). Sometimes these veins are relatively large and are arranged in a row around the artery (as in *Lagenorhynchus obliquidens*). Where they are arranged in several rows, they are small and numerous. Most *Cetacea* have a more or less developed nonskeletal dorsal fin, which resembles the tail flukes except for the absence of tendon layers. There are complex blood vessels in the center and on the periphery.

The sebaceous and sweat glands are absent. Specific skin glands are represented only by a twin anal gland at the sides of anus (Kükenthal, 1889) which is formed by a complex of multilobal sebaceous glands opening into the cavities in a single excretion duct.

There is no pelage, though there are individual vibrissae on the head of *Mysticeti* and the head and body of *P. gangetica*. Most species of *Odontoceti* have vibrissae on the head at various stages of embryonic development, but these soon disappear.

SUBORDO ODONTOCETI

The skin varies from very thin to very thick. Transverse and longitudinal epidermal septa are equal or the longitudinal septa are taller. Usually many processes project into the tissue of the dermal papillae from the epidermal septa, which are quite large in some *Odontoceti*. In transverse section, the dermal alveoli are oval, round, or polygonal. The dermis of most *Odontoceti* has a fat-free zone right under the epidermis. The network of elastin fibers in *Odontoceti* skin is less developed than in *Mystaceti,* though in some (e.g., dolphins) elastin fibers are very numerous. The great majority of *Odontoceti* have only prenatal vibrissae (*Delphinapterus leucas* and *Mondon monoceros* do not even have hair in the embryonic stage).

Vibrissae on the muzzle and occasionally on the body are typical in adult animals.

Pontoporia blainvillei

Material. Skin near eye (table 45).

The structure of the epidermis is typical (fig. 90) (Sokolov, 1974). The stratum spinosum is 50 to 55 cells deep. Highly deformed nuclei are retained in the stratum corneum. In the stratum basale, cells carry some pigment.

TABLE 45

FRESHWATER DOLPHINS: SKIN MEASUREMENTS

	Thickness			Height of dermal papillae (mm)	Thickness (mm)	
	Skin (mm)	Epidermis (mm)	Stratum corneum (mm)		Dermal subpapillary layer	Subcutaneous fat tissue
Pontoporia blainvillei near eye	5.89	1.09	38	0.6	0.3	4.5
Inia geoffrensis periorbital	8.5	0.71	51	0.6	7.8	—
Platanista gangetica side of head	9.28	0.76	21	0.5	2.2	6.3

There is less pigment in the stratum spinosum cells. In horizontal sections, the epidermal alveoli are oval and occasionally highly elongated (fig. 91). The processes adjoining the dermal papillae are few and small. In vertical section (fig. 90) the epidermal walls are of uniform thickness through their length. The large projections, which are present in *Inia geoffrensis* and *Platanista gangetica,* are absent. In the latter two dolphins, the epidermal walls are wedge-shaped, gradually thickening to the surface.

The dermal papillae height is 55% of the epidermal thickness, much less than in *Inia geoffrensis* (84.5%) and *Platanista gangetica* (65.7%). The subpapillary dermal layer is thin. It comprises a rather compact collagen fiber plexus with mostly horizontal bundles. Elastin fibers are numerous. The thickest are also horizontal. Fat cells are absent.

The subcutaneous fat tissue is made up of large fat cell accumulations, averaging $67 \times 72\mu$ in the upper parts of the subcutaneous fat tissue and $73 \times 132\mu$ in the lower (fig. 92). Big, sparse collagen fiber bundles run in different directions in the subcutaneous fat tissue. Elastin fibers are much smaller than in the dermis. There are a few cross-striated muscles at acute angles to the skin.

Inia geoffrensis

Material. Skin sample from the periorbital region (table 45).

The outer surface of the epidermis is microundulate (fig. 93). The height of the waves is 30 to 40μ, with the distance between their apexes being 80 to 120μ. The epidermis is typical of *Cetacea.* The stratum spinosum contains up to forty-three cell rows. The stratum corneum is up to 51μ thick. The

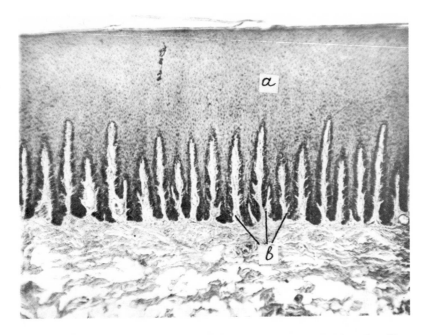

Fig. 90. General structure of *P. blainvillei* skin (x 24.5): *a*, epidermis; *b*, dermal papillae.

boundaries of the highly squamous cells are almost indistinguishable. The stratum corneum has deformed and compact cell nuclei. The epidermis is practically nonpigmented. Individual pigment cells occur only in the stratum basale or in the inner layer of stratum spinosum.

As can be seen in the vertical section, the walls of the epidermal alveoli are broken and dismembered (fig. 93). In the horizontal section level with the upper parts of the dermal papillae, the epidermal alveoli are oval or round with individual large processes reaching the papillae tissue (fig. 94). Collagen bundles run at different angles to the skin. Fat cells are absent. In the median parts of the subpapillary layer of the dermis, the collagen bundle plexus is sparser and the bundles are thicker. Most bundles run parallel to the skin and there are small groups of fat cells averaging 38 x 63μ. Occasionally there are bundles of cross-striated muscle fibers. In the inner part of the dermis, the compactness of the collagen fiber plexus increases again; there are no fat cells. The number of cross-striated muscles running in different directions increases (fig. 95).

Elastin fibers are few in the dermis. In the dermal papillae, thin fibers pass along the long axis of the papillae and in the subpapillary layer, parallel to the skin or at acute angles. The number of elastin fibers decreases with distance from the epidermis. The subcutaneous fat tissue is absent.

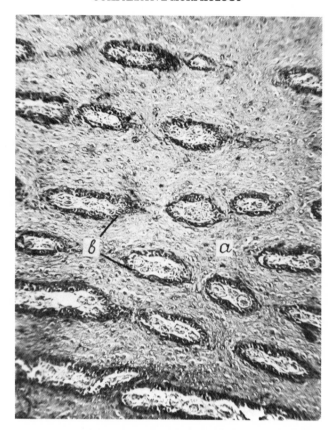

Fig. 91. Section of epidermal alveoli of *P. blainvillei* (x 140):
a, epidermis; *b*, dermal papilla.

Platanista gangetica

Material. Skin sample from side of the head (table 45).

The epidermis is typical of *Cetacea.* In the stratum spinosum there are 48 to 50 cell rows. In the stratum corneum deformed nuclei remain. Pigment granules, few in number, in the basal layer cells are fewer in the stratum spinosum. In horizontal sections, the epidermal alveoli are polyhedral, square, round, or irregularly round (fig. 96). Large processes reach up to the tissue of the dermal papillae from the walls of epidermal alveoli. Vertical sections shows that the epidermal septa of *Inia geoffrensis* have similar but fewer lateral outgrowths (fig. 97).

Collagen fiber bundles in the subpapillary layer of the dermis form fairly compact plexuses at different angles to the skin. Elastin fibers are few and run in various directions, mostly at acute angles to the skin. In the inner parts of the dermis, fat cells averaging 98 x 102 μ and a small number of

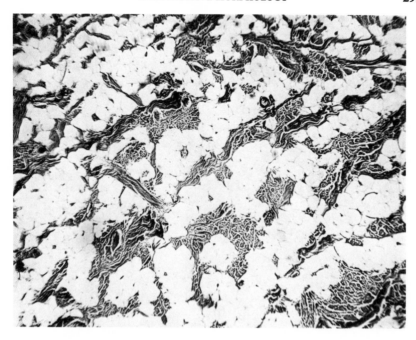

Fig. 92. Plexus of collagen fiber bundles in subcutaneous fat tissue of *P. blainvillei* (x 140).

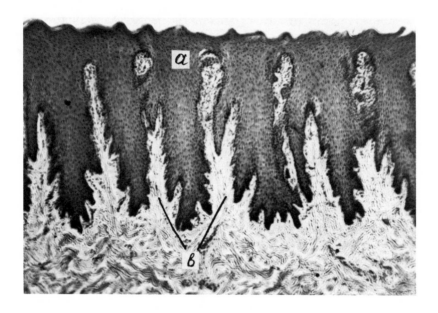

Fig. 93. General structure of *I. geoffrensis* skin (x 2.45): *a,* epidermis; *b,* dermal papillae.

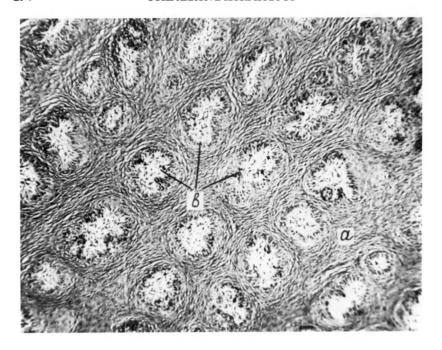

Fig. 94. Section of epidermal alveoli of *I. geoffrensis* (x 140): *a,* epidermis;
b, dermal papillae.

cross-striated muscles are found (fig. 98). The dermis changes into subcutaneous fat tissue without a pronounced boundary. The number and size of
fat cell groups increase. Collagen fiber bundles become relatively sparse.
Elastin fibers and cross-striated muscles also occur.

FAMILIA DELPHINIDAE

Stenella coeruleoalbus

Material. Skin from flippers over the body, side at the flipper, breast
between the flippers, surface of caudal peduncle, dorsal surface of flippers
and tail flukes, and the lateral surface of dorsal fin from three adult dolphins (table 46). Specimen (1), female (body length 218 cm); specimen (2),
female (217 cm); specimen (3), male (239 cm) taken November 1967, near
Honshu (Japan).

No general pattern of epidermal, dermal, and subcutaneous fat thickness
around the body has been revealed, but differences are small except in the
lateral part of the caudal peduncle.

The epidermal pigmentation is irregular and heavy on the back, the dorsal
surface, and the anterior edge of tail flukes and flippers, the lateral surface

TABLE 46

STENELLA: SKIN MEASUREMENTS

	Thickness (mm)			Height of dermal papillae	Thickness (mm)	
	Skin	Epidermis	Stratum corneum		Subpapillary layer of dermis	Subcutaneous fat tissue
Stenella coeruleoalbus						
Back						
(1)	12.2	0.9	40	0.45	0.6	10.7
(2)	11.0	0.7	30	0.40	0.9	9.4
(3)	13.1	1.4	40	0.72	1.1	10.6
Side						
(1)	11.8	1.2	65	0.75	0.3	10.3
(2)	8.6	0.9	40	0.42	0.6	7.2
(3)	8.2	1.3	30	0.70	0.6	6.3
Breast						
(1)	14.2	1.4	65	0.71	0.1	12.7
(2)	9.9	1.0	60	0.43	0.4	8.5
(3)	9.0*	1.4	35	0.75	0.3	7.3*
Flipper, top						
(1)	2.8	0.7	30	0.40	0.1	2.0
Dorsal fin, side						
(1)	—	1.2	45	0.60	0.2	—
Tail flukes, top						
(1)	—	0.5	25	0.31	0.2	—
Stenella frontalis						
Caudal peduncle, side	5.1	0.6	70	—	0.19	4.3
Stenella longirostris						
Back	13.2	1.0	30	0.57	0.9	11.3
Breast	9.9	0.9	30	0.57	0.5	8.5
Caudal peduncle	6.7	0.9	30	0.63	0.7	5.1
Flipper, front	—	0.7	*	0.51	0.5	—
Flipper, top	2.2	0.6	30	0.33	0.3	1.3
Tail flukes, front	—	1.1	*	0.54	0.2	—
Tail flukes, top	—	0.6	90	0.32	0.1	—
Dorsal fin, front	—	0.7	120	0.60	0.30	—

Inner layers of the subcutaneous fat tissue not preserved.
*Not preserved.

being more lightly pigmented (fig. 99). On the body sides few epidermal stratum basale cells contain pigment, and no pigment is recorded in the breast epidermis. Narrow epidermal alveoli are characteristic, especially on the side and breast. They are slit-shaped, some being irregular triangles in transverse section, their narrow side toward the head. Small processes and cells project from the septa of the epidermal alveoli into the tissue of the dermal papillae. In the dermis of specimen (1) numerous dermal papillae

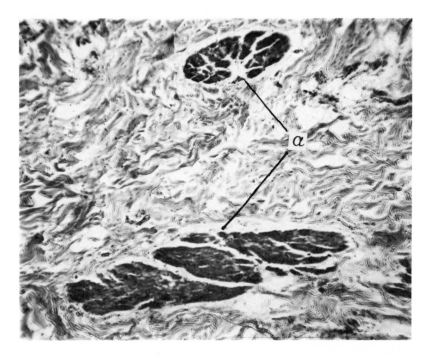

Fig. 95. Collagen fiber bundle in the inner parts of the dermal subpapillary layer in
I. geoffrensis (x 140): *a,* cross-striated muscles.

contain fat cells. The subpapillary layer of the dermis is formed by thin (35
to 80μ) bundles of collagen fibers arranged longitudinally (on the side, the
bundles run in various directions). The bundles form an intricate plexus
right by the epidermis.

In most parts of the body, a fatless layer occurs which is 180 to 300μ thick
(fig. 100). In some places (especially the abdomen) the fatless layer is absent
and the fat cells are contiguous to the epidermis. The collagen bundles in the
subcutaneous fat tissue are 80 to 150μ thick. The largest fat cells (120 x 190;
120 x 200μ) are in the middle of the subcutaneous fat tissue.

Stenella frontalis

Material. Skin from lateral side of caudal peduncle (table 46) of an adult
dolphin.

The pigmentation of the epidermis is moderately heavy. The stratum
corneum is compact. The epidermal papillae are irregularly oval or round in
section with small processes. The bundles of collagen fibers in the dermis
are mostly horizontal. The fatless dermal layer is very thin (70μ). Numerous
collagen fiber bundles (maximum thickness 120μ) are arranged in various

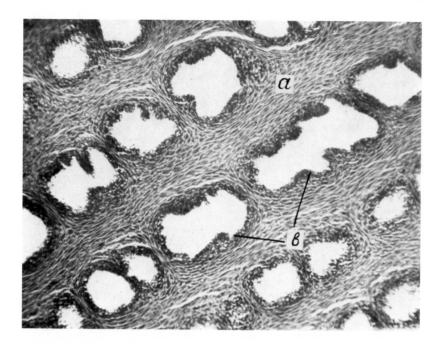

Fig. 96. Section of *P. gangetica* epidermal alveoli (x 140): *a*, epidermis; *b*, dermal papillae.

directions in the subcutaneous fat tissue, and there is a continuous mass of horizontal collagen bundles 600 to 800μ thick next to the subcutaneous muscles.

Stenella longirostris

Material. Skin from back over the flippers, breast, between flippers and lateral surface of the tail flukes, from the anterior edge and the dorsal surface of the flippers and tail flukes, and from the anterior edge of the dorsal fin (table 46) of an adult male taken near Hawaii.

The epidermis is heavily pigmented on the back, the dorsal surface of the tail flukes, and the lateral surface of the dorsal fin. On the side of the caudal peduncle and on the dorsal surface and anterior edge of the flipper, the pigmentation is moderately heavy. There is no pigment in the breast epidermis. The epidermal thickness is almost uniform in the body, flippers, dorsal fin, and tail flukes (table 34), except the anterior edge and lateral surface of the dorsal fin. The stratum corneum is rather thin, except on the dorsal surface of the tail flukes and the lateral part of the dorsal fin. Most of the epidermal alveoli are irregularly oval in section; some have angular hollows; a very few are round. Small processes project into the lumen of the alveoli and into

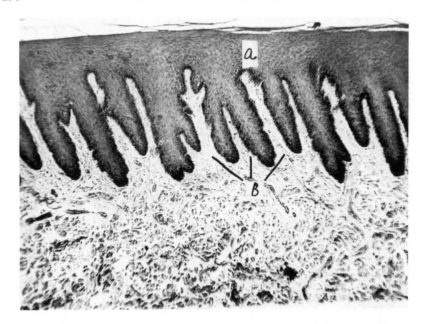

Fig. 97. General structure of *P. gangetica* skin (x 24.5): *a,* epidermis; *b,* dermal papillae.

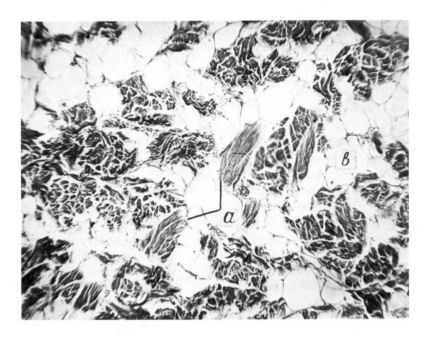

Fig. 98. Plexus of collagen fibers in dermis of *P. gangetica* (x 140): *a,* cross-striated muscles; *b,* fat cells.

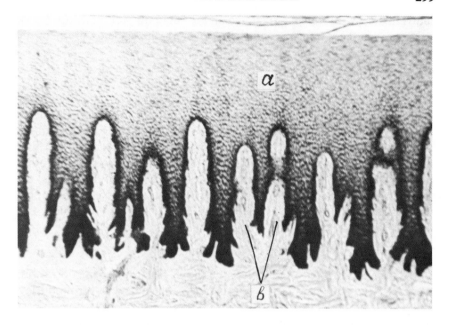

Fig. 99. *Stenella coeruleoalbus*. Skin of the back: *a*, epidermis; *b*, dermal papillae.

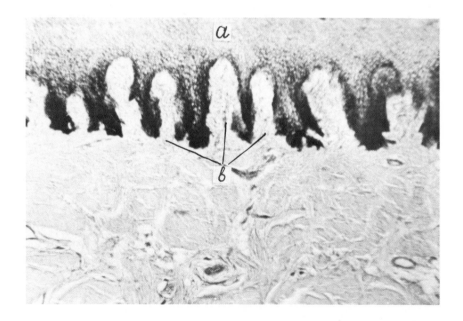

Fig. 100. *Stenella coeruleoalbus*. Skin of tail flukes: *a*, epidermis; *b*, dermal papillae.

individual cells. More epidermal alveoli are round in section in the breast epidermis. In the skin of the anterior side of the flippers and tail flukes the epidermal alveoli have peculiar, very narrow septa.

The dermis has a fine subpapillary layer (table 34). There is a 300 to 570μ fatless layer in the dermis over most of the body, thickest on the lateral surface of the tail fluke. Bundles of collagen fibers 40 to 80μ thick are mainly horizontal in the subpapillary layer (on the back and breast several run in different directions). Numerous fat cells are characteristic in the dermal papillae of the body and tail flukes, though not in all the papillae of the breast or on the lateral surface of the caudal peduncle (but in most of the other parts of the body under study). In the subpapillary layer of the dermis, the few fat cells are small (30 x 90; 50 x 70μ). The collagen bundles in the subcutaneous fat tissue are 120 to 150μ, the fat cells 70 x 90μ; 120 x 200μ.

Delphinus delphis

Material. Skin from the back over the flipper, in front of the dorsal fin, the side of the body near the flipper and below the dorsal fin, from the lateral surface of the caudal peduncles, and from the belly beneath the flippers and dorsal fins of two adult females (body lengths 160 cm and 163 cm) taken July 1953, near Novorossiisk, the Black Sea (table 47).

The skin is thickest on the belly and under the dorsal fin, and thinnest on the lateral parts of the body (Sokolov, 1962). The thickness of the epidermis is about the same all over the body. The thickest epidermis (2.6 mm) is at the rostrum's end (Surkina, 1973). The stratum corneum is thickest on the back and side of the caudal peduncle. The epidermis on the back and on the side of the body under the dorsal fin has medium pigmentation in both specimens and light pigmentation on the side of the body near the flipper in specimen (1). The pigment is distributed evenly through the thickness of the epidermal layer, in which melanocytes are also found. There is no pigmentation on the lateral side of the body near the flipper of specimen (2), or on the belly of either. The epidermal septa project small processes into the dermal papillae, and the height of the latter varies little in different parts of the body. The dermal ridges and rows of dermal papillae (the epidermal alveoli) at the front of the sides are at an angle to the longitudinal axis of the body and are directed upward toward the back (Sokolov et al., 1968). On the upper part of the back, the ridges pass along the body axis (up to the dorsal fin). The dermal ridges on the caudal peduncle are at a greater angle with respect to the longitudinal axis of the body than in the anterior part of the body. On the dorsal and ventral surfaces of the flipper, the ridges pass from its anterior edge to the apex (Purves, 1963).

The fatless part of the subpapillary layer of the dermis is present only on the belly below the dorsal fin of specimen (1), where it is 0.2 mm thick. The thickness of the subpapillary layer of the dermis varies considerably in different parts of the body (table 35). Collagen fiber bundles 30 to 55μ thick run mostly at an angle to the skin or parallel to it. The network of elastin fibers in the dermis of the whole area of the body is well developed. The majority of large fibers lie parallel to the skin surface.

The thickness of the subcutaneous fat tissue varies. It is least on the belly beneath the dorsal fins and flippers. The collagen fibers are 90 to 145μ, the size of the fat cells 90 x 170; 120 x 225μ. On the border with the subcutaneous muscles, collagen fibers form layers 155 to 450μ thick. Elastin fibers in most parts of the body are numerous, running along the collagen fiber bundles and independently between the fat cells. On the border with the subcutaneous muscles, elastin fibers form intricate plexuses and even small bundles up to 50μ thick.

Tursiops truncatus

Material. Skin from the back over the flippers and from the breast between the flippers from two specimens. Specimen (1), female (body length 20 cm) and specimen (2), male (body length 230 cm) taken in the Black Sea, June 1962. The subcutaneous fat tissue and lower parts of the dermis were cut off when the sample was taken (table 48).

In the septa of the epidermal alveoli, well-developed processes project into the dermal papillae. There are many melanocytes in the epidermis (Japha, 1910). Every cell at the surface zone of the epidermis is polyhedral with almost even sides (Surkina, 1973). A large nucleus of irregular shape contains numerous vacuoli. Smaller vacuoli are in the cytoplasm. All nuclear and cytoplasmic vacuoles are fat drops. There are more dermal papillae per cm^2 on the back—male (1) (504); female (1) (415)—then on the belly —male (1) (061); female (1) (238) (fig. 101).

The dermal ridges and rows of dermal alveoli on the body are mostly set lengthwise to the body, though set steeply upward around the caudal peduncle (Sokolov et al., 1968).

Skin innervation is greater and more specialized than in the equivalent parts of human skin (Palmer and Weddell, 1964). The nerve terminals form a peculiar complex in the septa of dermal papillae, and many are encapsulated. Some adjoin the epidermis; others are in it. Individual nerve fibers pierce the stratum basale, their terminals being separated only by two to three outer layers of the epidermis. Newborn *T. truncatus* have a few hairs on the upper lip which soon drop off. The holes remaining are richly innervated, and evidently act as specialized sense organs.

Lagenorhynchus obliquidens

Material. Skin from the back over the flippers and from the side of the body near the flipper, the breast between the flippers, the lateral surface of

COMPARATIVE MORPHOLOGY

TABLE 47

DELPHINIDAE: SKIN MEASUREMENTS

	Thickness			Height of dermal papillae (mm)	Thickness (mm)	
	Skin (mm)	Epidermis (mm)	Stratum corneum (μ)		Subpapillary layer	Subcutaneous fat tissue
Delphinus delphis						
Back above flippers						
(1)	19.2	1.4	45	0.84	1.0	16.8
(2)	18.5	1.1	63	0.67	0.8	16.6
Back at flipper						
(1)	17.8	1.5	63	0.88	0.7	15.6
(2)	17.5	1.4	54	0.75	0.7	15.4
Side at flipper						
(1)	14.0	1.3	36	0.71	0.5	12.2
(2)	16.9	1.4	36	0.75	0.5	15.0
Side under dorsal fin						
(1)	14.0	1.4	36	0.80	0.8	11.8
(2)	13.7	1.3	45	0.80	1.0	11.4
Between flippers						
(1)	15.1	1.3	**	0.71	0.3	13.5
(2)	21.8	1.3	45	0.79	0.6	19.9
Belly below dorsal fin						
(1)	21.0	1.4	36	0.75	0.8	18.8
(2)*	24.0	1.6	30	0.85	0.8	*
Caudal peduncle, lateral part						
(1)	12.2	1.3	54	0.67	0.4	10.5
Lagenorhynchus obliquidens						
Back	14.37	1.8	36	1.0	1.47	11.1
Body side	11.96	1.9	36	1.2	1.26	8.8
Breast	10.89	1.7	36	0.9	0.29	8.9
Caudal peduncle	8.50	1.7	36	1.0	0.50	6.3
Flipper, front	4.62	1.6	99	0.9	3.02	—
Flipper, top	2.34	1.1	49	0.7	0.84	0.4
Tail flukes, front	2.23	1.6	69	0.7	0.63	—
Tail flukes, top	1.48	1.2	39	0.7	0.18	0.1
Orcinus orca						
Side	42.6	5.0	**	3.7	3.6	34
Breast	40	5.6	117	4.0	2.4	32
Flipper	12.7	4.5	162	2.7	1.2	7
Tail flukes	7.5	3.1	171	2.0	1.4	3
Neophocaena phocaenoides						
Back above flippers	15.0	1.5	70	0.9	0.6	12.9
Back over anus	10.7	1.5	50	1.1	0.4	8.8

TABLE 47—*continued*

	Thickness			Height of dermal papillae (mm)	Thickness (mm)	
	Skin (mm)	Epidermis (mm)	Stratum corneum (μ)		Subpapillary layer	Subcutaneous fat tissue
Side	9.3	1.2	**	0.7	0.3	7.8
Breast	17.9	1.6	160	0.8	0.3	16.0
Caudal peduncle	5.3	1.2	40	0.7	0.3	3.8
Flipper	2.4	1.1	15	0.6	0.2	1.1

Fat cell has a pathological character.
**Poor condition.

TABLE 48

TURSIOPS TRUNCATUS: EPIDERMIS MEASUREMENTS

Sex	Thickness (mm)		Height of dermal papillae (mm)
	Epidermis	Stratum corneum	
Male (back)	2.15	0.03	1.37
Female (back	2.09	—	1.10
Male (breast)	2.42	0.05	1.32
Female (breast)	2.36	0.10	1.32

the caudal peduncle, the anterior edge and the dorsal surface from the flippers and tail flukes (table 36) of a female (body length 220 cm) taken September 1951, near the Kuril Islands.

The structure of the skin resembles that of *Delphinus delphis,* the differences being only in the indexes (table 47) (Sokolov, 1955).

The epidermal septa on the belly are smooth. Individual epidermal cells project into the tissue of the dermal papillae from the epidermal septa of the back, body side, the dorsal side of the flippers and tail flukes, and the anterior edge of the tail flukes. Small protuberances project into the epidermis of the lateral part of the caudal peduncle and the anterior edge of the flipper.

In the dermis the highest papillae are found in the side of the body, those on other parts of the body and fins being somewhat smaller. The fatless part of the dermal layer is thickest on the anterior edge of the flipper (1.2 cm)

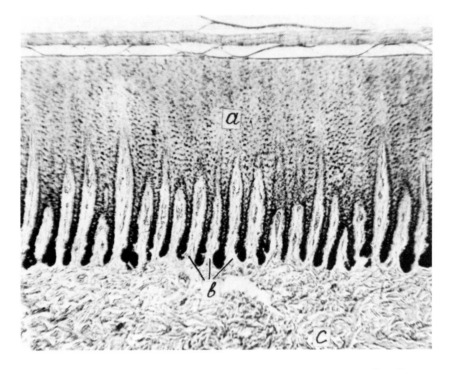

Fig. 101. *Tursiops truncatus.* Outer parts of back skin: *a,* epidermis; *b,* dermal papillae; *c,* subpapillary layer of dermis.

and the thinnest on the belly (0.04 mm). In the dermis of back, belly, and dorsal surface of the tail flukes, the bundles of collagen fibers (18 to 54μ thick) are mainly horizontal, their angles varying in other parts of the skin. The fat cells are 18 x 40; 36 x 72μ. In the subcutaneous fat tissue (fat cells 54 x 90; 126 x 144μ) of the dorsal surface of the flippers, the collagen bundles (54 to 294μ) are mainly horizontal, while on the dorsal surface of the tail flukes they are arranged at various angles.

Globicephala melaena

In the adult, stratum malpighii is 2.5 to 3.5 mm thick, stratum corneum is 0.3 to 0.4 mm (Fjelstrup, 1888). The stratum corneum cells retain nuclei and pigment granules. The structure of the subepidermal parts of the skin is typical of *Cetacea.*

Pseudorca crassidens

Material. Skin from the upper part of the head and lateral surface of caudal peduncle of a young animal, taken in summer, in the South Pacific.

The skin is not very thick (9.2 mm on the head and 10.3 mm on the caudal peduncle). The epidermis is heavily pigmented on the caudal peduncle and very lightly pigmented on the head. Both epidermis and stratum corneum are thin (head 0.63 mm and 60μ, caudal peduncle 0.70 mm and 50μ). The epidermal alveoli are large, oval, round, slit-shaped, or irregular in the head, and oval or round in the caudal peduncle. Small processes and individual cells project to the intradermal papillae from the septa of the alveoli.

The subpapillary layer is 1.2 mm thick on the head and 0.6 mm on the caudal peduncle. The fatless layer is rather thick in the dermis of the head (450μ) and thin in the caudal peduncle (90μ). Interweaving bundles of collagen fibers (60 to 90μ thick) form a compact plexus in the skin of the head and a considerably looser one in the skin of the caudal peduncle. The largest dermal fat cells are 40 x 50; 60 x 80μ. Occasional collagen bundles occur, up to 180μ thick in the subcutaneous fat tissue of the caudal peduncle, and the fat cells measure 120 x 180μ. Collagen bundles (up to 150μ thick) form a dense plexus in the subcutaneous fat tissue of the head, where fat cells (up to 100 x 140μ) occur in relatively small groups.

Orcinus orca

Material. Skin from the side near the flipper, the breast between the flippers, and the dorsal surfaces of the flippers and tail flukes of a male (body length 7 m) taken in June 1951, off Kuril Islands (table 47).

The stratum corneum is thickest on the dorsal surface of the tail flukes (Sokolov, 1955). The epidermis is highly pigmented on the dorsal surface of the flipper and also on the side of the body, where the pigment is distributed evenly in all the cells. On the dorsal side of the flipper, narrow strips of more heavily pigmented cells extend toward the stratum corneum over the apexes of the dermal papillae. Pigmentation is moderately heavy on the dorsal surface of the tail fluke, with vertical strips of less pigmented cells over the dermal papillae. The epidermis on the belly is colorless. Processes project into the dermal papillae tissue from the epidermal septa of the belly, the lateral part of the body, and the dorsal surface of the flipper (fig. 86). Protruding cells are found on the septa of the dorsal surface of the tail fluke epidermis.

The fatless part of the dermal subpapillary layer of the body is 840μ (as compared with 585μ on the dorsal surfaces of the flippers tail flukes and 450μ on the flipper). The dermal collagen bundles measure 36 to 108μ where the fat cells are 27 x 75; 45 x 63μ. They measure 180 to 324 where the fat cells are 72 x 90; 162 x 207μ in the subcutaneous fat tissue.

Phocaena phocaena

Material. Skin from various parts of the body of specimen (1), female (body length 138 cm). Specimens (2) and (3), males (body lengths 105 cm

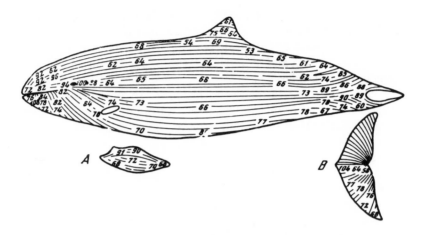

Fig. 102. *Phocaena phocaena.* Direction of dermal ridges in skin. Dorsal view of (A) foreflipper and (B) tail flukes. Figures indicate number of dermal ridges in 10-millimeter perpendicular section.

and 127 cm) taken in the Black Sea in 1968, with the permission of the USSR Ministry of Fisheries. A sample of skin from the tail fluke of specimen (4) an adult male, taken in April 1969, has been used for electron microscopy.

The skin is of uniform thickness (17 to 22 mm) throughout (up to 53 mm, Japha, 1910), being thinner only on the lateral surface of the caudal peduncle (14 to 15 mm) and near the middle of the flipper's dorsal surface (5 to 6 mm). The dermal and epidermal ridges are located throughout the body and limbs in regular rows, having the same direction as in *Delphinus delphis* (fig. 102). The thickness of the epidermis ranges from 0.6 to 0.9 mm (in front of the eye and on the flipper). The epidermal stratum germinativum per cm² of the skin on the side of the body beneath the dorsal fin is assessed at 9.2 per cm². The epidermal alveoli are round in section near the top, and oval or oblong lower down. The thickness of the stratum corneum varies considerably, ranging from 28μ on the posterior edge to 830μ on the anterior edge of the dorsal fin.

On the present specimen, a number of corneous protuberances (10 to 11) are found on the anterior edge of the dorsal fin (this is confirmed by Tsalkin, 1938). Kükenthal (1889) also finds these protuberances on the anterior edge of the flippers and tail flukes in Atlantic *P. phocaena.* The largest are up to 3 mm long, 2 mm wide and 0.9 mm high. The protuberances are histologically similar to the epidermal stratum corneum, of which they are a part. The thickness of the subpapillary layer (dermis) varies considerably: from 13μ on the ventral surface of the mandible to 124μ on the ridges of the caudal peduncle in specimen (3), the subpapillary layer being considerably thicker (up to 250 to 300μ) in specimens (1) and (2). The number of dermal

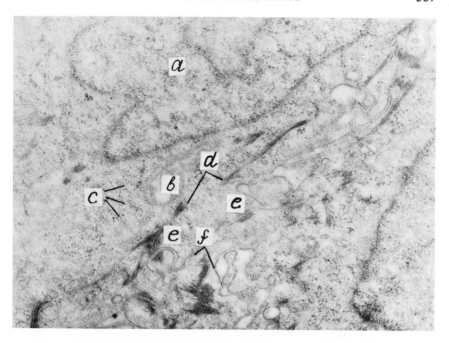

Fig. 103. *Phocaena phocaena*. Fragments of two cells of basal layer of epidermis: *a*, nucleus; *b*, mitochondrion; *c*, ribosome; *d*, tonofobrils; *e*, desmosomes; *f*, cellular membrane.

papillae (height 0.3 to 1.4μ) per 1 mm^2 of the horizontal sections is the same in all specimens (14 to 20 body; 15 to 19 head; 14 to 17 caudal peduncle; 15 to 21 flippers). The thickness of subcutaneous fat tissue varies with the skin thickness, of which it averages 90%.

Electron microscopy of the epidermis reveals that the stratum basale cells are oval and freely arranged on a rather thin basal membrane (Sokolov et al., 1971; Sokolov and Kalashnikova, 1971) with processes that interdigitate with neighboring cells. Desmosomes are noted at the root of individual cell membrane processes (fig. 103). The stratum basale cells have rather large, often irregular nuclei at the numerous nuclear membrane processes and in-ward projections. In the larger nucleoli, nucleolonemes are clearly distin-guishable and there are numerous free ribosomes in the ctyoplasm. Oval mitochondria have a light matrix and occasional cristae. Frequently, mito-chondria are in close contact with large fat globules. Lipids and pigment granules are present in rather large numbers in the cell cytoplasm in the Golgi apparatus zone. The Golgi apparatus, comprising two flat cisterns and a few vesicles, is in the immediate vicinity of the nuclei.

There is little basal structure in the stratum basale cells. Tonofibrils are shaped in a narrow strand mainly along the cellular membrane (fig. 103). In a few places small fragments of tonofibrils have entered the cytoplasm. The cells in stratum spinosum are notably more tightly interconnected than in

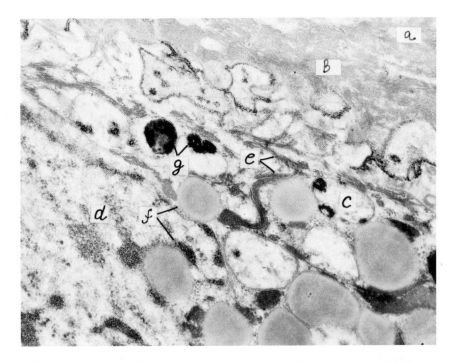

Fig. 104. *Phocaena phocaena*. Fragments of cells of stratum spinosum of epidermis on border with stratum corneum: *a*, fully cornified cell; *b*, cell with coarse, densely arranged tonofibrils; *c*, noncornified cell; *d*, nucleus; *e*, tonofibrils; *f*, ripid granule; *g*, pigment granule.

stratum basale, where there are more desmosomes in the cells, large and forming regular rows. There are many more tonofibrils. They form a heavy circular layer around the nucleus and a dense network in the cytoplasm, the bundles connecting with the fibrillar part of the desmosomes. There are fewer ribosomes in the cytoplasm of the stratum spinosum cells than in the stratum basale. They are mostly in the perinuclear zone and between the tonofibril strands. Oval mitochondria with a light matrix and individual cristae occur only in the ctyoplasm of the lower stratum spinosum cells. There are lipids and pigment granules in the stratum spinosum cytoplasm, although the pigment granules are fewer and smaller (and not in every cell) than in stratum basale (fig. 104). Electron-opaque granules surrounded by a single-layer membrane are also found near the desmosomes and tonofibrils, and small vacuoles and multivesicular bodies occur in the cytoplasm of stratum spinosum cells. There is a layer of cells rich in tonofibrils on the border between the stratum spinosum and stratum corneum (not as dense as in stratum corneum) which are connected with stratum corneum cells by desmosomes.

High electron opacity makes the cellular borders in the stratum corneum clearly visible. Desmosomes typical of the underlying layers of the epidermis are not seen in the stratum corneum. Inside the flattened cells are solid layers of fibrous substance, with the fibrillar structure less pronounced than in the inner layers. Ribosomes sometimes occur between fibrillar layers, as well as lipid and individual pigment granules, and irregularly shaped electron-opaque particles. Vacuoles are frequent. A similar description is given by Harrison and Thyrley (1974). These authors, curiously enough, do not reveal any electron-microscopic species distinctions in the epidermis of *Delphinus*, *Tursiops*, *Phocaena*, and *Orcinus*.

Neophocaena phocaenoides

Material. Skin from the back over the flippers and anus, from the side near the flippers, the breast between the flippers, the caudal peduncle laterally, and on the dorsal surface of the flipper from an adult female (body length 106 cm) (table 47).

The epidermis and stratum corneum are thickest on the breast. The epidermis has no pigment. The epidermal alveoli are round and small in section. On the sides of the epidermal-alveoli septa are oddly shaped, with processes and hollows.

Rows of corneous protuberances are characteristic of the skin. They run along the central dorsal surface in a band three to six cm wide from cervix to anus. The protuberances near the flippers are few and are larger. These protuberances are formed by a corneous bulge and are similar in consistency to the ordinary stratum corneum.

The nonfat layer of the dermis is 150 to 250μ thick. The plexus of collagen fiber bundles is loose in the subpapillary layer, the bundles being mostly horizontal or at an acute angle to the surface. In the subcutaneous fat tissue the bundles are thicker (70 to 190μ) than in the dermis (20 to 60μ). The size of fat cells in the dermis is $30 \times 50\mu$; $60 \times 80\mu$, and in the subcutaneous fat tissue 60×90; $90 \times 120\mu$.

Delphinapterus leucas

The structure of the skin has been studied by Weber (1886), Kükenthal (1889), Harmer (1930), Bonin and Vladykov (1940) and, in the greatest detail, by Belkovich (1959*a*, *b*, *c*, 1960, 1961, 1964). The epidermis is 5 to 12 mm thick. The epidermal alveoli are oval or round in section. Large processes from the alveoli septa interdigitate with the dermal papillae. The stratum corneum is usually highly developed where the epidermis is thickest and also near the blow hole. The stratum corneum is thicker near the oral, anal, and sexual orifices as well as on the flippers, dorsal fin, and tail flukes.

The epidermal pigmentation is peculiar and goes through several stages during ontogenesis. In embryos, black or brown granules of melanin in the cytoplasm of cells of the basal layer are especially numerous in a layer of flattened epidermal cells parallel to the skin surface. In the newborn, the pigment is most concentrated in the basal layer of the epidermis where it meets the dermis. In a suckling, 235 mm long, the pigment is no longer concentrated in the inner epidermis. It is distributed through the whole thickness. The pigmentation is greater in the outer cells. At 345 cm, there is already very little pigment. When the epidermal cells begin to cornify, what little pigment they contain is distributed unevenly. At 470 cm, there is no longer any pigment in the epidermis, which is now white. The 3 to 5 cm dermal papillae comprise half the thickness of the epidermis. The subpapillary layer is 2 to 36 mm thick. The nonfat dermal layer remains almost unchanged with age. It is 4.5 mm in a specimen 165 cm long, and it is 4.7 mm in a specimen 423 cm long. The dermal collagen fiber bundles mostly run longitudinally down the body at a slight angle. They may be 220 or even 990μ thick.

The subcutaneous layer on the body varies from 4 to 27 cm. In animals of similar size taken at the same time, the thickness of the subcutaneous fat tissue may differ by several centimeters. It also varies seasonally, being thickest in winter and thinnest (in adult animals) during the reproduction season; sex-specific differences are slight. The size of the fat cells is 130 to 200 x 110 to 150μ. The collagen fiber bundles running in various directions are 10 to 15μ in diameter.

FAMILIA PHYSETERIDAE

Kogia breviceps

Material. Skin from the back over the flippers, the side near the flippers, the lateral surface of the caudal peduncle, the anterior edge and dorsal surface of the flipper, the dorsal surface of tail flukes, and the anterior edge of the doral fins from an adult specimen (body length 109 cm) taken off Honshu Island (table 49).

The skin is not adequately preserved and the stratum corneum is damaged in most samples. The epidermal alveoli are oval (in the skin of the caudal peduncle) or ovally-round, with small processes and a few cells protruding into the alveolar cavity. The alveoli are very small. In the skin of the back and the dorsal surface of the flipper, the epidermal alveoli are at oblique angles to the skin. Their greatest vertical height is 110 to 120μ, and their greatest length is 160 to 240μ. The collagen fiber bundles are mainly horizontal. The 30μ nonfat layer is sometimes absent altogether. The collagen bundles are 15 to 30μ thick, and the fat cells are 30 x 60; 40 x 70μ. The dermal papillae of the lateral side of the tail flukes and the anterior edge of the tail flukes contain fat cells. They are plentiful in the caudal peduncle papillae, but are rare in other places and are not found in all dermal papillae.

Loose bundles of collagen fibers (30 to 80μ) are characteristic in the sub-

TABLE 49

PHYSETERIDAE AND ZIIPHIDAE: SKIN MEASUREMENTS

	Thickness			Height of dermal papillae (mm)	Thickness (mm)	
	Skin (mm)	Epidermis (mm)	Stratum corneum (μ)		Subpapillary layer	Subcutaneous fat tissue
Kogia breviceps						
Back	15.1	0.2	50	—	0.09	14.8
Side of body	13.5	0.1	*	0.1	0.09	13.3
Breast	18.1	—	*	—	0.09	18.0
Caudal peduncle	7.4	0.4	*	0.3	0.09	6.9
Flipper, front	4.1	0.5	*	0.4	0.12	3.5
Flipper, top	1.5	0.2	*	0.1	0.06	1.3
Tail flukes, top	0.5	0.2	*	0.1	0.30	—
Dorsal fin, front	—	0.6	*	0.5	—	—
Physeter catodon						
Head, front	58.3	6.3	45	5.4	30	22
Head, back	35.8	3.8	36	2.7	32	—
Head, side	63.8	4.8	36	3.8	19	40
Back	101.8	5.8	24	4.4	15	81
Side	114.8	4.8	45	3.9	15	95
Breast	135	*	*		16	119
Caudal peduncle	63.8	4.8	36	3.8	19	40
Flipper, front	21.3	3.3	*	2.8	18	—
Flipper, top	11.2	3.2	36	2.7	8	—
Tail flukes, front	17.3	4.3	36	3.3	13	—
Tail flukes, top	11.1	3.1	18	2.2	8	—
Berardius bairdii						
Head	68.4	3.8	45	2.1	0.63	64
Back	215.1	2.7	*	1.8	0.46	212
Side of body	137.4	3.5	27	2.0	0.96	133
Breast	142.7	3.3	24	1.9	0.42	139
Foreflipper, Upper part	16.1	2.9	36	1.5	1.26	12
Tail flukes, Upper part	9.9	2.1	45	1.1	0.84	7

*absent.

cutaneous fat tissue, with the fat cells somewhat bigger here than in the dermis (70 x 100; 90 x 120μ).

Physeter catodon

Material. Skin from anterior, upper, and lateral parts of the head, from the back over the flippers, the side near the flipper, the breast between the flippers, the lateral surface of the caudal peduncle, the anterior edge and dorsal surface of the tail flukes and flippers (table 49) of female (body length 11.15 m) taken August 28, 1951, near the Kuril Islands.

In large, fat males the skin on the belly may be 50 cm thick, but usually it is less than 20 cm. On the side, level with the dorsal fin, it rarely exceeds 11 or 12 cm. In epidermal structure, the sperm whale differs from other whales in that the lower edges of the longitudinal and transverse septa are level, so that polygonal, elongated alveoli (fig. 86) (no longitudinal septa as in other *Cetacea*) are seen in a tangential section through the inner layers of the epidermis. The septa of these alveoli divide into vertical and horizontal. The epidermal septa here are much thinner than in other *Cetacea*.

In tangential sections of the median parts of the dermal papillae, most of the alveoli are regular polygons (fig. 86). The layers of alveoli are less regular than in other *Cetacea*. They are so scattered that it is impossible to distinguish between horizontal and vertical sections. This epidermal septa pattern remains unchanged throughout to the apexes of the dermal papillae. The processes on the epidermal septa have a thickened distal and narrow proximal sector. Many processes are ball-shaped in tangential section, the ball being set on a thin stalk. The epidermis is highly pigmented only on the back. Thicker pigmented cells are in that part of the body over the dermal papillae. The epidermis in the other parts of the body except the dorsal surface of the tail flukes is moderately pigmented; all the cells above the dermal papillae are regularly pigmented.

The whale has a characteristic nonfat zone in the upper dermis (fig. 105), 13.1 to 11.3 mm thick on the anterior edges of tail flukes and flippers, to 4.5 to 4.9 mm on the side of the body and on the belly.

Collagen bundles run at various angles in the dermis of all the samples. Only in the inner dermis of the upper part of the head and the dorsal surface of the flipper a majority of the bundles are horizontal (0.29 to 0.45 mm) in the dermis, the fat cells being 45×126; $189 \times 360\mu$. There are no fat cells in the tail fluke skin. In the subcutaneous fat tissue the collagen bundles are 0.36 to 1.13 mm, the fat cells are 175×258; $420 \times 588\mu$. The middle of the subcutaneous tissue contains the most fat (Kolchev and Morieva, 1953).

The skin of an 8.5 m sperm whale embryo has 2 to 3 layers of sparse epithelial cells (Osipuk, 1975). The skin of embryos 13, 14.5, and 15 mm long includes epidermis and dermis. The epidermis consists of 5 to 11 layers of sparse cells with large nuclei. In the superficial layers, the cells are thickened. The dermis is formed by sparse process mesenchymal cells. In embryos 25.6 and 35 mm long, the epidermis is somewhat thicker. In embryos 54 and 72 mm long its thickness attains 30μ, with collagen fibers starting to form in the dermis. These fibers are especially pronounced near the subcutaneous muscle. In embryos 95 and 105 mm long, the epidermis basal layer is distinguished by highly basophilic nuclei. In the superficial layers of the dermis, the collagen bundles are arranged in various directions. In the deep parts they are parallel to the skin surface. In embryos 149, 220, 320, 520,

Fig. 105. *Physeter catodon.* Fat-free part of subpapillary layer.

and 650 mm long, the following epidermal layers are distinguished: corneum, spinosum, and basale. The spinosum layers are of irregular polygonal shape, disposed in 8 to 12 rows. The basal layer of the embryo 520
mm long first shows pigment cells.

The histogenesis of the skin of *P. catodon* has been studied by us in four
embryos: a three-month male (body length 79 cm) (according to Laws,
1959), a male (105 cm), and a female (114 cm) (both embryos about four
months old), and a female of unknown size with a more developed skin.
Skin samples have been taken from various parts of the body and flippers,
dorsal fins and tail flukes. The skin consists of epidermis, dermis, and the
subcutaneous fat tissue (Sokolov, 1963). There are no hair or skin glands.
In the three younger embryos, the relative thickness of the epidermis is
about the same as in adult whales. In the older embryos the relationship is
different; the epidermis is relatively much thicker. At later stages of embryogenesis, the epidermis develops fast (in particular, its alveolar structure is
formed) whereas the subepidermal layers (mostly the subcutaneous fat tissue) are slightly developed and their thickness appears to increase greatly
immediately before birth, or, more likely, during the first days after birth.

In *Balaenoptera physalus* embryos the relative thickness of the epidermis is less than in adult whales, which suggests a different rate of epidermal development from *P. catodon*. The epidermis in different body areas varies in more and less advanced *P. catodon* embryos. In the four month embryos, the epidermis is thicker on the fins than on the body, and in older embryos (as in adults) this is reversed—the epidermis presumably develops earlier on the fins. Three epidermal layers are noted: basale, spinosum, and corneum. Most stratum corneum cells still have nuclei.

In the three month *P. catodon* embryo, the inner surface of the epidermis lacks the alveolar structure typical of the adult. This begins to develop on the fins at four months, a month and a half earlier than in *B. physalus* (Naaktgeboren, 1960). Unlike *B. physalus, P. catodon*'s dermal papillae anlagen appear without the formation of dermal ridges. In the most developed *P. catodon* embryos, the epidermal alveoli are present throughout the body. The relative height of the dermal papillae is greater than in the adult. In the older embryos the alveoli are the same in section as in adult animals (round or polygon). The epidermal septa penetrate into the dermis to about the same depth as in the adult, but the processes of the septa are small and still lack distal bulges.

During embryo development, the epidermal thickness increases considerably. In fact, in the most advanced *P. catodon* embryo the back skin is twice as thick as in a five-month embryo, whereas the thickness of the epidermis is about twenty-five times that of the latter. The thickness of the subepidermal parts increases greatly immediately upon birth. A comparison would show the adult *P. catodon*'s back skin to be some forty-three times as thick as that of an advanced embryo and the epidermis only four times as thick. The absolute thickness of stratum corneum increases with the age of the embryo, but its relative thickness decreases. But even the absolute thickness of an advanced embryo's stratum corneum is greater than in an adult. A high development of stratum corneum is also recorded in newborn *Delphinapterus leucas* (Belkovich, 1956c) and *Globicephala melaena* (Starett and Starett, 1955). Possibly the thick stratum corneum provides insulation for the newborn (Belkovich, 1961). The absolute and relative thickness of stratum corneum on the flippers, dorsal fin, tail flukes, and on the body of the embryo sperm whale is almost the same as in the adult.

There are pigment granules in the epidermal cells (at early stages they are evenly spread through the cytoplasm, later around the nuclei) and in melanocytes. There is no clear division in the three- to four-month embryo between dermis and subcutaneous fat tissue. This begins to appear in a more advanced embryo, but it is not too well defined even in the adult whale. Their relative thickness approximately resembles that of the adult. The plexus of collagen fiber bundles is similar to the adult, and elastin fibers are few. Fat cells are present in the skin of all the embryos. The coronary vessels in the flippers, dorsal fin, and tail flukes are similar to those of the adult.

An examination of sperm whale embryos from 1 to 1.5 months to prebirth has led Veinger (1974) to distinguish five stages in skin histogenesis:

First stage: embryonal vesicular epithelium is characteristic of the earliest skin development period. In the caudal part, this stage is completed before the age of 1.5 months, somewhat later on the head. At this stage, epithelial cells are vesicular in shape and the dermis is syncytial.

Second stage (2.5 months): vascularization and early melanogenesis. A network of capillaries can be seen under the basal membrane. Melanin granules appear in the epithelial cells. Bundles of collagen fibers are being formed in the mesenchyme.

Third stage (4 months): the dermal reticular layer is being formed. The first elastin fibers develop. Cornification starts in the epidermis.

Fourth stage (7 months): anlage of the dermal papillary layer. Near the artery, an extensive proliferation of connective tissue cells forming cupola-like projections to the basal membrane is taking place near the artery. By this time the dermal papillae are better developed in the skin of the tail than on the head skin.

Fifth stage (11 to 13 months): formation of the subcutaneous fat tissue. Small insulae of fat cells appear in the inner layers on the boundary of the muscular layer between collagen fiber bundles. Subsequently, they merge and are crossed by collagen fibers. By this time, all constituents of the embryo skin are in evidence.

In histogenesis, the anlagen of skin constituents and structures is first in the caudal part. It is solely at the late stage of embryogenesis (by the age of 13 months) that formation processes level out.

Familia Ziphiidae

Berardius bairdii

Material. Skin from head, back over the flippers and side near the flippers, the breast between the flippers, the upper part of the flippers and tail flukes of a male (body length 10.5 mm) taken September 9, 1951, near the Kuril Islands (table 49).

The epidermis is heavily pigmented only on the belly. On the back, side of the body, head, and the dorsal surfaces of the flippers and tail flukes, pigmentation is moderate. In the epidermis of all these samples, narrow strips of heavily pigmented cells extend the stratum corneum over the apexes of the dermal papillae. On the side of the body, pigmentation is light and regular in all the cells above the dermal papillae level. Numerous processes project into the dermal papilla tissue from the epidermal septa (fig. 86). A fat-free dermal layer is found only in the skin of the dorsal surfaces of the flippers, dorsal fin, and tail flukes. In the flipper it is 360μ thick; in the tail flukes 180μ. In the upper dermis on the back and dorsal surfaces of the tail flukes, the bundles of collagen fibers are horizontal (fig. 106). In the inner parts of these two samples and through the whole of the other samples, bundles of collagen fibers up to 27 to 90μ are arranged at various angles to the

Fig. 106. *Berardius bairdii.* Dermal network of collagen fiber bundles: *a,* in subcutaneous fat tissue; *b,* epidermis.

skin surface. The fat cells are $54 \times 81\mu$ and the thickest collagen bundles in
the subcutaneous tissue are 162 to 546μ; the fat cells are 144×190; $351 \times$
477μ.

SUBORDO MYSTICETI

The skin is medium thick to very thick. Numerous longitudinal skin plicae
extend down most of the ventral side of the body. In the epidermis, the longitudinal alveoli septa are higher than those which are transverse. In the

transverse section, the epidermal alveoli are oval (rarely, round). The epidermal processes are usually small or absent. Sometimes they are very large. The dermal fat cells usually approach the epidermis, though occasionally there is a fat-free zone. In the dermis and the subcutaneous fat tissue, a network of elastin fibers, usually forming large bundles, is well developed. In some members of the suborder, bundles of muscle fibers in the subcutaneous muscles are surrounded by numerous parallel elastin fibers. Vibrissae are found on the muzzle.

FAMILIA BALAENOPTERIDAE

Large skin plicae extend down the body from chin to navel, over the throat, breast, and belly, and are separated by grooves. The epidermal alveoli are oval. The epidermal processes are small or absent. There is usually no dermal fat-free zone. The network of elastin fibers and bundles is highly developed. Vibrissae are arranged on the end of the mandible and on the dorsal surface of the head (Japha, 1916; Yablokov and Klevezal', 1964).

Balaenoptera acutorostrata

Material. Skin from the back over the flippers, from the bands on the breast between the flippers, from the belly near the navel, the anterior edge and the dorsal surface of the flippers and tail flukes of a female (body length 7.08 m) taken August 2, 1951, off the Kuril Islands.

B. *acutorostrata* is very similar to B. *physalus* in skin structure, though all the indexes are smaller (table 50). The epidermis is heavily pigmented on the dorsal surface and anterior edge of the tail flukes, and lightly on the back. All the cells above the dermal papillae apexes are uniformly pigmented. The epidermis near the navel, on the apexes of the bands, and on the dorsal surface and edge of the flipper is unpigmented. Small processes protrude into the tissue of dermal papillae from the epidermal septa on the body (fig. 86). Their size increases considerably on the flippers, dorsal fin, and tail flukes. The dermal fat-free layer is present only on the belly, where it is 120μ thick. In the dermis of the back, the belly near the navel, and the belly bands on the anterior edges of the flipper and tail flukes most of the collagen bundles are arranged at various angles to the surface. On the side of the body and on the dorsal surfaces of the fins they are horizontal. The dermal collagen fibers are 20 to 72μ thick, the fat cells are 48 x 72; 130 x 144μ. In the subcutaneous fat tissue of the dorsal surface and the anterior edge of the flipper, the collagen bundles run at various angles, while in the subcutaneous tissue of the dorsal surface of the tail flukes, they are horizontal, 90 to 360μ thick; the fat cells being 49 x 72; 120 x 400μ. There are very thick (375μ) bundles of elastin fibers in the subcutaneous tissue of the belly bands (900μ) and around the navel.

In the skin of the belly, small elastin bundles appear 2.8 cm inward from the epidermis. In the inner layers of the subcutaneous tissue they predominate in number over the collagen bundles. Gigantic horizontal bundles of elastin fibers, oval in section, run along the long axis of the body.

Balaenoptera borealis

Material. Skin from belly near the skin bands of an adult taken in summer off the Kuril Islands.

There are forty to sixty-four skin bands on the belly which do not reach the navel. The skin thickness ranges from 5 cm on the belly to 10 cm on the back (Japha, 1907). The median thickness of the epidermis is 1.5 mm. It is thicker where the skin is not pigmented (up to 2 mm) (Japha, 1905, 1907). The sample under study (Sokolov, 1962) is 47 mm from the apex of the band to the subcutaneous muscles, and the grooves between bands are 13.5 mm deep.

The epidermis, which is thickest at the apex of the band (1 mm), gradually thins laterally toward the bottom of the groove, where it is only 0.3 mm thick. The stratum corneum is 110μ thick at the band apex and 25μ at the bottom of the groove. The epidermis is very slightly pigmented. The dermal papillae are up to 0.6 mm high at the apex of the band and down to 0.2 mm on the lateral parts and the bottom of the groove. The subpapillary layer of the dermis is 0.5 mm thick. Collagen bundles up to 20μ thick run in various directions. There are many elastin bundles, and numerous fat cells (30×54; $45 \times 64\mu$) approach the epidermis itself.

Around the hair, the subcutaneous fat tissue is 45.5 mm thick and near the grooves it is 32 mm. The collagen fiber bundles run in various directions, to a maximum thickness of 400μ. The fat cells are larger (90×190; $110 \times 130\mu$) than in the dermal layer and elastin fibers, are numerous. The first thin bundles of elastin fibers, which lie 10.1 mm from the band apex, become thicker as they approach the subcutaneous muscles, the largest being 440μ thick.

There are about one-hundred vibrissae on the head. Their roots are 10 to 15 mm long (Japha, 1905, 1907), the outer part of the hair is about 8 mm long and 0.16 mm thick (Yablokov and Klevezal', 1964). The blood cavities surrounding the roots of the hair are broad and asymmetric. The hair bursae are richly innervated (Japha, 1910; Nakai and Shida, 1948).

Balaenoptera physalus

Material. Skin from the back over the flipper, the side near the flipper, the bands between the flippers, the belly near the navel, the flipper (back and top), and the top of the tail flukes of a male (body length 15.65 mm) taken September 21, 1951, off the Kuril Islands (table 50).

TABLE 50

BALAENOPTERIDAE AND BALAENIDAE: SKIN MEASUREMENTS

	Thickness			Height of dermal papillae (mm)	Thickness (mm)	
	Skin (mm)	Epidermis (mm)	Stratum corneum (μ)		Subpapillary layer	Subcutaneous fat tissue
Balaenoptera acutorostrata						
Back	28.6*	1.7	120	1.0	0.9	26*
Bands on breast	46.8	1.9	60	1.0	0.9	44
Belly	42.9	***	—	0.8	0.9	42
Flipper, front	18.1	3.0	345	1.7	1.1	14
Flipper, upper part	8.2	2.4	300	1.1	0.8	5
Tail flukes, upper part	2.5	1.4	180	0.7	0.8	0.3
Tail flukes, anterior part	4.9	2.4	405	1.2	2.5	—
Balaenoptera physalus						
Back	47.7	3.0	60	1.8	0.7	44
Side	45.3	3.3	45	1.9	0.9	41
Bands on breast	58.5	3.9	92	2.7	0.6	54
Belly	45.6	2.9	60	1.7	0.7	42
Flipper, anterior part	13.9	3.4	144	2.1	10.5	—
Flipper, upper part	11.5	3.3	54	2.0	1.2	7
Tail flukes, upper part	5.2	2.8	72	1.9	0.4	2
Eubalaena glacialis						
Specimen (44), side	236	12.2	60	6.7	8.8	215
Flipper, top**	—	8.6	110	5.2	18.0	—
Specimen (40), back	160	13.2	100	8.2	7.8	139
Tail flukes, top	38	9.7	100	6.3	6.6	22
Specimen (41), side	250	9.0	330	7.4	8.3	233
Specimen (50), side	272	10.2	130	7.7	7.0	255
Specimen (98), side	153**	—	—	—	7.3	146
Side	192***	—	—	—	7.5	185
Belly	209	11.6	70	7.7	6.0	191

*Inner subcutaneous fat tissue, not preserved.
**The sample not taken through full thickness of skin.
***Outer epidermis and the upper section of dermal papillae, not preserved.

In the 16-meter-long fin whale studied by Giacometti (1967), the epidermis on the belly (3 mm) is thicker than on the back (2.5 mm). The epidermis is highly pigmented on the back, around the navel, and on the dorsal side of the tail flukes. The most heavily pigmented cells form vertical bands running upward above the top of the dermal papillae. The epidermal septa have no protrusions (fig. 86). There is no fat-free dermis layer except on the anterior edge of the flipper, where its thickness is 1428μ, and on the dorsal sur-

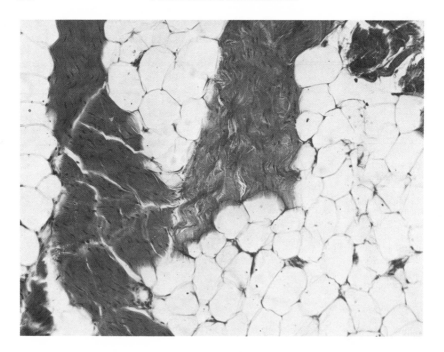

Fig. 107. *Balaenoptera physalus.* Collagen fiber bundles and fat cells in subcutaneous tissue
of back.

face of the tail flukes (180μ). In the upper subpapillary layer occur Vater-
Pacini corpuscles, weakly reactive for cholinesterase (Giacometti, 1976).
The collagen fiber bundles are horizontal or at an acute angle to the surface
in the dermis on the back, the belly bands, around the navel, and on the
dorsal side of the flipper. On the side they are all at an acute angle, and on
the anterior edge of the pectoral and on the dorsal surface of the tail flukes
they run at various angles. The bundles are 28 to 420μ thick, the fat cells are
27 x 27; 63 x 72μ. In the subcutaneous fat tissue, the collagen bundles are
0.17 to 1.34 mm thick, the fat cells are 92 x 128; 100 x 168μ (fig. 107).

The greatest accumulation of fat is in the middle layer of the subcutaneous fat tissue (Kolchev and Morieva, 1953). The subcutaneous tissue is
absent on the anterior edge of the front flipper. The plexus of elastin fibers
is very well developed in the subcutaneous tissue of the back and the dorsal
surface of the fins. There are elastin fibers in the subcutaneous tissue of the
body. These reach a great thickness on the belly bands (750μ) and near the
navel (515μ), but are considerably thinner on the back (225μ) and sides
(135μ). They are very thick in the subcutaneous tissue at the hairs on the
belly (750μ) and near the navel (515μ), but they are considerably thinner on

the back (225μ) and on the side of the body (135μ). There are no bundles of elastin fibers in the subcutaneous tissue of the flippers.

The first bundles of elastin fibers in the skin of the bands of the belly appear 9.7 mm from the lower border of the epidermis, and both number and size increase toward the subcutaneous muscles. In the lower layers of the subcutaneous tissue, bundles of elastin fibers appear in larger numbers than the collagen fiber bundles to which they are parallel.

The subcutaneous muscles are typically cross-striated. However, along the muscle bundles are large (up to 8μ in diameter) elastin fibers (up to 40 fibers around every muscle bundle) surrounding the muscles on all sides (Sokolov, 1960).

Unlike the sei whale, the fin whale often has vibrissae on the upper side of the head between the blow hole slits (Yablokov and Klevezal', 1964). The number and distribution of vibrissae between the mandible and the maxilla is different in male and female. The vibrissa root is up to 11.6 mm long, the outer hair 10.1 mm. Vibrissae are 0.15 mm thick.

The histological differentiation of the epidermis in ontogenesis begins in a two-month embryo. By six months the layers of the epidermis are the same as in an adult (Naaktegeboren, 1960). The first anlagen of the dermal ridges are noted in the five- and-a-half-month embryo, developing earlier in the cranial than the caudal parts of the body. The dermal papillae appear in an eight-month embryo. At nine months there are few differences from the adult animal's skin.

The skin from the various parts of the body of the embryos under investigation (a male, body length 502 cm, about 9 months, according to Laws, 1959; and a female of unknown age with a less developed skin) consists of epidermis, dermis, and subcutaneous fat tissue (Sokolov, 1963). Hair anlagen and skin glands are not found. The embryo epidermis is in three layers as in adult whales: the strata basale, spinosum, and corneum. In most of the embryos the stratum corneum cell nuclei are similar to the adult's, except that they are flattened. But on the back and flippers skin of a nine-month embryo there are no nuclei in the cells, and the stratum corneum resembles that of land mammals. The thickness ratios between the epidermis in various parts of an embryo with a less developed skin are the same as in adult whales: the epidermis is much thicker on the flippers than on the body. Unlike Naaktgeboren (1960), who claims that the alveolar structure of the epidermis is formed first in the cranial parts of the body, we have found that the first anlagen are on the flippers and belly bands, and that dermal ridges are formed before the dermal papillae. In a nine-month embryo, epidermal alveoli are found over the whole body, and the dermal papillae are relatively higher in relation to the epidermis than in the adult, indicating a relatively poor development of the outer epidermis, especially of the stratum spinosum. The oval section of the adult whale's epidermal alveoli is also typical of our more advanced embryos. The epidermal septa forming the alveoli are similarly arranged; the adult's longitudinal septa are considerably deeper than the transverse.

As the embryo develops, the epidermis gets much thicker than its subepidermal layers, far more so than most of the skin. The nine-month embryo's back skin is 60% thicker than the back skin of the less developed embryos, and the epidermis is 3300% thicker. In the transition to postnatal life, the rate of development changes in various parts of the skin. Subepidermal layers begin to grow much thicker. On the back, an adult's skin is three times as thick as a nine-month embryo's, though the epidermis is only 1.7 times as thick.

While the embryo's stratum corneum does get thicker with age, its thickness in relation to the whole epidermis decreases. It is interesting to note that in some body areas the nine-month embryo's stratum corneum is thicker than the adult's. Both the absolute and relative thickness of stratum corneum on a younger embryo is greater on the body than on the flippers, dorsal fin, and tail flukes. But in a nine-month embryo this is reversed. Here, as in the adult, the stratum corneum is thicker on the flippers, dorsal fin, and tail flukes than on the body. And in a nine-month embryo, the epidermis and stratum corneum are thicker at the apexes of the belly bands than in the grooves between, while in the earlier stages the relationship is reversed—the epidermis and stratum corneum are thicker in the grooves. This is especially interesting, for it shows the profound changes that have taken place in the *Cetacea*'s skin—the thicker epidermis and stratum corneum in the areas subject to the greatest wear (from water while swimming) manifests itself in the embryo.

The epidermis of both embryos under study is pigmented, with melanocytes in the stratum basale and pigment granules in the epidermal cells. At early stages, these granules are evenly distributed through the cytoplasm; later they are concentrated above the nucleus of the cell as in adults. In both embryos, the subepidermal layer of the skin can be divided into dermis and subcutaneous fat tissue, though there is no clear-cut boundary as in the adult. A progressive decrease in the relative thickness of the dermal layer and an increase in the subcutaneous fat tissue is noted in the course of embryonic development. In fact, in the embryo with a less developed skin, the dermis on the side of the body is 2.64 mm thick (38.8% of skin); in the nine-month embryo, it is 1.1 mm (3.1% of skin); and in an adult, it is 0.9 mm (2.1% of skin).

The plexus of collagen fiber bundles usually has the same pattern as in adults, and the elastin fibers peculiar to the adult fin whales are present in both embryos and most developed in the ventral side in both. However, in the subcutaneous muscles, the elastin fibers that surround every muscle bundle in the adult are absent in embryos. There are no fat cells in the younger embryo's skin. The coronary vessels (large arteries surrounded by a complex of veins) peculiar to the adult whale's fins are also found in the embryo fins.

FAMILIA BALAENIDAE

The skin is very thick, but the folds on the ventral side are absent. The section of the epidermal alveoli is round or polygonal. The epidermal processes are large. Usually there is a well developed fat-free zone in the dermis. The network of elastin fibers and bundles is less developed than in Balaenopteridae.

Eubalaena glacialis

Material. Skin from five whales taken in summer (the numbers are from the whale ship registers). Female (44) (body length 16.30 m); male (40) (body length 10.75 m); female (41) (body length 17.40 m); male (50) (body length 17.96m); female (98) (body length 11.35 m). The whales were taken with special license No. 0113/2010, Ministry of Fisheries of the USSR, dated March 3, 1955.

Unfortunately, the samples are not taken from the same part of the body in all the whales: the side of the body, approximately where the lines from the pectoral flippers and dorsal fins cross (41), (44) and (98); the middle of the back (40); the side of the body near the flippers (50) and (98); the belly near the navel (98); the dorsal surface of the flippers (44); and the tail flukes (40) (table 50).

In the epidermis of some samples, larger, rounder column-shaped cells reach from the dermal papilla apex to the surface (Sokolov, 1962). Similar formations are found in the skin of Steller's sea cow (Haffner, 1957). In most of the samples, the epidermal pigmentation is of medium intensity. Large processes project from the epidermal septa into the dermal papilla tissue. Unlike the other *Mysticeti* studied, *E. glacialis* has a well-developed dermis (table 38). There is a fat-free layer (1.7 to 5.3 mm thick), except in (98). Collagen fiber bundles (230 to 440μ thick) run in various directions forming a dense plexus in the dermis, where elastin fibers are few in number (especially in the flippers, dorsal fin, and tail flukes), and fat cells (45 × 70; 120 × 160μ) occur in large groups. Collagen fiber bundles in the subcutaneous fat tissue reach 330 to 660μ in thickness, and on the body skin there are bundles of elastin fibers 165 to 550μ thick in the inner subcutaneous tissue (they are not found in the flippers, dorsal fin, and tail flippers and are more developed on the belly than the side). Note, for instance (98): 350μ belly; 250 to 300μ on the side. They occupy a large area in the skin (the zone of elastin fibers on the belly is 8.1 cm and in the body skin 3.3 to 4.9 cm). Common elastin fibers in the subcutaneous tissue are rather numerous. The fat cells are 110 × 130; 160 × 240μ.

There are peculiar skin bonnets on the head with hollows on the surface caused by parasites (*Cyamus*). The shape, size, and distribution of the bonnets is fairly regular, although they vary considerably in some whales. Most of *E. glacialis* have large bonnets on the anterior edge of the maxilla, the lateral sides of the anterior edge of the mandible, above the eyes, and in front of the blowhole (Allen, 1908; Matthews, 1938; Tomilin, 1957; Klumov, 1962). There is a double (sometimes single) row of smaller bonnets on the maxilla between its anterior edge and the blowhole, and also a row of small bonnets on the lateral sides of the mandible. Small bonnets are 10 to 30 cm, and large ones are 50 to 135 cm in diameter (Klumov, 1962). The bonnets are noted on embryos and babies and they grow with age. There are hairs on the bonnets (32 to 35 mm long). Any number between one and two hundred may be found, varying in size, location (Klumov, 1962), and from whale to whale. The function of these bonnets is not known. Allen (1908) believes they are a result of a permanent skin irritation from whale lice. Tomilin (1957) also regards this as likely. On the other hand, the author (Sokolov, 1957) holds that they might have grown out of the protuberances on which the hairs grew under the action of the whale lice. Matthews (1938) does not accept these theories, pointing out that the bonnets also occur in embryos, and that parasitic crustaceans are also found on other parts of the body. G. Allen (1916) assumes that the largest bonnet on the head serves as a weapon, and Howell (1930) believes that it corresponds functionally to the fat layer on the toothed whales.

We have studied the bonnets of two *E. glacialis*: the upper part of the head, the chin, and a bonnet whose label had been lost from specimen (1); and the top of the head and the mandible from specimen (2). In longitudinal section, it can be seen that the bonnet is really a thickening of the epidermis, in a place where the epidermis is 2.6 to 5.6 times as thick and the dermis surface is also thickened. The epidermis on the bonnets as in other parts of the body consists of strata basale, spinosum, and an outer layer that could be called corneus, for the borders are retained between cells and cellular nuclei. The thickness of these layers varies considerably. The stratum corneum is 1.5 to 40 mm thick; the stratum spinosum is 3.5 to 12 mm. This high bonnet of stratum corneum is an epidermal reaction to the whale lice. The stratum corneum has numerous corneous papillae extending vertically downward from the outer skin surface to the apexes of the dermal papillae. The cells in the corneous papillae are in various stages of degeneration. The inner cells of the papillae barely differ from the surrounding cells, but nearer the surface they become rounder. They are vacuolized, with their nuclei becoming deformed and hypochromatic. Farther on, the cells become still flatter, the border between them almost disappears, and the cellular nuclei break down. The corneous papillae have a layered pattern characteristic of land mammals' epidermal stratum corneum.

As has been noted, the skin of *Cetacea* in the normal state has no genuine stratum corneum. Potentially, however, the capacity for cornifying appears to be retained. It is manifested as a result of a reaction to irritation produced by whale lice. In some rodents and birds a similar reaction of skin is noted in response to mechanical and chemical irritation (by saliva) which is caused by mites parasitizing these animals. An intensive cornification at the surface of the epidermis takes place there and around the tongue of the sucking mite, and there is an even formation of corneous "cones" around the tongue (Pawlowsky and Alfeeva, 1941, 1949; Berlin, 1957; Pokrovskaya, 1957; Lototsky et al., 1959).

The dermis near the area of the bonnet, as in other parts of the body, consists of dermal papillae rising into the epidermal alveoli and the subpapillary layer in which the upper part adjoins the epidermis. The fat-free layer is distinguished by a more compact plexus of collagen bundles. The dermal papillae in the bonnets are 1 to 2 mm higher, and the fat-free and the subpapillary layers are thicker (the first layer is 2 to 3.4 times, and the second layer is 1.2 to 2.4 times as thick as the normal body skin).

The structure of the dermal papillae of the bonnets is the same. In the subpapillary layer of the bonnet dermis, the plexus of collagen fiber bundles is firmer than in the normal skin, and the bundles themselves (which run in various directions) are thicker and sometimes form regular layers. The elastin fibers in the subpapillary dermis of the bonnet are few, as in the normal skin. The subcutaneous fat tissue under the bonnets is normal.

The histological structure of the bonnet epidermis shows pathological characteristics that do not fit with the functions attributed to the bonnets by Allen (1916) and Howell (1930). We believe (Sokolov, 1961) that the best explanation of the origin of the bonnets is given by Tomilin (1957), who suggests that the bonnets are the protuberances on which the hair grew. Possibly these hairs on the bonnets function better as vibrissae, so that a small thickening of dermis and epidermis in these places in the embryo would be natural. The whale lice, which have passed all the stages of evolution from a freeliving organism through the commensal stage to parasitic animals (Klumov, 1962), have adapted to these places, as it is easier for them to stick to the skin there. A very high outgrowth of the epidermis, the cornification to its papillae, and the pitted pattern of its surface are secondary formations appearing postnatally in reaction to the parasite.

ORDO CARNIVORA

The skin and, especially, the hair of certain *Carnivora* in the USSR with commercially important fur has been studied rather well; notably *Nyctereutes procynoides* (Kogteva, 1963), *Alopex lagopus* (Tserevitinov, 1951; Komarova, 1968), *Vulpes vulpes* (Tserevitinov, 1951, Leschinskaya, 1952a, b), *Martes zibellina* (Pavlova, 1951; Gerasimova, 1958), *Martes martes* (Pavlova, 1951), *Mustela erminea* (Marvin, 1958a; Pavlova, 1959), *Mustela sibirica* (Marvin, 1958*a*), *Mustela vison* (Marvin, 1958*a*; Eremeeva, 1967).

The pattern of distribution of the thickness of skin and pelage has been studied in many *Carnivores* (Kuznetsov, 1952; Tserevitinov, 1958, 1969), and specific skin glands are described in many *Carnivores* (Schaffer, 1940; Spannholf, 1960; Azbukina, 1970).

We have examined the skin of *Canis aureus, Vulpes corsac, Vulpes vulpes, Cuon alpinus, Ursus arctos, Thalassarctos maritimus, Martes zibellina, M. foina, Mustela nivalis, M. erminea, M. sibirica, M. lutreola, M. putorius, Vormela peregusna, Lutra lutra, Meles meles, Viverra civetta, Herpestes griseus, Felis ocreata,* and *F. caracal.*

The skin is thin, or medium thick, the epidermis and stratum corneum being thin. The stratum granulosum and lucidum are usually absent on the body skin. The epidermis is usually not pigmented. The papillary layer borders on the reticular layer at a level with the hair bulbs or the bottom of the sweat glands. The sebaceous glands are commonly large; the secretion cavity of two sweat glands is sausage-shaped, slightly or highly twisted, or tubular. The hairs are markedly differentiated into guard, pile, and fur, the last being the most numeous. The medulla is not very well developed in any category. The hairs grow in tufts and groups. There are large sole glands on the feet. Many species have anal glands and sometimes other specific skin glands.

FAMILIA CANIDAE

The border between the papillary and reticular layers of the dermis is level with the hair bulbs. The sebaceous glands are one- , two- , and three-lobed. The sweat glands are tubular and slightly twisted. The pile hairs may or may not have a granna.

Nyctereutes procyonoides

The outer surface of the skin is irregular, plicated, and covered with hollows (Kogteva, 1963). The skin is thickest on the middle of the back and the shoulder blades, thinner on the sides, and thinnest on the belly (Tserevitinov, 1958). The epidermal thickness varies from 18 to 33μ over the year (Kogteva, 1963) (table 51), its inner surface being regular with neither alveoli nor processes. The stratum malpighii is two to three cell layers deep in winter and five to six in summer. The stratum corneum, which is not very highly developed, is at its thickest in winter. In summer and autumn, there is an epidermal stratum granulosum. The dermis varies from 1.3 to 4.1 mm over the year. Bundles of collagen fibers form a compact plexus in the dermis (looser in summer) (fig. 108).

The sebaceous glands vary from 128μ (in May) to 700μ (in September). There are two or three glands at every hair (fig. 108). The sweat glands have

TABLE 51

SEASONAL VARIATIONS IN THE SACRAL AREA SKIN OF NYCTEREUTES PROCYONOIDES (KOGTEVA, 1963)

Month	Thickness of skin* (μ)	Thickness of epidermis (μ)	Number of cell rows in stratum malpighii and its thickness (μ)		Stratum corneum (μ)	Length of sebaceous glands (μ)	Thickness of arrectores pilorum muscles (μ)
December	2,000	24	2-3	12	Well developed	420	220
January	1,800	18	2	—	Highly developed	400	200
March	1,300	19	—	—	—	185	—
April	1,520	18	—	9	9	200	—
May	1,700	18	Thickened		Very thin	128	100
June	1,800	32	3	—	Slightly developed	300	—
July	2,200	32	5-8	22	10	600	80
August	3,860	38	3-5	19	19	400	198
September	4,100	25	3-4	16	Well developed	700	260
October	1,900	25	2	12	Well developed	400	260
November	1,900	25	2	—	Very highly developed	—	220

*Without subcutaneous fat tissue.

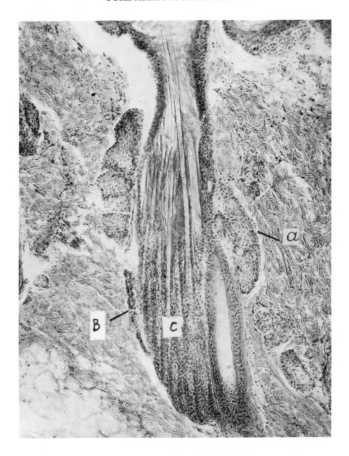

Fig. 108. *Nyctereutes procyonoides.* External layer of skin:
a, sebaceous gland; *b,* sweat gland; *c,* hair follicles.

a slightly twisted tubular secretion cavity, their ducts opening into the funnels of the hair bursae above the sebaceous gland ducts. The well-developed
arrectores pilorum muscles achieve their maximum thickness in autumn and
winter (260μ), and are thinnest in July (60 to 80μ). The well-developed subcutaneous fat tissue is thickest from autumn to spring.

Seasonal skin changes are shown on table 39 (Kogteva, 1963). In the first
stage (end of November to February) the skin remains relatively unchanged.
The skin and epidermis are moderately thin and the stratum malpighii is
very thin, only two cells deep. The stratum corneum is well developed, the
papillary layer making up 60% of the dermis. The sebaceous glands are
large and active. The arrectores pilorum muscles are well developed, and the
subcutaneous fat tissue is thick.

The second period (March to July) is the molting season, and the first half

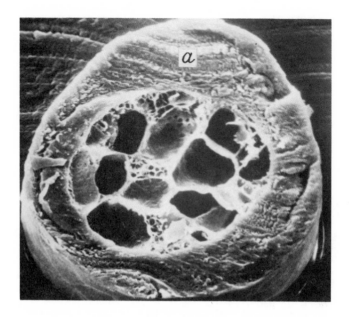

Fig. 109. *Nyctereutes procyonoides*. Section of pile hair:
a, cortical layer.

reveals processes of atrophy. The whole skin, especially the stratum corneum, gets thinner, and the sebaceous glands shrink and stop secreting. The
hair begins to drop out, leaving large numbers of empty sheaths. The second
half is a period of regeneration: the skin thickens and the number of stratum
malpighii cell rows increases. May brings the beginnings of new hair. In the
third period (August to mid-November) the new hair is growing. It reaches
full length by the end of November and stops growing. During the first half,
the skin still manifests processes of morphogenesis. The skin thickens during hair growth. By the end of November, the skin activity is over.

According to Tserevitinov (1958), *N. procyonoides* has three categories of
hair: guard, pile (three orders), and fur (see table 52). According to Kogteva
(1963) there are four orders (see table 53) of pile and two of fur. All the categories of hair have a medulla of 69% in the granna. The guard hairs have
63% in the base; the pile hairs (first order) have 69% in the granna and 66%
in the base; and the fur hair has 29% in the base (with no medulla in the distal part) (Kogteva, 1963) (fig. 109).

The length of the hairs in various parts of the body differs greatly. The
hairs are three times as long on the back as on the belly, and pile and fur
hair include seven different length zones (Tserevitinov, 1958). In guard and
pile hairs, the granna is poorly differentiated and takes up only 50% of the
length (guard) or 40 to 50% of the length (pile).

TABLE 52

HAIR LENGTH AND THICKNESS OF NYCTEREUTES PROCYONOIDES (MALES FROM LENIN DISTRICT, KHABAROVSK TERRITORY) WINTER (AFTER TSEREVITINOV, 1958)

Hair category	Withers			Sacral region		
	n	M±m	lim	n	M±m	lim
Length (mm)						
Guard	16	82.20±0.51	79-86	15	97.80±0.55	95-101
Pile						
1st order	48	75.60±0.38	70-81	50	90.78±0.70	81-97
2d order	41	63.40±0.44	59-70	43	80.26±0.90	71-92
3d order	43	58.0 ±0.44	50-63	42	67.80±1.00	57-82
Fur	50	48.40±0.75	38-56	50	62.60±0.52	52-68
Thickness (μ)						
Guard	18	131.20±2.03	120-152	15	133.60±3.10	116-156
Pile						
1st order	50	99.60±2.06	80-132	56	112.30±2.00	84-136
2d order	41	68.00±1.45	56-92	43	86.80±1.52	72-108
3d order	43	41.50±1.00	32-60	45	60.00±1.40	40-76
Fur	50	15.00±0.29	12-20	50	17.00±0.24	16-20

TABLE 53

NYCTEREUTES PROCYONOIDES: HAIR LENGTH AND THICKNESS (WINTER PELAGE, LENINGRAD ZOO) (AFTER KOGTEVA, 1963)

Hair category	Withers			Sacral region		
	n	$M \pm m$	lim	n	$M \pm m$	lim
Length (mm)						
Guard	75	90.6 ± 0.15	86-94	25	102.0 ± 0.26	99-107
Pile						
1st order	75	87.0 ± 0.2	74-91	25	99.7 ± 0.38	96-103
2d order	75	76.3 ± 0.36	70-86	25	83.8 ± 1.05	74-92
3d order	75	61.8 ± 0.21	53-79	25	69.2 ± 1.36	56-78
4th order	35	47.0 ± 0.38	35-61	25	55.0 ± 0.38	53-58
Fur						
1st order	75	31.7 ± 0.19	27-36	20	46.6 ± 0.90	33-51
2d order	75	51.9 ± 0.6	45-57	43	65.5 ± 0.90	48-78
Thickness (μ)						
Guard	75	161.0 ± 0.14	—	25	152.0 ± 0.23	—
Pile						
1st order	75	130.8 ± 0.18	—	25	129.0 ± 0.21	—
2d order	75	130.6 ± 0.16	—	25	84.5 ± 0.37	—
3d order	75	70.0 ± 0.39	—	25	71.0 ± 0.43	—
4th order	35	46.4 ± 0.21	—	25	62.5 ± 0.35	—
Fur						
1st order	75	38.4 ± 0.23	—	43	24.9 ± 0.30	—
2d order	75	28.8 ± 0.38	—	43	24.9 ± 0.30	—

The guard hairs are slightly wavy, oval in section in the granna section, and round in the main section. The pile hairs have a wavy shaft. The hairs in the skin grow in tufts and groups (fig. 110) with one pile and twenty-one to thirty-one fur hairs to a tuft, and three or even four tufts grouped around a single guard or pile hair. The average hair group contains one guard hair, two to three pile hairs, and sixty-five to sixty-eight fur hairs. There are 5,500 to 7,287 hairs per 1 cm² on the withers, and 8,514 to 8,600 hairs on the sacrum (Tserevitinov, 1958; Kogteva, 1963) (table 54).

There is one long molt only from the end of March to November (Kogteva, 1963), during which time the old winter pelage is replaced. Growth is from June to mid-November. There is little or no fur hair in the summer, the guard and pile hairs are short and growing, and the number of fur hairs is increasing toward winter. By September, there is an average of 5,400 hairs per 1 cm² of sacrum, 7,360 by October, and 8,400 by December.

Canis aureus

Material. Skin from withers, breast, and front and hind soles of a male (body length 90 cm) taken February 3, 1961, at Kopet-Dag, near Firjuza (table 55).

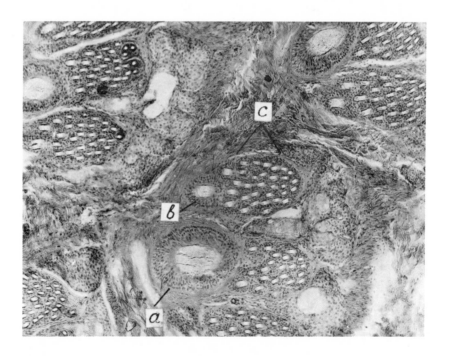

Fig. 110. *Nyctereutes procyonoides.* Withers skin, groups and bundles of hair in horizontal section: *a,* guard hairs; *b,* pile hairs; *c,* fur hairs.

TABLE 54

NUMBER OF HAIRS PER 1 CM² ON WINTER N. PROCYONOIDES, TAKEN LENIN REGION, KHABAROVSK TERRITORY (TSEREVITINOV, 1958)

Hair category	Withers	Sacrum
Guard hairs	10	10
Pile hairs	212	148
Fur hairs	7,075	8,356
Total	7,248	8,514

The skin is thicker on the withers than on the breast, though epidermis and stratum corneum are thinner. The stratum malpighii is one to two cells deep. The stratum corneum is loose. On the hind soles, the epidermis is very thick, with a strong, solid stratum corneum. The inner surface of the epidermis has alveoli.

There is virtually no reticular layer, as subcutaneous fat tissue cells (45 x 56; 50 x 62μ) adjoin the hair bulbs and sweat glands, the border between dermis and subcutaneous fat tissue being irregular. Between the hair tufts, dermal processes project into the subcutaneous tissue, and at the hair tufts the subcutaneous tissue protrudes into the dermis. The dermal network of collagen bundles is not very dense; horizontal bundles predominate; many bundles are set at different angles. There are no fat cells in the dermis. There are few collagen bundles in the subcutaneous fat tissue, which is 120μ thick on the withers, 35μ on the breast.

The sebaceous glands vary in size. There is a large two- or three-lobed gland behind a smaller gland at every fur hair. Sweat glands are poorly developed, with a tubular secretion section (diameter 50 to 60μ) extended along the hair roots. They are slightly twisted. Some of the sweat glands reach down below the bulbs. There are large eccrine sweat gland glomi (0.7 to 0.8 mm; 300 x 650; 330 x 1000μ) beneath the epidermis, 200 to 500μ apart. The diameter of the secretion gland parts is 34 to 40μ. The excretion ducts of the glands open onto the skin. The jackal has an anal gland (Schaffer, 1940). The arrectores pilorum muscles are well developed, 120μ thick in the withers and 35μ in the breast. They have several parts. The pelage is coarse and rough. Its general coloration in winter is dull reddish gray with a touch of black on the back; it is redder in summer.

There are two categories of hair: pile (two orders) and fur (three orders). The pile hairs (first order) are slightly wavy, long (81.6 mm), and 12.6μ thick along the whole length, gradually tapering toward the tip. There is no granna. The medulla is well developed down the whole length (74.4%), except the base and tip. The second order is shorter (54.5 mm) and thinner (64μ). The medulla is less developed (65.7%). The three orders of fur hairs

TABLE 55

CANIDAE: SKIN THICKNESS MEASUREMENTS

	Thickness						
	Skin (mm)	Epidermis (μ)	Stratum corneum (μ)	Dermis (mm)	Papillary layer (mm)	Reticular layer (mm)	Sebaceous glands (μ)
Canis aureus							
Withers*	1.88	28	14	1.2	1.2	—	120 × 385
Breast*	1.43	45	28	0.9	0.9	—	62 × 123
Vulpes corsac							
Withers	0.66	62	45	0.6	0.4	0.2	56 × 101
Breast	0.43	34	22	0.4	0.3	0.1	39 × 90
Vulpes vulpes							
Withers	0.92	28	17	0.9	0.6	0.3	56 × 168
Cuon alpinus							
Withers	2.056	56	34	2.0	0.9	1.1	77 × 130; 110 × 363
Breast*	1.98	45	28	1.0	1.0	—	56 × 140

*Thickness of subcutaneous fat tissue is indicated in the text.

are wavy and of even thickness (38, 27, and 22μ) the entire length (45.8; 35.1 and 22.9 mm), with a rather well developed medulla down the whole shaft (57.9; 25.0 and 45.5%). This is broken to a line of dashes in some hairs. The hairs grow in tufts, one pile hair (second order) and nine to thirteen (occasionally up to 16) fur hairs. The first order pile hairs grow singly, 83μ pile hairs (first order), 166μ pile hairs (second and third orders) and 2,000 fur hairs per 1 cm^2.

Alopex lagopus

Material. The skin of *A. lagopus*, Norwegian type, kept on a fur farm.

The thickness of the skin averages 900μ in winter, thinning down to 702μ in spring molt (Komarova, 1968). When the summer hair begins to grow, the skin thickens up to 1218μ. The epidermis also thickens from 17μ in winter to 30μ in summer, and the number of cell layers in stratum malpighii increases to four or five. The sebaceous glands are much bigger (87 x 155μ) in winter than in summer (96 x 168μ).

In blue fox ontogenesis, the first anlagen of sebaceous glands appear on withers on the thirty-third day of embryonal development (Eremeeva, 1957). They are individual glandular cells among the basal layer cells of the epidermis. The development of sebaceous glands on the sacrum is more rapid than on the withers, and by birth their size on the withers and on the sacrum becomes identical. Cubs are born with fully formed sebaceous glands, though these are smaller than in adults.

The anlage of the first sweat glands appears later than that of sebaceous glands (on the thirty-seventh to thirty-eighth day of embryonic development). In newborn cubs, sweat glands are most developed and numerous on the belly, where they are well divided into secretion cavity and duct. On the withers and sacrum, the differentiation into the secretion cavity and the duct takes place when the cubs are three to five days old.

At the base of the tail (level with the fifth tail vertebra) at the dorsal surface is the so-called violet gland (Petskoi, Pichugin, and Sinyavina, 1971). It comprises very large sebaceous and apocrine sweat glands. The latter are about as large as in the skin of the trunk, but they are more numerous. The hairs in the gland region are similar to guard hairs, but are shorter and thicker and have a wide medulla. The cuticle and cortical layers of the hair have microslits through which the secretion penetrates into the medulla.

The winter hairs are divided into guard, pile (four orders) (Komarova, 1968) (or four to six orders) (Tserevitinov, 1951) and fur (two orders) (table 56). In summer there is only one order of fur hairs (Komarova, 1968). Hair measurements on different parts of the body are given in table 57. Seasonal variations in hair length and thickness are pronounced (table 58).

There are five distinct periods of the year for the state of the fur (Kaletina et al., 1957). The first period (the end of December to February) is a state of

COMPARATIVE MORPHOLOGY

TABLE 56

ALOPEX LAGOPUS, WINTER HAIR MEASUREMENTS (TSEREVITINOV, 1951)

	Length (mm)		Thickness (μ)	
	n	M$\pm$m	n	M$\pm$m
Withers				
Guard	13	70.46$\pm$1.11	13	74.15$\pm$1.16
Pile				
1st order	23	69.13$\pm$0.80	23	74.47$\pm$0.76
2d order	30	58.73$\pm$0.46	28	78.85$\pm$0.67
3d order	36	54.80$\pm$0.29	34	71.29$\pm$0.70
4th order	36	52.44$\pm$0.25	36	63.12$\pm$0.87
5th order	36	50.67$\pm$0.18	36	51.56$\pm$1.30
6th order	36	50.62$\pm$0.18	35	38.86$\pm$0.92
Fur				
1st order	36	49.62$\pm$0.23	36	27.89$\pm$0.48
2d order	36	43.77$\pm$0.33	36	14.45$\pm$0.48
Sacral region				
Guard	22	64.40$\pm$0.29	20	81.40$\pm$1.45
Pile				
1st order	27	58.48$\pm$0.43	26	87.23$\pm$0.96
2d order	36	52.25$\pm$0.35	36	83.22$\pm$0.92
3d order	36	49.06$\pm$0.31	36	74.22$\pm$1.00
4th order	36	47.41$\pm$0.30	36	64.22$\pm$0.77
5th order	36	47.02$\pm$0.24	36	44.44$\pm$1.16
Fur				
1st order	36	46.41$\pm$0.22	36	31.45$\pm$1.08
2d order	36	40.22$\pm$0.59	36	16.33$\pm$0.43
Belly				
Guard	27	65.93$\pm$0.92	28	68.28$\pm$0.85
Pile				
1st order	35	51.80$\pm$0.41	35	41.60$\pm$0.85
2d order	36	48.67$\pm$0.51	36	35.45$\pm$0.69
3d order	36	48.13$\pm$0.25	36	30.78$\pm$0.97
4th order	36	38.23$\pm$0.26	36	25.11$\pm$0.54
Fur				
1st order	36	31.33$\pm$0.25	36	15.60$\pm$0.76
2d order	36	28.19$\pm$0.28	36	15.44$\pm$0.23

relative rest. The second period (the end of February to June) is the spring molt, when the old hair falls out and new summer fur begins to grow. In the third period (July and the first half of August) there is a slow growth of summer pelage. The fourth period (the second half of August to September) is the autumn molt, when the summer hairs fall out and the winter hair begins to grow. And in the fifth period (October, November, and December) the winter pelage is growing.

TABLE 57

ALOPEX LAGOPUS: HAIR LENGTH AND THICKNESS IN DIFFERENT SEASONS (KOMAROVA, 1969)

Hair		Length (mm)		Thickness (μ)	
		Winter	Summer	Winter	Summer
Guard	80	63.0±0.7	—	108.0±0.2	—
Pile					
1st order	200	55.6±0.1	32.0±1.9	98.1±0.7	57.4±1.5
2d order	200	45.8±0.3	26.0±0.4	69.4±1.9	41.2±2.2
3d order	200	44.1±0.4	23.8±0.1	38.7±0.6	29.1±0.8
4th order	200	41.3±0.2	22.4±0.3	24.4±0.8	20.8±0.7
Fur					
1st order	200	33.2±0.2	19.5±0.3	17.0±0.3	19.0±0.5
2d order	200	24.4±0.2	—	16.6±0.3	—

TABLE 58

ALOPEX LAGOPUS: HAIR DENSITY IN DIFFERENT SEASONS (AFTER KOMAROVA, 1968)

Hair	Hairs per 1 cm^2 of skin			
	Winter		Summer	
	n	%	n	%
Guard	5	0.02	—	—
Pile				
1st order	105	0.42	77	0.94
2d order	174	0.69	113	1.38
3d order	262	1.04	189	2.31
4th order	1,481	5.86	298	3.65
Fur				
1st order	14,555	57.66	7,491	91.71
2d order	8,660	34.31	—	—
Total	25,242	100.00	8,168	100.00

Vulpes corsac

Material. Skin from withers, breast, and front and hind soles from specimen (1) (body length 45 cm) taken in winter in Turkmenia (table 55) and pelage from withers of specimen (2) taken in summer in Kalmyk steppes near Terekli-Mekteb.

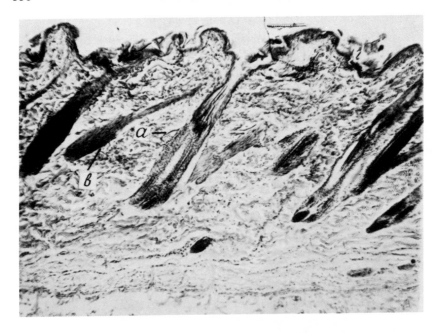

Fig. 111. *Vulpes corsac.* Papillary layer of withers: *a,* sebaceous glands; *b,* arrectores pilorum muscle.

The skin is thickest around the withers and sacrum, and thinnest on the ventral side of the body (Tserevitinov, 1958). Four zones of thickness can be distinguished. The epidermis and stratum corneum are relatively thick, and thicker on the withers than on the breast. The stratum malpighii is one to two cells deep. The stratum corneum is rather loose. In the dermis, the papillary layer is thicker than the reticular. Collagen fiber bundles form a dense plexus in which the horizontal predominate (fig. 111). There are groups of fat cells (22 x 34; 17 x 34μ) stretching in horizontal strings in the upper reticular layer. The sebaceous glands are unilobal, small, anterior and posterior to every tuft of hair (the anterior glands are not present at every tuft).

The sweat glands do not appear to be active and are very shrunken in the withers (diameter 22μ). There are large tubular gland glomi in the fat tissue of the front and hind soles (front foot 220 x 500μ, hind foot 330 to 550μ, diameter of secretion cavity of both 39μ). The ducts open into the hair funnels and onto the skin surface where there are no hairs. Arrectores pilorum muscles are thin (17 to 22μ).

There are three categories of hair. The guard hairs (30.9 mm long) thicken very gradually from the base to the distal two-thirds of the shaft (to 101μ), then taper toward the tip. The medulla (87.2%) runs down the whole hair except base and tip. The pile hairs (first order, length 25.4 mm, thickness 102μ, medulla 82.3%) have a weak granna for about one-third of their length. The second order is shorter (21.6 mm), thinner (85μ), with a granna for about one-fourth of its length and a well-developed medulla (81.2%). The fur hairs (21.6 mm long) are almost uniform the whole length (18μ thick, medulla 66.7%).

The summer withers pelage divides into the same categories and orders as in winter. These guard hairs are few, straight, and long (26.4 mm), thickening gradually from the proximal to 80μ at two-thirds up the shaft. The medulla (82.5%) runs through the whole hair. The pile hairs (first and second orders) differ from the guard hairs only in size: they are shorter (15.3 and 13.1 mm) but thicker (101 and 109μ), with a well-developed medulla (74.3 and 79.9%). The fur hairs, which have up to five waves, are short (8.2 mm) and evenly thin (26μ) down the whole length. The medulla is well developed (61.6%).

Vulpes vulpes

Material. Skin from withers and hind sole of adult specimen (body length 60 cm) taken in January, in Repetek, Kara Kum (table 55).

The outer surface of the skin is plicated; the epidermis is thin. The stratum malpighii is one to two cells deep. The papillary layer is thicker than the reticular. The collagen fiber plexus is more compact in the papillary than the reticular layer. In the papillary layer many bundles rise to the surface at various angles, although horizontal bundles predominate in the dermis (fig. 100). Elastin fibers (mainly horizontal) are numerous in the papillary and the inner reticular layers, where fairly large groups of fat cells may be found (17×34; $28 \times 39\mu$).

The sebaceous glands (mainly with one lobe) are well developed, anterior to each hair tuft (fig. 112) and sometimes also posterior. The sweat glands are few and small (diameter up to 30μ), tubular and slightly twisted (fig. 112), apparently shrunken for the winter. On the hind sole the whole skin is covered in fur. The sebaceous glands are larger than on the body (100×140; $112 \times 200\mu$), and multilobal. Large sweat eccrine gland glomi are linked to form a continuous accumulation (fig. 113) (diameter of secretion cavity 80μ).

On the dorsal surface of the tail is a violet gland 5 to 6 cm to its base (Schaffer, 1940). It is formed of huge, multilobal sebaceous glands opening into the hair bursae. On the ventral side of the tail is a subcaudal gland (Schaffer, 1940) formed of large sebaceous glands with big tubular gland glomi underneath. Between the outer and inner anal sphincters is a twin anal glandular sac with a short duct opening ventrally into the anus. Its walls are made up of tubular glands twisted into complex glomi (Schaffer,

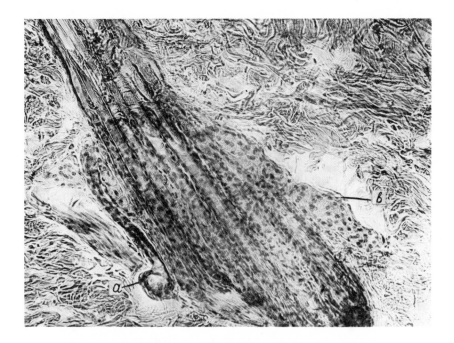

Fig. 112. *Vulpes vulpes.* Hair tuft in withers skin: *a,* sweat gland; *b,* sebaceous gland.

1940; Spannhof, 1960). It secretes most actively in March and April, rather less in August and September, and very little during the rest of the year. The arrectores pilorum muscles are 30μ thick.

The structure of the skin changes during three distinct annual cycles. (Leschinskaya, 1952*a, b*). During the first period (late November to February), the skin, the dermal layer, and the epidermis are all thin, and the stratum corneum is well developed. The sebaceous glands are large and very active. The sweat glands are slightly developed. The arrectores pilorum muscles are powerful.

A molt occurs during the second period (March to June) in which old hairs drop out and new ones grow, and the dermis and stratum malpighii of the epidermis thicken. The sebaceous glands shrink, and the sweat glands grow in size. The arrectores pilorum muscles become thin, and the subcutaneous tissue becomes thinner. In the third period (August to early November) there is a full molt. All the old hair drops out and new hair grows. The hair is short at the beginning of the period but attains normal length by the end. The skin also gradually returns to a relative state of rest, the epidermis and dermis grow thinner, the sebaceous glands and arrectores pilorum muscles expand, the number of sweat glands decreases, and the subcutaneous fat tissue thickens.

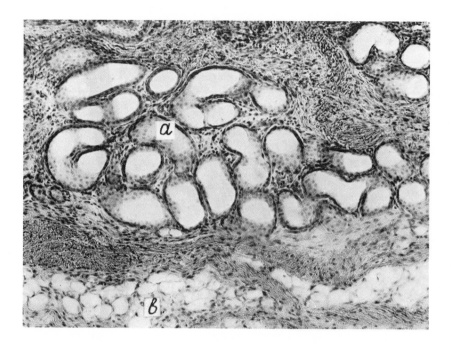

Fig. 113. *Vulpes vulpes.* Hind foot sole gland: *a,* sweat gland; *b,* fat cells.

The fox has one lasting molt during the year. In winter pelage, summer hairs may be found (Pichugin and Sudakov, 1973).

There is a high geographic variability in pelage. The pelage of *V. v. karagan erxleben,* to which this fox belongs, is short and sparse compared with that of other foxes. There are four hair categories: guard, pile (two orders), intermediate, and fur. The guard hairs are straight, 51.3 mm long and 87μ thick down the whole length. The medulla (77.1%) is well developed. The first order of pile hairs (length 42.6 mm, thickness 43μ) is slightly wavy and has a small thickening near the tip that cannot be treated as a granna. The medulla (72.1%) runs all the way except for base and tip. The second order pile hairs are shorter (30.4 mm), and they thicken to 86μ in the distal two-thirds (medulla 74.5%). The intermediate hairs (25.4 mm long and 54μ thick) are peculiar. The lower half of the shaft is straighter and thicker than the wavy upper half, with a slight distal thickening. The medulla is well developed (80.4%). The fur hairs are 35.5 mm long, slightly wavy, and 22μ thick through the length. The medulla is poorly developed (54.6%). There are 12,807 hairs per 1 cm² on the back, 9,762 on the withers, and 6,553 on the belly (Tserevitinov, 1958).

The length of the pile hairs varies in seven different zones. The longest are in three pairs of areas: patches 3 cm long and 2 cm wide on either side of the

TABLE 59

VULPES VULPES: HAIR LENGTH AND THICKNESS (WINTER, KAMCHATKA) (TSEREVITINOV, 1951)

Hair	n	Length (mm)	n	Thickness (μ)
Withers				
Guard	44	99.32±1.26	44	86.00±1.30
Pile				
1st order	50	80.58±0.95	50	83.44±1.05
2d order	50	72.50±1.12	50	71.12±1.26
3d order	50	66.52±1.09	50	54.60±1.32
4th order	20	61.75±0.36	20	53.40±1.90
Fur hairs*	40	48.12±0.41	39	15.00±1.33
Sacral region				
Guard	38	71.63±1.22	33	110.54±1.56
Pile				
1st order	50	61.42±0.92	50	114.48±2.07
2d order	50	52.84±0.96	50	102.76±2.34
3d order	50	50.63±0.68	50	85.84±2.44
Fur	50	43.64±0.74	50	15.12±0.34
Belly				
Guard				
1st order	49	59.30±0.74	50	70.32±1.14
2d order	50	49.80±0.89	50	57.76±1.15
3d order	50	46.86±0.50	50	47.36±0.26
Fur*	50	33.84±0.36	50	19.00±0.31

*Fur hairs do not divide into orders.

withers; 8 cm long and 3 or 4 cm wide on the shoulder blade and on the edge of the sacrum at the sides of the caudal base. The second zone surrounds these patches and also covers large areas of the sides near the thighs. The third zone is on the back, except the middle, and surrounding the fur patches on the withers and shoulder blades, reaching down to the groin. The fourth zone surrounds the third one near the withers and the sides on to the groin, plus a patch right in the middle of the back. The fifth zone extends from the forelegs up the legs to breast and belly. The sixth zone covers the antebrachium and thighs. The seventh zone with the shortest pelage is on the muzzle and front and hind legs. The Kamchatka specimen has longer hair than does the Turkmenian (table 59).

Cuon alpinus

Material. Skin from withers, breast (table 55), and hind sole of an adult male (body length of 95 cm). Died at Moscow Zoo, February 16.

The epidermis and stratum corneum on the withers are thicker than on the breast. The stratum malpighii is two to four cells deep. The epidermis is very thick on the hind sole with alveoli on the inner surface and a well-developed compact stratum corneum.

The reticular layer is thicker than the papillary on the withers but absent on the breast, owing to hair growth. Bundles of collagen fibers form a compact plexus in the papillary layer. They are predominantly horizontal, although many rise to the surface at different angles. There are sizable groups of fat cells (62×84; $56 \times 78\mu$) in the inner papillary layer near the hair bulbs. The reticular layer on the withers is loose, and horizontal layers of collagen bundles alternate with thick layers of fat cells, the same size as those in the papillary layer. The subcutaneous fat tissue on the breast is well developed, up to 0.9 mm thick. The outer part is a continuous mass of fat cells (45×62; $56 \times 95\mu$), though there are few in the inner part, which is made up of loose horizontal collagen bundles. On the withers, the sebaceous glands are large, with twin lobes (fig. 114). In a horizontal skin section they are usually seen posterior to hair tufts. In the breast skin, small sebaceous glands with a single lobe are seen anterior and posterior to every hair tuft. The sweat glands are not visible. On the hind sole, large eccrine sweat gland glomi (up to 0.38×1.43 mm) are found in the subcutaneous fat tissue 1 to 1.5 mm below the epidermis (diameter of secretion section up to 55μ). The arrectores pilorum muscles are poorly developed (55 to 66μ thick).

The hair is growing and divides into three orders of pile and two of fur. The pile hairs (first order) are long (73.3 mm), slightly wavy, without granna, and of uniform thickness down the length, tapering only toward the apex. The medulla is well developed all the way (62.0%). The rather more wavy second and third orders are not so long (59.1 and 49.2 mm), are thinner (75 and 71μ), and with a weaker medulla (57.4 and 55.0%). The fur hairs are of uniform thickness their whole length, and their shaft is sinuate. The first order is longer than the second, 23.4 mm against 44.1 mm (they may be growing hairs). They are of uniform thickness (39μ), and the medulla is equally developed (38.5%). The hairs grow in tufts of one (sometimes two) pile hairs (second order) and six to twelve fur hairs, three tufts to a group, one in the middle and one each side. The pile hair in the middle tuft is first order. There are 83 pile hairs (first order), 250 pile hairs (second order) and 2,583 fur hairs, per 1 cm².

Familia Ursidae

The stratum spinosum of the epidermis consists of a few layers of cells, and stratum granulosum (sometimes absent) of one or two layers. The epidermis may be pigmented (usually lightly). The papillary and reticular layers adjoin the hair bulbs or the lower sections of the sweat glands. The plexus of (mainly horizontal) collagen fiber bundles in the dermis is fairly compact,

Fig. 114. *Cuon alpinus*. Withers skin: *a,* sebaceous gland; *b,* hair follicle.

and there are a few elastin fibers. The fat cells may be in the dermis and the subcutaneous fat tissue may be well developed. The sebaceous glands on the body, which are mostly unilobal, more rarely multilobal, are arranged singly, rarely in twos, at every pile hair. The sweat glands (body) have twisted secretion cavities sometimes forming glomi. One or two sweat glands run into the hair channel at every hair tuft. The sweat glands are also in the skin of the soles. There are relatively small arrectores pilorum muscles at the hair tufts. The hairs divide into vibrissae, guard, pile, and fur. The pile hairs have no granna. The hairs grow in groups of tufts, the fur hair in the tuft tending to be behind the pile hairs.

Ursus arctos

Material. Skin from withers and breast of specimen (1), an adult male taken October 24, 1959, on the Nalchik hunting preserve in the Caucasus. Skin from body sides below withers, breast, sacrum, belly beneath sacrum, and hind and foot soles of specimen (2), an adult female (6 to 8 years) (body length 161 cm) taken in lair, March 20, 1971, in Kostroma region. Skin from withers, body side below withers, breast, sacrum, and belly beneath sacrum of specimen (3) an adult female (body length 147 cm) taken August 27, 1971, on Kamchatka. Skin from the withers, body sides beneath the sacrum and belly under sacrum of specimen (4), a young male (three years) (body length 144 cm) taken August 23, 1971, on Kamchatka. Skin from withers, breast, and sacrum of specimen (5), a female cub (one year and two or three months old) (body length 102 cm) sacrificed May 11, 1971, in a laboratory 54 days after it was taken from its lair (table 60).

In autumn, specimen (1) is very fat before hibernation. The skin is thick, thicker on the withers than on the breast. The stratum malpighii is three to four cells deep. There is a single-cell-layer stratum granulosum in the withers epidermis. The stratum corneum is solid.

The papillary layer is thicker than the reticular. The dermal collagen fiber bundles are rather loosely arranged at various angles. Fat cells begin to appear in the middle of the papillary layer, increasing in number with depth. A small number of collagen bundles and a large accumulation of fat cells make up the reticular layer. The fat cells increase in size with depth, varying from 75×90; $55 \times 75\mu$ in the papillary layer to 90×100; $74 \times 120\mu$ lower down. Elastin fibers are numerous only under the epidermis, with just a few in the rest of the dermis. The subcutaneous fat tissue is well developed, 4.5 mm on the breast. The fat cells are uniform in size throughout the subcutaneous tissue.

The withers sebaceous glands are large, but their boundaries are difficult to determine. No sebaceous glands are found in the breast. The sweat glands (withers) are very large (fig. 115), with secretion cavities twisted into gigantic glomi stretching from above to 3.3 mm; 440μ in diameter. It is hard to say whether each of these glands is formed by one or several glands. In some glands the diameter of the secretion cavity is very large, up to 500μ. Other actively secreting glands are smaller. In the breast, the sweat glands are relatively small, glomi $110 \times 660\mu$, the diameter of the secretion cavity 66μ. There is an anal gland (Schaffer, 1940). The arrectores pilorum muscles are small, 45μ thick, and rare, and are not found at all in the breast.

All the hairs are of the same type, varying only in size. Three hair categories different from similar categories in other mammals should be noted. The guard hairs are slightly wavy, and evenly thick (138μ) along the whole length (60.6 mm), except that they taper toward the tip. Their medulla is rather poorly developed (40.5%). The pile hairs (first order) are thinner

TABLE 60

URSUS ARCTOS: SKIN MEASUREMENTS (THICKNESS OF THE SUBCUTANEOUS FAT TISSUE IS INDICATED IN THE TEXT)

	Skin (mm)	Epidermis (μ)	Stratum corneum (μ)	Dermis (mm)	Papillary layer (mm)	Reticular layer (mm)	Sebaceous glands (μ)
				Thickness			
Ursus arctos							
(1) Autumn, Caucasus, male							
withers	9.5	56	22	9.4	7.6	1.8	275 × 1,100; 440 × 1,400
breast	8.1	56	—	3.5	2.9	0.6	not recorded
(2) Winter, Kostroma region, female							
side	3.1	105	75	3.0	1.5	1.5	55 × 187
breast	1.9	105	75	1.5	1.0	0.5	60 × 195
sacral region	2.0	170	150	1.8	1.5	0.3	40 × 200
belly	1.9	104	80	1.5	1.1	0.5	56 × 108
(5) Spring, cub, Kostroma, female							
withers	3.4	45	30	1.4	1.4	—	48 × 200
breast	3.2	135	120	1.4	1.4	—	60 × 195
sacral region	3.1	59	44	1.3	1.3	—	45 × 225
(3) Summer, Kamchatka, female							
withers	7.7	78	40	7.6	7.5	0.1	50 × 300
side	3.4	37	12	2.6	2.6	—	50 × 350
breast	3.9	307	183	3.0	2.7	0.3	408 × 816
sacral region	6.3	50	20	5.8	5.8	—	50 × 250
groin	3.9	102	60	2.9	2.9	—	70 × 450
(4) Summer, Kamchatka, male							
withers	6.0	58	30	5.9	5.9	—	32 × 315
shoulder	2.5	120	105	2.4	1.9	0.5	45 × 255
breast	2.1	56	30	2.1	1.8	0.3	30 × 450
flank	3.3	96	72	3.2	2.9	0.3	32 × 320
belly	3.1	110	90	2.3	2.3	—	45 × 405

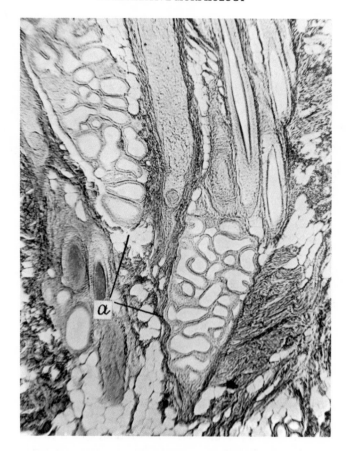

Fig. 115. *Ursus arctos*. Withers skin: *a,* sweat glands (taken in
Caucasus, autumn).

(95μ), but longer (65.2 mm), with a still less developed medulla (30.9%).
The pile hairs (second order) are altogether smaller (length 34.0 mm, thickness 57μ), with a thin medulla (22.8%), which is sometimes broken. The fur
hairs (20.7 mm long and 50μ thick) are somewhat different in shape in that
they thicken slightly on the distal section. The medulla is weaker than in
guard and pile hairs (22.0%), absent in the proximal, and usually broken in
the distal section.

Specimen (2), taken at the end of winter during hibernation, is rather
thin. The skin is thickest on the withers, and much the same over the rest of
the body (table 55). The epidermis is relatively thick, with a loose stratum
corneum comprising over 70% (fig. 116). The stratum corneum is thickened
in the sacrum skin. The stratum malpighii is three to six cells deep. The

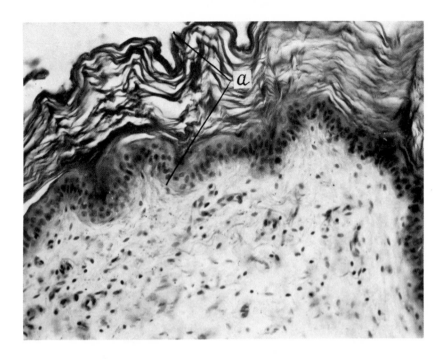

Fig. 116. *Ursus arctos.* Belly skin: *a,* epidermis (winter).

outer and inner surface of the body epidermis is slightly sinuate. The unpigmented epidermis is very thick on the soles. Deep alveoli on its inner surface
interdigitate with dermal papillae. The stratum malpighii is twenty-one cells
deep. The stratum granulosum is formed by two layers of flattened cells.
The stratum corneum is very compact. It is thicker in the skin of the front
foot sole (3.2 mm) than in the hind foot (2.8 mm).

In the dermis, the papillary layer adjoins the reticular layer level with the
young hair bulbs. In all samples the papillary layer (51 to 86% of the dermis
thckness) is thicker than the reticular. A compact plexus of mainly horizontal collagen bundles, right under the epidermis, becomes looser and
multidirectional deeper down, though deeper still the bundles are again
mainly horizontal. There are many blood vessels in the papillary layer. Thin
elastin fibers are found mainly at the hair tufts, near the glands, and below
the epidermis where the papillary and reticular layers meet. The glands are
round, diameter 20 to 30μ. The upper half of the reticular layer is more
compact than the lower with mainly horizontal collagen bundles. The reticular layer contains some larger blood vessels, and fat cells and elastin fibers
form compact plexuses next to the subcutaneous muscles.

The subcutaneous fat tissue has been lost in some samples. It is somewhat
thicker on the belly (0.25 mm) than on the breast (0.22 mm). The sebaceous
glands, which are very small and apparently not active, are about the same

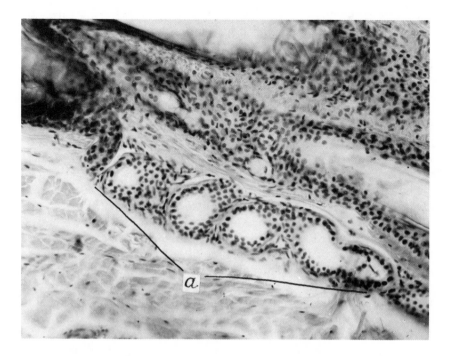

Fig. 117. *Ursus arctos.* Breast skin: *a,* sweat gland (winter).

size in all the samples. The glands look like elongated, unilobal sacs arranged lengthwise at an acute angle to the hair roots. Their short ducts run into the hair channels, with one or two sebaceous glands found at each pile hair. The sweat glands have large, highly twisted secretion cavities (diameter 40 to 60μ) in all samples. Most hair tufts have one sweat gland, though some tufts have two arranged laterally (fig. 117). In the front and hind soles are eccrine sweat glands whose secretion sections (diameter 60μ) form glomi deep in the subcutaneous fat tissue. The arrectores pilorum muscles are found only on the body side and the sacrum is wide (40 to 60μ). These are thick bands.

The morphology of the hairs is similar to that of specimen (1), except that there is only one order of pile hairs. The hairs are longest on the side of the body and sacrum, and densest on the withers and the side of the body (table 61). The medulla is very well developed in all categories and the fur hairs are long. The hairs grow in tufts and groups, usually one guard hair, one or two pile hairs, and ten to twelve fur hairs in each tuft, usually three tufts to a group. The fur hairs are behind the pile hairs in the tufts.

In spring the skin of specimen (5) is equally thick on the withers, sacrum, and breast (table 60). The epidermis is pigmented on the breast, where it is considerably thicker than in other places mainly owing to a loose stratum corneum.

The roots of growing hairs cut through the whole dermis into the subcutaneous tissue, so it is impossible to distinguish a reticular layer. There are small dermal papillae. The upper 20μ of the papillary layer is a dense plexus of thin collagen bundles running in various directions. Deeper down mainly horizontal bundles form a looser plexus.

Well-developed subcutaneous fat tissue is 1.92 mm thick on the withers, 1.72 mm on the breast, and 1.80 mm on the sacrum. The fat cells are large (up to $90 \times 105\mu$). Groups of fat cells protrude from the subcutaneous fat tissue to the bulbs of old (not growing) hairs. There are a few horizontal collagen bundles. Young, pigmented hair bulbs lie in the subcutaneous tissue. The active sebaceous glands are small and pressed up against the hair sheaths. The sweat glands have twisted secretion cavities (35 to 40μ), which do not form glomeruli. Every hair tuft has two sweat glands. The arrectores pilorum muscles are from 35μ thick on the withers to 75μ on the breast.

The hairs are similar to those of the female specimen (2) except that there are only two orders of pile, with a more marked distal thickening of the shaft of guard and pile hairs. Tufts of one guard, one pile, and three or four fur hairs (or one pile, and three to five fur hairs) are arranged in groups, three or four in a row, with the central tuft usually containing a guard hair. On the withers the hair is only starting to grow. There are more young than old hairs on the breast, and more old than young hairs on the sacrum.

In summer, specimen (3), a nursing female, has a very poor subcutaneous fat tissue. The hair follicles are active, and on the body side, sacrum, and belly, hair roots extend through the whole dermis, nearly reaching the subcutaneous muscles on withers and breast. The skin thickness is considerable, especially on withers and sacrum. The epidermis is rather thick on the ventral side of the body, mostly owing to an enlarged stratum malpighii (the stratum corneum is no more than 59.6% of the epidermis). A single-cell-layer stratum granulosum is found in the epidermis of withers, sacrum, and belly; it is two to five cells deep on the breast and absent from the epidermis of the side. The epidermis is lightly pigmented, with pigment granules arranged like a cap over the nuclei. The structure of the epidermis is peculiar. On its inner surface it has small hollows or numerous irregular alveoli up to 150μ deep. In the walls of the alveoli, the long axis of stratum spinosum cells in the alveolar walls is perpendicular to the skin surface, but horizontal where there are no alveoli.

The outer dermal subpapillary layer, about 0.3 to 0.6 mm thick is comprised of mostly horizontal, compact collagen-bundle networks. Minute collagen bundles are upright in the breast skin dermal papillae, but throughout the dermis the collagen bundles are mostly horizontal. The collagen bundles reach 10 to 15μ in thickness in the outer and 50 to 80μ in the inner dermis. Fat cells ($50 \times 100\mu$) surround some of the hair bulbs. The subcutaneous fat tissue is formed by loose connective tissue, with a small number of fat cells nearly all localized near blood vessels and hair bulbs. It is 0.6 mm

TABLE 61

URSUS ARCTOS: HAIR MEASUREMENTS (FEMALE, WINTER)

Areas	Length (mm)			Thickness (μ)			Medulla (%)			Hairs per 1 cm^2
	Guard	Pile	Fur	Guard	Pile	Fur	Guard	Pile	Fur	
Withers	111.1	95.3	73.2	39	39	11	43.5	28.2	no data	1840
Side	129.8	143.2	123.0	71	59	28	50.0	44.0	39.2	1400
Breast	79.0	67.7	36.0	143	144	56	32.8	36.8	28.5	1000
Sacral region	129.2	107.1	106.7	85	123	37	24.7	43.2	35.1	728
Belly	116.4	81.7	56.0	73	53	36	30.1	28.3	30.6	628

TABLE 62

FEMALE URSUS ARCTOS (SUMMER): HAIR MEASUREMENTS

	Withers	Breast
Length (mm)		
Guard hairs	98.5	78.8
Pile hairs		
1st order	44.5	45.1
2d order	32.8	35.5
Fur hairs	13.2	24
Thickness (μ)		
Guard hairs	93	101
Pile hairs		
1st order	103	142
2d order	82	76
Fur hairs	55	55
Medulla (%)		
Guard hairs	19.3	16.8
Pile hairs		
1st order	10.6	11.2
Number of hairs per 1 cm^2	2,000	600

thick on the breast, 0.8 mm on the body side, 0.5 mm on the sacrum and 0.9 mm on the belly. The fat cells, which are withered, with sinuate walls, are only $10 \times 20\mu$. The sebaceous glands, inactive in the withers, body side, and sacrum skin, are small, with single lobes pressed up against the hair sheath. On breast and belly they are large, multilobal and active. The lobes of the gland encircle the hair bursae, onto which they open in large ducts.

Sweat glands are slightly developed in the skin of body side, sacrum, and belly. Their secretion cavities are slightly sinuate and stretched along the hair roots. Small glomi (up to $180 \times 200\mu$) may occur in the breast skin. In the withers skin, the sweat glands are large, forming $150 \times 350\mu$ glomi. The diameter of the secretion cavity of the sweat glands is the greatest in the skin of the breast (80μ). In other parts of the body it ranges from 40 to 50μ. There is one sweat gland at every hair tuft. The arrectores pilorum muscles are thin (40 to 50μ), and none are found in the sacrum or belly.

The hairs of the withers and the breast are similar to specimen (1), except that the pile hairs divide into two orders (table 62).

A short pile hair is characteristic, as is the slight development of the medulla. The greatest hair density is recorded on the withers. The hair tufts usually contain one to three pile hairs and six to thirteen fur hairs, the number of fur hairs being smallest on the belly. The tufts are grouped in twos and threes, sometimes singly. The hair tufts are sometimes arranged in

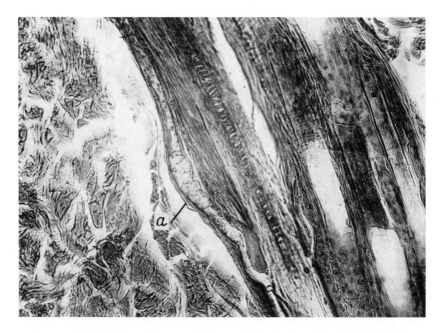

Fig. 118. *Ursus arctos*. Withers skin: *a,* sebaceous gland (summer male from Kamchatka).

curved lines. In some tufts, near the pile hair is another hair with cell nuclei of the inner epithelial sheath, intensely stained with hematoxylin, the outer epithelial sheath being common to both—an old hair, apparently, that has not yet dropped out, together with a growing young hair.

Specimen (4), a young male, is very fat. The hair is growing and the skin is thickened. The skin is thickest in the withers. The epidermis is relatively thin, being thickest on the body side due to the stratum corneum (75 to 87% of the epidermal thickness). The stratum malpighii is usually two to three cells deep. The epidermis, which is lightly pigmented, has a two-layered stratum granulosum and a very solid stratum corneum in the soles. The epidermal alveoli are 1.3 to 1.5 mm high. The papillary layer takes up nearly the whole thickness of the body dermis. The bulbs of the growing hairs lie deep in the dermis—in the skin of the belly they reach the subcutaneous tissue. The dermal structure is similar to specimen (3). The subcutaneous fat tissue was cut off when the samples were taken, except on the belly, where it is 0.75 mm thick.

The one-lobed sebaceous glands are active (fig. 118). They are multilobal on the belly. Two sebaceous glands are usually found at the pile hairs. The sweat glands are large, with slightly twisted secretion cavities (50 to 60μ in diameter) with no glomeruli. In the hind soles the sweat glands are twisted into glomi of 0.45 × 1.2 mm on the front feet and 0.45 × 1.05 mm on the

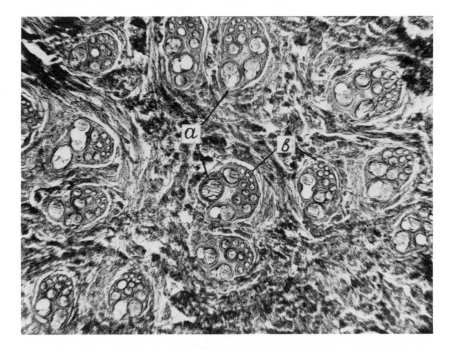

Fig. 119. *Ursus arctos*. Withers skin (hair tufts): *a,* pile hair; *b,* fur hair (summer male from Kamchatka).

hind feet (secretion section diameter 52μ). There are no arrectores pilorum muscles.

The hair is the same as in specimen (3), except that only one order of pile hairs is found in the withers skin. The hair is densest on the withers (2,900 hairs per 1 cm²) and on the body side (2,016 hairs), and sparsest on the belly (500 hairs). As a rule, there are up to three pile and eight to thirteen fur hairs in a tuft. The old fur hairs have given way to new ones, whereas old and new pile hairs in common outer sheaths still occur (fig. 119).

Thalassarctos maritimus

Material. Skin from withers, middle back, sacrum, middle of the belly, groin, jowl, elbow, and knee of an adult female, specimen (1) (body length 225 cm). Skin from middle of back, sacrum, breast, and middle of belly of female, specimen (2) taken in lair, May 1969, on Wrangel Island (table 63).

The hair is not growing. The skin is thickest on the sacrum and jowl, and thinnest on the groin (table 63). In specimen (1) the skin on the dorsal is thicker than the ventral, which is reversed in specimen (2). The thickness of

TABLE 63

THALASSARCTOS MARITIMUS: SKIN MEASUREMENTS

Indexes	Thickness							Sebaceous glands (μ)
	Skin (mm)	Epidermis (μ)	Stratum corneum (μ)	Dermis (mm)	Papillary layer (mm)	Reticular layer (mm)		
Specimen (1)								
Scapula	3.0	102	62	2.99	2.59	0.40		120 × 300
Midback	3.3	214	204	3.08	1.88	1.20		80 × 240
Sacral region*	5.2	102	51	4.20	3.38	0.82		51 × 307
Midbelly*	3.1	51	30	1.94	1.94	—		41 × 164
Groin	2.1	63	42	2.08	1.53	0.55		51 × 266
Jowl	5.2	180	168	5.02	1.32	3.70		30 × 200
Ulnar joint*	3.6	306	296	1.90	1.20	0.70		40 × 100
Patellar joint	4.4	92	61	2.00	2.00	—		72 × 256
Specimen (2)								
Midback	4.1	54	33	4.04	2.15	1.89		31 × 256
Sacral region	4.2	153	111	4.04	2.24	1.80		80 × 200
Breast*	5.0	67	46	4.05	2.25	1.80		42 × 190
Midbelly*	5.0	54	37	3.28	3.28			71 × 300

*Thickness of subcutaneous fat tissue is indicated in the text.

the epidermis varies considerably from zone to zone. It is thickest on the elbow, the dorsal body, and the jowl. A lax stratum corneum makes up to 96.7% of the epidermal thickness. The stratum malpighii is one to three cells deep. The sacrum, middle belly, groin, and knee have a one-cell-deep stratum granulosum. The epidermis is heavily pigmented on shoulder blade, middle back, and sacrum. In the other samples the epidermal pigmentation is slight. On the front and hind soles the epidermis is thick, about five to seven mm. The stratum corneum is tough, 4 to 4.1 mm thick, and very compact. The outer surface of the epidermis is plicated, wrinkled; the inner surface has alveoli of up to 1.6 mm. A two- to three-layered stratum granulosum is marked. The epidermis is lightly pigmented.

In the dermis of the body, the papillary and reticular layers meet level with the hair bulbs. The reticular layer is absent on the middle belly as large accumulations of fat cells protrude from the subcutaneous fat tissue to the hair bulbs; the reticular layer is also absent from the knee. The plexus of collagen bundles is rather loose in the dermis. In the outer papillary layer, horizontal collagen bundles predominate; in the other parts of the papillary layer there are also many bundles at various angles. In the reticular layer the collagen bundles run in several directions, the horizontal predominating. The elastin fibers are few in number in the dermis.

There are large groups of fat cells (75 x 96; 80 x 110μ) in the reticular layer under the hair bulbs. The subcutaneous fat tissue varies greatly in thickness with location on the body (Ristland, 1970). In polar bears of the migrating part of the population the maximum thickness (11 cm) is on the hind end of the body. In our animals, the subcutaneous fat tissue measures as follows: specimen (1) 0.92 mm on the sacrum, 1.13 mm on the belly, 1.4 mm on the elbow, 2.4 on the knee; specimen (2) 0.92 mm on the breast, 1.69 on the belly. In polar bear females taken in late winter, the thickness of the subcutaneous fat tissue ranges from 5 to 15 mm on the neck and 30 to 35 mm in the pelvis region (Grushetskaya et al., 1973). In the subcutaneous fat tissue fat cells alternate with sparse, loose bundles of collagen fibers. The sebaceous glands are small, with one lobe in specimen (1) and multiple lobes in specimen (2). They are slightly active (fig. 120). Two appear at every hair tuft, their lobes stretching along the hair roots.

In the jowl skin of specimen (1), the glands are unilobal, completely inactive on the elbow. In specimen (1) sweat glands are found only on the jowl, where their almost straight secretion cavities (20μ in diameter) stretch along the hair bursae. The glands are arranged singly at hair tufts (but by no means at every tuft), the ducts opening into the hair funnels. On the sacrum the sweat glands are inactive, seen as a strip of cells with nuclei intensively stained with hematoxylin. The strip stretches along the hair roots. In specimen (2) the sweat glands are slightly twisted, with very thin secretion cavities (20μ), apparently inactive. There are small, twisted sweat glands on

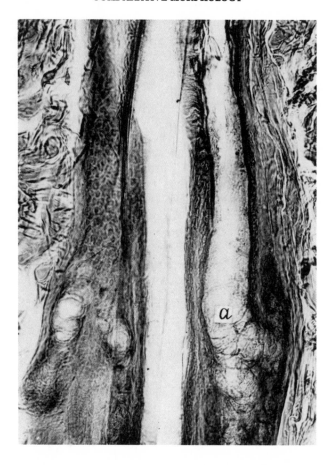

Fig. 120. *Thalassarctos maritimus.* Jowl skin: *a,* sebaceous glands.

front and hind soles. The arrectores pilorum muscles are very thin (20 to 30μ on the shoulder blade and 50μ on the sacrum. Possibly this is why they are not found in most of the samples.

All categories of hair are nearly uniform in length and have a relatively well-developed medulla. All the hairs are similar in shape, and the division into pile, guard, and fur is academic. On the middle of the back, the length of the guard hairs in specimen (1) is 57.8 mm, and in specimen (2) it is 53.4 mm. The pile hairs are 61 and 52.6 mm respectively, the fur hairs 37.6 and 45.6 mm (thickness 116 and 102μ, 83 and 106μ, 40 and 66μ, respectively; medulla 38.7 and 41.1%, 30.1 and 32.0%, 27.5 and 51.5%. On the breast of specimen (2) the guard hairs are 83.9 mm long; the pile hairs (first order) 67.3 mm; pile hairs (second order) 50.4 mm; pile hairs (third order) 45.3

mm; fur hairs 40 mm, thickness 149μ, 98μ, 92μ, 86μ and 56μ, respectively; medulla (in %) (none in pile hairs, second order): 37.5%, 43.8%, 36.9%, 19.6%.

The guard hairs have a thin base, which gradually gives way to a thickening. The scales of the cuticle are not ring-shaped. They are larger at the base (height about 11μ) and are smaller toward the distal (about 7μ). Sometimes the pile hairs (second order) are slightly wavy. Unlike the guard and pile hairs, the fur hairs are sinuate and the distal thickening is slight, or nonexistent. The cuticle scales are ring-shaped, the cuticles being large at the base and small toward the tip. The density of hair is much the same in all the areas of the skin under investigation, but is highest in the shoulder blade area and on the sacrum, and much lower on the groin. In specimen (1) there are 1,100 hairs per 1 cm^2 on the middle of the back, 1,410 hairs on the shoulder blades, 1,350 on the sacrum, 1,050 on the belly, 150 on the groin. In specimen (2) there are 2,350 hairs on the sacrum, 1,850 on the middle of the back, 1,510 on the breast, and 1,800 hairs in the middle of the belly. The hairs grow in tufts and groups, each tuft having four to five pile and six to ten fur hairs, the fur hairs growing behind the pile hairs. Sometimes a few tufts have a common guard hair. On body and jowl, three or sometimes five tufts form a group, though some of the tufts are arranged singly. On elbow and knee the tufts usually grow singly in curved lines.

Familia Mustelidae

In the dermis, the reticular and papillary layer usually meet level with the lower parts of sweat glands. The sebaceous glands are large, usually deep in the skin, almost at the hair bulbs. Most of the sweat glands have sausage-shaped, slightly twisted secretion cavities. The pile hairs have a clear granna, and there is a characteristic, highly developed specific anal gland.

Martes zibellina

Material. Skin from withers and breast (body length 46 cm). The animal died July 6, at Moscow Zoo.

The outer surface of the skin is irregular on the withers. The skin is thicker on the breast (0.82 mm) than on the withers (0.77 mm). So are the epidermis (34 and 34μ) and stratum corneum (22 and 24μ). The stratum malpighii is one to two cells deep. The stratum corneum is fairly compact. In the dermis (0.73 mm on withers and 0.78 mm on breast), the reticular layer (0.22 and 0.33 mm) is thinner than the papillary (0.51 and 0.45 mm). They meet level with the lower parts of the sweat glands. The collagen plexus, mostly horizontal bundles, are rather compact in the dermis. Elastin fibers, few in number, are found in the upper papillary and the inner reticular layer, where they form a loose plexus. The sebaceous glands (56 x 168;

TABLE 64

SEX AND SEASONAL VARIATIONS IN SABLE INGUINAL GLANDS (AFTER MONAKHOV AND LEONTYEV, 1968)

Age	Summer			Autumn		
	Weight of testicles or uterus and ovaries (g)	Length of inguinal glands (cm)	Weight of inguinal glands (g)	Weight of testicles or uterus and ovaries (g)	Length of inguinal glands (cm)	Weight of inguinal glands (g)
	Males					
Mature	n(4) 0.868	n(4) 5.70	n(4) 9.300	n(2) 0.120	n(2) 4.45	n(2) 3.400
Immature	n(1) 0.090	n(1) 3.50	n(1) 3.600	n(2) 0.095	n(2) 3.25	n(2) 1.075
	Females					
Mature	n(2) 2.450	n(2) 3.65	n(2) 1.065	n(4) 0.720	n(3) 3.15	n(3) 0.590
Immature	—	n(2) 3.20	n(2) 1.885	n(4) 0.188	n(1) 0.750	n(1) 0.750

$67 \times 140\mu$) have one or two lobes and are characteristically deep in the skin, with the lower part almost reaching the hair bulbs. There are two anterior and posterior glands at every hair. The sweat glands ($20 \times 28\mu$ in diameter), which reach into the skin deeper than the hair bulbs, have short, tubular, slightly twisted secretion cavities. The ducts open into the hair funnel (Matveyev, 1942). There are large glomi of eccrine sweat glands in the soles (Matveyev, 1942).

Paired inguinal glands are made up of sebaceous glands (Monakhov and Leontyev, 1968). Inguinal glands occur both in males and females, much larger in the former (table 64). In the reproductive period, the inguinal gland becomes larger (table 64).

In the first few weeks of life the cubs have a functioning neck gland made up of large apocrine glands (Petskoy and Kolpovsky, 1970).

No arrectores pilorum muscles are found in withers skin, though they are found in sacrum and breast.

There are four categories of hairs: guard, pile, intermediate, and fur. The guard hairs (19.3 mm long, 105μ thick) are straight and at three-quarters they thicken very gradually from the base to the distal quarter, tapering gradually toward the tip. The medulla is well developed through the whole hair, except base and tip. The pile hairs have a well developed granna and a slightly wavy club (two to three waves). The pile hairs have two size orders. The first order is 16.6 mm long, 100μ thick, medulla 80.0%, granna 8.0 mm. The second order is 14.3 mm, 66μ, 81.9%, and 6.8 mm. The intermediate hairs (11.4 mm long, 30μ thick) have a slightly developed granna, which differs from the proximal part in the morphology of the cortical layer but not in thickness. The medulla is slightly developed (53.3%). The shaft of the intermediate hair has up to three waves.

The fur hairs (11.4 mm long, 18μ thick) have no granna but up to five slight waves. The distal shaft differs somewhat in structure from the proximal. The medulla is more developed than in the intermediate hairs (55.5%). In winter the hair is much longer, though the thickness remains almost unchanged (Pavlova, 1951; Gerasimova, 1958) (table 65). The hairs grow in tufts, three tufts to a group (Matveyev, 1942), with six or seven fur hairs to a tuft, and a guard hair or a first order pile hair in the middle bundle, in lateral tufts—with a pile hair of the second to fourth order each. On the sacrum of a winter Barguzin sable taken in Buryatia are 353 guard and pile hairs and 19,635 fur hairs per 1 cm² (Gerasimova, 1958). Paired anal sacs have nonglandular walls and large sebaceous and tubular glands in the duct area (Stubbe, 1969).

TABLE 65

MARTES ZIBELLINA (BARGUZIN VARIETY): SACRAL REGION HAIR MEASUREMENTS
(WINTER, BURYATIYA) (AFTER GERASIMOVA, 1958)

	n	Length (mm)		Thickness (μ)	
		M±m	lim	M±m	lim
Guard	30	49.0 ±0.18	42-56	101.49±1.25	85.1-114.7
Pile					
1st order	129	45.58±0.24	39-50	101.31±0.62	85.1-114.7
2d order	225	47.18±0.16	35-47	94.28±0.78	62.9-111.0
3d order	245	37.38±0.24	31-42	73.78±1.24	25.9-107.3
4th order	210	33.01±0.14	29-38	37.7 ±0.77	25.9- 66.6
Fur	276	30.0 ±0.17	25-38	16.4 ±0.35	11.1- 22.2

Martes martes

The hair of *M. martes* is similar to that of *M. zibellina*. In summer the pile hairs are 23 to 27 mm long, 100 to 150μ thick. In winter they are 36 to 38 mm and 75 to 90μ, respectively (Pavlova, 1951). The fur hair is (summer) 11 to 12 mm long, 15 to 16μ thick, and (winter) 22 to 22.5 mm and 14 to 15μ, respectively.

Martes foina

Material. Skin from breast (thickness 0.77 mm) and hairs from withers of an adult (body length 50 cm) taken on August 5, 1961 in the Caucasus.

The hair is growing. The epidermis is 45μ thick, the stratum corneum 23μ. The stratum malpighii is two or three cells deep. The dermis is 0.72 mm thick, the papillary and reticular layers meet level with the hair bulbs. The papillary layer (0.47 mm) is thicker than the reticular layer (0.25 mm). The plexus of the collagen fiber bundles in the dermis is compact, and the bundles are predominantly horizontal (especially in the reticular layer) in a characteristic wavy pattern. Elastin fibers are few, horizontal, mostly in the upper papillary layer. The sebaceous glands are large (66×242; $143 \times 240\mu$) and level with the middle of the hair roots. They have one lobe, which is more or less oval. There are one or two glands at every tuft. The sweat glands are visible only in nongrowing hair tufts. Their secretion cavity (34μ in diameter) is sausage-shaped and slightly twisted. Anal sacs are similar to those of *M. foina erxleben* (Stubbe, 1969, 1970). The arrectores pilorum muscles are thin (17μ), but visible at every hair tuft.

The hair of the withers falls into four categories: guard, pile, intermediate, and fur. Guard hairs (25.1 mm long) are straight, thickening gradually from the base for the first quarter of the shaft to 112μ, then thinning toward the tip. The medulla is not highly developed (60.8%), although it runs down the whole hair except for base and tip. There are two orders of pile hairs. They are 19.9 and 17.3 mm long, 154 and 80μ thick. They have a highly pronounced granna, 10.5 mm long in the pile hair (first order) and 7.1 mm in the pile hair (second order), and have a slightly wavy (two to four waves) club. The medulla (57.8 and 75.0%) is absent only at the very base and tip. The intermediate hairs, 13.2 mm long and 48μ thick, have a small granna (4.4 mm long), a wavier club (up to seven waves), and a less developed medulla (50.0%). The fur hairs are short (9.7 mm) and thin (22μ) as compared with other hair categories. They do not have a granna and are wavy (up to six waves). The medulla, which passes through the whole hair shaft (except base and tip), is less developed than in other hair categories (27.2%).

Mustela nivalis

Material. Skin from withers, breast, and hind sole of specimens taken in summer, specimen (1) near Pskov, June 28 (body length 15 cm); specimen (2) near Gellenor, August 12 (body length 15 cm); specimen (3) in Turkmenia (body length 17 cm); and specimen (4) in the northern part of the European USSR near Vyborg, in winter (body length 15 cm) (table 66).

In summer the skin is thickest in June, because the hair is growing. In the two other specimens the skin thickness is about equal. The epidermis and stratum corneum are thinner on the breast than the withers. The stratum malpighii is one to two cells deep. In all three specimens the papillary layer is thicker than the reticular. In specimens (2) and (3) the hair has finished growing and the two layers meet level with the lower parts of the sweat glands. In specimens (2) and (3) in the papillary layer, and in specimen (1) in the upper part of this layer, the plexus of collagen fibers is dense. In the other parts of the dermis the plexus is mainly horizontal, especially in the reticular layer, where collagen bundles form a loose plexus. Elastin fibers are not numerous in the dermis. Near the hair bulbs, fat cells (28×28; $22 \times 34\mu$) may occur. In specimen (1) (molting) the lower part of the papillary layer and the whole reticular layer are filled out by fat cells (28×50; $34 \times 56\mu$). The sebaceous glands (one lobe) are about the same size in all three specimens. In the growing hairs of specimen (1) (molting) they are about level with the upper third of the hair roots. In the other two, their upper sections are halfway up the hair root, with the lower sections almost reaching the hair bulb. In horizontal section, one or two glands are seen at every hair tuft. The sweat glands have sausage-shaped, slightly twisted secretion cavities along the hair roots. In specimens (2) and (3) these are heading down below the hair bulbs. Highly developed anal glands are formed by large sebaceous and tubular glands (Schaffer, 1940). The arrectores pilorum muscles are thin (17 to 22μ).

There are three categories of hair: guard, pile, and fur. The guard hairs are straight, length 9.5 mm. They gradually thicken to 70μ in the distal quarter, then taper gradually toward the tip. The well-developed medulla (80.0%) runs through the whole hair, except base and tip. The two size orders of pile hairs are 9.0 and 7.6 mm long, 55 and 27μ thick, medulla 80.0 and 63.1% with a well-developed granna (first order is 5.1 mm long; second order 2.7 mm) and a wavy club (first order, one to three waves; second order, up to four waves). The fur hairs have no granna and have up to three waves; length is 6.0 mm, thickness 17μ, with a less developed medulla (53.0%) than the other hair categories.

For winter, the skin proper is studied only on the breast of specimen (4). The hair follicles are resting, but the skin is rather thick (table 66). The epidermis and stratum corneum do not differ in thickness from those of the summer specimens. The dermal reticular layer is rather well developed but is thinner than the papillary layer. They border level with the lower parts of the sweat glands. The plexus of the collagen bundles is the same as in summer specimens (2) and (3). In the upper reticular layer and lower papillary layer there is an almost continuous layer of fat cells (28×56; $39 \times 50\mu$). The sebaceous glands have single lobes, are larger than in summer, and are horizontally elongated. The glandular cells have large vacuoles; there is a gland at every hair tuft; the sweat glands are sausage-shaped, straight or slightly twisted, lying at an acute angle to the skin surface with their lower end deeper into the skin than the hair bulbs. The secretion cavity diameter is smaller than in summer. The arrectores pilorum muscles have a smaller diameter (10μ) than in summer.

The winter hair categories are the same as in summer, but the guard hairs are longer (9.8 mm), with a marked granna, which comprises over half the

TABLE 66

MUSTELA NIVALIS AND MUSTELA ERMINEA: SKIN MEASUREMENTS

	Thickness (μ)						Sebaceous glands (μ)	Sweat glands diameter (μ)
	Skin	Epidermis	Stratum corneum	Dermis	Papillary layer	Reticular layer		
Mustela nivalis (summer)								
Withers								
(1)	660	28	18	632	599	33	56 × 224	34
(2)	440	34	23	406	351	55	84 × 140	45
(3)	495	22	11	273	207	66	78 × 235	39
Breast								
(1)	616	28	14	586	586	absent	39 × 50	34
(2)	440	22	11	418	363	55	67 × 146	34
(3)	638	22	11	616	506	110	62 × 140	34
(4) (winter)	495	22	14	473	330	143	67 × 252	28
Mustela erminea								
Withers								
(1) (summer)	429	28	17	401	291	110	84 × 179	50
(2) (winter)	275	22	14	253	176	77	39 × 123	22
Breast								
(1) (summer)	330	22	7	308	253	55	34 × 118	50
(2) (winter)	253	22	16	231	187	44	56 × 168	28

hair length (6.8 mm), reaching a thickness of 84μ. The medulla is thicker than in summer (85.5%). Both orders of pile hairs have a wavy club (one to two waves) and differ only in size: length 9.4 and 8.8 mm, thickness 82 and 62μ, medulla 85.4 and 70.7%, the length of the granna 4.7 and 3.9 mm. The fur hairs are the same as in summer, except that they are somewhat shorter (5.7 mm) and thinner (14μ), though the medulla is thicker than in summer (57.2%). The hairs grow in tufts of one large hair and four or five fur hairs, though many tufts contain only fur hairs. Three tufts commonly form a group: the central tuft with a large hair, with tufts of fur hairs at the side. There are 1,000 pile and guard hairs and 23,000 fur hairs per 1 cm².

The hind sole skin is different enough to merit separate consideration. In summer, in fur-covered parts of the sole, the sweat glands do not form glomi, although they are longer and more twisted than on the body. The diameter of their secretion cavities is 39μ. The sebaceous glands have two to three lobes and are large (84 × 224μ). The skin areas without fur usually bulge like cushions, have very large tubular gland glomi, and are tightly pressed together so that their boundaries are almost invisible. The diameter of the secretion cavities is 39μ. The epidermis here is rather thick, its outline is curved inside and out. The gland ducts open into the hollows of the epidermis. In winter the skin of the sole has the same structure as in summer except that the glomi of the tubular glands are much smaller. The diameter of the secretion cavities is 34μ.

Mustela erminea

Material. Skin from withers and breast of specimen (1) (body length 23 cm) taken August 20, at Altai; and specimen (2) (21 cm), taken December 10 in the village of Rysevo (table 66).

In summer the skin, epidermis, and stratum corneum are thicker on the withers than the breast. The hair is not in growth. The stratum malpighii is two cells deep. The papillary layer is considerably thicker than the reticular, and they meet level with the lower part of the sweat glands. In the dermis, horizontal bundles of collagen fibers predominate; the plexus is dense in the upper and middle papillary layers and loose in the lower papillary and reticular layers. There are groups of fat cells (22 × 45; 28 × 39μ) near the hair bulbs and sweat glands. Horizontal elastin fibers are few, and are found only in the papillary layer. The sebaceous glands, which have single lobes, stretch horizontally level with the lower third of the hair roots, almost reaching the hair bulbs. There are two glands at every hair tuft. The sweat glands are sausage-shaped, horizontally very twisted secretion cavities with a rather large diameter. The sweat glands are deeper in the skin than the hair bulbs. The anal glands are formed of highly enlarged sebaceous and tubular glands (Schaffer, 1940). The arrectores pilorum muscles are thin (8 to 12μ).

There are three categories of hairs: guard, pile, and fur. The guard hairs

TABLE 67

Mustela erminea: sacral region hair measurements (summer, Moscow region)
(after Pavlova, 1959)

	Guard	Pile (orders)				Fur
		1st	2d	3d	4th	
Length (mm)	13.9	12.1	11.3	10.5	9.9	8.1
Thickness (μ)	102	97	73	61	50	18
Medulla as % of granna	67	63	62	58	50	49
Hairs per 1 cm^2 (total of 14,544)	20	108	96	92	72	14,156

are straight, 11.7 mm long, with a club. They gradually thicken to three-quarters of the way up the hair (74μ). The medulla (73.0%) runs through the whole of the hair, except base and tip. The pile hairs have a marked granna, and up to two waves. They divide into two size orders, 12.1 and 10.9 mm long, 75 and 54μ thick, medulla 78.7 and 81.5%. The medulla is absent at the base and tip. The fur hairs have no granna, but have three to four waves, 7.4 mm long, 19μ thick. The medulla is less developed (68.5%) than in the other categories. E. Pavlova (1959) has obtained somewhat different data on the summer fur (table 67). She distinguishes four size orders of pile hairs in the sacrum skin.

In winter the hair follicles are resting. The skin is thinner than in summer (table 66), and its outer surface is irregular. The winter skin is uniformly thick in all parts of the body (Tserevitinov, 1958), but thickest (over 0.3 mm) on the sacrum. The second thickest areas (0.2 to 0.3 mm) are head, neck, and part of the sides and belly. The third (to 0.2 mm) is a band down the center of the back. The latter is not consistent with our findings, according to which the skin is thicker on the withers than the breast. The epidermis and stratum corneum of the withers are thinner than in summer. The stratum malpighii is only two cells deep. The stratum corneum is loose.

The sebaceous glands have single lobes, are smaller than in summer and horizontally elongated, two at every hair tuft. The glandular cells have large vacuoles. The sweat glands have elongated, sausage-shaped (almost tubular) secretion cavities. Some of them are slightly (some highly) twisted (their long axes being horizontal to the skin surface). They lie somewhat deeper than the hair bulbs and their diameter is smaller than in summer. The arrectores pilorum muscles are thin (8 to 12μ).

The hair is similar to summer M. erminea, except that it is without pigment and has a marked granna in the guard hairs (7.3 mm, about half the length of the hair) and some pile hair tips differ morphologically from the

TABLE 68

Mustela erminea: sacral region hair measurements (winter, Moscow region)
(E. Pavlova, 1959)

	Guard	Order of pile				Fur
		1st	2d	3d	4th	
Length (mm)	15.75	13.7	12.2	11.9	11.5	9.1
Thickness (μ)	96	98	79	65	55	14
Medulla as % of granna	68	76	75	74	71	55
Hairs per 1 cm^2 (Total of 19,232)	8	204	144	116	212	18,548

TABLE 69

Mustela erminea: hair measurements (#25) (Karelian ASSR, March)
(after Marvin, 1958)

Hair	Body area				
	Withers	Midback	Sacral region	Side	Belly
Length (mm)					
Guard	14.52±0.76	17.00±0.64	18.37±0.62	16.68±1.13	14.88±0.77
Pile	11.44±0.61	13.00±0.64	14.72±0.68	12.12±0.56	11.42±0.55
Fur	9.24±0.80	10.8 ±0.78	12.12±0.49	10.12±0.72	9.40±0.66
Thickness (μ)					
Guard	92	100	92	87	85
Pile	39	39	40	40	38
Fur	14	14	15	14	14

lower part of the shaft. The hair indexes for length are: guard hair 13.5 mm; pile hair (first order) 12.2 mm; pile hair (second order) 9.9 mm; fur 9.1 mm. For thickness: 111μ, 91μ, 51μ and 16μ; for medulla: 82.0%, 80.4%, 80.4% and 62.5%; for granna length; pile hair (first order) 5.5 mm, (second order) 3.8 mm.

The indexes obtained by E. Pavlova (1959) and Marvin (1958) differ somewhat from ours (tables 68 and 69).

On the belly, the density of the pelage is lower than on the back. In fact, in the Sayan animal there are 17,020 hairs per 1 cm² (16,480 of them fur), and 15,916 on the belly (15,340 fur) (E. Pavlova, 1959). The hairs are arranged in tufts and groups. Eight to fourteen (most commonly 11 or 12) hairs to a tuft, three tufts to a group. There is one guard or pile hair in one of the tufts which is always on the edge, never in the middle of the tuft

(E. Pavlova, 1959). There are five zones for pile hair length (Tserevitinov, 1958). The longest pile hairs are in a 4 cm wide band on the back, more than halfway down the body. The shortest pile hairs are on the head, and front and hind feet. Fur hairs have four length zones resembling the pile zones.

Mustela sibirica

Material. Skin of withers, breast, and hind sole of specimen (body length 33 cm) taken September 11, on Shanganda River.

The outer surface of the skin is irregular. The hair follicles are active, and the skin is rather thick, more so on the withers than the breast (1.65 mm *vs* 1.26 mm), though the epidermis (23μ) and stratum corneum (17μ) are thinner on the withers than the breast (39 and 28μ). The stratum malpighii is two cells deep. There are three skin-thickness zones (Tserevitinov, 1958). The thickest skin is on the shoulder blades and sacrum. The dermis on the withers is 1.6 mm thick and 1.2 mm on the breast; the papillary layer (1.0 mm on the withers and 1.1 mm on the breast) is considerably thicker than the reticular, the two meeting at the hair bulbs.

The plexus of collagen fiber bundles in the outer papillary layer is dense. It gets looser in the middle and inner part, and still looser in the reticular layer. In the outer papillary layer the collagen bundles are arranged in various directions. In the middle and inner layers horizontal bundles predominate, and in the reticular layer they are almost exclusively horizontal. In the inner papillary layer, near the hair bulbs, fat cells (28×50; $34 \times 56\mu$) are concentrated. There are a few horizontal elastin fibers in the papillary layer and rather more in the reticular. The sebaceous glands are rather large (100×275; $100 \times 153\mu$) with single lobes, arranged in pairs at every hair tuft, level with the upper third or half of the hair roots. The sweat glands have sausage-shaped, slightly twisted secretion cavities 45 to 50μ in diameter running along the hair roots but not apparent at every hair tuft. Their lower end does not reach the hair bulbs. In the first four to five weeks of life, the cubs have a functioning neck gland formed of large apocrine glands (Petskoy and Kolpovsky, 1970). The arrectores pilorum muscles are 15μ thick.

There are five categories of hair: guard, pile (two orders), intermediate, fur, and near-bristles. The guard hairs (16.2 mm long and 122μ thick) are straight, thickening gradually in the upper two-thirds of the shaft. The medulla (70.5%) is well developed through the whole hair except at base and tip. The pile hairs, 15.3 and 12.0 mm long, 94 and 66μ thick, have a medulla of 76.6 and 66.7% with a marked granna that takes up about half the hair length (7.4 mm in the first order, and 5.8 mm in the second). In winter, there are six length zones of pile hairs (Tserevitinov, 1958). The longest hairs are on the lower end of the back and on the sacrum, and the shortest are on the head and feet. The intermediate hairs (9.1 mm long and 35μ thick) have a very small granna (3.7 mm long) and up to five waves.

TABLE 70

MUSTELA SIBIRICA: HAIR MEASUREMENTS (#25), BODY LENGTH 48 CM
(OCTOBER, SVERDLOVSK REGION) (AFTER MARVIN, 1958)

Hair	Withers	Midback	Sacral region	Side	Belly
Length (mm)					
Guard	22.31 ± 9.47	28.01 ± 0.63	35.01 ± 0.25	28.00 ± 0.31	23.37 ± 0.68
Pile	21.20 ± 0.34	24.00 ± 0.28	25.21 ± 0.61	22.80 ± 0.19	18.30 ± 0.41
Fur	12.00 ± 0.15	16.00 ± 0.24	18.00 ± 0.73	14.00 ± 0.20	13.00 ± 0.25
Thickness (μ)					
Guard	10	104	106	98	95
Pile	85	85	86	85	83
Fur	14	14	14	14	14

The medulla is less developed than in guard and pile hairs (60.0%). The fur hairs are longer than the intermediate hairs (10.2 mm), but thinner (19μ) and granna-free. The medulla is poorly developed (36.8%). The fur has up to eight waves.

There are six fur length zones on the body of the winter *M. sibirica,* their distribution being much like the pile zones (Tserevitinov, 1958). The bristle hairs (6.7 mm long) are very odd. They seem a bit like the grannae of the pile hairs, but are definitely not growing pile hairs as there is no medulla in the base. A thin, medulla-free club thickens rather sharply at first, then gradually thickens to about its half (to 103μ), then gradually tapers down. The medulla is better developed than in other hair categories (78.7%). Marvin (1958) distinguishes only guard, pile, and fur hairs. His evidence does not agree with ours (table 70).

The hairs grow in tufts: (1 guard, 1 pile, and 14 to 18 other hairs) or (1 pile and 14 to 18 other hairs) or (14 to 20 other hairs with neither guard nor pile). The tufts form groups. The tuft with guard hair is slightly forward of center in the group; those without a guard are slightly behind or to the side. There are 340 guard, 1,000 pile and 27,660 other hairs per 1 cm². There are other data on hair density of the *M. sibirica* which differ from the above (e.g., there are 292 guard, 282 pile, and 8,040 fur hairs per 1 cm² of the withers of October Siberian *M. sibirica*) (Marvin, 1958).

Seasonal variability of hair in *Mustela sibirica* Pall. (central and south Sikhote-Alin, Far East) is characterized by a 1.5-fold increase in hair length in winter and 2-3-fold in hair density (Bromley, 1956) (table 71).

The structure of the hind sole skin will be treated separately. On the hair-covered areas of the sole (the hair follicles are active there), the sebaceous glands are multilobal and considerably larger (up to $220 \times 550\mu$) than on the body skin. The sweat glands are also somewhat larger than on the body. Nearer the hairless areas, the size of the sweat glands increases. They be-

TABLE 71

MUSTELA SIBIRICA: SEASONAL VARIABILITY OF HAIR

	Males		Females	
	summer (Aug.)	winter (Oct,)	summer (Aug.)	winter (Oct.)
Length (mm)				
Guard and Pile	21	31	18	27
Fur	9	16	8	15
Hair per 1 cm^2				
Guard	16	66	32	64
Pile	113	224	144	320
Fur	3,344	9,576	4,048	13,312

come twisted and small glomeruli appear. On the hairless sole areas, the epidermis is very thick—up to 605μ, with alveoli of up to 242μ long. A thick stratum granulosum is most pronounced, but there is no stratum lucidum. The stratum corneum is very thick (308μ) and looks like stratum lucidum. There are no sebaceous glands in these areas. There are large eccrine gland glomi ($275 \times 495\mu$) in the dermis, surrounded by fat cells. The diameter of their secretion cavities is 34μ.

Mustela lutreola

Material. Skin from withers and the breast (body length 26 cm) taken in summer in the Tatar SSR.

The hair follicles are active and the skin is thick, more so on the breast than on the withers (1.70 mm *vs* 1.48 mm), while the epidermis and stratum corneum are thicker on the withers (34 and 18μ) than on the breast (22 and 11μ). The stratum malpighii is one to three cells deep. In the dermis (1.45 mm thick on withers and 1.68 mm on the breast) the papillary layer (1.06 mm and 1.13 mm) is considerably thicker than the reticular layer (0.38 and 0.55 mm) because of hair growth. They meet level with the hair bulbs. Horizontal collagen fiber bundles predominate in the dermis. There are also many rising bundles. The plexus of the collagen bundles is rather loose. In the middle and inner papillary layers are small fat cells. The reticular layer has large groups of round fat cells 34×39; $28 \times 39\mu$. The plexus of collagen bundles is loose in the reticular layer. The sebaceous glands, which have single lobes, are oval and elongated. They are larger in the breast skin ($112 \times 224\mu$) than in the withers ($40 \times 242\mu$). Sweat glands in the middle of the papillary layer have slightly twisted, sausage-shaped secretion cavities 28μ in diameter. They are rather sparse. There are no arrectores pilorum muscles.

The hair divides into three categories: guard, pile, and fur. The guard hairs are straight, 20.3 mm long and 91μ thick, a granna taking up the upper half (10.8 mm). The shaft thickens into the granna gradually. The medulla is moderately developed down the whole hair (60.5%) except for the tip and a relatively short segment at the base. The pile hairs have a long, wavy club (three to five waves) and a weak 3.8 mm granna. They are much shorter (13.5 mm) and thinner (31μ) than the guard hairs. Their medulla (46.7%) is less developed than in the guard hairs, especially in the club. The fur hairs (9.7 mm long and 18μ thick) have no granna, with up to seven slight waves. The medulla, which runs through the whole hair, is thin (33.3%). In horizontal section, we have not been able to distinguish between pile and fur hairs. The guard hairs grow singly or are pressed closely against the tufts of four to seven pile and fur hairs. There are also tufts without guard hairs containing six to nine pile and fur hairs. There are groups of one tuft with guard hairs and three or four tufts without. There are 670 guard and 26,330 pile and fur hairs per 1 cm².

Mustela vison

The thickness of the body skin in winter does not vary (Tserevitinov, 1958). The thickest skin (0.4 to 0.5 mm) is on the sacrum, between the shoulder blades, and on the neck. The second zone (0.3 to 0.4 mm) is on the shoulder blades, back, head, and part of the sides. The third zone (0.2 to 0.3 mm) is the belly. The ontogenesis of epidermis of mink has five periods (table 72) (Utkin, 1969).

TABLE 72

MUSTELA VISON: AGE CHANGES IN THE THICKNESS OF EPIDERMIS AND STRATUM CORNEUM (μ)
(UTKIN, 1969)

Age in days	n	Epidermis		Stratum corneum	
		Lim	M±m	Lim	M±m
Embryos					
32-36	5	16-30	21.3±0.4		
44-47	10	5-63	31.0±0.6	2-26	9.5±0.3
Over 48 days	5	59-98	72.6±1.2	2-28	15.0±0.7
Cubs					
Newborns	4	20-80	43.6±0.6	8-58	28.5±0.7
10	2	25-60	39.6±0.8	8-86	41.7±1.2
20	2	18-53	32.4±0.6	13-63	35.1±1.2
30	2	16-50	31.1±0.6	15-75	32.3±1.3
45	4	10-35	21.5±0.6	5-43	17.6±0.8
75	15	8-43	17.9±0.3	2-47	16.1±0.4
105	15	6-48	15.8±0.4	2-35	10.9±0.6
135	12	8-41	18.9±0.5	2-38	14.9±0.6
165	6	8-50	18.8±0.2	3-80	20.9±1.2
225	6	3-21	11.7±0.4	2-20	8.7±0.4

TABLE 73

MUSTELA VISON: AGE CHANGES IN DERMIS THICKNESS (μ)

Age in days	Embryos				Cubs								
	32-36	44-47	Over 48	New-borns	10	20	30	45	75	105	135	165	225
	5	10	5	4	2	2	2	4	15	15	12	6	5
	71-392	54-324	307-571	126-450	234-900	306-846	504-990	450-826	360-882	360-954	342-953	324-918	414-1,116
M$\pm$m	162.1±7.3	180.9±3.1		278.2±6.4	495.0±17.3	528.5±12.2	693.0±11.1	835.2±16.6	582.3±8.4	599.4±10.8	620.3±11.3	627.7±15.5	647.5±1.8

TABLE 74

MUSTELA VISON: AGE CHANGES IN THE WIDTH OF SWEAT AND SEBACEOUS GLANDS (μ)

Age in days	n	Sweat glands (width)		Sebaceous glands (width)	
		Lim	M$\pm$m	Lim	M$\pm$m
Newborns	4	43-150	84.0$\pm$2.3	—	—
10	2	36-71	58.5$\pm$1.6	—	—
20	2	36-100	65.6$\pm$1.3	17-35	23.3$\pm$0.7
45	4	29-57	44.3$\pm$0.7	36-100	67.0$\pm$1.6
75	6	29-50	42.4$\pm$0.5	29-107	67.7$\pm$0.7
105	12	—	—	36-143	63.2$\pm$1.4
135	6	29-57	41.6$\pm$2.5	43-128	75.8$\pm$0.9
165	6	—	—	43-121	82.5$\pm$1.4
225	5	36-50	36.4$\pm$2.4	29-100	60.1$\pm$1.1

The thickening of the dermis is closely related to hair growth. The dermis becomes thicker with the formation and development of hair; when the hair stops growing the dermis becomes thinner (table 73).

The anlage of hairs in mink begins at the age of thirty-seven to forty-three days. The first rudiments of sweat glands are recorded in embryos at the age of forty-eight days, those of sebaceous glands are recorded in newborn cubs. With age, their size changes (table 74).

The anlage of the neck gland appears at the beginning of the fetal period, some ten days before birth (Kolpovsky, 1969, 1972). No glands are distinguishable on the back throughout the fetal period.

Newborn cubs have numerous specific skin glands: sole, anal, withers, gluteal, and caudal (Azbukina, 1970), which are rather large (table 75).

The withers, caudal, and gluteal glands are formed of large apocrine sweat glands. The tubular glands of the neck gland have apocrine-type secretion (Petskoy and Kolpovsky, 1970). The glandular secretion mecha-

TABLE 75

MEAN DATA (MM) FOR SPECIFIC SKIN GLANDS OF 10 NEWBORN MUSTELA VISON (AZBUKINA, 1970)

	Males	Females	Indexes	Males	Females
Body weight (g)	9.1	7.4	Buttock glands		
Body length (cm)	8.7	7.8	Length	7.8	5.6
Withers gland			Thickness	3.3	4.0
Length	25.3	21.0	Anal glands		
Thickness	20.0	17.0	Length	1.8	1.7
Caudal gland			Thickness	1.1	1.0
Length	4.5	3.6			
Thickness	2.8	2.6			

TABLE 76

M̲USTELA VISON: AGE CHANGES IN THE ANAL GLAND (IN MM)̲

Months	Male (n=131)		Female (n=134)	
	Gland diameter	Inner cavity diameter	Gland diameter	Inner cavity diameter
1	6.8	4.1	6.9	4.8
2	12.1	6.5	9.9	5.9
3	11.2	8.6	9.8	6.9
4	11.1	4.0	12.9	5.3
5	10.9	4.3	10.2	4.4
6	12.0	7.3	10.3	7.1
7	15.5	9.9	10.3	7.9
8	14.3	5.6	14.4	5.6
9	13.2	5.9	10.6	5.7
10	16.1	9.7	11.1	8.3

nism is characterized by separation of globular electron-transparent vacuoles surrounded by a plasma membrane from the top of the microbristles of the epithelial cells (Kolpovsky, 1969; Kolpovsky and Zaitsev, 1972). The anal glands are not fully formed at birth. When the cubs are fifteen days old, alveolar holocrine glands develop in the anal glands (Azbukina, 1973). The alveolar glands grow intensively from the moment of their appearance until the cubs are twenty-five to thirty days old. At six weeks the withers, caudal, and gluteal glands have shrunk considerably while the anal glands have grown.

According to Petskoy and Kolpovsky (1970) the cervical gland (on the withers) begins to shrink in three-week-old cubs, usually disappearing in six-week-old cubs. In six-month-old cubs the first three glands have disappeared completely, while the anal glands are well developed. The anal glands reach maximum size and development by the time of sex maturity (Azbukina, 1973) (table 76). Maximum diameters of the lumens in the tubular glands secretion cavities in the anal gland, maximum height of the glandular epithelium, and the number of ducts are noted during the rut. The anal glands of mink are oval subcutaneous twin structures lateral to the anal orifice (Azbukina, 1967, 1972). In adult males they are 2.5 cm long and 1.5 cm wide, and they are somewhat smaller in females. In males they are on a level with the fourth to fifth caudal vertebrae; in females with the third to fourth caudal vertebrae. Only dot orifices of the ducts can be seen exteriorly.

The greatest part of the anal gland is a cavity where the secretion accumulates (fig. 121). Its size may vary, as is seen from the walls' plicate pattern. The cavity walls are made up of connective tissue and muscular membrane and are lined with a multilayered, squamous, cornifying epithelium. Exte-

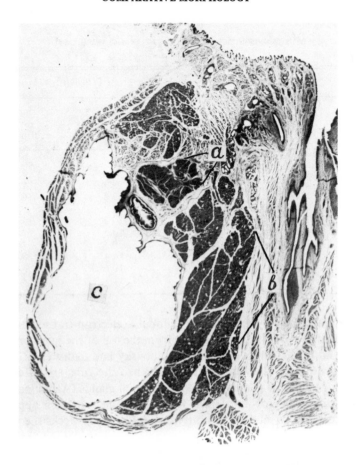

Fig. 121. *Mustela vison*. Anal gland structure: *a*, sebaceous glands;
b, tubular glands; *c*, anal gland cavity.

riorly, the cavity opens as a central duct, up to 4 mm long in females. A
sphincter formed of bundles of collagen fibers is in the upper part.

In the lower part of the central duct, six ducts of sebaceous holocrine
glands open on two levels: three ducts at first and another three ducts some
300 mm deeper. The ducts are arranged in a chessboard pattern. The alveolar multilobal glands are similar to common skin sebaceous glands but are
much larger. There are tubular apocrine glands peripheral to the sebaceous
glands. They consist of individual packets pressed tight against one another
dorsoventrally. There are up to fifty of them in the dorsal part of the anal
gland. The tubular glands have two orifices—one in the dorsal and the other
in the ventral part of the anal gland near the central duct.

An electron-microscopic examination confirms an apocrine-type secre-

TABLE 77

WINTER HAIR ON THE SACRUM OF MUSTELA VISON (STANDARD COLORATION) FROM
SALTYKOVO BREEDING FARM (EREMEYEVA, 1967)

Category	Length (mm)	Thickness (μ)
Guard	27.0	133
Pile		
1st order	24.5	123
2d order	23.0	107
3d order	21.0	98
Intermediate	18.5	47
Fur	15.1	15

tion in the anal gland tubular portion cells and a holocrine-type secretion in the alveolar portion (Afrosin, 1975a, b).

Guard, pile, and fur hairs are noted (table 77) (Marvin, 1958). The guard hairs, which are longest, are straight, spindle-like, and flattened in the upper half (medulla 70.37%). The pile hairs have a granna, at the base of which the shaft is bent. The pile hairs are longest on the sacrum (30 to 38 mm). The second zone (25 to 30 mm) embraces the median part of the back, scapulae, and the sides. The third zone (20 to 25 mm) is on the withers and surrounds the scapulae, the area between the forelegs, and the edge of the belly. The fourth zone (15 to 20μ) takes in part of the head, neck, the lower part of the belly, and the legs. The shortest pile hairs are on the muzzle (up to 15 mm). 67.9% of the hair is medulla thickness. The fur hairs are short and thin, with a shaft that is wavy down the whole length. At the tip is a small lancet-shaped thickening (Marvin, 1958).

There are five length zones for fur hairs (Tserevitinov, 1958). The first zone (16 to 17 mm) is located in two regions on the sides of the spinal area. The second zone (13 to 15 mm) is on the back surrounding the first zone and part of the sacrum. The third zone (10 to 12 mm) is on the withers of the anterior part of the back, sides, and sacrum. The fourth zone (7 to 9 mm) is on the head, lower part of the neck, the belly, and the feet. The fifth zone (3 to 6 mm) is on the muzzle. The medulla makes up 49.5% of the shaft. Marvin (1958) has different indexes, divides the hairs into somewhat different categories, and obtains results at variance with Eremeyeva (1967) (table 78).

The hairs grow in groups (Marvin, 1958) both simple and complex. The simple groups have two to three fur hairs, the complex groups contain several categories. There are groups of a single pile hair and 8 to 28 fur hairs and groups with one guard, two to four pile hairs, and 36 to 80 fur hairs.

The density of the pelage varies in different parts of the body (table 78). According to Kaszowski, Rust, and Shackelford (1970), live mink taken

TABLE 78

NUMBER OF HAIRS PER 1 CM² OF SKIN SURFACE OF WINTER MUSTELA VISON (MARVIN, 1958)

Category	Withers	Middle of back	Sacrum	Belly	Side of body
Guard	40	40	36	30	40
Pile	225	150	130	143	220
Fur	9,600	9,900	9,100	6,833	8,500
Total	9,865	10,090	9,266	7,006	8,760

in winter (n = 9) average 14,712 hairs per 1 cm². The female pelts (n = 5) average 380 pile hairs and 24,163 fur hairs; the male pelts (n = 20) average 295 pile and 2,083 fur hairs (table 79).

Mustela putorius

There are four thickness zones of skin on winter *M. putorius* (Tserevitinov, 1958). The thickest skin (0.58 to 0.8 mm) is on the front of the body—head, neck, and withers. The second zone (0.3 to 0.4 mm) comprises the lower half of the neck, the shoulder blades, and the whole back down to the sacral region. The third zone (0.2 to 0.3 mm) begins at the front feet and runs along the sides to the sacral region. The fourth zone (up to 0.1 mm) is on the belly and hind legs.

There are guard, pile, and fur hairs (Tserevitinov, 1958). The pile hairs form six zones for length; the fur hairs form five. There are about 8,500 hairs per 1 cm² on the back and about 6,000 on the belly.

Mustela eversmanni

Material. Skin from withers, breast, and hind sole of male specimens (1) to (3) taken March to April in Askania-Nova (body lengths 39 cm, 37 cm, and 40 cm) and from female specimens (4) to (5), taken in January near Zaisan (body lengths 37 cm and 38 cm) (table 80).

In spring, the hair follicles are inactive in specimen (1), are active in specimen (2), and are in initial growth in specimen (3). The thickness of the withers skin in specimen (2) is greater than in the other two specimens. The breast skin is thinner than the withers skin. The stratum malpighii of the epidermis is two to three cells deep. The stratum corneum is rather compact. In the dermis of specimens (1) and (3) the papillary and reticular layers meet level with the lower part of the sweat glands, and in the molting specimen (2) they meet at the hair bulb level. The papillary layer is thicker in the latter.

The dermal collagen bundle plexus is compact except in the inner reticular

TABLE 79

MUSTELA VISON: HAIR MEASUREMENTS (n = 25) WINTER (AFTER MARVIN, 1958)

Hair	Withers	Midback	Sacral region	Side	Belly
Length (mm)					
Guard	18.3 ± 0.84	28.1± 1.23	27.9 ± 1.56	27.08± 1.95	19.62± 1.42
Pile	13.18±47.1	19.8± 1.70	21.9 ± 2.82	17.66± 2.07	14.28± 1.53
Fur	10.5 ± 0.14	16.4± 0.80	13.33± 0.56	13.34± 0.55	10.37± 0.32
Thickness (μ)					
Guard	96.8 ±10	99.2±16.3	100.9 ±12.3	110.2 ±14.7	101.9 ±14.46
Pile	84.0 ± 9.7	55.4±13.6	47.3 ±12.7	64.0 ±13.9	57.3 ±10.8
Fur	14.0 ± 0.8	14.6± 0.5	15.3 ± 0.6	16.4 ± 0.7	15.96± 0.73

layer. The bundles are mainly horizontal. Thick, horizontal elastin fibers are sparsely arranged on the border with the subcutaneous muscles. In specimen (2), with growing hair, large groups of fat cells (22 x 50; 170 x 40μ) are in the inner papillary layer. The sebaceous glands are very large, mostly with single lobes (though some have several lobes), and with two glands enveloping each hair tuft at front and back. The glands are characteristically deep in the skin; the upper section is about halfway down the hair roots, and the lower section is almost at the bulbs. In specimen (2), with growing hair, the sebaceous glands are considerably smaller and are level with the middle of the hair roots, a pair at every hair tuft. The sweat glands are well developed and have tubular, highly twisted secretion cavities deeper in the skin than the hair bulbs. In specimen (2), with growing hair, the sweat glands are smaller and less twisted. The arrectores pilorum muscles are well developed, 30 to 77μ in diameter. In the molting specimen, they begin about halfway down the hair roots.

The hind sole skin was studied only in the spring specimens. On the furred areas of the soles, the epidermis has no alveoli, is 45μ thick, has a stratum malpighii five to six cells deep, and has a well-developed stratum granulosum. The stratum corneum is compact. The epidermis is not pigmented. The sebaceous glands are not very big (121 x 220μ). The sweat glands are better developed there than on the body (diameter of the secretion section 55μ) but more twisted. The sweat gland glomi get bigger close to the hairless parts of the sole, where the glomi become very large indeed (specimen 1: 110 x 242; 120 x 340μ), though the diameter of their secretion cavities is rather small (22 to 23μ). The glomi are surrounded by fat cells. The ducts open straight onto the skin surface. The epidermis is much thicker (550μ) (stratum corneum 220μ) in the hairless areas. The inner surface of the epidermis has alveoli up to 275μ.

There are three hair categories on the body: guard, pile, and fur. The guard hairs, 11.6 mm long, are typical of *Mustelidae*. They gradually thicken from the base to 85μ at the upper two-thirds of the shaft, then taper gradually toward the tip. The medulla (67.2%) is found only in the upper, thickened part of the hair. The two size orders of pile hairs are 9.8 and 7.2 mm long and 67 and 50μ thick. They have a marked granna, 6.6 mm in the first order and 3.9 mm in the second. The medulla, 79.2 and 64.0%, respectively, is found only in the granna and the upper part of the club. The fur hairs (6.8 mm long, 18μ thick) have three waves. Their upper third or half differs from the lower in the character of the cortical layer and in the presence of a medulla. In the lower section of the fur, the weakly developed medulla (33.3%) is broken or absent.

In winter the hair follicles in specimens (4) and (5) are inactive. The skin is as thick or slightly thinner than in the spring animals (table 80). There are three zones for skin thickness (Tserevitinov, 1958). The epidermis is thinner than in spring. The stratum malpighii is two cells deep. The dermis and the

TABLE 80

MUSTELA EVERSMANNI: SKIN MEASUREMENTS

Indexes	Thickness						Sebaceous glands (μ)	Sweat glands diameter (μ)
	Skin (mm)	Epidermis (μ)	Stratum corneum (μ)	Dermis (mm)	Papillary layer (mm)	Reticular layer (mm)		
Withers (spring)								
(1)	1.21	11*	—	1.19	0.68	0.51	232 × 64	67
(2)	1.42	34	17	1.38	1.05	0.33	110 × 220	45
(3)	1.21	28	11	1.18	0.74	0.44	209 × 253	50
(4) (winter)	1.21	22	14	1.18	0.74	0.44	132 × 143	22
(5) (winter)	0.82	22	14	0.80	0.42	0.38	88 × 187	20
Breast (spring)								
(1)	0.98	28	14	0.95	0.79	0.16	198 × 330	77
(2)	0.66	28	14	0.63	0.56	0.07	99 × 220	45
(3)	1.04	22	11	1.02	0.69	0.33	132 × 374	77
(4) (winter)	0.66	22	11	0.63	0.41	0.22	132 × 253	28

*Without stratum corneum.

papillary and reticular layers meet level with the lower sweat gland sections, these layers being more nearly equal than in the spring animals. The plexus of collagen fiber bundles is the same as in the spring animals. In the upper part of the reticular layer, large accumulations of fat cells (56×56; $50 \times 73\mu$) may occur, and in some places they form a continuous layer. The sebaceous glands are unilobal and are smaller than in the spring animals. There is a pair at every hair tuft. The sweat glands are inactive and much smaller than in the spring animals. They are slightly twisted and go deeper than the hair bulbs.

The arrectores pilorum muscles are well developed but are thinner than in spring (20 to 28μ). The hair categories are the same. Guard hairs (16.8 mm long and 110μ thick) gradually thicken to the top quarter of the shaft. The medulla (72.8%) is missing only at base and tip. The two orders of pile hairs (14.1 and 11.6 mm long, 111 and 82μ thick, medulla 82.0 and 75.7%, granna 7.6 and 5.1 mm long) are without medulla only at base and tip. There are five zones of pile hair length on the body (Tserevitinov, 1958). The fur hairs are wavy (up to four waves), 8.8 mm long, and 13μ thick. In the lower and middle sections of the fur hair shaft, the medulla runs in a series of dashes. In the upper section it is continuous, though thin (34.8%). There are four fur hair length zones (Tserevitinov, 1958). The hairs grow in tufts of one guard or pile hair and seven to eleven fur hairs, or ten to fourteen fur hairs only.

Vormela peregusna

Material. Skin from withers, breast, and the front and hind soles of a male adult (body length 32 cm) taken June 6 in Azerbaijan.

The breast hair is in growth, so the skin is thicker (1.43 mm) than the withers skin (0.77 mm), where the hair follicles are resting. The epidermis and stratum corneum of the breast skin (56 and 34μ) are also thicker in the withers (50 and 28μ). The stratum malpighii is three to four cells deep. The stratum corneum is loose. In the skin of the soles the epidermis is very thick (660 to 792μ). The inner surface of the epidermis contains large alveoli. The height of the alveoli is 275 to 330μ. The stratum granulosum and stratum lucidum are marked. The stratum corneum is compact (the stratum corneum and stratum lucidum are 264 to 451μ, respectively, and still thicker on the front paw). In the body dermis, the papillary layer (0.50 mm withers and 1.27 mm breast) is thicker than the reticular (0.22 and 0.10 mm). They meet at the lower level of sweat glands.

The plexus of collagen fiber bundles in the dermis of the withers is compact except in the middle and inner reticular layer. In the dermis it is loose. The collagen bundles are mostly horizontal, with a few elastin fibers arranged along them. Under the hair bulbs a few rows of fat cells (39×45; $34 \times 56\mu$) stretch along the secretion cavities. The sebaceous glands are large

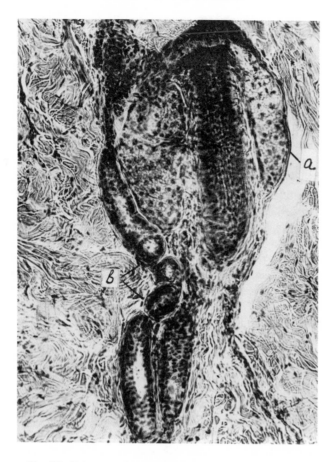

Fig. 122. *Vormela peregusna.* Withers skin: *a,* sebaceous gland;
b, sweat glands.

(up to 112 x 140μ withers, 154 x 220μ breast) and deep. In the withers their
upper sections are about halfway down the hair roots, the bottom almost
reaching the hair bulbs (fig. 122). The glands have one, two, or three lobes,
a pair at every hair tuft. In the breast, the unilobal sebaceous glands lie a
little higher than halfway down the hair roots.

The sweat glands, well developed in the withers, can be seen at every hair
tuft. Their secretion cavities are sausage-shaped or tubular, 39μ in diameter. Slightly twisted, they are arranged at an angle of 45° to the skin surface
(fig. 121). They go deeper into the skin than the hair bulbs. There are large
pigment cells at some glands. Most of the sweat glands are surrounded by
fat cells. The glands in the breast are inactive, and only small remnants of
their secretion cavities are visible at the hair tufts. In the dermis of the soles
are large (385 x 440μ front, 165 x 605μ hind sole) eccrine sweat gland glomi

pressed close together, interspersed with bundles of collagen fibers abundant in elastin fibers and forming continuous networks. Their secretion cavities are 45 to 56μ in diameter. The transition from the highly twisted but small sweat gland glomi in the hair-covered areas of the sole to the large, highly twisted glomi in the hairless areas is very noticeable. The arrectores pilorum muscles are thin in the body skin (17 to 22μ).

The hair falls into three categories: guard, pile, and fur. The guard hairs (11.9 mm long) gradually thicken from the base to 75μ in the upper quarter of the shaft and then taper very gradually to the tip. The medulla does not run through the lower quarter of the hair shaft. There are two size orders of pile hairs (11.1 mm and 9.8 mm long, 65μ and 50μ thick). The granna length is 5.3 mm (first order) and 4.4 mm (second order). In the first order the medulla is well developed only in the granna (87.7%) with just a narrow strip reaching the upper club. In the second order (medulla 80.0%) it reaches farther into the thin part of the hair. The fur hairs are 7.6 mm long, 13μ thick, and have three to four waves. The medulla is underdeveloped, although it runs through the whole hair shaft (38.4%). The hairs grow in tufts, some of only five or six fur hairs, some of five or seven fur and one big pile or guard hair. Three tufts form a group: the middle tuft, usually with a big hair, and two tufts on either side of fur hairs only. There are 660 big hairs and 6,000 fur hairs per 1 cm^2.

Meles meles

Material. Skin from withers, breast, soles, and anal region of adult specimen (body length 70 cm) taken June 17, 1971, near Oxford, England.

Some hair is growing, some is not. The withers skin is 5.8 mm thick, the breast skin 2.6 mm, the epidermis 126 and 105μ, the stratum corneum 84 and 42μ, respectively. The stratum malpighii is three to four cells deep. The stratum corneum is loose. The stratum granulosum is absent. The epidermal cells are not pigmented. The withers dermis is 5.7 mm thick, the breast 1.3 mm, the papillary layers 3.0 mm and 1.3 mm, respectively. In the skin of the breast the hair follicles reach through the whole dermis, so there is no reticular layer. The collagen bundles are mainly horizontal in the dermis. In the inner reticular layer, fat cells form horizontal strands. There are deposits of fat cells in the subcutaneous fat tissue of the breast (1.1 mm thick). Multilobal sebaceous glands are 72 x 307; 51 x 154μ in size. They are arranged in pairs at every hair tuft, but many are inactive. The sweat glands (fig. 123), which are tubular and slightly twisted, 38 to 42μ in diameter, are not found at every tuft of hair. The sweat gland ducts open into the hair funnel. The arrectores pilorum muscles are well developed, 147 to 153μ in diameter.

The hairs are arranged in tufts of one pile and three (more rarely four to five) fur hairs and, on the breast, of one pile and two or three fur hairs. The tufts are arranged in slightly curved lines. There are 550 hairs per 1 cm^2 of skin on the withers, 650 on the breast.

The subcaudal gland is a twin-lobed sac under the tail that opens as a narrow slit above the anus. It is formed of very large, intricately branching, superficial sebaceous glands averaging 1.2 mm in diameter, with large, abundant, highly twisted apocrine glands underneath (fig. 124). The sweat gland excretion ducts open into the bottom of the hair funnels. Our data agrees with Schaffer, 1940; Stubbe, 1971. Twin anal glands (with gland-free walls) have apocrine tubular glands in the area of their ducts and "sebaceous glands" (Stubbe, 1971). There are sole glands in the front and hind soles formed of large eccrine sweat gland glomi (256 x 410μ), their secretion cavities averaging 42μ in diameter (fig. 125).

Lutra lutra

Material. Skin from withers and breast of two adult females taken in summer (table 81).

The hair in both is in growth, so the skin is thick. The stratum malpighii is two to three cells deep. The stratum corneum is compact. The papillary layer is thicker than the reticular, which it meets at the hair bulb level.

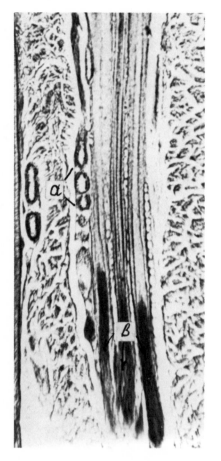

Fig. 123. *Meles meles.* Withers skin: *a,* sweat gland; *b,* hair follicles.

Horizontal collagen fiber bundles predominate in the dermis, although there are quite a few oblique bundles. The collagen bundles form a moderately compact plexus. Small groups of fat cells, 27 x 28; 22 x 34μ in specimen (1); 28 x 45, 34 x 45μ in specimen (2), occur near hair tufts in the inner papillary layer. The sebaceous glands are about the same size in both, with single (sometimes elongated) oval lobes, one at every hair tuft. There is also a small gland at every hair in the tuft. The sweat glands are found only in specimen (2), sausage-shaped and almost straight. The diameter of the secretion cavity is small (22 to 28μ). There is a caudal gland (highly enlarged tubular gland) proximally lateral on the tail (Schaffer, 1940). The walls of twin anal sacs are gland-free; however, very large sebaceous and tubular glands open into their ducts (Stoddart, 1972). The arrectores pilorum muscles are not found in specimen (2) and they are very thin in specimen (1).

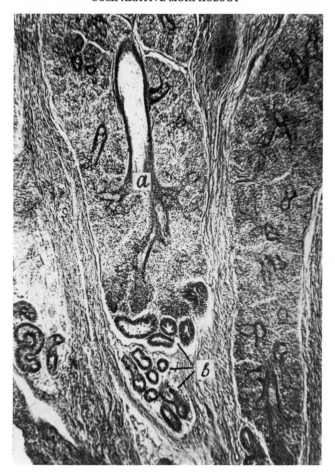

Fig. 124. *Meles meles.* Subcaudal pocket wall: *a,* sebaceous glands;
b, sweat glands.

There are four categories of hair on the withers: guard, pile (three orders), intermediate, and fur. The guard hairs are straight, 19.5 mm long and 149μ thick. Unlike the guard hairs of most of other mammals, the granna is very pronounced (its length is 0.1 mm). Although there is no clear neck, there is a narrowing of the medulla before it. The medulla is poorly developed (46.9%). The pile hairs are straight. A clearcut granna has a neck in front of it in which there is no medulla. The three orders of pile hairs are 16.0, 14.0, and 12.1 mm long; 7.7, 5.4, and 3.9μ thick; medulla 49.1, 51.2, and 42.2%; the length of the granna 7.7, 5.4, and 3.9 mm. The intermediate hairs (length 13.1 mm, thickness 25μ) have up to ten waves, a weak granna, 3.1 mm long; no medulla in the club, weakly developed in the granna (28.0%), and broken in some hairs. The fur hairs 11.6 mm long, 10μ thick, have up to

TABLE 81

LUTRA LUTRA: SKIN MEASUREMENTS

Indexes	Skin (mm)	Epidermis (μ)	Thickness Stratum corneum (μ)	Dermis (mm)	Papillary layer (mm)	Reticular layer (mm)	Sebaceous glands (μ)
Withers							
(1)	1.54	34	22	1.50	1.01	0.49	56 x 224
(2)	2.14	50	34	2.09	1.21	0.88	56 x 168
Breast							
(1)	2.09	56	39	2.03	1.59	0.44	73 x 880
(2)	1.76	39	28	1.72	1.34	0.38	34 x 168

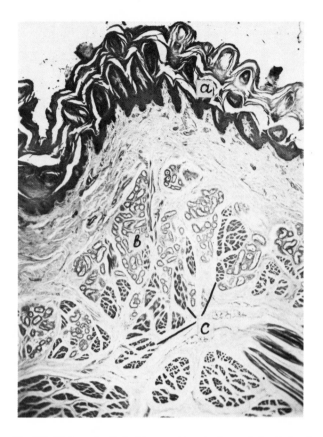

Fig. 125. *Meles meles*. Sole skin: *a*, epidermis; *b*, sweat glands glomi;
c, bundles of muscle fibers.

eleven waves, no granna or medulla. The tip of the fur hair is different from the rest of the hair.

The hair categories on the breast are the same as on the withers. The guard hairs are 20.5 mm long and 120μ thick, granna length 10.4 mm. The medulla is thin (50.0%). Unlike those in the withers, the guard hairs have a neck with no medulla. The three orders of pile hairs are 17.1, 13.3, and 11.4 mm long; 106, 90, and 28μ thick; medulla 49.0, 40.0, and 28.5%; granna length 4.3, 4.5, and 3.8 mm. The intermediate hairs (length 9.6 mm, thickness 13μ, medulla 23.0%, length of granna 2.2 mm) have up to seven waves. The fur hairs are the same as on the withers, length 8.6 mm, thickness 10μ with up to fourteen waves and no medulla. The hairs grow in tufts, usually of one guard or pile and seventeen to twenty-two fur (and intermediate) hairs. The fur hairs are posterior to pile and guard hairs in the tuft. There are tufts of fur hair only. Six to seven tufts of pile or fur hairs grow around the guard hair tufts. There are 1,333 guard, 2,333 pile, and 37,666 fur hairs per 1 cm^2.

In winter the hairs are longer than in summer (table 82).

TABLE 82

HAIR LENGTHS (MM) OF A WINTER L. LUTRA
TAKEN IN THE NORTH (TSEREVITINOV, 1958)

Hairs	Back	Belly
Guard	24.2	21.0
Pile	18.4	17.2
Fur	14.6	12.2

There are 50,668 hairs on the belly of a winter animal, more than on the back (34,972) (Tserevitinov, 1958). Similar hair measurements are reported by Ermilov and Marvin (1971), in an otter taken in October in the southern Urals (table 83).

TABLE 83

LUTRA LUTRA (URALS): HAIR MEASUREMENTS (ERMILOV AND MARVIN, 1971)

Hair types		Body areas				
		Withers	Middle of back	Sacrum	Side	Belly
		Hair length (M±m) in mm				
Guard	25	24.53±0.12	22.52±0.11	26.00±0.11	25.95±0.13	22.55±0.11
Pile	25	19.71±0.09	17.29±0.13	21.54±0.11	22.88±0.12	19.35±0.14
Fur	25	17.78±0.17	12.77±0.16	16.03±0.17	15.62±0.37	17.01±0.22
		Hair thickness (M±m) in mm				
Guard	25	209±1	204±3	204±4	181±5	155±2
Pile	25	197±10	188±4	187±5	141±4	132±2
Fur	25	13	13	12	13	13

Winter otters from Sakhalin and Western Siberia show somewhat similar hair measurements (Vshivtsev, 1972) (table 84).

TABLE 84

L'UTRA LUTRA (SAKHALIN AND WESTERN SIBERIA): HAIR MEASUREMENTS (VSHIVTSEV, 1972)

	Sakhalin (n=6) $M \pm m$	Western Siberia (n=6) $M \pm m$
Pile hairs		
Length (mm)	25.3 ± 0.78	24.38 ± 0.51
Granna thickness (μ)	101 ± 5	170 ± 1
Thickness of medulla in		
granna (μ)	57 ± 9	116 ± 4
Fur hairs		
Length (mm)	12.2 ± 0.53	15.82 ± 0.58
Thickness in middle of		
shaft (μ)	8	11

Enhydra lutris

The pelage comprises guard, pile, intermediate, and fur hairs (Barabash-Nikiforov et al., 1968). Pile and fur hairs are most abundant. The pile hairs are slightly twisted, round in proximal cross section, highly flattened and dilated in middle and distal portions, and round in cross section at the tip. The pile hair length is 15.6 to 42.0 mm, maximum thickness 40.8 to 149μ (table 87). Fur hairs are wavy and round in cross section. They are 8.3 to 27.2 mm long and 4.8 to 17.2μ thick (tables 85-87). The guard hairs are similar to the pile hairs but longer than the latter (26.5 to 48.4 mm). The intermediate hairs contribute to a transition category between pile and fur. The hairs grow in tufts of one pile and an average of seventy other hairs (Kenyon, 1972). In winter, the belly is more densely haired than the back (Barabash-Nikiforov et al., 1968). In summer, the difference in hair density is reduced but still remains. In a sea otter female carcass found on April 5, 1932, on Medny Island, the number of hairs per 1 cm² of the skin surface amounted to 32,875 (middle of back) and 44,979 (middle of belly).

The vibrissae are maximum length on the upper lip (50 to 70 mm) (thickness up to 180μ). There are five to ten bundles of small vibrissae 30 mm long. Short, relatively thin vibrissae are also on the breast and front feet. Sea otter hair dimensions vary with age. The maximum absolute length is recorded at the age of one month.

FAMILIA VIVERRIDAE

The papillary and reticular layers meet level with the lower parts of the sweat glands or at the bulb level. The sweat glands have sausage-shaped or tubular, highly twisted secretion cavities. The hairs differ from those of

TABLE 85

MEAN LENGTH OF HAIR (MM) OF VARIOUS-AGED SEA OTTERS FROM THE KOMANDORSKIE ISLANDS
(BARABASH-NIKIFOROV, ET AL., 1968)

Age	Withers		Middle back		Rump		Throat		Breast		Belly	
	Pile	Fur	Pile	Fur	Pile	Fur	Pile	Fur	Pile	Fur	Pile	Fur
About 1 month	37.0	18.0	41.0	17.0	42.0	22.0	20.0	13.0	—	17.0	37.0	20.0
3-5 months	25.7	17.0	28.7	24.0	27.3	23.0	27.5	19.0	23.0	12.7	25.7	19.7
Over 5 months	27.8	22.5	29.1	24.4	28.5	22.8	27.8	17.8	22.8	12.8	25.8	20.8
About 1 year	30.5	21.9	29.8	25.9	28.0	23.7	30.6	17.5	23.7	12.4	24.4	19.6
Adult	30.1	22.5	30.0	25.0	28.0	22.0	30.5	19.0	27.2	13.0	28.1	21.7
Old	27.7	20.6	31.5	26.0	29.8	23.1	26.6	16.6	24.8	12.0	24.1	18.0

TABLE 86

LENGTH OF HAIRS (MM) OF ADULT SEA OTTERS FROM THE KOMANDORSKIE ISLANDS IN WINTER
(BARABASH-NIKIFOROV, ET AL., 1968)

Hairs	Female					Male				
	M±m	σ	Cv	Lim	n	M±m	σ	Cv	Lim	n
Guard										
Withers	37.2±1.26	2.18	5.86	35.5-48.4	14					
Back	32.1±0.41	0.93	2.90	31.0-39.5	16					
Rump	36.9±0.14	0.24	6.50	36.6-37.2	14	29.3±6.13	8.65	38.79	28.6-30.1	13
Throat	28.9±0.29	0.66	2.28	27.9-29.5	16	34.4±1.49	2.10	6.10	31.0-46.0	13
Breast	27.9±0.71	1.28	4.59	26.5-29.6	14	25.9±0.13	0.22	0.85	26.7-29.2	14
Belly	29.8±1.20	0.17	0.60	29.2-30.3	13					
Pile										
Withers	29.5±0.27	1.51	5.12	27.1-32.2	30	27.0±0.24	1.34	4.96	22.5-28.5	30
Back	28.0±0.17	0.90	3.21	26.2-29.5	27	24.6±0.25	1.37	5.57	21.8-27.5	30
Rump	25.9±0.43	2.34	9.03	20.5-29.0	30	23.0±0.26	1.38	6.00	19.9-26.4	30
Throat	23.0±0.24	1.33	5.77	21.2-26.8	30	18.8±0.36	1.68	8.94	16.0-22.2	22
Breast	19.3±0.37	2.04	5.36	15.6-23.7	30					
Belly	22.3±0.27	0.93	4.18	20.5-23.8	31	36.0±0.26	1.42	5.46	22.5-28.0	30
Fur										
Withers	18.5±0.39	1.95	10.54	12.2-21.9	26	20.8±0.93	4.64	22.31	12.8-29.8	26
Back	16.7±0.73	3.67	21.98	10.5-22.6	26	19.4±0.53	2.63	13.56	14.2-23.8	26
Rump	16.7±0.89	4.44	26.27	9.8-23.5	26	16.7±0.42	2.12	12.69	11.8-20.1	26
Throat	11.9±0.58	2.94	24.70	8.6-19.3	26	10.9±0.32	1.61	14.77	8.3-14.5	26
Breast	12.9±0.47	2.35	18.22	8.4-20.5	26					
Belly	20.4±0.48	2.40	11.76	13.1-24.0	26	22.7±0.29	1.45	6.39	20.1-24.8	26

TABLE 87

GREATEST THICKNESS OF HAIRS (μ) OF ADULT SEA OTTERS FROM THE KOMANDORSKIE ISLANDS IN WINTER
(BARABASH-NIKIFOROV ET AL., 1968)

Hairs	Female					Male				
	M±m	σ	Cv	Lim	n	M±m	σ	Cv	Lim	n
Guard										
Withers	126.6±6.06	10.48	8.28	111.9-136.1	14					
Back	124.0±3.12	6.98	5.63	113.1-132.2	16					
Rump	128.8±5.03	8.71	6.76	119.5-136.4	14	106.4±6.58	9.28	8.72	95.8-113.2	13
Throat	136.8±3.77	8.44	6.17	123.5-144.2	16	132.0±8.76	12.35	9.36	119.5-144.2	13
Breast	149.7±1.67	2.78	1.78	145.2-150.0	14	141.0±3.13	5.41	3.83	135.5-146.8	14
Belly	134.5±3.08	4.34	3.23	129.5-137.5	13					
Pile										
Withers	86.5±3.00	20.16	23.31	48.5-123.1	45	85.7±2.80	14.30	16.68	52.8-108.1	26
Back	72.4±2.52	16.11	22.24	54.4-111.0	41	89.3±1.21	5.93	6.64	58.4-129.6	25
Rump	86.2±0.33	20.40	23.67	43.8-127.1	38	108.8±3.89	19.05	17.51	51.2-143.2	25
Throat	102.3±3.96	26.88	26.26	40.8-138.1	46	108.2±4.65	22.78	21.05	49.2-137.4	25
Breast	109.0±3.18	21.36	29.59	62.0-149.0	45	89.5±2.25	11.93	13.33	70.2-108.2	29
Belly	79.0±3.27	18.50	23.40	42.8-115.2	32					
Fur										
Withers	8.9±0.27	1.37	15.39	6.2-12.8	26	9.0±0.27	1.33	14.78	7.0-12.0	26
Back	9.1±0.69	3.43	37.69	4.8-17.2	26	7.9±0.23	1.14	14.43	5.4-10.5	26
Rump	7.7±0.29	1.43	18.57	5.4-11.1	26	8.2±0.19	0.98	11.95	6.8-10.1	27
Throat	8.6±0.32	1.59	18.49	6.7-14.6	26	8.3±0.21	1.03	12.41	6.6-10.0	26
Breast	9.0±0.24	1.25	13.89	6.9-11.1	29	8.3±0.25	1.27	15.30	6.9-11.0	26
Belly	9.4±0.52	2.58	27.45	6.0-16.2	26					

other *Carnivora* in shape. The specific skin glands are highly developed in the anal region.

Viverra civetta

Material. Skin of breast and withers, hairs of withers of specimen (body length 23 cm) taken in Cameroun, West Africa.

Although the hair follicles on the breast are active, the skin is thin (473μ). The epidermis is 28μ thick; the stratum corneum 14μ. The stratum malpighii is two to three cells deep; the dermis is 445μ. The papillary layer of the dermis (313μ) is more than twice as thick as the reticular layer (132μ). Horizontal collagen bundles predominate in the dense plexus in the dermis and there are a few elastin fibers in the papillary layer. The sebaceous glands at the first category hairs are larger ($34 \times 140\mu$) than those at other hairs ($56 \times 73\mu$). The sweat glands at the first category hairs have highly twisted secretion cavities 40μ in diameter. The arrectores pilorum muscles are thin (22μ).

Three categories of withers hair do not correspond in their morphology to any hair categories in other *Carnivora*. It seems reasonable to give them only numbers. The first category hairs (length 23.7 mm, thickness 53μ) have three to four waves and a small thickening in the upper shaft. The moderately developed medulla (58.5%) runs through the whole hair. The three size orders of second category hairs have no thickening and are 20.0, 15.8, and 7.6 mm long; 30, 27, and 22μ thick; medulla 32, 61.8, and 55.6%. They have several waves. The first order has three to five, the second has three to six, and the third has two to four. In some of the third order hairs the upper medulla is a series of dashes. The third category hairs are very different from the first and second. They are short (6.1 mm), straight, flattened, thickening gradually from the base to middle (up to 51μ) after which they taper toward the tip. The medulla runs through all but the base and tip. It is less developed than in the other two categories (37.2%).

Herpestes griseus

Material. Skin from withers, breast, and hind soles of two adult specimens (both body lengths 31 cm). In specimen (1) the hair follicles are active, so the skin is thicker than in specimen (2) (table 88).

Epidermis and stratum corneum are thinner on the breast than on the withers. The hind sole epidermis is thick, with an alveolar inner surface, a stratum granulosum, stratum lucidum, and a very compact stratum corneum. The body dermis consists of a papillary layer only. In specimen (1), roots of the growing hair cross the entire dermis deep into the subcutaneous fat tissue (0.55 to 0.66 mm from the lower border of the dermis), and in

TABLE 88

HERPESTES GRISEUS: SKIN MEASUREMENTS (μ)

	Skin	Epidermis	Stratum corneum	Dermis	Papillary layer	Subcutaneous fat tissue	Size of sebaceous glands (μ)
(1) Withers	1,870	28	15	770	770	1,072	220 × 330
(2) Withers	1,210	56	45	604	604	550	50 × 168
(1) Breast	1,980	10*	—	770	770	1,200	55 × 176
(2) Breast	1,100	28	17	550	550	522	39 × 122

*Without stratum corneum.

394COMPARATIVE MORPHOLOGY

specimen (2), the sweat glands reach the border between the dermis and subcutaneous fat tissue or into the latter. The collagen fiber bundles form a
compact plexus in the dermis, the structure of which is uniform through the
dermis. Perpendicular horizontal bundles predominate. Elastin fibers are
abundant, and a loose network of horizontal elastin fibers lies under the
subcutaneous fat tissue. The subcutaneous fat tissue is thick, and collagen
fiber bundles are very few. The fat cells are 56 x 112; 67 x 113μ. Sebaceous
glands (with one or two lobes) are larger in the withers, one at every hair
tuft, usually posterior. No sweat glands are found in specimen (1) (they
appear to be absent during hair growth). In specimen (2) their sausage-
shaped secretion cavities are 30 to 44μ in diameter. They are very twisted.
The lower sections are usually level with the hair bulbs and sometimes reach
into the subcutaneous fat tissue.

Large eccrine sweat gland glomi in the lower parts of the dermis of the
entire sole are often surrounded by fat cells (fig. 126). The number and size
of the glomi increase sharply wherever the skin is thicker (diameter of secretion cavities 12μ).

Large sebaceous and apocrine sweat glands are in the walls of the anal
sacs of *Herpestes auropunctatus*. The secretion of these glands (Gorman,
Nedwell and Smith, 1974) gets into the anal sac where it is influenced by
bacteria (four bacterial species were identified) with the resultant production of a series of saturated carboxylic acids; acetic, propionic, isobutyric,
isovaleric, and valeric.

The arrectores pilorum muscles in the skin are highly developed on the
withers. In specimen (2) they begin at the hair bulbs and in specimen (1)
(with growing hair) halfway up the hair roots. In the skin of the breast the
muscles are small (25μ thick) as compared with 50 to 56μ on the withers.

The three categories of hair are designated, for convenience, guard, pile,
and fur. The guard hairs, 26.5 mm long, are straight, thicken gradually in
the proximal third of the shaft (to 136μ), and then taper toward the tip. The
medulla is well developed (61.8%). The pile hairs (12.6 mm long, 32μ thick)
have a wavy shaft (two or three waves). There is a slight thickening and a
weak granna in the upper dermis. The medulla is less developed (55.6%)
than in the guard hairs. The fur hairs, which are wavy, are of even thickness
through most of the shaft and taper in the distal sixth. The medulla is developed through the whole hair. There are two orders: length 9.1 and 6.5 mm,
thickness 35 and 32μ, medulla 54.3 and 56.2%. The guard hairs grow singly,
the pile and fur hairs in tufts. Some tufts contain only three to seven fur
hairs. There are 166 guard, 416 pile, and 5,508 fur hairs per 1 cm^2.

Familia Felidae

The papillary and reticular layers meet at the hair bulb level. The sebaceous
glands are small. The sweat glands have tubular, slightly twisted secretion

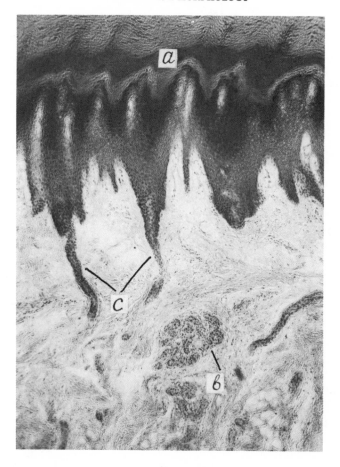

Fig. 126. *Herpestes griseus*. Sole gland of hind foot: *a*, epidermis;
b, secretion cavity of gland; *c*, ducts.

cavities. The pile hairs have a weak granna. The guard hairs and first order
pile hairs are straight. The other pile hairs are wavy. The fur hairs are curly.

Felis silvestris

There are four skin thickness zones. The thickest skin is on the anterior part
of the head, withers, and half of the back (Tserevitinov, 1958). The hair is
shorter on the belly than the back (first order pile hairs on the back are 41.3
mm long, and on the belly 22.6 mm long; Tserevitinov, 1969). The hair
indexes are given in table 89. There are five zones of pile-hair length, the
longest (first zone) hairs (50 to 60 mm) being on the back. The second zone
(45 to 50 mm) starts at the shoulder blades, reaching down the anterior sec-

tion of the back, the sides, and the sacral region. The third zone (34 to 45 mm) is on the thighs. The fourth zone (25 to 35 mm) is on the withers, neck, and hind legs. The fifth zone (under 25 mm) includes head and feet. The fur hair zones are about the same. There are 8,900 hairs per 1 cm² on the back, and 2,000 on the belly. There are ten fur hairs to one guard or pile hair.

Felis ocreata

Material. Skin from withers and breast of an adult specimen taken in winter in Repetek (Kara Kum).

The skin is thicker on the withers (0.93 mm) than on the breast (0.55 mm); the epidermis (28μ) and the fairly compact stratum corneum (18μ) are the same in both. The stratum malpighii is one to two cells deep. The dermis is 0.90 mm on the withers and 0.52 mm on the breast, the papillary layer being thicker (0.46 and 0.30 mm) than the reticular layer (0.43 and 0.22 mm). They meet at the hair bulb level. In the papillary and the upper reticular layers, collagen fiber bundles form a compact plexus (fig. 111). In the upper and middle papillary layers the collagen bundles run at various angles, but they are mainly horizontal in the rest of the dermis. Elastin fibers are numerous in the papillary layer and on the border with the subcutaneous muscles where they form smooth plexuses and small bundles, always predominantly horizontal. There are fat cells (34 x 39; 28 x 45μ) in the reticular layer, the inner part of which has a loose plexus of collagen bundles. One-lobed sebaceous glands (39 x 101; 56 x 112μ in the withers and 34 x 67; 34μ 78μ in the breast) lie singly, anterior or posterior to the hair tuft (fig. 127), though usually there is only one posterior gland. The secretion cavities of the sweat glands, shrunken and inactive (apparently for the winter), are tubular and slightly twisted. Their diameter is 14 to 28μ. The ducts are behind the hair tufts. The arrectores pilorum muscles are 28 to 62μ thick.

The hair indexes are given in table 89. There are five length zones for pile hairs (Tserevitinov, 1969). The first zone with the longest hairs (over 50 mm) takes in the back half of the belly and part of the neck. The second zone (45 to 50 mm) covers the front half of the belly. The third zone (40 to 45 mm) runs down the middle of the hind part of the back, reaching the sacrum, and takes in the sides and top of the neck. The fourth zone (35 to 40 mm) is on the sacral area, up along the sides and shoulder blades. The fifth (30 to 35 mm) and the sixth zones (up to 30 mm) cover the legs and head.

Felis euptilura

The indexes of the hairs on back and belly are given in table 89. The hair on the belly is shorter and much thinner than on the back (Tserevitinov, 1969). The first zone (50 mm and over) is in the middle of the back. The second zone (40 to 55 mm) surrounds it and reaches down the sides and belly. The third zone (40 to 45 mm) is on the sacrum and withers. The fourth zone (35

TABLE 89

FELIDAE: HAIR MEASUREMENTS (WINTER) (AFTER TSEREVITINOV, 1969)

		Hair				
		Pile (orders)				
Indexes	Guard	1st	2d	3d	4th	Fur
1	2	3	4	5	6	7
Felis silvestris						
Back						
Length (mm)	44	41	39	37	36	34
Thickness (μ)	81	87	71	62	43	20
Felis ocreata						
Back						
Length (mm)	51	43	39	37	35	35
Thickness (μ)	92	102	76	56	42	23
Belly						
Length (mm)	58	54	49	46	44	38
Thickness (μ)	59	48	42	35	31	21
Felis euptilura						
Back						
Length (mm)	47	49	47	39	35	34
Thickness (μ)	108	106	75	66	34	18
Belly						
Length (mm)	45	43	37	31	—	28
Thickness (μ)	60	56	40	41	—	16
Felis chaus						
Back						
Length (mm)	61	56	47	43	39	30
Thickness (μ)	89	74	52	40	30	20
Belly						
Length (mm)	50	37	30	26	—	20
Thickness (μ)	48	30	28	27	—	21
Felis lynx						
Back						
Length (mm)	51	38	39	38	35	31
Thickness (μ)	94	88	74	63	42	27
Belly						
Length (mm)	88	70	53	55	50	41
Thickness (μ)	76	58	46	36	30	22
Acinonyx jubatus						
Back						
Length (mm)	35	33	32	32	31	25
Thickness (μ)	87	77	73	64	44	29
Belly						
Length (mm)	116	77	73	54	46	40
Thickness (μ)	82	70	54	38	28	25

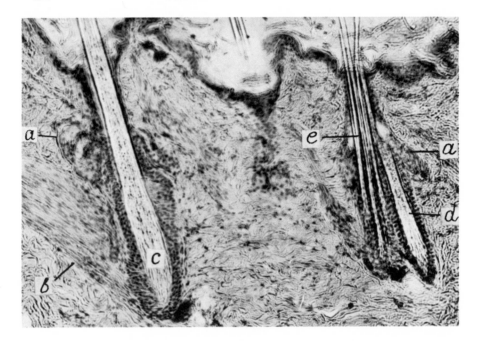

Fig. 127. *Felis ocreata*. Withers skin: *a*, sebaceous gland; *b*, arrector pili muscles; *c*, guard hairs; *d*, pile hairs; *e*, fur hairs.

to 40 mm), the fifth zone (30 to 35 mm), and the sixth zone (shorter than 30 mm) take in parts of the neck, head, and legs. There are 7,000 hairs per 1 cm^2 on the back and 1,950 on the belly. There are up to thirty fur hairs to one guard or pile hair.

Felis chaus

Hair indexes on the back and on the belly are given in table 89. There are five pile hair length zones (Tserevitinov, 1969). The first zone (50 mm and over) is on the back. The second zone (40 to 50 mm) surrounds the first, covers the sacrum, part of the sides, the lower back, and goes around to the breast. The third zone (30 to 40 mm) takes in the shoulder blades, part of the neck, sides and thighs. The fourth zone (20 to 40 mm) includes the neck, and the front and hind legs, and the fifth zone (shorter than 20 mm) comprises the feet and head. There are about 4,000 hairs per 1 cm^2 on the back and 1,700 on the belly. On the back there are twelve and on the belly four or five fur hairs to one guard or pile hair.

Felis lynx

The hair indexes are given in table 89. There are six pile hair length zones (Tserevitinov, 1969). The first, the long hair zone (80 to 90 mm), is on the belly. The second zone (70 to 79 mm) surrounds the first one, and covers the front part of the belly and lower neck. The third zone (60 to 69 mm) is a narrow band on the sides. The fourth zone (40 to 49 mm) occupies the whole back except the sacral region. The fifth zone (35 to 39 mm) is on the sacrum and thighs, and the sixth zone (shorter than 30 mm) is on the head and feet. There are 9,000 hairs per 1 cm² on the back, and 4,600 on the belly. There are twelve to thirteen fur hairs to one guard or pile hair.

Felis caracal

Material. Skin from withers of adult male that died on February 25 at the Planernaya Zoo Center, near Moscow.

The skin is thick (2.14 mm), the epidermis thin (28μ). The epidermis is almost wholly comprised of a loose corneous layer (21μ). The stratum malpighii is mostly one cell deep. In the dermis (2.11 mm), the papillary layer (0.68 mm) is thinner than the reticular layer (1.43 mm). The plexus of collagen fiber bundles in the dermis is dense (fig. 128). In the inner reticular layer the collagen bundles are horizontal and much looser. Occasionally they form a continuous layer. Large numbers of elastin fibers are found, but only in the inner reticular layer. There are no fat cells in the subcutaneous tissue. The sebaceous glands are small (28 × 62; 34 × 62μ) with single lobes, one at each hair. Sweat glands are not recorded, probably because it is winter. The arrectores pilorum muscles are thin (28μ).

Fig. 128. *Felis caracal.* Withers skin: *a,* pile hair; *b,* fur hair.

The hair follicles are resting. There are three categories of hair: guard hairs 17.5 mm long, 113μ thick, straight, with a weak granna down half the shaft and a well-developed medulla (76.8%) through the whole hair except base and tip. The pile hairs (first order) 14.9 mm long, and 88μ thick, have a more pronounced but shorter granna than the guard hairs (one-third of the hair length). The medulla is well developed (79.6%). The pile hairs (second order) differ from the first order in size (length 11.8 mm, thickness 56μ) and are slightly wavy in the shaft below the granna (3.8 mm, about a third of the hair length). The medulla is thick (78.6%). The third category hairs are peculiar. Short (5.6 mm) and straight, they thicken from base to middle then taper in the distal half toward the tip. At their thickest point (73μ) they are thicker than the second order pile hairs. The medulla (67.2%) is less developed than in the other categories. Tserevitinov (1959) gives different indexes from ours (table 89). There are about 2,500 hairs per 1 cm^2 on the back, and 1,000 on the belly (Tserevitinov, 1969).

Felis manul

The indexes for back and belly hairs are in table 89. The long hairs are on the back. There are 9,000 hairs per 1 cm^2 on the back, and only 800 on the belly (Tserevitinov, 1969). There are ten fur hairs to one guard or pile hair on the back and six on the belly.

Felis tigris

The tiger has anal glandular sacs (Schaffer, 1940). The indexes for the hairs are in table 89. The hairs are straight. Even the fur hairs are only slightly coarse and wavy (Tserevitinov, 1969). The yellow hairs of the tiger (except fur hairs) are shorter and thicker than the black. There are 2,500 hairs per 1 cm^2 on the back, and only 600 on the belly. There are only 1.4 fur hairs to one guard or pile hair.

Felis pardus

The hair indexes are in table 89. The black hairs are longer and thinner than the yellow hairs (Tserevitinov, 1969). There are 3,500 hairs per 1 cm^2 on the back. There are four fur hairs to one guard or pile hair.

Felis uncia

The hair indexes are in table 89. The longest hair is on the belly (Tserevitinov, 1969). There are about 4,000 hairs per 1 cm^2. There are eight fur hairs to one guard or pile hair.

Acinonyx jubatus

The hair indexes are in table 89. The hairs are mostly straight, or slightly wavy. The longest hairs are on the belly (Tserevitinov, 1969). *A. jubatus* has sparser pelage than other Felidae. There are 2,000 hairs per 1 cm^2 on the back, and 600 hairs on the belly (Tserevitinov, 1969). There are six fur hairs to one guard or pile hair on the back, and 2.5 to one on the belly.

ORDO PINNIPEDIA

There are a number of works on the pinnipeds, including a description of the skin structure and pelage of *Eumetopias jubatus* (Sokolov, 1959, 1960), *Callorhinus ursinus* (Sokolov, 1959, 1960a, b, c; Scheffer, 1962, 1965; Bychkov, 1968), *Odobaenus rosmarus* (Sokolov, 1960a, b), *Phoca vitulina* (de Meijere, 1894; Troll-Obergfell, 1930; Montagna and Harrison, 1957; Sokolov, 1960a, b; Mohr, 1952), *Pusa sibirica* (Belkovich, 1964b), *Pagophilus groenlandicus* (Bergensen, 1931; Belkovich, 1964b), *Cystophora cristata* (Belkovich, 1964a, b; Mohr, 1952), *Mirounga leonina* (Ling, 1965a, b, 1968).

We examined the skin of *E. jubatus, Callorhinus ursinus, Odobaenus rosmarus, Phoca vitulina, Pusa caspica,* and *Pusa sibirica.*

Pinnipedia's skin is thickest on the belly. The epidermis is usually thick and has either a regular inner surface or numerous alveoli (*O. rosmarus*). The epidermis is commonly multilayered: basale, spinosum, granulosum, lucidum, and corneum; and may be pigmented. The dermis is highly developed. The papillary and reticular layers meet on a level with the lower part of the sweat glands. In the papillary layer, thin bundles of collagen fibers run in many directions and form a compact plexus. There are only few elastin fibers in this layer. In the reticular layer, bundles of collagen fibers arranged at various angles form a compact plexus. Bundles of smooth muscle fibers in the reticular layer are characteristic, the upper border coinciding roughly with the border between the papillary and reticular layers, and they reach down to the subcutaneous fat tissue. Most of these bundles are horizontal or at acute angles to the surface. Thin in the upper reticular layer, they become much thicker nearer to the subcutaneous fat tissue. The network of elastin fibers is denser in the reticular layer than in the papillary. On the border with the subcutaneous fat tissue, the elastin fibers form vast and rather compact networks.

The well-developed subcutaneous fat tissue, consisting of large accumulations of fat cells interlayered with bundles of collagen fibers, is mainly responsible for the thickness of the skin. In these interlayers, elastin fibers are fairly numerous, and there are also some blood vessels. The fat cells are about the same size throughout the subcutaneous tissue. There are capillaries between the fat cells, but few large blood vessels. The big arteries are of the muscular type, even the smallest having a thick middle envelope and highly developed elastin elements. The sebaceous glands commonly lie in pairs at every hair tuft, the duct of one opening into a hair sheath of the pile hair, the other (the larger one) opening into the common sheath posterior to the tuft of fur hairs.

All pinnipeds have well-developed sweat glands (except in hairless skin areas) with twisted secretion cavities, and ducts opening into the hair bursae. Arrectores pilorum are absent. There are no sweat glands in the hairless skin. The pile, intermediate, and fur hairs are flattened and consist of a cuticle and a cortical layer. Only *C. ursinus* and *E. jubatus* have the medulla. The cuticle of pile and fur hairs is not usually ring-shaped. The hairs grow in tufts of one pile and several fur hairs. The number varies in different species and genera and, with hair indexes, serves as a species and genus marker (table 90). The arrangement of the hair in tufts is common to all, with fur hairs growing behind pile hairs. The pile hair bulbs lie deeper in the skin than the fur bulbs. The hairs have no arrectores pilorum muscles.

FAMILIA OTARIIDAE

The skin is thicker than in *Phocidae* but thinner than in *Odobaenidae*. The epidermis is medium thick (*E. jubatus*) or thin (*C. ursinus*) and slightly pigmented. In the dermis, the papillary layer may be thicker (*C. ursinus*) or thinner (*E. jubatus*) than the reticular. There are more bundles of collagen fibers in the reticular layer than in other *Pinnipedia*. The elastin fibers form a denser network than in other *Pinnipedia*. The sebaceous glands are not highly developed. The sweat glands are very large, their secretion cavities longer and more twisted than in *Phocidae,* but similar to *Odobaenidae.* The ducts of the sweat glands open into hair bursae over the sebaceous gland ducts, except in the limbs, where this relationship is reversed (Ling, 1965).

Unlike other *Pinnipedia, Otariidae* have a well-developed medulla. The number of fur hairs per tuft varies from three in *Eumetopias jubatus* to thirty in *Callorhinus ursinus*.

Eumetopias jubatus

Material. Skin from back over the forelimbs, the breast between them, the belly just before the anus, and the side of the body by the forelimbs of specimen (1), a young female (body length 143 cm) taken July 6, 1961, off Sakhalin Island, skin from the breast between the forelimbs and belly of specimen (2), a young female taken July 30, off the Kuril Islands (table 91).

The epidermis has pronounced layers despite a considerable pelage development: strata basale, spinosum, granulosum, lucidum, and corneum. The basal layer is formed of typical prism-like cells with large oval nuclei. The long axis of the cells and the nuclei is perpendicular to the skin. The stratum spinosum is formed of three to five polyhedral cell layers. Their nuclei are round or oval (the long axis is horizontal). In the stratum granulosum there

TABLE 90

Pinniped Hair Measurements (After Scheffer, 1964)

Species	Sex	Age, years	Pile hairs — Number of hairs per 1 cm²	Pile hairs — Maximum length (mm)	Pile hairs — Mean main diameter (μ)	Pile hairs — Ratio of diameters max/min	Fur hairs — Number of hairs in tuft, Min.	Fur hairs — Number of hairs in tuft, Max.	Fur hairs — Number of hairs in tuft, Mode	Fur hairs — Mean main diameter	Fur hairs — Ratio of diameters
Otaria byronia	F	13	530	11	76	1:9	1	3	1	10	1:5
Eumetopias jubatus	M	ad	380	31	73	1:5	1	4	1	11	4:7
Zalophus californianus	F	ad	540	9	104	1:9	1	4	1	8	2:3
Neophoca hookeri	M	ad	540	30	161	1:9	1	4	3	20	5:3
Arctocephalus forsteri	M	6-8	730	25	113	2:6	—	—	45	7	—
Callorhinus ursinus	M	12	840	24	102	2:1	—	—	53	11	—
Callorhinus ursinus	F	12	1,150	18	68	3:0	—	—	45	7	—
Odobaenus rosmarus	M	ad	180	7	146	1:4	0	1	0	15	2:7
Phoca vitulina	M	12-15	360	9	158	1:4	3	6	5	28	8:4
Pusa hispida	M	1	500	16	63	2:7	2	4	3	12	2:0
Halichoerus grypus	F	5	370	14	108	2:1	4	7	5	18	9:3
Histriophoca fasciata	M	ad	290	8	144	1:4	2	6	4	17	7:3
Pagophilus groenlandicus	M	ad	280	13	136	1:4	3	5	5	29	11:5
Erignathus barbatus	F	ad	350	10	71	2:0	3	8	7	13	10:7
Monachus schauinslandi	M	20	450	7	125	1:9	0	0	0	0	0
Lobodon carcinophagus	M	1	940	11	137	1:9	1	4	2	13	2:4
Ommatophoca rossi	?	?	1,030	13	83	2:8	0	2	1	14	3:3
Hydrurga leptonyx	F	?	990	8	180	1:6	1	1	1	30	1:9
Leptonychotes weddelli	F	ad	340	13	91	1:7	2	5	3	31	5:4
Cystophora cristata	?	ad	230	15	175	1:5	4	6	5	20	7:7
Mirounga angustirostris	M	3	370	7	204	1:3	0	0	0	0	0

are one to two layers of flattened cells. The stratum lucidum is very thin. The stratum corneum is well developed and is thickest on the belly. The epidermis is slightly pigmented. The cells of the stratum basale contain the most pigment. The inner surface of the epidermis has small processes.

Bundles of collagen fibers, which form a compact plexus in the dermis, lie in all directions with horizontal bundles predominating only in the deep inner dermis, where they border on the subcutaneous fat tissue. The collagen fiber bundles in the papillary layer get thicker inward from the epidermis, remaining much the same all over the body. Numerous elastin fibers are mainly horizontal or oblique and also run along the hair bursae, sweat, and sebaceous glands. The fat cells are found only near the sweat glands (only on the back in specimen 1). The network of blood vessels is highly developed, though no large vessels are found.

In the reticular layer the largest bundles of collagen fibers are on the back. The network of elastin fibers is denser in the reticular than in the papillary layer, and the fibers are much thicker. In the upper reticular layer the fibers lie in various directions (fig. 129), but in the inner part they are mainly horizontal, thick, and grouped into loose bundles. There are no fat cells in the reticular layer except in the sample from the belly of specimen (2). In the

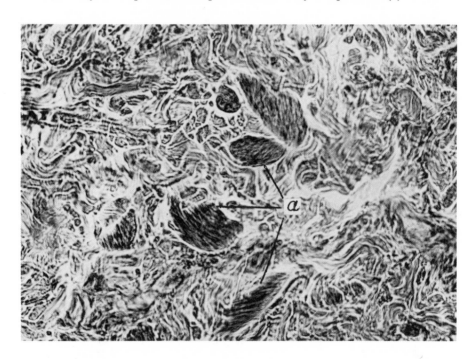

Fig. 129. *Eumetopias jubatus.* Network of collagen fiber bundles in reticular layer: *a,* bundles of smooth muscle fiber.

TABLE 91

EUMETOPIAS JUBATUS: SKIN MEASUREMENTS (MM)

Indexes	Skin	Epidermis	Stratum corneum	Dermis	Papillary layer	Reticular layer	Subcutaneous fat tissue	Zone of smooth muscle bundles
Specimen (1)								
Back	30.7	0.145	0.07	6.3	1.8	4.5	24.3	4.4
Breast	33.5	0.225	0.08	8.4	2.6	5.8	24.9	6.1
Belly	46.1	0.14	0.085	7.1	2.3	4.8	38.9	5.1
Side	23.3	0.11	0.055	9.9	2.8	7.1	13.3	5.2
Specimen (2)								
Breast	24.5	0.11	0.05	5.4	2.1	3.3	19.0	3.5
Belly	17.2	0.17	0.085	6.2	2.1	4.1	10.9	4.7

inner papillary layer, thin (up to 45μ) bundles of smooth muscle fibers are found arranged at an acute angle to the surface, becoming less steep toward the inner side, where both size and number increase considerably (fig. 129). However, the inner reticular layer (0.8 to 1.3 mm) contains practically no smooth muscles and consists only of tightly interwoven bundles of collagen fibers and thick elastin fibers. The thickest bundles of muscle fibers (420μ) are found on the back—in other parts of the body they are considerably thinner (55 to 130μ). The subcutaneous tissue cells are of the same size (45 x 90; 72 x 108μ) in all the samples, only slightly bigger than the dermal fat cells. Numerous blood vessels cross the subcutaneous tissue in various directions.

In the subcutaneous fat tissue thick bundles of nerve fibers branch obliquely toward the surface, straightening to the horizontal as they approach the inner side of the dermis. In the inner dermis are thick nerve bundles made up of thick, medium, and thin nerve fibers, usually beside blood vessels. Nerve terminals branch dichotomously and become much thinner as they approach the epidermis. Tiny fibers penetrate between the epidermal cells and the lower layers of the epidermis. The receptors also approach the hair sheaths, sebaceous, and sweat glands. The sebaceous glands are large with single lobes and are located singly posterior to the hairs (fig. 130). The sweat glands are well developed and have a slightly twisted upper and a more twisted lower cavity (fig. 131).

The pelage consists of coarse, straight, smooth pile hairs and very sparse, thin, short fur hairs. The longest fur is on the neck and shoulders, gradually decreasing toward the tail. Fur is almost absent on the limbs. The hair tufts consist of one pile and one or two, occasionally three (and sometimes four, Scheffer, 1964) fur hairs. The hairs grow at an angle of 45°. The pile hairs (up to 31 mm long; Scheffer, 1964) are lancet-shaped (table 91). The thickened distal two-thirds has a continuous medulla, which thins in the lower section and disappears at the root. There is no medulla in the tip. The pile hair cuticle is not ring-shaped. The borders of the cuticle cells have irregular outlines and are sinuate. Near the tip, the cuticle cells are largely transverse. There are 380 pile hairs per 1 cm² (Scheffer, 1964). The cuticle of the fur hairs is usually not ring-shaped, the outline of the cells is not sinuate, and the cells themselves are tall. There is no medulla in the fur hairs, which are flattened (table 83).

Callorhinus ursinus

Material. Skin from the back above the forelimbs. Specimens (1), (2) and (3) are two-, three-, and four-year-old bachelors; specimen (4) is a six-year-old male; specimen (5) is an adult male over six years.

The epidermis is badly preserved in all samples, so it is impossible to find details of the layers. The stratum corneum alone is clearly visible. The inner

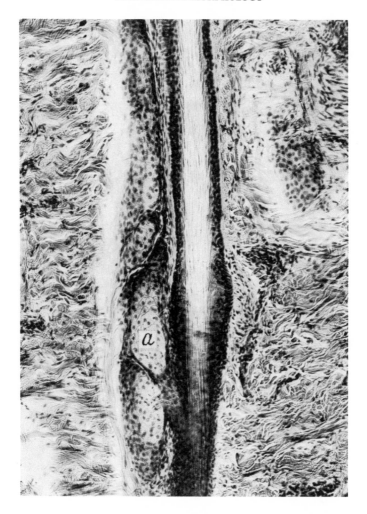

Fig. 130. *Eumetopias jubatus*. Back skin: *a,* sebaceous gland.

structure of the epidermis is completely smooth. There is no pigment in the epidermis (Sokolov, 1959, 1960). In the upper parts of the papillary layer the bundles of collagen fibers are thin, mostly horizontal, and slightly twisted. Elastin fibers, relatively thick at this level, lie along the bundles of collagen fibers; in the middle papillary layer most of the collagen bundles and elastin fibers are perpendicular to the hair sheaths, with some oblique bundles of collagen and elastin fibers arranged along the hair sheaths. In the lower parts of the papillary layer, bundles of collagen fibers rise to the surface at various angles, continuing to twist like worms, but considerably thicker. The elastin fibers also become thicker. Fat cells, found only in the

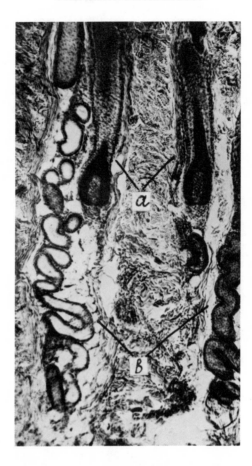

Fig. 131. *Eumetopias jubatus*. Back skin: *a,* hair
follicle bulbs; *b,* sweat glands.

lower papillary layer, are few in number and are grouped between the curves
of the sweat glands.

The numerous blood vessels that run through the outer dermis are small,
their number decreasing greatly with depth, though larger vessels do occur.
In the reticular layer of the dermis, most of the collagen fiber bundles lie at
various angles to the surface (fig. 132); some are horizontal or rise steeply
upward. Their thickness is the same or slightly greater in the inner papillary
layer. The bundles continue to be twisted. The very numerous elastin fibers
mostly run along the side of collagen bundles, increasing in size and number
as they approach the subcutaneous fat tissue near which they are loosely
grouped (fig. 132).

There are no fat cells in the reticular layer and few blood vessels, though
these are rather large. The inner reticular layer adjoining the subcutaneous

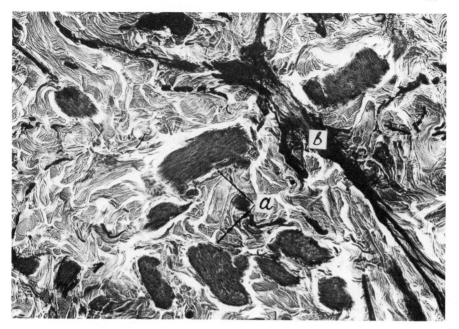

Fig. 132. *Callorhinus ursinus.* Network of collagen fiber bundles in reticular layer of adult back skin: *a,* smooth fiber bundles; *b,* elastin fiber bundles.

fat tissue has the most compact plexus. Its upper border protrudes into the inner papillary layer of the dermis. The lower border almost reaches the subcutaneous tissue. The bundles of smooth muscle fibers are either horizontal or at an acute angle to the surface. Most of the bundles are wavy in shape, and their thickness increases with depth (fig. 132). The subcutaneous fat tissue is absent in the samples. The sebaceous glands are few and small. Their ducts open into the upper section of the pile hair sheaths. The sweat glands are relatively better developed than in other *Pinnipedia* (fig. 133). They have a very well-developed, twisted secretion cavity and straight ducts that open into the hair follicles.

Sex and age dimorphism is pronounced in the pelage. There are three categories of hair: pile, intermediate, and fur. The pile hairs have a granna in the upper third of the shaft. The medulla is highly developed in the lower section of the hair. It disappears at the beginning of the granna and reappears only at the very end. The medullary cells are large (fig. 134). Their membranes have a wavelike pattern. The fibers branch out dichotomously to form feltlike structures at the cell wall surfaces. The cuticle is not ring-shaped. The cuticle cells in the lower section of the hair are spear-shaped and firmly flattened in the thickened and distal sections. The intermediate hair resembles the pile hair in shape and structure.

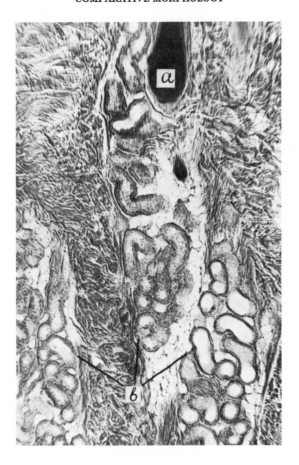

Fig. 133. *Callorhinus ursinus.* Back skin: *a,* hair bulb;
b, sweat glands.

The fur hair has a fairly uniform thickness. There is no medulla. The cuticle cells are spear-shaped, pointing upward. The hairs grow in tufts of one pile, two or three intermediate, and ten to thirty fur hairs (or 45 to 53; Scheffer, 1964) (some tufts have no intermediate hairs) (fig. 135). Every hair tuft has a common upper epidermal sheath, while deeper down each hair in the tuft has a separate sheath.

The thickness of the epidermis increases greatly with age. In the adult male it is four times as thick as in the two-year-old male. It is largely stratum corneum that thickens whereas stratum germinativum increases in thickness but negligibly. While in the two-year-old male the stratum corneum (10μ) is only 22.2% of the epidermis thickness. In the three-year-old male it is 42.8% (27μ). In the four-year-old it is 57.1% (36μ). In the five- to six-year-old male it is 66.6% (90μ), and in the adult male it is 75% (135μ).

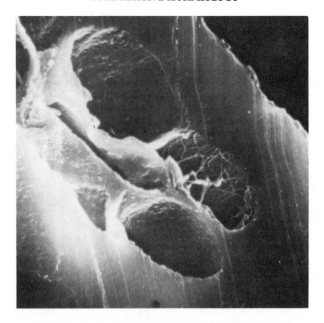

Fig. 134. *Callorhinus ursinus.* Cells of medulla.

The dermis, too, thickens with age. As the lower parts of the reticular layer were cut off when the sample was taken, only changes in the papillary layer can be traced. In four age groups of seals (2-, 3-, 4-, and 5- to 6-year-old males) the papillary layers differ very little in thickness (table 92). In the five- to six-year-old male it is only 0.9 mm thicker than in the two-year-old male. But in the adult male the papillary layer is 6.4 mm thicker than in the five- to six-year-old male. The pattern of the collagen fiber plexus in the dermis is the same for all ages, though the thickness of the bundles in the inner dermis increases with age. In two- and three-year-old males, the bundles are 27μ thick; in the four-year-old male, 45μ; in the five- to six-year-old male, 72μ; and in the adult male, 92μ.

The elastin fibers thicken considerably with age, but decrease in numbers. The changes in the first four age groups are not striking, but the adult male differs a great deal. The slightly developed network of thick elastin fibers completely contradicts the data of Zubin and Pchelina (1951) on high development of the elastin-fiber network in the adult male compared with young animals.

The fat cells rarely differ in size or numbers in different age groups, except for the three-year-old with no fat cells. The plexus of smooth muscle fiber bundles in the dermis in the first four age groups is more or less equally developed, but it is sparser in the adult male than in the younger animals. The thickness of the bundles increases with age (table 92). The sebaceous

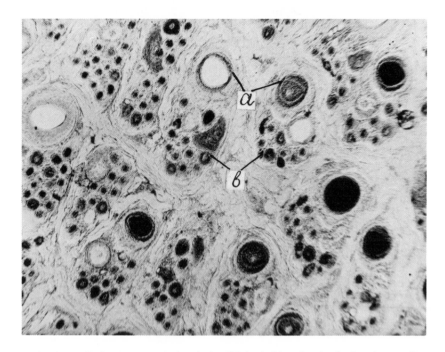

Fig. 135. *Callorhinus ursinus.* Hair tufts in adult back skin at longitudinal section: *a,* pile hair; *b,* fur hair.

glands in animals of all ages are poorly developed. The five- to six-year-old male and the adult male have more and bigger sebaceous glands than the young seals, but the sweat glands are equally numerous at all ages, though their size increases considerably with age, as does the diameter of the secretion section. In a two-year-old male, it is 18μ; in a three-year-old, it is 27μ, in a four-year-old and a five- to six-year-old male, it is 45μ; and in an adult male, it is 54μ. The depth at which the hair bulbs lie and the length of the hair roots increase considerably with age.

The pile hair length increases unequally in males (Scheffer, 1962; Bychkov, 1968) (table 93). In gray cubs the pile hairs are longer over the whole body than in bachelors of up to four years. In bachelors under five years the average pile hair length differs insignificantly. It becomes much longer on the withers and back and a bit longer on other parts of the body when they are five years old. In the gray cubs the average length for different parts of the body varies little, and in the two- to four-year-old bachelor group, the mean length of pile hair shafts is only slightly longer on the belly. In males of five years and over, the average length varies greatly in different parts of the body.

The length of the fur varies less in the *C. ursinus* males (Scheffer, 1962;

TABLE 92

CALLORRHINUS URSINUS: SKIN MEASUREMENTS ON THE BACK OVER THE FORELIMBS

Age	Thickness					Zone of smooth muscle bundles (mm)	Thickness of smooth muscle bundles (μ)
	Epidermis (μ)	Stratum corneum (μ)	Dermis (mm)	Papillary layer (mm)	Reticular layer (mm)		
2-year-olds	45	10	2.9	1.5	1.4	1.5	35
3-year-olds	63	27	3.3	2.1	1.2*	1.3*	45
4-year-olds	63	36	4.2	2.1	2.1	2.1	45
Semimature	135	90	6.6	2.4	4.2	4.0	80
Mature	180	135	15.5	8.8	6.7	8.0**	90

*The lower border of smooth muscles in the mature animal is approximately 1 mm short of the subcutaneous fat tissue.

**The lower layers of the reticular layer are not in the sample.

Bychkov, 1968) (table 94). The mean length of the fur hairs is similar in different age groups and different parts of the body, excepting the belly, where the fur hairs are 30 to 40% shorter in all age groups. Females have shorter pile and fur hairs than males on withers and back (Bychkov, 1968) (table 95). These differences are revealed as early as in gray cubs. Other age and topographical variables in the pelage are little different from those in males.

There are no differences in bachelors taken on Tjuleny, Sakhalin, Komandorskie, and Pribilof Islands (Bychkov, 1968). A medium adult female has about 52 hairs per tuft, and 303,000,000 hairs on the whole fur-covered area of 0.62 m² (Schaffer, 1964). In both *C. ursinus* and *Eumetopias jubatus* a molt is recorded during embryonic development (Belkin, 1964). In the seven-month *C. ursinus* embryo, the initial pelage and skin glands begin to be replaced by new pelage and glands, the process taking two to two-and-a-half months. This is accompanied by a loss of stratum corneum and a shrinkage in stratum spinosum.

FAMILIA ODOBAENIDAE

The skin is very thick. The epidermis is thicker in absolute terms and in percent relative to the skin thickness than in *Otariidae* or *Phocidae*. The inner surface of the epidermis is alveolar. The stratum corneum is very well developed. The epidermis is not pigmented. In the dermis, the reticular layer is thicker than the papillary. The collagen fiber bundles are thicker than in any other *Pinnipedia* under investigation, and bundles of smooth muscles and elastin fibers in the dermis of the walruses are least numerous. The absolute thickness of the subcutaneous fat tissue in walrus is the thickest, and the percent relative to the thickness of the skin is the least. The sebaceous glands are small. The sweat glands are very large but few.

In adult specimens the hair is sparse, with only one to three fur hairs per tuft and no medulla in the hairs. In the adult male there are characteristic skin bulges in the cervicothoracic region, with a very durable plexus of collagen bundles in the dermis, a thick epidermis, and a compact stratum corneum.

Odobaenus rosmarus

Material. Skin from belly near the navel, breast near the forelimbs, and the bulge on the neck of specimen (1), a male (body length 366 cm). Skin from belly near the navel, breast, and back of specimen (2), a young male (body length 232 cm). Skin from breast of specimen (3), a female (body length 310 cm) (table 96), all taken late September, early October, 1951, in the Chukotski Sea.

The skin is ventrally thickest (Sokolov, 1960). In accordance with published data, adult (especially old) males have the thickest skin in the cervico-

TABLE 93

CALLORHINUS URSINUS: LENGTH OF PILE HAIRS (MALES; JUNE-JULY, TJULENY ISLAND) (BYCHKOV, 1968)

Age in years	n	Withers		Back		Sacral region		Side		Belly	
		M	Lim	M	Lim	M	Lim	M	Lim	M	Lim
½	6	20.2±0.44	19-22	18.6±0.57	16-20	18.0±0.41	16-19	18.2±0.49	16-20	15.2±0.49	13-17
1	5	17.7±0.43	16-19	18.2±0.52	16-19	15.7±0.10	15-17	15.6±0.07	15-16	13.0±0.94	11-16
2	25	19.3±0.28	16-22	16.7±0.26	15-19	14.9±0.26	12-18	15.9±0.19	15-18	10.3±0.19	8-13
3	67	19.7±0.20	16-25	16.9±0.18	13-20	14.9±0.13	12-18	15.8±0.12	14-18	10.2±0.14	7-13
4	39	22.8±0.28	18-30	18.1±0.23	15-21	15.4±0.18	13-19	16.3±0.21	13-20	10.5±0.22	7-13
5	18	26.2±1.04	21-36	19.2±0.29	17-21	15.8±0.22	15-17	17.0±0.24	15-18·	10.9±0.29	9-14
6	13	30.7±1.12	24-41	19.5±0.46	17-22	16.2±0.36	14-19	17.2±0.44	15-20	11.0±0.30	9-13
7	6	38.6±1.75	32-44	24.6±0.43	20-29	17.2±0.70	15-20	18.8±0.64	17-21	13.2±0.54	11-16
Over 7	5	58.8±4.53	42-70	27.5±2.70	20-37	17.7±0.27	17-19	18.5±0.71	17-21	13.0±0.79	10-15

TABLE 94

CALLORRHINUS URSINUS: LENGTH OF FUR HAIRS (JUNE-JULY, TJULENY ISLAND) (BYCHKOV, 1968)

Age in years	n	Withers		Back		Sacral region		Side		Belly	
		M	Lim	M	Lim	M	Lim	M	Lim	M	Lim
½	6	14.6±0.37	13-16	14.0±0.41	12-15	13.3±0.37	12-15	13.4±0.50	11-15	10.4±0.26	9-11
1	5	14.0±0.45	12-15	13.5±0.59	11-15	12.0±0.79	10-14	13.0±0.63	11-15	8.5±0.05	8-9
2	25	14.9±0.19	13-17	13.6±0.16	12-15	11.6±0.29	8-14	12.8±0.21	10-14	7.1±0.28	5-9
3	67	14.7±0.13	12-17	13.4±0.16	10-16	11.6±0.16	9-16	12.4±0.14	10-15	7.0±0.19	4-10
4	39	15.4±0.19	13-19	13.9±0.20	10-16	11.5±0.19	9-15	12.3±0.18	10-15	6.7±0.23	3-9
5	18	15.5±0.36	13-20	14.2±0.22	12-15	11.0±0.33	9-13	12.7±0.28	11-15	7.2±0.31	5-10
6	13	14.8±0.30	13-17	14.2±0.38	11-16	12.8±0.39	11-15	13.6±0.33	12-16	7.2±0.36	5-9
7	6	11.8±0.72	10-15	13.6±0.51	12-16	12.6±0.62	10-15	13.2±0.55	11-15	7.4±0.41	6-9
Over 7	5	18.5±1.53	13-22	15.0±1.22	10-18	11.2±0.65	9-13	12.7±0.65	11-15	7.5±0.59	5-9

TABLE 95

CALLORRHINUS URSINUS: LENGTH OF PILE AND FUR HAIRS (FEMALES; JUNE-JULY, TJULENY ISLAND) (NYCHKOV, 1968)

Age in years	n	Withers		Back		Sacral region		Side		Belly	
		M	Lim	M	Lim	M	Lim	M	Lim	M	Lim
½	2	18.5±1.06	17-20	17.5±0.35	17-18	18.5±0.35	18-19	16.5±1.06	15-18	14.0±0.70	13-15
1	3	18.0±0.15	18-20	17.7±0.28	17-18	18.0±0.15	17-19	17.7±0.55	17-19	13.7±0.28	13-14
2	2	17.5±0.35	17-18	16.5±0.35	16-17	16.0±0.70	15-17	16.5±0.35	16-17	10.5±0.35	10-11
3	9	18.9±0.48	17-22	16.5±0.06	17-17	16.4±0.33	15-18	16.1±0.53	14-19	11.6±0.30	10-13
4	18	19.5±0.39	16-23	17.3±0.36	15-21	16.2±0.31	13-19	15.5±0.30	13-18	10.5±0.33	8-13
5	20	20.7±0.29	17-23	17.7±0.28	15-20	15.8±0.23	13-18	16.2±0.30	14-20	10.2±0.27	8-13
6	7	22.7±0.72	20-26	19.3±0.33	18-21	17.9±0.47	16-20	17.7±0.44	16-19	13.9±0.37	12-15
7	4	22.7±1.51	19-27	19.0±0.62	17-20	18.0±0.36	17-18	17.3±0.54	16-19	12.7±0.36	11-13
Over 7	5	23.5±1.18	21-28	20.5±0.59	18-22	15.3±0.76	12-17	13.4±0.61	11-15	11.0±0.28	10-12
½	2	13.5±1.06	12-15	13.0±0.00	13	12.0±1.00	11-13	12.5±0.35	12-13	9.0±0.70	8-10
1	3	13.7±0.55	13-14	13.0±0.39	12-14	12.7±1.00	11-15	11.7±0.04	11-12	9.3±0.73	8-10
2	2	15.0±0.00	15	14.5±0.35	14-15	13.5±0.35	13-14	13.5±0.35	13-14	7.5±0.35	7-8
3	9	14.0±0.09	13-16	14.3±0.43	12-16	12.4±0.31	11-14	12.1±0.47	11-16	7.3±0.21	6-8
4	18	14.2±0.30	11-16	13.5±0.29	11-15	11.2±0.06	10-13	12.3±0.27	10-14	6.8±0.32	5-10
5	20	12.3±0.17	11-14	11.1±0.15	10-12	10.6±0.18	9-12	10.5±0.29	8-13	7.3±0.34	4-10
6	7	11.8±0.59	10-15	12.3±0.39	11-14	10.3±0.33	9-12	11.0±0.35	10-13	7.3±0.44	5-9
7	4	14.3±0.89	12-17	13.5±0.43	12-14	11.7±0.07	11-12	11.3±0.41	10-12	5.5±0.56	4-7
Over 7	5	11.2±0.33	10-12	10.8±0.52	10-13	8.8±0.69	6-11	10.7±0.59	9-13	5.7±0.53	5-6

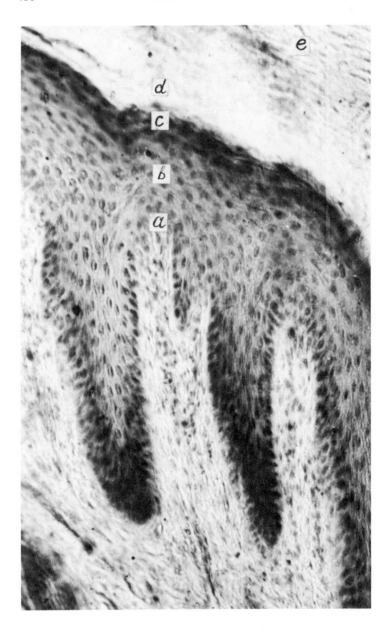

Fig. 136. *Odobaenus rosmarus*. Epidermis of back skin layers: *a,* basale; *b,* spinosum; *c,* granulosum; *d,* stratum lucidum; *e,* stratum corneum.

TABLE 96

ODOBAENUS ROSMARUS: SKIN MEASUREMENTS (MM)

	Skin	Epidermis	Stratum corneum	Dermis	Papillary layer	Reticular layer	Subcutaneous fat tissue	Zone of smooth muscle bundles
Specimen (1)								
Belly	72.0	0.88	0.44	32.0**	8.2	23.8	39.2	23.8
Breast	88.0	1.10	0.44	28.3***	7.0	21.3	58.5	20.7
Near limb*	—	3.30	2.70	14.2***	—***	—***	—*	13.6
"Knot" on neck*	—	1.76	0.44	50.4***	8.6	41.8	—*	31.2
Specimen (2)								
Belly	56.4	1.10	0.55	19.2***	5.5	13.7	36.1	14.3
Breast	74.0	0.66	0.33	25.2***	6.9	18.3	48.2	19.0
Back	53.6	1.05	0.53	18.7***	5.3	13.4	33.9	14.6
Specimen (3)								
Breast	126.8	0.77	0.55	25.4	4.9	20.5	100.7	12.3
Back	56.2	0.55	0.33	22.2**	5.8	16.4	33.5	15.2

*Without subcutaneous fat tissue.
**Without dermal papillae.
***The dermis does not divide into layers.

thoracic area, which is covered with bulges. Epidermal thickness varies widely in the three specimens (table 96). The stratum basale, stratum spinosum, stratum granulosum, stratum lucidum, and stratum corneum can be clearly distinguished in the epidermis (fig. 136). The stratum corneum accounts for 50% of epidermal thickness in most of the specimens, and it is several times thicker near the forelimbs. The septa of the epidermal alveoli are formed by the stratum basale and the stratum spinosum cells. In a horizontal section the alveoli have varied outlines.

The dermis is thickest on the ventral part of the walrus body and on the "bulge." The various body parts of all three walrus specimens vary little in dermal thickness. The papillary layer subdivides into two parts: the dermal papillae and the subpapillary layer under it. In the papillae, very thin bundles of collagen fibers rise steeply upward, as do very fine elastin fibers. According to the season, there may be numerous fat cells in the dermal papillae or none at all. Blood vessels come near the apexes of the papillae. The height of the dermal papillae varies widely in different body parts and in different animals, both in absolute and relative terms. No pattern has been found in the distribution of the highest papillae, except that they are highest in the bulges. In the outer and inner papillary layers the collagen fiber bundles twist like worms and run toward the surface at various angles, while in the middle they are mainly perpendicular to the hair sheaths. They are thickest in the papillary layer in the bulge and near the forelimbs in specimen (1) (220μ). The fat cells, which vary with the season, may be found only around the sweat glands, or they may be the main component of the papillary layer, as in specimen (3). They are about the same size (60 x 70; 80 x 110μ) in all the samples. There are many small blood vessels in the papillary layer.

The reticular layer of the neck bulge of specimen (1) has the greatest absolute thickness. In specimens (2) and (3) the reticular layer is thicker on the belly than on the back. The relative thickness of the reticular layer (percentage of dermis) is about the same in all samples. Bundles of collagen fibers in the reticular layer form a compact plexus in all directions, but in the inner dermis they are mainly horizontal. The thickness of the bundles varies in different animals but is greatest in the bulge in specimen (1). There are a few elastin fibers in the reticular layer, the number increasing where it meets the subcutaneous fat. They form loose bundles arranged at various (mostly acute) angles to the surface. Fat cells, which are present only in the upper reticular layer, are arranged under the sweat glands in strands extending down into the subcutaneous fat tissue. They are only slightly bigger than those in the papillary layer. They are, on the whole, similar in all the samples.

The network of collagen fiber bundles, 80 to 120μ thick, is poorly developed in the dermis. In most of the samples its upper border lies deep in the upper reticular layer. Blood vessels are numerous but small in the reticular

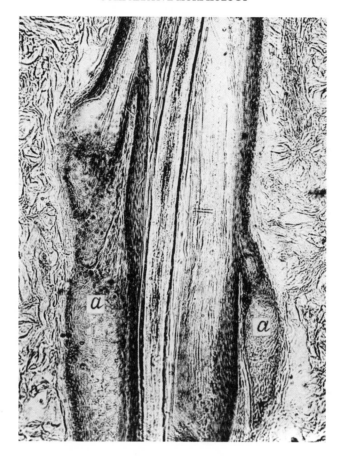

Fig. 137. *Odobaenus rosmarus.* Belly skin: *a,* sebaceous glands.

layer. The subcutaneous fat tissue, in absolute terms and as relative percentage of the skin, is thickest on the ventral side of the body. The subcutaneous tissue, which becomes thinner from head to tail, is obliquely interspersed with bundles of collagen fibers. These vary in thickness in the samples, but are always thicker on the ventral side. In the subcutaneous fat tissue, the fat cells are larger than in the dermis, and uniformly so in all samples (90 x 145; 110 x 145μ). The sebaceous glands are small and stretch along the hair bursae (fig. 137). The sweat glands are highly developed, with very twisted secretion cavities (fig. 138). Some sweat glands in the skin of specimen (3) are variably shrunken (fig. 139), their lumen being much reduced and their cells and nuclei very flattened, though in some of the glands their ducts remain wide. Pile and fur hairs grow in tufts of one pile and one or two fur hairs. The cuticle is not ring-shaped.

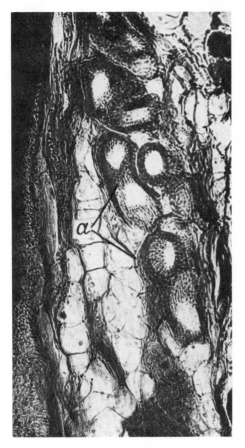

Fig. 138. *Odobaenus rosmarus.* Back skin:
a, sweat gland.

The skin of specimen (3) differs considerably from the other two in that it has huge quantities of fat cells in the dermal layer. Almost the whole space between the epidermal alveoli is filled with minute fat cells (fig. 140). The collagen fiber bundles in the papillae are very thin and far apart. The papillary layer of the dermis contains large accumulations of fat cells and a very loose plexus of collagen bundles, with the density of the plexus increasing somewhat in the reticular layer.

The skin on the neck bulge in specimen (1) has a thicker epidermis and somewhat higher dermal papillae than the skin on other parts of the body. But the main distinction is a very thick dermal layer with a much more compact collagen bundle plexus than in other parts of the body. There are fat cells in the reticular layer, but only in small numbers. Elastin fibers and bundles of smooth muscle fibers are few.

The skin near the forelimbs of specimen (1) has neither glands nor hair. The epidermis, especially the stratum corneum, is very thick. There is no papillary and reticular layer division in the dermis. The upper dermis, 1.2 mm thick, has no fat cells, and collagen fiber bundles form a compact plexus. Nearer the subcutaneous fat tissue a few fat cells appear, which then increase and form large accumulations.

Comparing the skin of the young specimen (2) (body length 232 cm) with the more mature specimen (1) (body length 365 cm) indicates that the dermal layer and the subcutaneous fat tissue become thicker with age. This is confirmed by the following indexes (table 97).

The relative thicknesses of the papillary layer and the smooth muscle zone in the dermis (as percentages of the dermis) is greater in the young than in the adult animal, and the collagen fiber plexus is better developed.

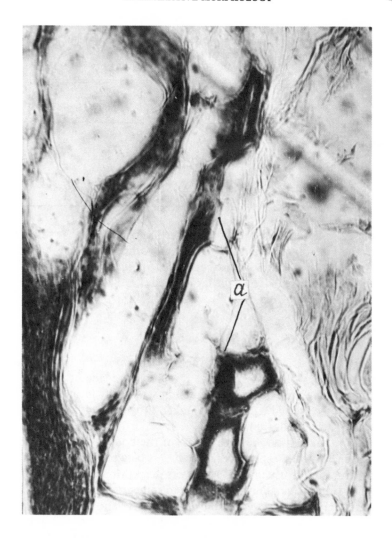

Fig. 139. *Odobaenus rosmarus.* Specimen 3: *a,* sweat gland.

FAMILIA PHOCIDAE

The absolute thickness of the skin is less than in other *Pinnipedia.* The papillary layer is thicker than the reticular layer. For development of the smooth muscle bundles and elastin fibers, *Phocidae* is next to *Otariidae.* The relative thickness of the subcutaneous fat tissue (as percentage of the skin) is greater in *Phocidae* than in other *Pinnipedia,* and the fat cells in the subcutaneous fat tissue are a little larger than in *Otariidae* and *Odobaenidae.*

The sebaceous glands are very large, while the sweat glands are less devel-

TABLE 97

MEAN THICKNESS (IN MM) OF DERMIS AND SUBCUTANEOUS FAT TISSUE IN VARIOUS AGE-SEX GROUPS
OF O. ROSMARUS (CHAPSKY, 1936)

Age	Dermis	Subcutaneous fat tissue
Under one year	8	26
Yearlings	13.5	37
Older but still immature	20	47
Mature males	26	64
Mature females	22	54.5

oped than in other *Pinnipedia,* the lower section penetrating the skin slightly beyond the hair bulbs. The sweat gland ducts open into the hair bursae below the sebaceous gland duct (except in *Pagophilus groenlandicus,* where the position is reversed). The hair has no medulla. The number of fur hairs in a tuft ranges from five to ten in various species.

Phoca vitulina

Material. Skin from the back over the forelimbs, the back near the sacrum, the side of the body near the forelimbs, the side of the body near the hind limb, the chest between the forelimbs, the belly just before the anus of a young male (body length 135 cm) taken October 1, 1951, on Shikotan Island, the Kurils (table 98).

The skin is thickest on the ventral part of the body, as are the epidermis and stratum corneum (especially on the breast between the forelimbs). The stratum spinosum is usually ten cell layers deep (Montagna and Harrison, 1957). The stratum lucidum is very thin and sometimes cannot be seen at all. The stratum granulosum is found only on the belly. The stratum corneum is rather compact. The pigmentation of the epidermis is heavy or moderate. The inner surface of the epidermis is slightly wavy, so that small epidermal processes protrude into the dermis. The stratum corneum contains a large number of sulfhydryl groups (Montagna and Harrison, 1957).

Thin collagen fiber bundles form a compact plexus in the outer papillary layer. They are arranged at various angles with very thin, oblique elastin fibers, no fat cells, and numerous capillaries running along the hair bulbs and sweat glands. But most are twisted like worms and are horizontal. The bundles are about the same thickness in all the parts of the body (30 to 45μ), and large elastin fibers lie alongside them. A few fat cells (65 × 80; 90 × 90μ) similar in size in all samples, occur in the curves of the sweat glands. There are none in the papillary layer of the abdomen on the back near the sacrum.

The reticular layer gets thicker toward the tail. Two parts are clearly dis-

TABLE 98

PHOCIDAE: SKIN MEASUREMENTS

Indexes	Skin (mm)	Epidermis (μ)	Stratum corneum (μ)	Thickness Dermis (mm)	Thickness Papillary layer (mm)	Thickness Reticular layer (mm)	Subcutaneous fat tissue (mm)
Phoca vitulina							
Back over forelimbs	37.0	90	30	5.2	3.2	3.0	31.8
Back at sacral region	25.2	100	50	6.1	2.4	3.7	19.0
Side at forelimbs	31.0	90	30	4.9	2.4	2.5	26.0
Flank	29.0	110	50	5.7	2.3	3.4	23.2
Breast between forelimbs	40.0	150	100	5.2	2.9	2.3	34.7
Belly at anus	50.0	100	50	5.6	2.7	2.9	44.3
Pusa sibirica							
Withers	23.1	138	96	2.65	1.86	0.79	20.34
Breast	39.9	92	50	3.45	2.79	0.66	36.35
Pagophilus groenlandicus (after Belkovich, 1964)							
Side	—	88-181	16-27	5.4	1.6	3.8	30-70
Cystophora cristata (Belkovich, 1964)							
Side	—	87-187	44-132	14.8	3.8	11.0	35-45

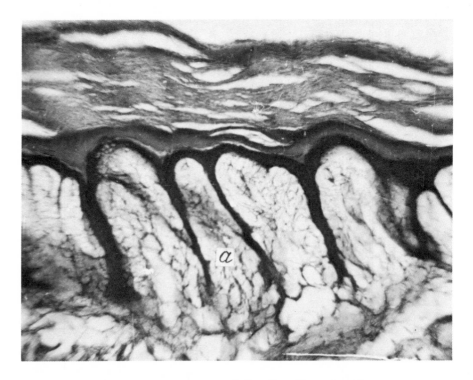

Fig. 140. *Odobaenus rosmarus*. Dermal papillae: *a,* fat cells.

tinguishable—the outer, which is looser and contains vertical strands of fat cells extending from the lower parts of the sweat glands, and the inner, which is more compact. Wavy bundles of collagen fibers run at all angles to the skin surface, mainly in small bundles (35 to 60μ in diameter). Many thick elastin fibers are arranged at acute angles, and bundles of smooth muscles appear in the lower parts of the papillary layer, running to the skin surface at very sharp angles. They are thickest in the middle of the reticular layer. Closer to the subcutaneous fat tissue, there are fewer smooth muscles. These, however, retain their thickness. The thickest bundles are lateral and on the abdomen (100 to 200μ). The smooth muscle fiber zone thickens caudally and attains its maximum on the back near the sacrum. The reticular layer fat cells are a little bigger than the papillary layer fat cells and are similar to all other samples. Mast cells are not recorded in the dermis (Montagna and Harrison, 1957).

The subcutaneous fat tissue is typical in structure, being thickest on the ventral side of the body. On the dorsal and lateral parts of the body the absolute and relative (% of skin) thickness of the subcutaneous fat tissue decreases caudally and increases abdominally. The largest collagen bundles are on the lateral and abdominal sides of the body (120 to 235μ). Fat cells, of uniform size in all the samples, are somewhat larger than in the dermis (80 × 120; 110 × 165μ). There are not many large blood vessels, but a huge number of capillaries run between the fat cells. There are thick bundles of nerve fibers near the blood vessels. In the dermis, the nerve bundles become thinner and follow the blood vessels. Fibers of various thickness form the bundles. Receptors coming up to the epidermis, hair sheaths, and sebaceous and sweat glands are seen on the preparations. The sweat glands are less innervated than the sebaceous glands. There are multilobal sebaceous glands, two and three at every hair tuft (fig. 141). The larger glands are behind it. The glands are nearly oval in section. Pigment cells lie on the periphery of the glandular cells and also between the cells of the duct walls. Most of the glandular cells contain pigment granules.

The sweat glands are slightly twisted (fig. 142) as in *P. vitulina vitulina* (Leydig, 1859; de Meijere, 1894; Troll-Obergfell, 1930). Their lower sections reach down below the hair bulbs (as in *P. v. vitulina,* Troll-Obergfell, 1930). The diameter of the gland lumen ranges from 30 to 40μ. The ducts open into the hair bursae exactly beneath the sebaceous gland duct (Montagna and Harrison, 1957).

There are pile, intermediate, and fur hairs growing in tufts of one pile and two to five fur hairs (fig. 143). One or two intermediate hairs are found in some tufts. This hair tuft pattern is also noted in *P. v. vitulina* (de Meijere, 1894; Troll-Obergfell, 1930; Mohr, 1952; Montagna and Harrison, 1957). The hair consists only of a cortical layer and cuticle (as in *P. v. vitulina,* Troll-Obergfell, 1930). In the pile and intermediate hairs, the cuticle cells of the lower part of the shaft resemble arrowheads. In the middle and upper shafts, the cells extend across the hair and their edges become sinuate. The cuticle of the fur hairs is usually ring-shaped. The histochemical properties of the hair follicles are similar to those of the hair follicles of land mammals (Montagna and Harrison, 1957).

Pusa caspica

Material. Skin of the withers, shoulder, sacrum, flank, belly, and fore-limbs (upper and lower surface), four-year-old female of *P. caspica* taken in mid-February, 1972, north of Caspian Sea (by special permit).

Body skin thickness of *P. caspica* gradually increases both front to back and dorsoventrally (table 99) (Sokolov and Sumina, 1974). The epidermis is thickest on the sacrum and belly, thinnest on the lower part of the forelimbs and on withers (table 99).

The epidermis is three layers deep: stratum basale, stratum spinosum, and stratum corneum. The basal layer is one cell deep. The stratum spinosum is two to three cells deep, topped by a compact stratum corneum. The cells of stratum basale are cylinder-shaped and ovoid; their nuclei are arranged vertically. Large cells of the stratum spinosum are rhomboid in

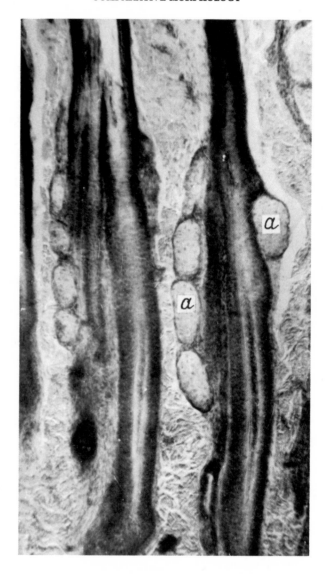

Fig. 141. *Phoca vitulina.* Back skin: *a,* sebaceous glands.

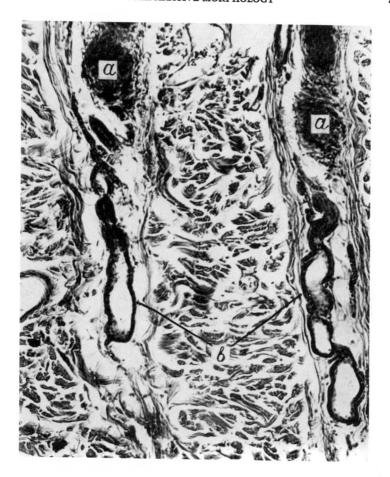

Fig. 142. *Phoca vitulina*. Back skin: *a,* hair bulbs; *b,* sweat glands.

section, with oval nuclei, each with a long axis parallel to the skin surface. The stratum corneum is fairly thick and compact; its relative thickness (as a percentage of the epidermis) ranges from 28 to 49%, with the thickest part being on the forelimbs, belly, and sacrum (table 92). Small processes are noted on the inner surface of the epidermis. The basal layer cells are filled with pigment granules. In the stratum spinosum, cell pigmentation is less. There is some pigment in the corneous layer.

The dermis is formed of the papillary and reticular layers. Their boundary is level with the terminal parts of the sweat glands. The papillary layer is thicker than the reticular, with this thickness varying in different parts of the body (table 92). The papillary layer thickens from front to back and dorsoventrally.

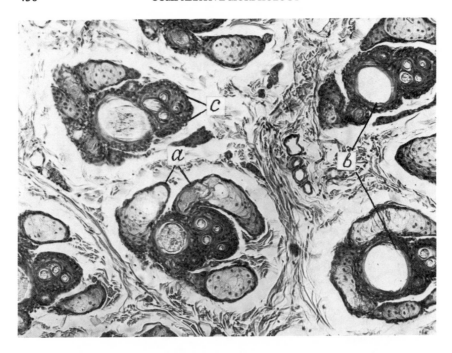

Fig. 143. *Phoca vitulina.* Back skin (tufts of hair in longitudinal section): *a,* sebaceous glands; *b,* pile hair; *c,* fur hair.

Nearing the epidermis, in the upper part of the papillary layer, thin (4μ) collagen bundles form a compact, horizontally wavy plexus. A similarly compact collagen bundle plexus surrounds the hairs and the sebaceous and sweat glands. Collagen bundles are sparser in the inner parts of the papillary layer but thicken up to 15μ. In sections, numerous collagen bundles are cut crosswise, probably because they are arranged in circles around the hairs.

The papillary layer of the dermis is rich in large (19 x 19; 26 x 26; 74 x 56μ) blood vessels filled with blood. They are mostly vertical throughout the papillary layer. The elastin fibers in the papillary layer are rather few, differing in thickness and arrangement pattern in various parts of this layer. Thin elastin fibers are below the epidermis and around the sebaceous glands. They run parallel to the skin between hair bursae. Not infrequently, fairly thick fibers merge with one another forming a network. Many elastin fibers are in the connective tissue hair bursae, where they alternate with collagen bundles.

Tongue-shaped accumulations of fat cells occasionally occur near the sweat and sebaceous glands in the reticular layer and also around the hair bulbs. These accumulations project to the papillary layer. Fat cells range from 33 to 70μ.

TABLE 99

MEASUREMENTS OF SKIN OF PUSA CASPICA

| Indexes | Skin without subcutaneous fat tissue (mm) | Thickness | | | | | Size of sebaceous glands (two dimensions, mm) | Diameter of secretion cavities of sweat glands (μ) |
		Epidermis (μ)	Stratum corneum (μ)	Dermis (mm)	Papillary layer of dermis (mm)	Reticular layer of dermis (mm)		
Withers	3.000	76	23	2.924	1.574	1.350	0.825 × 0.165	52
Shoulder	4.940	81	26	4.859	1.959	1.900	0.600 × 0.150	51
Breast	5.314	81	26	5.233	2.700	2.533	0.675 × 0.157	52
Sacrum	4.170	106	36	4.064	2.850	1.214	0.817 × 0.180	76
Flank	4.462	92	26	4.370	2.280	2.090	0.855 × 0.190	53
Belly	6.093	93	37	6.000	3.150	2.850	0.950 × 0.135	56
Forelimb, top	2.592	92	37	2.500	1.995	0.505	0.523 × 0.190	56
Hind limb, bottom	2.475	75	37	2.400	1.875	0.525	0.600 × 0.150	48

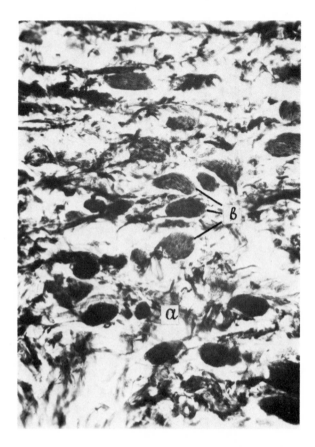

Fig. 144. *Pusa caspica.* Reticular layer in vertical section of skin
(x 140): *a,* collagen bundles; *b,* bundles of smooth muscles.

In the lower part of the papillary layer and also in the underlying reticular
layer are many free bundles of smooth muscles unconnected with hairs (fig.
144). They are numerous in all skin samples except in the skin from the forelimbs, where they are few. Muscles are arranged in a pattern. The bundles in
horizontal sections of the reticular layer form a peculiar network with the
collagen bundles and elastin fibers (fig. 144). Accumulations of fat cells,
blood vessels, and collagen bundles are arranged in the diamond-shaped
alveoli of this network. In vertical sections of the skin, the smooth muscles
are mostly parallel to the skin. They are spindle- or strip-shaped and are
covered with fascias, which relate to the elastin fibers of the reticular layer
(fig. 145). The diameter, length, and number of free bundles of smooth
muscles increase with their depth in the dermis. There are more muscular
bundles in the belly skin than in other parts of the body. In the lower areas

of the papillary layer (vertical sections) the muscles average 27μ; in the upper parts of the reticular layer they average 38μ; in the middle area of this layer they average 76μ; and on the boundary with the subcutaneous fat tissue they average 95μ. There are 3,000 bundles of smooth muscles per 1 cm² of the reticular layer of the dermis (vertical section). The muscles in the reticular layer of the skin approximately equal that of the connective tissue.

On the border between the papillary and reticular layers, connective tissue shows frequent accumulations of fat cells. They are not in a compact layer but form individual clusters with regular intervals filled with collagen bundles, elastin fibers, and smooth muscles. At greater depth, the elastin fibers disappear, and the connective tissue assumes its usual appearance with a multitude of thick elastin fibers (thickness 7μ), which deviate somewhat from the horizontal direction. Together with bundles of collagen

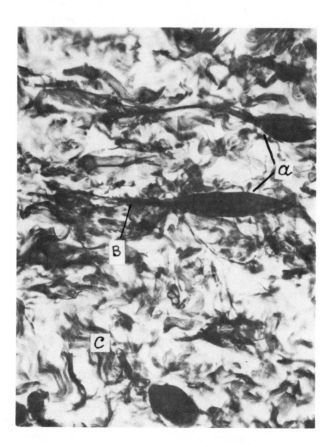

Fig. 145. *Pusa caspica*. Smooth muscles and their relation to elastin fibers of reticular layer (x 100): *a,* smooth muscles; *b,* elastin fibers; *c,* collagen bundles.

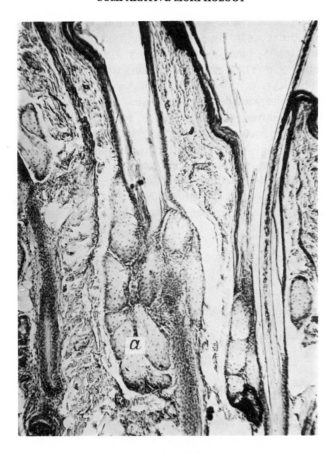

Fig. 146. *Pusa caspica*. Vertical section of skin (x 100):
a, sebaceous gland.

fibers they encircle the muscle bundle clusters. Some collagen fibers rise vertically in broad bands (70 to 90μ) to the hair bulbs, forming hair bursae around the hairs.

The collagen bundles in the reticular layer are from 15 to 26μ thick. Numerous blood vessels of various diameters (285 x 190; 155 x 63; 96 x 59; 37 x 37μ) pass through the reticular layer. Large holocrine sebaceous glands are on the sides of the pile hair bursae, the largest in the belly skin and the flank, the smallest in the limb skin.

The sebaceous glands are multilobal; lobes (from 5 to 11) are cluster-arranged (fig. 146). Glands are active. The secretion lobes are in the middle of the papillary layer of the dermis. They are pressed against the hair bursae or are at sharp angles to the hair. Some lobes resemble a broad or narrow sac. They differ in size: the largest secretion lobes may be 25μ long and 83μ

wide. An inverse relationship exists between the size and the number of secretion lobes in the sebaceous gland: the larger the lobe, the smaller their
number. The ducts of the secretion lobes in the sebaceous gland merge into
a common excretion duct. The latter opens into the upper part of the hair
sheath. The width of the duct increases with its length.

The secretion of the sebaceous gland first reaches the inner hair sheath of
a pile hair up to where the hairs form a tuft. It then moves up the hair into
the common hair sheath, lubricating the other hairs of the tuft. Horizontal
sections show two or (more rarely) three glands at each hair bundle. Skin
sections stained with hematoxylin-eosin show all the typical stages of secretion formation in the sebaceous gland. The epithelium of the apical part of
the secretion lobes and their ducts is cornified. It is eosin-stained, as is the
stratum corneum of the epidermis. The number of sebaceous glands per
1 cm² and the depth at which they lie vary in different parts of the body.
Assuming two sebaceous glands per hair bundle, the number of sebaceous
glands varies from 1,200 to 2,240. Their greatest number is on the sacrum
and in the lower part of the forelimbs. Their smallest number is on the belly
and breast. The glands are deepest in the sacrum skin and on the flank, followed in order by breast, withers, shoulder, and belly. In the skin of the
limbs, the sebaceous glands are close to the surface.

The tubular apocrine sweat glands are characterized by two secretion
stages: necrobiotic, involving a rupture of the cell's apex, which projects
into the lumen of the tubule as secretion; followed by regeneration of the
apical part of the cell, which produces ordinary secretion. The secretion cavity of the gland is lower than the hair bulbs and is spiral shaped. Farther
down it suddenly narrows, becoming a straight duct that occasionally loops
downward and rises to the surface. The sweat gland duct is thin (15μ), opening into the hair sheath above the duct of the sebaceous gland. There are
sweat glands at each hair tuft. They are nearer the connective tissue bursa of
the pile hair, and most ducts pass between the pile and other hairs of the
tuft. Active and inactive sweat glands may be seen in sections. The diameter
of the secretion cavities of sweat glands in the body skin of *P. caspica* increases from front to back. The diameter of the secretion cavities of sweat
glands is larger in the upper part of the forelimbs.

The longest pile and intermediate hairs are on the breast and on the belly.
Shorter ones are on the side of the trunk, and the shortest are on the withers, sacrum, and limbs (see table 100). The greater the density of the hairs,
the less is their length, and vice versa.

The body hair of *P. caspica* is in the intermediate stage between growth
and rest (catagen). The hair bulbs are shriveled and vesicle-free. The bulb
terminal, previously formed of live cells, is fully cornified and of the panicu-
liform shape characteristic of the so-called hair cone. It has broken off from
the vesicle and is gradually rising to the surface. Strands of cells of solid
connective tissue of the hair sac are still between the papilla and the hair

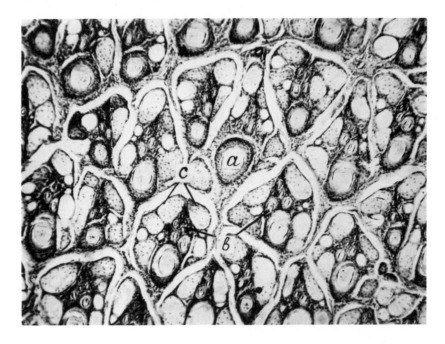

Fig. 147. *Pusa caspica*. Hair pattern in skin (x 50): *a*, pile hairs; *b*, fur hairs; *c*, sebaceous glands.

bulb. The distance between the papilla and the hair bulb in the skin of the seal ranges from 0.059 on the withers to 0.143 on the sacrum. On the fore-limbs it is invariably 0.41 mm.

P. caspica has pile, intermediate, and fur hairs growing in tufts (fig. 147). Pile hairs are anterior to the intermediate and fur hairs. Tufts may form groups of two to three, mostly in short, crosswise rows in relation to the body axis. The tufts are invariably of one pile, one to five intermediate (on the shoulder there may be tufts of five intermediate hairs) and two to seven fur hairs. Tufts with the greatest number of fur hairs are found on the shoulder and on the breast.

Pile hairs are straight; the shaft is flattened the entire length of the hair. The base of the pile hair is somewhat smaller than the middle. In section, the shafts of pile and intermediate hairs are like an elongated lens, convex anteriorly and flattened posteriorly. The pile and intermediate hairs are medulla-free. The cuticle is not ring-shaped throughout the shafts of these hairs. The scales are broad and short. At the base of the pile hair, the aver-age width of the scale is 78μ; the average height is 14μ. In the middle part the average width is 42μ; the average height is 12μ. Near the shaft's apex the average breadth is 30μ; the average height is 10μ. The scale edges are irregular. The pile hairs are unpigmented except for those at the withers and

TABLE 100

Pusa caspica: Hair measurements

| Sample | Hair length (mm)* | | | | Breadth of hairs in the middle part of the shaft (mm) | | | Number | | |
	Pile	Intermediate	Fur		Pile	Intermediate	Fur	Number of hairs in a tuft (n=10)	Number of tufts per 1 cm²	Number of hairs per 1 cm²
Withers	6.5	4.2	3.2		0.175	0.090	0.022	8.0	1,070	8,600
Shoulder	7.5	4.8	2.8		0.215	0.100	0.014	9.5	730	7,000
Breast	10.4	8.3	9.7		0.250	0.190	0.018	9.5	730	7,000
Sacrum	5.7	4.1	3.9		0.160	0.075	0.016	6.5	1,110	7,200
Flank	7.7	4.7	3.3		0.240	0.130	0.014	7.5	790	5,900
Belly	8.5	5.0	4.0		0.203	0.070	0.014	7.0	600	4,200
Forelimb, top	5.2	3.4	8.1		0.192	0.138	0.021	7.5	1,030	7,700
Forelimb, bottom	5.4	3.9	2.9		0.150	0.080	0.014	7.0	1,120	7,800

*Mean values for 6 to 10 hairs.
**Broken fur hairs.

sacrum, where both pigmented and unpigmented hairs occur. Occasionally, pigment is found at the base of some pile hairs.

There is little difference between intermediate and pile hairs. Intermediate ones are somewhat shorter and thinner (table 100). They are bent over the base at about 45 to 50°.

Fur hairs are thin and twisted (three to four curves). Most of the fur hairs have broken tips. In some body areas the fur hairs are longer than the pile hairs (forelimb, on the top) or equal to them (breast). The terminal part of the fur hairs is usually pigmented and is somewhat thicker than the other parts of the hair. The cuticle is either ring-shaped or not. The medulla in the fur hairs is well developed all through the shaft and thinner in the curves of the hair, improving the mechanical properties of the hair.

Hairs are longest in the skin of the withers, limbs, and sacrum, and shortest in the skin of the side, flank, and on the belly. Hair is denser on the limbs, due to an increase in tuft density rather than to an increase in the number of hairs in the tuft. The greatest number of hairs in a tuft is at the side of the body, on the shoulder, and on the breast (table 100).

Pusa sibirica

Material. Samples of the withers skin over the forelimbs and the breast between the forelimbs (tables 101-103) of an adult female taken in Baikal.

The hair follicles are active. The skin and epidermis are thicker on the breast than on the withers. The stratum corneum is loose. The stratum malpighii is three to four cells deep. The cells of the stratum basale have a tiny amount of pigment. Horizontal collagen bundles predominate in the dermis; their plexus is loose. The smooth muscle bundles, most of which run horizontally, are numerous in the reticular layer. There are innumerable fat cells in the reticular layer which form peculiar strands from the hair bulbs to the subcutaneous fat tissue.

In a *P. sibirica* yearling, fat cells were also found in the breast papillary layer (Belkovich, 1964a, b). The subcutaneous fat tissue is very thick (in October females, 5 to 9 cm in the back, 3 to 6 cm on the breast, and 4 to 5 cm on the belly; and in males, 4.5 to 8 cm, 3.5 to 5 cm, and 3 to 5 cm) (Shepeleva, 1971). The sebaceous glands are small (102 x 256; 71 x 256μ) and sometimes multilobal, arranged in pairs at every hair tuft (fig. 148). In the yearling, the sebaceous glands have up to ten lobes (Belkovich, 1964). The sweat glands have a long, slightly twisted secretion cavity 30 to 51μ in diameter (up to 70μ in the yearling). The ducts open into the hair funnels.

The pelage of the yearling has pile, intermediate, and fur hairs (Belkovich, 1964b). The pile hairs, up to 15 mm long and 231μ wide, are flattened and lancet-shaped with a medulla only at the tip. The cuticle cells are cobble-shaped. The intermediate hairs are also flattened. They grow to 8.7 mm long, 33μ wide, and have no medulla. The cuticle is similar to that of the

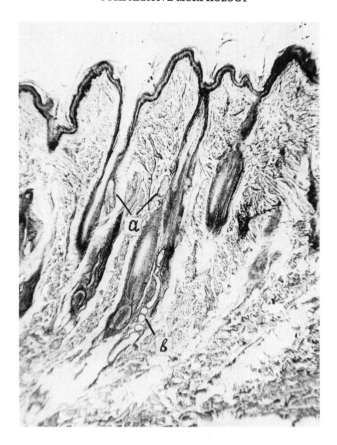

Fig. 148. *Pusa sibirica.* General structure of withers papillary layer:
a, sebaceous glands; *b,* sweat glands.

pile hair. The fur hairs are 8.9 mm long, 22μ thick, and very curly with no medulla. The cuticle is not ring-shaped. The hairs grow in tufts of one pile, one intermediate, and four fur hairs. The tufts grow in rows across the body, or sometimes in close groups of two or three tufts. There are about 1,200 pile and intermediate hairs and 4,800 fur hairs per 1 cm^2.

In the adult female *P. sibirica* the distribution of pelage is the same as in the yearling, with 2,800 hairs per 1 cm^2 on the withers and 3,300 on the breast. Skin and hair age variability has been studied in a one-month-old female (body length 70 cm), in a one-and-a-half-month-old female (81 cm), in two two-month-old males (77 and 81 cm), and a fifteen-year-old male taken in April (1971 and 1972) at the Baikal Lake (Sokolov, Sumina, and Dyachkova, in press). With age, the indexes of all the skin structures increase (table 101). In adult animals, smooth muscle bundles get thicker.

TABLE 101

SKIN MEASUREMENTS IN PUSA SIBIRICA OF VARIOUS AGES

| Indexes | Skin (mm) | Epidermis (μ) | Stratum corneum (μ) | Thickness | | | Subcutaneous fat tissue (mm) | Size of sebaceous glands (μ) | Diameter of secretion sections of sweat glands (μ) |
				Dermis (mm)	Papillary layer (mm)	Reticular layer (mm)			
1	2	3	4	5	6	7	8	9	10
Female, 1 month old									
Withers	23.4	100	60	3.1	2.8	0.3	20.2	92 × 410	42
Shoulder	23	84	42	3.2	3.2	—	19.7	20 × 307	42
Breast	25.3	84	42	3.9	3.6	0.3	21.3	105 × 378	42
Sacrum	21.5	84	42	2.8	2.5	0.3	18.6	63 × 273	50
Flank	24.4	84	42	2.5	2.1	0.4	21.8	82 × 256	63
Belly	23.4	84	42	2.4	2.1	0.3	20.9	102 × 307	42
Forelimb, top	9.0	100	60	3.3	2.8	0.5	5.6	82 × 307	55
Forelimb, bottom	2.7	63	20	2.1	1.5	0.6	—	103 × 379	38
Hind limb, top	4.5	84	42	2.8	2.3	0.5	1.6	231 × 307	50
Hind limb, bottom	18.6	100	50	3.6	3.1	0.5	14.9	105 × 357	50
Female, 1.5 months old									
Withers	23.1	100	60	2.6	1.9	0.7	20.3	71 × 256	30
Shoulder	25.8	100	60	3.9	3.4	0.5	21.8	133 × 410	51
Breast	39.9	90	50	3.4	2.8	0.6	36.3	102 × 256	51
Sacrum	22.6	84	42	3.3	2.6	0.7	19.2	410 × 502	63
Flank	30.0	100	60	3.3	2.8	0.5	26.6	87 × 380	42
Belly	23.4	100	60	3.4	2.6	0.8	19.9	103 × 257	72
Foreflipper, top	5.0	100	60	3.1	2.6	0.5	1.8	205 × 410	50
Foreflipper, bottom	4.6	88	50	2.8	2.2	0.6	1.7	95 × 293	57
Hind flipper, top	4.9	84	42	2.4	1.9	0.5	2.4	109 × 294	46
Hind flipper, bottom	3.2	84	42	3.1	1.6	1.5	—	174 × 226	42

TABLE 101—*continued*

Indexes	Skin (mm)	Epidermis (μ)	Stratum corneum (μ)	Thickness Dermis (mm)	Papillary layer (mm)	Reticular layer (mm)	Subcutaneous fat tissue (mm)	Size of sebaceous glands (μ)	Diameter of secretion sections of sweat glands (μ)
1	2	3	4	5	6	7	8	9	10
Male, 2 months old									
Withers	28.96	110	70	2.85	2.18	0.66	26.0	167 × 570	52
Shoulder	28.58	104	63	2.98	2.18	0.79	25.5	114 × 456	41
Breast	30.76	129	85	3.13	2.12	1.00	27.5	130 × 475	48
Sacrum	27.81	97	56	2.71	1.76	0.95	25.0	133 × 380	44
Belly	27.78	129	89	2.66	1.90	0.76	25.0	170 × 437	53
Forelimb, top	3.29	88	44	3.21	1.78	1.42	—	144 × 304	41
Forelimb, bottom	3.15	111	44	3.04	1.33	1.71	—	152 × 324	37
Hind limb, top	3.43	103	59	3.33	1.61	1.72	—	133 × 304	41
Hind limb, bottom	20.80	104	52	2.70	1.75	0.95	18.0	130 × 300	41
Male, 2 months old									
Withers	33.03	191	134	2.84	1.86	0.98	30.0	121 × 421	48
Shoulder	33.20	115	70	3.09	2.20	0.89	30.0	147 × 414	56
Breast	35.68	125	80	2.55	1.82	0.73	33.0	129 × 430	53
Sacrum	32.36	178	109	2.18	1.52	0.66	30.0	130 × 373	47
Flank	30.33	147	95	2.19	1.53	0.65	28.0	97 × 339	46
Belly	39.62	181	105	3.44	2.94	0.50	36.0	127 × 559	73
Forelimb, top	2.37	112	72	2.25	1.44	0.81	—	159 × 420	52
Hind limb, top	2.98	180	121	2.80	2.36	0.43	—	175 × 470	82
Hind limb, bottom	9.81	132	67	2.68	1.76	0.92	7.0	139 × 309	73

TABLE 101—*continued*

Indexes	Skin (mm)	Epidermis (μ)	Thickness Stratum corneum (μ)	Thickness Dermis (mm)	Thickness Papillary layer (mm)	Thickness Reticular layer (mm)	Subcutaneous fat tissue (mm)	Size of sebaceous glands (μ)	Diameter of secretion sections of sweat glands (μ)
1	2	3	4	5	6	7	8	9	10
Male, 15 years old									
Withers	—	292	150	7.06	3.44	3.62	absent	231 × 693	72
Shoulder	—	351	160	10.64	4.92	5.71	absent	168 × 795	109
Breast	—	393	189	7.90	4.20	3.70	absent	253 × 821	87
Sacrum	—	294	133	7.08	3.72	3.36	absent	268 × 907	96
Flank	—	249	132	6.18	3.63	2.55	absent	270 × 823	78
Hind limb, top	—	291	108	6.07	3.47	2.60	absent	306 × 819	90
Hind limb, bottom	—	228	102	4.05	1.93	2.11	absent	274 × 414	67
	—	231	139	5.65	2.11	3.54	absent	100 × 432	88

TABLE 102

PUSA SIBIRICA: HAIR LENGTH ON WITHERS (MM)

Age and sex	Hair categories and orders														
	Embryonic			Pile									Fur		
				1st order			2d order			3d order					
	n	Lim	M±m	n	Lim	M±m	n	Lim	M±m	n	Lim	M±m	n	Lim	M±m
Female 1 month	43*	11.2-37.6	23.2±2.33	10	7.6-12.6	10.0±1.0	18	7.1-10.9	8.5±0.49	2	5.6-6.6	—	54	1.3-6.4	3.9±0.32
Female 1.5 months	44*	13.8-26.4	20.6±1.06	10	12.3-17.5	14.6±0.9	15	8.5-14.5	10.8±0.9	15	4.8-8.7	7.7±0.55	53	4.8-8.6	6.9±0.24
Male 2 months	—	—	—	13	12.8-17.6	15.2±1.02	25	7.1-14.1	10.6±0.74	10	6.6-9.8	8.2±0.71	48	2.6-10.5	6.8±0.28
Male 15 years	—	—	—	10	8.6-10.8	9.6±0.42	19	5.9-8.5	7.3±0.33	17	4.9-7.4	5.9±0.3	53	2.2-5.9	4.2±0.27

*For all skin samples.

TABLE 103

Pusa sibirica: Maximum hair thickness on the withers (·)

Age and sex	Hair categories and orders														
	Embryonic			Pile									Fur		
				1st order			2d order			3d order					
	n	Lim	M±m	n	Lim	M±m	n	Lim	M±m	n	Lim	M±m	n	Lim	M±m
Female, 1 month	43*	0.028-0.100	0.093±0.001	10	—	0.200	18	0.100-0.200	0.133±0.022	2	—	0.056	54	—	0.014
Female, 1.5 months	44*	0.028-0.150	0.090±0.007	10	0.100-0.200	0.180±0.008	15	0.056-0.200	0.090±0.027	15	0.042-0.070	0.051±0.005	53	0.014-0.028	0.016±0.001
Male, 2 months	—	—	—	13	0.100-0.200	0.170±0.007	25	0.100-0.200	0.100±0.015	10	0.056-0.100	0.079±0.015	48	0.014-0.028	0.017±0.001
Male, 15 years	—	—	—	10	0.200-0.300	0.240±0.039	19	0.100-0.300	0.120±0.026	17	0.028-0.084	0.055±0.007	53	0.014-0.028	0.017±0.001

*For all skin samples.

Their maximum thicknesses are 87μ (one-year-old female), 70μ (one-and-a-half-month-old female), 82 to 83μ (two-month old males), 186μ (fifteen-year-old male).

The pelage of *P. sibirica* under study includes three orders of pile hairs and fur hair (the one- and one-and-a-half-month-old females also having embryonic hair (table 102, 103).

Pagophilus groenlandicus

The thickness of the epidermis varies considerably in different parts of the body (Belkovich, 1964*b*) (table 91). The epidermis is strongly pigmented. The inner surface of the epidermis is even and without processes (fig. 149). The outer surface of the papillary layer (140 to 330μ thick) is formed of a compact plexus of small collagen fibers. In the middle and inner papillary layers are fat cells (single and in groups) and the collagen fiber plexus becomes looser. In the reticular layer the plexus of collagen fibers, mainly perpendicular to the surface, is loose and has fat cells. There are many blood vessels in the dermis. The bundles of collagen fibers become considerably thinner in the subcutaneous fat tissue, which varies considerably with the season (Bergensen, 1931). In *P. groenlandicus,* taken in March, the subcutaneous fat tissue is 5.6 cm thick in males and 4.5 cm in females; in May it is 2.1 cm and 2.4 cm. The sebaceous glands have six to nine lobes and are up to 1 mm long (Bergensen, 1931). Their ducts open into the upper hair sheath. Sweat glands of *P. groenlandicus* are almost shriveled like the sweat glands of the land mammals they resemble (fig. 150), and they vary seasonally (Bergensen, 1931). In March they are inactive, and in May they actively secrete.

The hairs divide into pile, intermediate, and fur (the hair classification for *P. groenlandicus* suggested by Bergensen is very complicated and of little use). The pile hairs are 12.5 mm long and 203μ wide. They are flattened (oval with one concave side in secretion) and have no medulla (fig. 151*a*). Their cortical layer has lengthwise fibers (fig. 151*b*). The cuticle scales stretch along the length of the hair. The intermediate hairs (6.2 mm long and 69μ wide) are also flattened and without medulla. Their shafts are wavy. The length of the curly fur hairs is 5.5 to 7.7 mm, their thickness 22μ. They are round in section, and their cuticle is not ring-shaped. The hairs grow in tufts of one pile, one or two intermediate, and four or five fur hairs. The intermediate and fur hairs grow behind the pile hair. The tufts usually grow in rows across the body. There are about 11.9 thousand hairs per 1 cm^2.

Ommatophoca rossi

In an adult male, taken on January 26, 1963, the skin on the back is thinner than on the belly (Polkey and Bonner, 1966). No mitoses are revealed in the

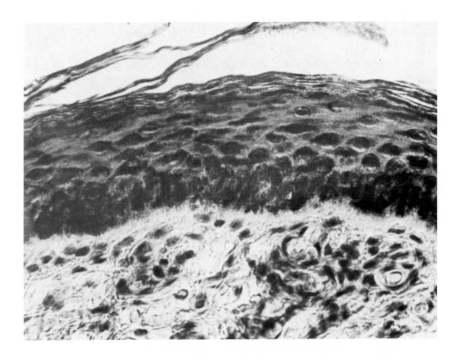

Fig. 149. *Pagophilus groenlandicus.* Epidermis of belly skin.

epidermal cells. Melanin-containing epidermal cells make up a layer 120μ thick on the back and 80μ thick on the belly. The epidermal stratum corneum is thin. Sebaceous glands are small, paired, opening into the upper part of the hair bursae. No sweat glands are found. There are two hair categories: primary and secondary (evidently, pile and fur hairs—V. S.), growing in tufts of one pile and one (rarely two) fur hairs. The pile hair is always anterior to the fur hair. Pile hair diameter is 120μ, that of fur hair is 50μ. The average length of the pile hair on the belly (10.6 mm) is greater than on the back (6.5 mm). On the top of the head, there is an average of 3.6 pile hairs per 1 cm²; 3.0 on midback; 2.6 mm² on the middle of the belly, and 1.9 mm² on the breast between the limbs.

Cystophora cristata

The inner surface of the epidermis is smooth (Belkovich, 1964), its thickness is variable (table 98), and it is lightly pigmented. In the papillary layer, only the outer surface, 165 to 250μ thick, contains a compact plexus of collagen fibers. In the middle and inner papillary layer, the collagen bundles run at various angles and blood vessels are numerous. The reticular layer has a compact collagen fiber plexus, fat cells, and smooth muscle bundles. The

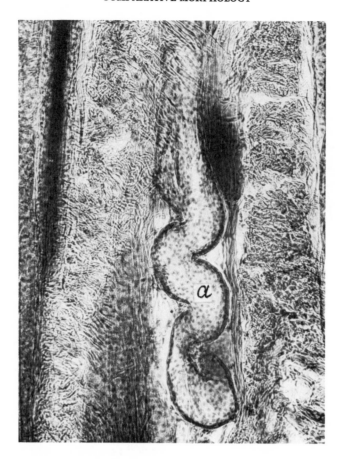

Fig. 150. *Pagophilus groenlandicus.* Belly skin: *a,* sweat gland.

subcutaneous fat tissue is 3.5 to 4.5 cm on the side, 5 to 6.5 cm on the back, and 4 to 7 cm on the breast. It is thicker in males than in females. Sweat glands are not found in our specimens and the sebaceous glands are not well developed and are almost inactive.

The pelage consists of pile, intermediate, and fur hairs. The pile hairs are 18 mm long and 230μ wide, straight and flattened, with no medulla. The cuticle is cobble-shaped. The intermediate hairs are wavy, 7.8 mm long and 44μ thick, and flattened. The cuticle in the distal and proximal sections is not ring-shaped, and it is cobble-shaped in the middle section. The first hairs are curly and slightly flattened, 5.20 mm long, and 22μ wide. The hairs grow in tufts of one pile, two or three intermediate, four or five (sometimes six) fur hairs (eight or nine, according to Mohr). The tufts grow in fairly regular rows across the body. There are 800 pile, 1,600 to 1,800 intermediate, and 4,000 to 5,000 fur hairs per 1 cm^2.

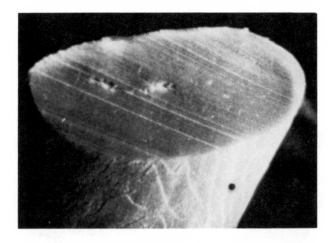

Fig. 151*a*. *Pagophilus groenlandicus*. Cross-section of pile hair.

Mirounga leonina

The thickness of the epidermis varies with the season from 150 to 500μ (Ling, 1965, 1968), changing mainly in the stratum corneum. Melanocytes are common in the epidermis. The stratum spinosum is six to nine cells deep. The stratum granulosum and the stratum lucidum are not pronounced, but cells with granular inclusions occur in the upper stratum spinosum that can be regarded as part of it. The outer stratum corneum contains a fatty secretion from the sebaceous glands. The inner parts of the stratum corneum react actively to bound calcium and phospholipids and moderately to the sulfhydryl groups, a reaction that is considerably weaker in the outer parts of the stratum corneum (Spearman, 1968).

In the papillary layer, collagen bundles lie at various angles. Elastin fibers, which are more numerous in younger animals, are commoner, thicker, and shorter in the reticular than in the papillary layer. There are fat cells in the upper reticular layer, and collagen bundles form a diamond-shaped plexus. The reticular layer contains smooth muscles. The subcutaneous fat tissue is 2.2 to 2.4 cm thick (Bryden, 1964). There are apocrine sweat glands at every follicle, with ducts opening into the hair bursae some 0.4 mm lower than the sebaceous gland duct. The outer diameter of the sweat gland's special secretion cavity averages 38μ, its lumen averages 14μ. The sebaceous glands (one at every hair follicle) have two lobes. The sebaceous gland attains 0.25 x 1.5 mm in size. The lipid content of the sebaceous gland consists of free cholesterol (9%), triglycerides (38%), and wax esters and phospholipids (53%) (Ling, 1965*a*, 1968). The hairs are like lancet-shaped bristles. There are only pile hairs.

Fig. 151*b. Pagophilus groenlandicus.* Cortical layer of pile
hair.

ORDO PROBOSCIDEA

A review of elephant skin (Frade, 1955) includes a bibliography (e.g.,
Smith, 1889; Eggeling, 1901, et al.) and some new data. A description of the
epidermis of the African elephant is given by Spearman, 1970.

There is only one family of this order, *Familia Elephantidae,* so the skin
characteristics refer to family and order. The skin is very thick. The outer
surface of the epidermis is irregular. In spite of its thickness, the epidermis
consists of only three layers: stratum basale, stratum spinosum, and stratum corneum. On its inner surface, alveoli interdigitate with dermal papillae. The dermis is not divided into papillary and reticular layers. The colla-

gen bundles in the dermis are mainly horizontal and form a compact plexus. Common sebaceous and sweat glands are absent. There is a specific skin temporal gland. Heavy, bristlelike hairs are found on the body. No arrectores pilorum muscles are recorded.

Elephas indicus

Material. Skin from tip of trunk, lower part of neck, breast, the groin, and near the foot of an adult female that died December, 1959, in the Moscow Zoo.

On the whole, our data agree with the evidence from other authors (Möbius, 1892; Neuville, 1917, 1918, 1923; Eales, 1925; Frade, 1955). The outer surface of the skin is covered with epidermal protuberances of varying size. These form regular rows on the neck and are scattered irregularly in the other parts. They are small on the groin (0.3 to 0.6 mm in diameter) but tall on the foot (up to 5 mm high), flattened, to an oval measuring on average 2 x 4 mm in section. Their apexes are rounded obtusely.

The skin of the samples varies considerably in thickness: 11 mm on the trunk, 14 mm on the breast, 16.5 mm in the groin, 21.5 mm on the neck, and 22 mm on the feet (20 mm over most of the body). The epidermis is 0.48 mm on the neck, 0.88 on the groin, 1.4 mm on the trunk, 1.6 mm on the breast, 5.7 mm at the hind foot, and 1.2 to 2 mm on the body. The epidermis is thinner between the epidermal protuberances: 0.13 mm, 0.44 mm, 0.60 mm, 0.20 mm, and 5.66 mm, respectively. The stratum spinosum passes directly into the stratum corneum, and the border between them is pronounced. In the outer stratum spinosum there are round cell nuclei absent in the inner part of stratum corneum. The section of the epidermal alveoli varies from round or polygonal to long and slit-shaped.

There are pigment cells in the basal layer of the epidermis on the body and numerous pigment granules are scattered in the protoplasm of the epidermal cells. The stratum basale is especially heavily pigmented where the epidermis protrudes deep into the dermis. The pigment granules continue into the stratum corneum. They are clearly visible on the upper part, although less pronounced. The epidermis of the breast, neck, and groin is moderately pigmented, while pigmentation is slight on trunk and foot.

The dermis comprises an intricate plexus of collagen fiber bundles (fig. 152) with a thickness ranging from 33μ near the epidermis to 260μ in the deep inner dermis. There are few elastin fibers, mainly near the border with the subcutaneous muscles, most of them oblique. The usual division of the dermis into papillary and reticular layers is impossible in this instance because of the few hairs, the absence of skin glands, and the unchanging pattern of collagen bundle plexus through the whole thickness of the dermis.

The dermal papillae, rising into the epidermal alveoli, are formed by thin bundles of collagen fibers and a few elastin fibers. These pass lengthwise along the papillae, which in the groin reach 0.45 mm in height, on the trunk

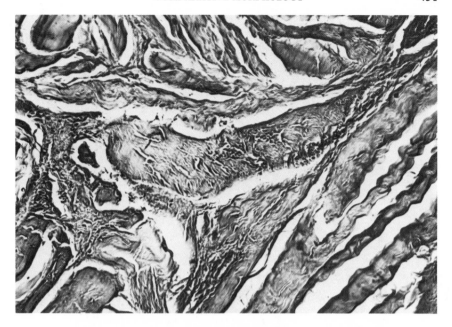

Fig. 152. *Elephas indicus*. Plexus of collagen fiber bundles in trunk skin.

0.70 mm, and on the foot 1.8 mm. Minute blood vessels forming small glomeruli rise to their apex. Fat cells are absent in the dermis. Blood vessels are few in the dermis. They are most frequent near the epidermis, especially at the hair follicles. There are also large arteries up to 0.8 mm in diameter. The sebaceous and sweat glands are absent in the body skin of adult specimens, and no rudiments are found in the embryos (Eales, 1925).

In the temporal region a glandular formation (a temporal gland) produces an abundant liquid secretion with an unpleasant odor during heat (Eggeling, 1901; Cnyrim, 1914; Eales, 1925; Schneider, 1930; Perry, 1953). This gland lies in the temporal hollow beneath the skin, bordering on the zygomatic arch at the lower edge. The opening of the gland is 20 cm from the outer angle of the eye and is very small (about 4 mm in diameter). A flattened tube runs across the opening, ending in three hemispherical cavities where the tubular gland ducts open. They merge with the lobes, which are separated from one another by connective tissue. The structure of this gland in the embryo and adult animal is similar (Eales, 1925). The secretion is pressed out of the gland reservoir by the contraction of *M. glandulae temporalis*.

There are separate, bristle-like, nonpigmented hairs up to 9.6 mm long in the groin which grow rather far apart (fig. 153). They are round in section and their thickness (154μ) is about the same down the whole length. Almost the entire thickness of the hair is filled with a cortical layer, the very thin medulla (about 10μ) being present only in spots. The hair cuticle is not ring-

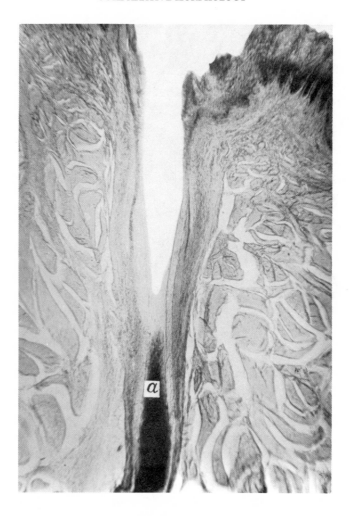

Fig. 153. *Elephas indicus*. Structure of breast skin: *a,* hair root.

shaped. The hair bulb is 4.3 mm down from the skin surface. The follicle is vertical. In the skin of the hind foot there are sometimes dark brown, bristle-like hairs 24 mm long. They are slightly oval (100 × 132μ) in section. These hairs are about the same diameter down the whole length, with no medulla, and with pigment granules filling out the thickness. The number of granules is less at the periphery.

Loxodonta africana

The epidermal surface is comprised of corneous papillae, each about 0.5 mm in diameter, and narrow troughs (Spearman, 1970); hair follicles with deepset necks are confined to the troughs. The stratum malpighii is 60 to 110μ thick in the area of the corneous papillae, and somewhat thinner in the troughs. The stratum granulosum of the epidermis, with keratohyalin granules, is well developed. The stratum corneum is up to 60μ thick, comprising half the epidermis in the corneous papillae area. It is made up of moderately flattened cells with a solid, eosinophilic, pink-stained cytoplasm. The cells of the stratum corneum in the corneous papillae contain accumulations of melanin granules marking the original position of the nucleus.

Dermal papillae project into the epidermis. Cytoplasmic disulfide bonds of cystine are found in the hair cortex and in the stratum corneum. Bound sulfhydryl groups are concentrated in the papillary stratum corneum at the base of the above layer in the troughs and in the hair follicle neck.

ORDO HYRACOIDEA

The skin of *Hyracoidea* has been studied only with regard to the specific skin gland (Schaffer, 1940; Monod and Dekeyser, 1968; Gabe and Dragesko, 1971).

The skin of *Procavia capensis, P. syriacus habessinica, P. s. syriacus,* and *Heterohyrax brucei* provided specimens for this study. There is only one family in the order (*Procaviidae*). The skin is rather thick and the epidermis and its stratum corneum are thin. The stratum granulosum and stratum lucidum are usually absent in the epidermis, which is commonly unpigmented. The papillary and reticular layers meet at the pile hair bulb level exactly where the pattern of the collagen fiber plexus changes.

On the whole of the body, sparse vibrissae roots penetrate twice as deep as pile hairs. The reticular layer is much thicker than the papillary, especially on the back. The continuous layer of pigment cells 20 to 60μ from the epidermis is characteristic. The small sebaceous glands have single lobes. No sweat glands are found in the trunk skin. In the sole are huge glomi of tubular glands. On the back a specific skin gland is made up of tubular glands. Arrectores pilorum muscles are typically thin.

The withers hair falls into three categories: vibrissae, pile, and fur. The pile hairs have a granna and a medulla. The vibrissae and fur hairs have no medulla. The hairs grow in tufts of one pile and seven to twelve fur hairs.

Procavia syriacus habessinica

Material. Skin from withers, breast, and the hind foot of a specimen (body length 30 cm) taken in Abyssinia on January 9).

The withers skin is 1.04 mm thick; the breast skin 0.77 mm. The withers epidermis is thin, 36μ. The greatest part of it is taken up by a 22μ stratum corneum. The epidermis and stratum corneum are somewhat thicker on the breast: 45 and 28μ, respectively. The stratum malpighii is two to three cells deep. The withers epidermis has a thin stratum lucidum. There is no stratum granulosum in withers or breast and no stratum lucidum in the breast. The stratum corneum is rather loose. The epidermis is not pigmented. It is very thick on the hind sole (825μ) and its inner surface has large alveoli. The stratum malpighii is 198μ thick.

The epidermal septa take up 132μ. The lower parts of the epidermal septa are pigmented. The stratum granulosum is pronounced. The stratum corneum is solid and similar in consistency to the stratum lucidum over it. Papillary and reticular layers are distinguishable in the dermis. In the withers, the vibrissae bulbs reach down as much as 0.65 mm, but they are sparse and the pattern of the plexus of the collagen bundles is such that it is more reasonable to draw the border between the two dermal layers level with the pile hair bulbs, 0.33 mm from the epidermis.

In the papillary layer, the mainly horizontal collagen bundles form a rather compact plexus. Elastin fibers are very sparse. Twenty to 60μ from the epidermis is a thin but continuous layer of large pigment cells. The hind sole dermis has no pigment cells. In the outer and middle parts of the reticular layer, bundles of collagen fibers form a rhombic plexus. They are horizontal on the inner side. Elastin fibers are present only on the border with the subcutaneous muscles, where they form networks for a zone of about 50μ. In the breast, the papillary layer is only a fraction thinner than in the reticular layer. The collagen fiber bundles run in various directions in the papillary and outer middle reticular layers. They are mostly horizontal in the inner reticular layer. Elastin fibers are rather numerous in the papillary layer, but are rare in the reticular, where they are found only on the border with the subcutaneous muscles. The fibers are mostly horizontal.

Arrectores pilorum muscles are very thin and can be seen only at some hair tufts. From the lower parts of the hair sheath posterior to the tuft, they pass obliquely back and up to the epidermis (fig. 154). The sebaceous glands are small, single-lobed, and singly posterior to the hair (but not at every hair). In the breast, the glands are somewhat larger. In the dermis of the hind foot sole, some 250μ from the epidermis, there are large tubular gland glomi with secretion cavities 11 to 13μ in diameter (fig. 155). Their ducts run upward, pass through every epidermal alveolus, and open at the surface. There is a specific skin gland on the back up to 9 mm wide and 22 mm long (Schaffer, 1940). At this point the skin is about 1 mm thick. The epidermis is thin; the stratum malpighii is two to three cell layers.

The spinal gland is formed by large, apocrine tubular gland glomi (Monod and Dekeyser, 1968; Gabe and Dragesco, 1971). Their ducts open into large cavities, the ducts of which open outward.

There are numerous vibrissae on the head: supraorbital, buccal, labial, and genial. There are three categories of hair on the withers: vibrissae, pile, and two orders of fur. There are no vibrissae in the sample, but they are normally found over the whole body, especially in those parts that fre-

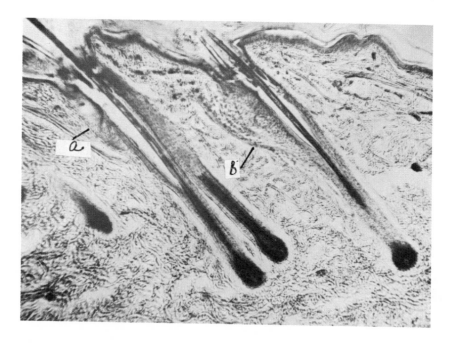

Fig. 154. *Procavia syriacus habessinica.* Skin of withers: *a,* sebaceous gland; *b,* arrector pili muscle.

quently touch things when the animal is moving (Sale, 1970). There are two size orders of pile hairs. Pile hairs are not numerous. In the first order, the granna is poorly developed and is thickest two-thirds up the shaft. It is 30.2 mm long and 101μ thick, with medulla all the way except for base and tip, up to 66.4% at the thickest point. Second order pile hairs have a pronounced granna extending for less than a third of the hair. The average length is 22.4 mm, thickness 65μ, medulla 69.3% of the granna thickness. In some hairs, the medulla is broken in the club. The fur hairs (first order) are 25μ thick down their whole length of 16.4 mm. They have up to eight waves. The medulla is absent in most of the hairs, and is simply a series of dashes in some hairs. The second order differs only in length (9.8 mm), and has up to five waves. Pile hairs grow in a common tuft with fur hairs, usually one pile and seven to twelve fur hairs in a tuft, though some tufts have fur hairs only. There are 660 pile and 12,000 fur hairs per 1 cm^2 and 11,666 hairs per 1 cm^2 on the breast.

Procavia syriacus syriacus

Material. Skin from withers, breast, and the hind sole (body length 33 cm).

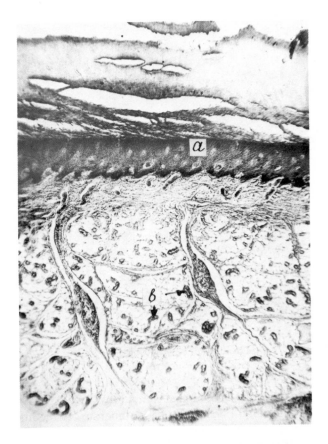

Fig. 155. *Procavia syriacus habessinica.* Skin of hind foot sole:
a, epidermis; *b,* glomi of tubular glands.

The skin is somewhat thicker than in *P. habissinica:* 1.90 mm on the withers and 0.90 mm on the breast. The epidermis is 44μ thick on the withers
and 45μ on the breast. The stratum corneum on the breast is 28μ. The stratum malpighii is two to three cells deep. The stratum granulosum and stratum lucidum are absent. The stratum corneum is compact. The epidermis is
lightly pigmented; its structure on the hind sole is the same as in *P. habessinica.* The epidermis is $1,045\mu$ thick, of which stratum malpighii takes up
220μ. The height of the epidermal alveoli reaches 150μ. The dermis on the
withers and breast has the same structure as in *P. habessinica.* The pigment
cells are more numerous in the withers papillary layer and form a band 30 to
400μ from the epidermis. The sebaceous glands are larger than in *P. habessinica.* Tubular glands form large glomi in the hind sole, their secretion cavities being 11 to 15μ in diameter. The gland ducts pass into every epidermal
alveole and open onto the skin surface.

There are three categories of withers hairs: vibrissae, pile (two orders), and fur. The few vibrissae are straight, bristlelike, and very long (54.7 mm). They are thick (168μ), tapering gradually from base to tip. They have no medulla. The pile hairs (first order) are 34.1 mm long (granna 7 mm long, 90μ thick, medulla 60.0%). In the long club, the medulla is broken into dashes and is often absent altogether. The pile hairs (second order) have a thicker granna (69μ) and a more poorly developed medulla (36.4%). They are also shorter (10 mm). The fur hairs are uniformly thin down the whole length. The hairs grow in tufts of one pile and seven to twelve fur hairs. The vibrissae grow singly and sparsely. There are 666 pile and 6,333 fur hairs per 1 cm^2.

Heteroxyrax brucei

Material. Skin of withers, breast, sacrum, dorsal gland, and soles of an adult male (body length 42 cm) taken in Kenya in December 1972.

The hair on the trunk is in growth. The skin on the withers and sacrum is of similar thickness (1.6 mm), and is somewhat thinner on the breast (1.3 mm). The thickness of epidermis and stratum corneum is similar on the sacrum (84 and 63μ, respectively) and on the breast (70 and 42μ) and thinner on the withers skin (42 and 13μ). The stratum malpighii is formed by 2 to 3 cell layers in the withers skin and by 3 to 5 layers in the skin of sacrum and breast. The stratum granulosum and stratum lucidum are absent. The stratum corneum is loose. There is no pigment in the epidermis.

In the soles, the thickness of epidermis and stratum corneum is great: 1,064 and 798μ on the front soles, and 957 and 665μ on the hind. The inner surface of the sole epidermis is alveolar, the alveoli having thick walls. The stratum granulosum is 2 to 3 cells deep. There is no stratum granulosum, or it is not pronounced. The stratum corneum is very solid. In section, it shows occasional large slits, with the long axis parallel to the skin surface. It appears that these slits are characteristic of a live animal. The deep epidermal layers are moderately pigmented and the upper layers are slightly pigmented.

The boundary between the papillary and reticular layers in the trunk skin is level with the bulbs of pile hairs. The thickness of the papillary layer (0.7 mm) in the breast skin is greater than that of the reticular layer (0.5 mm), this being reversed (0.7 against 0.8 mm; 0.6 against 0.9 mm) in the skin of the withers and sacrum.

In the upper papillary layer, collagen fiber bundles form a dense plexus. In the deep papillary and in the deep reticular layers the plexus is loose. The collagen fiber bundles are largely parallel to the skin surface. They are twisted and thin, up to 15μ thick. In the upper papillary layer there are pigment cells at a small distance from the epidermis. Their number is small, and they do not form a continuous layer. The arrectores pilorum muscles are well developed, up to 30μ in thickness. In the growing hairs they proceed from the medium parts of the lower follicle (below the sebaceous gland) crosswise, backward, and up to the epidermis.

The sebaceous glands are one-lobed, one anterior and one posterior to each bundle. In the breast skin, the size of the sebaceous glands ($42 \times 147\mu$) is somewhat larger than in the withers skin ($29 \times 105\mu$) and the sacrum ($34 \times 126\mu$). In the sole skin are numerous glomi of tubular glands. The twisted secretion cavities of the glands pass along the dermal papillae, rise to their apexes, pierce the entire epidermal thickness, and open outward.

The skin in the specific dorsal gland region is of greater thickness (2.5 mm). The epidermis (80μ) and stratum corneum (26μ) are not thicker than in other areas of the trunk. The epidermis is moderately pigmented. The hairs are set farther apart than in surrounding skin areas. The hair is in growth. The arrectores pilorum muscles are better developed than in other areas of the trunk, up to 70μ thick. The sebaceous glands have three to four lobes, $185 \times 308\mu$. Huge glomi of tubular glands form a nearly continuous layer. These are separated by connective tissue interlayers. The size of the glomi is $880 \times 1,300\mu$; $1,200 \times 1,400\mu$. Some tubes with cells that are finished with secretion are very large in diameter, up to 200μ. The inner parts of the glomi are not far from the skin surface at a distance of 300 to 400μ. The ducts may be sinuate, long, and very thin (13μ), or short and wide (up to 90μ). The ducts open into the hair funnels. The hairs on the trunk grow in tufts of six to eight. The vibrissae in the trunk skin are typical in structure. Their follicle is surrounded by blood lacunae. The length of the vibrissa root is about 2 mm. The bulb is at a distance of 1.2 mm from the skin surface (vertically). Two small, one-lobed anal glands are near the vibrissa's funnel, opening into the funnel. The arrectores pilorum muscles are not recorded in vibrissae.

ORDO SIRENIA

A description of the skin of the manatee (Matthews, 1929) is available in literature. The skin of *Dugong dugon* and *Trichechus inungius* provided specimens for this study. The skin has a smooth surface. Its absolute thickness is considerable. The epidermis is thick, though it is formed of only three layers: the strata basale, spinosum, and corneum. The inner surface of the epidermis is alveolar. In cross section the alveoli are round, sometimes oval. As in *Cetacea,* the longitudinal epidermal septa pass mostly along the long axis of the body, protruding into the dermis deeper than the transverse septa. The dermis is pronounced, formed by dermal papillae (fig. 156) and a subpapillary layer. In the dermal papillae, thin collagen fiber bundles follow the long axis of the papillae, which contain capillaries. The subpapillary layer, which is not divided, is formed by a dense collagen bundle plexus on the boundary with the subcutaneous fat tissue, which is well developed and

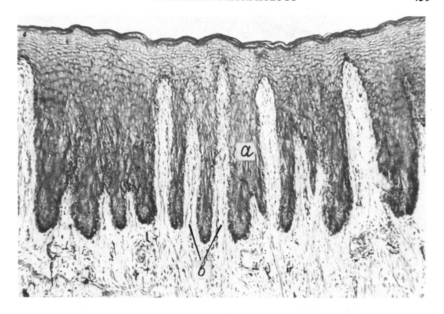

Fig. 156. *Dugong dugon*. Back skin: *a*, epidermal septae; *b*, dermal papillae.

clearly distinguishable from the dermis. Its structure is the same as in land mammals. There are no sebaceous or sweat glands and no arrectores pilorum muscles. There is little pelage, though there are vibrissae on the lips, body, and flippers (fig. 157). The structure of the roots and sheaths is typical of vibrissae in land mammals.

FAMILIA DUGONGIDAE

Dugong dugon

Material. Skin from the back over the forelimbs, the breast in between, and the anterior edge of forelimb of a newborn (body length 113 cm) taken off Western Australia (table 104).

The thickness of the skin and of the epidermis as a whole is greater on the back than on the breast (the same is noted in manatee, Matthews, 1929). The stratum malpighii of the epidermis is 28 to 30 cells deep on the back and 20 to 22 cells deep on the breast. The dermal papillae rise to 410μ on the breast, 420μ on the anterior edge of the forelimb, and 543μ on the back. In the cells of stratum basale (back) a few pigment granules surround the nuclei. Epidermis of the breast and the anterior forelimb edge is pigmented. The stratum corneum is thin and compact (slightly thicker on the anterior edge of the forelimb).

The dermal collagen fiber bundles form a pronounced, compact rhom-

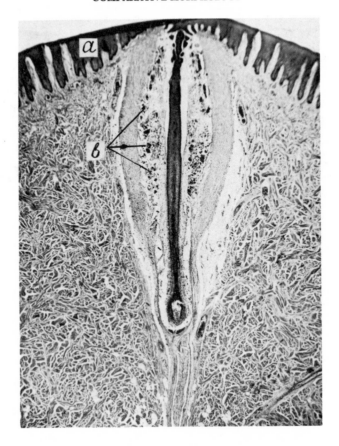

Fig. 157. *Dugong dugon.* Vibrissae root from back skin:
a, epidermis; *b,* blood lacunae.

boid plexus and, in the dermis of the skin, a loop-patterned plexus on fore-limb and breast. About one-fourth to one-third of the dermal thickness on the border with the subcutaneous tissue is formed by horizontal bundles. The back and breast dermis has small groups of fat cells 34 x 42; 41 x 51μ, and a large number of blood vessels, smaller vessels being numerous where it adjoins the epidermis. The subcutaneous fat tissue is thickest on the anterior edge of the forelimb. It is an accumulation of fat cells crossed by mainly horizontal collagen fiber bundles.

The whole body is covered by sparse individual vibrissae growing on skin protuberances, these being well developed on the back, reaching a diameter of 3 mm and a height of 0.5 mm. The vibrissae are 1.2 to 1.7 mm apart on the back, and about a centimeter apart on the belly. The lower lip is covered

TABLE 104

SIRENIA: SKIN MEASUREMENTS

Indexes	Thickness				
	Skin (mm)	Epidermis (mm)	Stratum corneum (μ)	Dermis (mm)	Subcutaneous fat tissue (mm)
Dugong dugon					
Back	19.54	0.64	31	16.6	2.3
Breast	15.49	0.49	31	6.81	8.2
Forelimb					
Anterior part	8.29	0.59	41	7.7	—
Trichechus inungius					
Back	21.80	1.00	123	8.5	12.3
Breast	16.54	0.84	90	6.6	9.1

with vibrissae, but there are none on the upper lip, although there are many protuberances close together. There are more vibrissae on the forelimb than on the hind limb, and they are present on the anterior edges of the limbs.

FAMILIA TRICHECHIDAE

Trichechus inungius

Material. Skin from the back over the forelimb and the breast between the forelimbs of a young specimen (table 104).

Skin, epidermis, and stratum corneum are thicker on the back than on the breast. The epidermal alveoli are 840μ high. The outer surface of the epidermis over the dermal papillae protrudes as a small tubercle. The epidermal pigment granules are mostly in epidermal cells on the back, as a cap over the nucleus, the pigmentation being moderate. The inner epidermis is more heavily pigmented. There is practically no pigment in the breast epidermis.

In horizontal sections, the upper and middle parts of the epidermal alveoli are round or oval. Deeper down they are star-shaped, with numerous processes protruding into the tissue of the dermal papillae.

In contrast to *Cetacea,* the inner parts of the epidermis do not form ridges but penetrate into the dermis as digitate processes. The dermal collagen fiber bundles form mostly rhomboid plexuses. In the inner dermis there are thick horizontal collagen bundles for about 0.5 mm. The subcutaneous fat tissue is common in structure. The body vibrissae are about 1 cm apart.

ORDO PERISSODACTYLA

FAMILIA EQUIDAE

The skin of *Perissodactyla,* especially *Equidae,* has been so poorly studied that it is not possible to offer details of the order or the family. Works on the skin of the domestic horse (Braun, 1935) and the white rhinoceros (Cave and Allbrook, 1959) are available. We studied the skin of *Equus hemionus.*

Equus hemionus

Material. Skin from withers, middle of the back, sacrum, the middle of the barrel, the side of the body under the sacrum, the middle of the belly, the belly beneath the sacrum, callosity (table 105) of an adult male (body length 190 cm) shot by poachers on the Badkhyz reservation on April 20, 1961.

The epidermis is thin and includes stratum basale, stratum spinosum, and stratum corneum. The epidermis of the callosity is very much thicker, has a two-cell stratum granulosum, and an alveolar inner surface (fig. 158). Melanocytes may occur in stratum basale. The pigmentation of the epidermis is moderate (heavy in the callosity).

The dermis is a compact plexus of collagen fibers, thin near the epidermis, gradually becoming thicker toward the subcutaneous nucleus. The papillary and reticular layers meet at the hair bulb level. The collagen fiber bundles are mainly horizontal in the papillary layer and run in all directions in the reticular. Elastin fibers are found only in the papillary layer, where they are very thin and oblique, and in the innermost part of the reticular layer, where they are horizontal and much thicker, often forming loose networks.

In the withers, thick elastin fibers are close to the lower border of the skin, forming loose horizontal networks. Some 0.2 to 0.4 mm of the inner reticular layer is free of elastin fibers. The most developed elastin fiber network is on the belly skin. Blood vessels are numerous in the papillary layer and rare in the reticular layer, where they are much larger. The only fat cells (45×70; $45 \times 55\mu$) in the dermis are on the belly beneath the sacrum, where small groups are found beside the subcutaneous muscles. Subcutaneous fat tissue is found only on the middle of the belly—an accumulation of fat cells and rare bundles of collagen fibers. The fat cells are uniform in size (45×84; $65 \times 95\mu$) all through the subcutaneous tissue.

The sebaceous glands are small in size (the largest are noted on the middle of the belly) and arranged caudal to the hairs (fig. 159). The sweat glands are very shrunken—this is probably seasonal. Only the ducts are pronounced; the secretion cavities are virtually absent. Arrectores pilorum

TABLE 105

EQUUS HEMIONUS: SKIN MEASUREMENTS

| Area | Thickness | | | | | | Sebaceous glands (μ) | Arrectores pilorum muscles (diameter) (μ) |
	Skin (mm)	Epidermis (μ)	Stratum corneum (μ)	Dermis (mm)	Papillary layer (mm)	Reticular layer (mm)		
Withers	4.2	56	39	4.1	1.2	2.9	50 × 27	15
Midback	4.5	84	50	4.4	1.1	3.3	55 × 220	60
Sacral region	4.7	97	32	4.6	1.6	3.0	90 × 275	55
Shoulder	2.3	40	17	2.3	0.9	1.4	80 × 220	35
Middle of side	2.8	56	22	2.7	0.8	1.9	90 × 220	40
Flank	3.1	44	16	3.1	0.8	2.3	70 × 220	30
Midbelly	5.5	56	12	3.4	1.2	2.3	80 × 285	45
Groin	2.2	55	22	2.1	0.8	2.2	55 × 195	20
Callosity	1.1	275	55	no measurements were taken			absent	

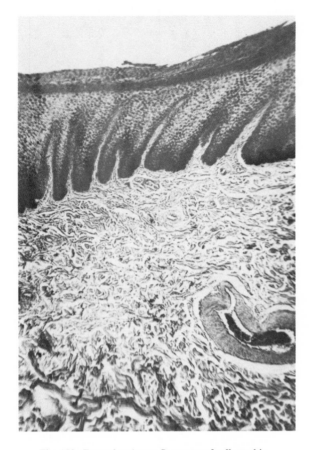

Fig. 158. *Equus hemionus.* Structure of callous skin.

muscles approach the anteriorly disposed hair bursae from the upper to lower end.

The hair was studied only on the withers, sacrum, and middle of the belly. There are two categories: pile and fur (table 106). The hairs grow singly, but sometimes form groups. The fur hairs are slightly wavy; the pile hairs are straight. The fur hairs are almost uniform in thickness throughout, with a small thickening in the middle. The pile hairs have a narrow neck with a sharp thickening above it, after which they taper gradually to the tip. The fur hairs are round in section. The pile hairs are oval with a slightly concave side and medulla along the whole hair, except base and tip. The cuticle cells are not ring-shaped, though ring-shaped cells do occur in the fur hairs.

On the withers there are some elongated fur hairs alongside the common ones. These are slightly wavy, with five to six waves (the common fur hairs have two or three). The hairs form loose groups of one to three pile and

Fig. 159. *Equus hemionus.* Withers skin: *a,* sebaceous glands;
b, hair root.

three fur hairs. There are 3,000 pile and 5,000 fur hairs per 1 cm². The medulla usually comprises 90% of the thickness. In the sacrum skin the fur hairs have three-and-a-half to four waves. The pile hairs have 88% medulla. Here the groups consist of one pile and two or three fur hairs, and there are 2,000 pile and 3,000 fur hairs per 1 cm². On the middle of the belly there are only pile hairs, some of them in growth. They form no groups, and average 9,500 per 1 cm² (including young hairs).

In winter, hairs are much longer (Rashek, 1972) (table 107).

The hair coat of newborn Asiatic wild asses is long and soft. It includes long, sparse pile hairs and shorter, softer fur hairs (Rashek, 1972). The hairs of the newborn are longer than those of adult animals taken in summer. For instance, the hairs on the back in the newborn are 5 to 16 mm long; on the sacrum, 10 to 20 mm; on the side of the trunk, 15 to 30 mm; on the middle of the belly, 28 to 50 mm; on the breast, 3.9 mm.

TABLE 106

EQUUS HEMIONUS: HAIR MEASUREMENTS

Hair	Length (mm)		Max. breadth (μ)	
	M	Lim	M	Lim
Withers				
Pile	9.5	7.4-11.6	80	70-93
Elongated fur	38.5	36.6-41.8	26	22-30
Fur	11.0	9.0-12.4	23	22-28
Sacral region				
Pile	33.7	30.7-36.7	88	70-100
Fur	25.9	20.2-31.0	34	28-39
Midbelly				
Pile	5.0	4.2-5.9	91	90-93

TABLE 107

EQUUS HEMIONUS: HAIR MEASUREMENTS (RASHEK, 1972)

Body area	Length of hairs (mm)				
	Winter				Summer
	Adult females (n=3)	Female 10 months old	Male 32 months old	Male 8 months old	Adult males (n=2)
Back					
(fur hairs)	40-45	—	43	45	12
Sacrum					
(fur hairs)	30-35	45	38	40	11
Side					
(fur hairs)	43-50	—	45	45	12
Middle of belly					
(pile hairs)	50-65	55	38	50	10
Breast					
(fur)	35-40	30	30	35	5

FAMILIA RHINOCEROTIDAE

Ceratotherium simum

The skin is very thick and inelastic, with sparse pelage (Cave and Allbrook, 1959). The epidermis is relatively thin. In the belly skin of an immature rhinoceros, it is about 1 mm thick. The stratum corneum is highly developed, 0.25 mm thick. The stratum granulosum (0.5 mm thick) is spotted. The stratum granulosum usually consists of two cell layers. The inner surface of the epidermis is alveolar. The dermis is extremely thick (up to 18 to 20 mm) with a very compact plexus of collagen fiber bundles running in all directions. In the skin of the occipital eminence, characteristic of white rhinoceros, the dermis is still thicker (up to 45 mm) with elastin fibers occurring only in the septa of the skin arteries. The blood vessels are numerous.

The subcutaneous fat tissue is well developed throughout the body, 2.5 cm thick on the back and 5 cm on the belly. The sebaceous glands are small and saccular, a pair at every hair follicle. The apocrine sweat glands are very large, forming twisted glomi around the base of the hair follicles. Their ducts are spiral. Their lumen is very constricted within the section that pierces the epidermis. The secretion and excretion cavities of the glands are surrounded by large myoepithelial cells. Relatively large arteries approach the glands.

There are hair follicles throughout the skin, though there are no hairs on the surface except for a tuft on the occipital eminence—the number drops off with age. The occipital eminence skin is not different in morphology from the skin on other parts of the body, except that the epidermis is thicker, and the dermis very thick and compact. The occipital eminence, absent in embryos, develops soon after birth.

ORDO ARTIODACTYLA

A lot of work has been done on the skin of domestic ungulates (e.g., Bogo-molova, 1933; Braun, 1935, and others), but the wild species have received much less attention. The skin of *Alces alces* was studied to some extent (Lopukhova and Kulikova, 1958; Sokolov, 1964), *Bison bonasus* (Sokolov, 1962), *Saiga tatarica* (Sokolov, 1961), *Rangifer tarandus* (Bol' and Nikola-yevsky, 1932; Braun and Ostrovskaya, 1933; Efimov, 1938, 1940). There are some data on the structure of the hairs in *Capriolus capriolus* (Flerov, 1928), *Saiga tatarica* (Adolf, 1959), mouflon (Ivanov and Belikhov, 1929), and a relatively large amount of work has been done on the skin specific glands of ungulates (Schaffer, 1940; Ortman, 1960; Quay, 1954, 1955, 1959; Epling, 1956, and others).

We studied the skin of *Sus scrofa, Tayassu tajacu, Moschus moschiferus, Cervus nippon, Cervus elaphus, Capreolus capreolus, Alces alces, Rangifer tarandus, Bison bonasus, Gazella subgutturosa, Saiga tatarica, Rupicapra rupicapra, Capra aegagrus, Capra caucasica, Ovis ammon musimon, Ovis ammon cycloceros,* and *Ovis ammon karelini.*

The skin is medium thick or medium thin. The epidermis and stratum corneum vary in thickness in different groups, most having neither stratum granulosum nor stratum lucidum. The epidermis of many species is pig-mented, the skin glands are well developed, hair is abundant with a highly developed medulla, or relatively sparse without a medulla. The pelage is subject to reduction. The hairs grow singly or in tufts. The specific glands are numerous and diversified.

SUBORDO SUIFORMES

(FAMILIA SUIDAE AND FAMILIA TAYASSUIDAE)

The skin is thick. The thickened epidermis on the inner side may have alveoli that interdigitate with dermal processes. The stratum granulosum is occasionally distinguished in stratum malpighii. The reticular layer is absent or very thin. There are very few elastin fibers, and the subcutaneous fat tissue is usually very thick. The sebaceous glands are large and multilobal. The sweat glands have tubular secretion cavities twisted into large glomi which pass deep into the skin almost down to the bristle bulbs. There are peculiar specific skin glands, carpal in *Suidae* and dorsal in *Tayassuidae*. The arrectores pilorum muscles are very large. The hair divides into bristles, pile, and fur, which grow singly, are not numerous, and have no medulla.

FAMILIA SUIDAE

Sus scrofa

Material. Skin from withers and breast of specimen (1), an adult male (body length 162 cm) and specimen (2), female (body length 130 cm), taken in June near Primorsko-Akhtarsk. Specimen (3), an adult male (body length 149 cm) and specimen (4), female (body length 170 cm), taken in November near Nalchik, specimen (5), female (body length 160 cm) taken in February in Kalinin district. Of specimens (6) and (7), yearlings, only the hairs were studied. Also studied were the carpal gland of specimen (1) and the skin between the hoofs of specimen (3) (table 108).

The epidermis of summer specimens (1) and (2) has small alveoli on the inner surface, from 88 to 110μ in height. The stratum granulosum, one to two cells deep, occurs only in spots. The stratum lucidum is absent. The stratum basale is moderately pigmented. The multirow stratum corneum is thick. The lower border of the dermis has large processes protruding into the subcutaneous fat tissue near the bristle roots, which go deep into the skin. The dermis is not divided into layers, because the roots of the bristles pierce the whole dermis, where predominantly horizontal collagen fiber bundles form a rather loose plexus.

In the dermis of the withers many thick elastin fibers run in the same direction as the collagen bundles. They are less numerous in the breast. In the subcutaneous fat tissue are very thick horizontal collagen fiber bundles (sometimes a continuous layer). The fat cells are the same size at different depths (maximum 80 x 130; 110 x 130μ) in the subcutaneous tissue. The sebaceous glands are peculiar. There are two huge, multilobal glands posterior and anterior to every bristle (fig. 160). The long duct opens into the hair bursa (about 0.6 to 0.7 mm) from the surface of the skin. The seba-

TABLE 108

SUS SCROFA AND TAYASSU TAJACU: SKIN MEASUREMENTS

Indexes	Skin (mm)	Epidermis (μ)	Thickness Stratum corneum (μ)	Dermis (mm)	Subcutaneous fat tissue (mm)	Arrectores pilorum muscles (diameter) (mm)	Sebaceous glands (mm)	Sweat glands diameter (μ)
Sus scrofa								
June								
Withers (1)	—	209	44	9.6	not preserved	0.7	0.55 × 3.9	143
Withers (2)	30.1	209	44	11.2	18.7	0.9	1.50 × 2.7	220
Breast (1)	16.5	154	33	4.1	12.3	—	0.22 × 1.1	165
November								
Withers (3)	26.6	396	88	7.2	19.0	0.9	0.55 × 2.2	495
Withers (4)	24.2	206	40	5.7	18.3	0.9	0.55 × 1.5	—
Breast (3)	—	144	50	5.2	—	—	0.27 × 0.88	142
Breast (4)	22.7	188	44	4.7	17.8	0.6	1.0 × 2.2	143
February								
Withers (6)	—	352	55	5.0	—	0.9	0.28 × 1.1	143
Withers (7)	—	162	17	4.7	—	0.4	0.33 × 1.3	100
Tayassu tajacu								
Withers	3.7	67	56	3.6	—	—	0.55 × 0.29	90
Breast	2.2	112	90	2.1	—	—	0.07 × 0.22	55

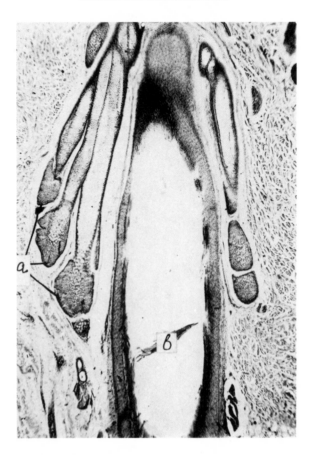

Fig. 160. *Sus scrofa*. Withers skin: *a,* sebaceous gland;
b, bristle.

ceous glands reach down 4.2 mm. At common hairs, the sebaceous glands
(one at each hair) have a single lobe.

There is a sweat gland at every bristle (fig. 161). Their long secretion cavities are twisted into glomi, usually arranged level with the bristle bulbs, with
fat cells near the glomi. The diameter of the secretion cavity of some glands
is very large. In domestic pigs (big white and Siberian North varieties) the
sweat glands are not recorded on the belly and are found on the sides only,
some 5 to 12 cm from the tail root (Sadovskaya, 1961). This finding, which
appears erroneous, contradicts Rzhanytsina (1960), who states that sweat
glands in pigs are developed over the whole skin surface.

Domestic pigs have a considerable number of specific skin glands: anal,
formed of alveolar glands (Ellenberger, 1960; Schaffer, 1940); perineal,
formed of tubular glands (Ellenberger, 1906); preputial (Schaffer, 1940);

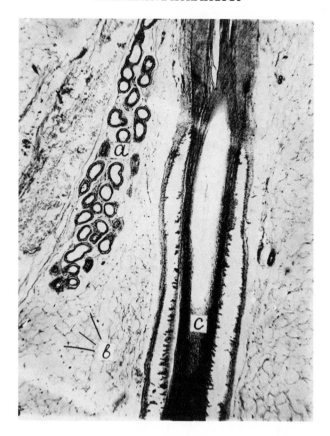

Fig. 161. *Sus scrofa*. Withers skin: *a*, sweat gland; *b*, fat cells;
c, root of bristle

auditory passage, formed of sebaceous glands (Hegewald, 1914); eyelids, formed by tubular glands (Zietschmann, 1904); preorbital (not forming a hollow or a pocket) and muzzle (Schaffer, 1940); carpal, formed of tubular glands (Schaffer, 1940; and others).

As for wild boar, it is briefly noted that they have preorbital and carpal glands (Schaffer, 1940). We studied the carpal glands of boar specimen (1) and the interdigital glands in boar specimen (3). The carpal gland is situated at about the middle of the carpus at its inner surface. The glands open exteriorly with a small aperture. In addition to large, typical sweat gland glomi (the diameter of their secretion cavities reaches 132μ), there are large accumulations of narrower tubular gland secretion cavities up to 100μ in diameter. These are highly twisted into large glomi (fig. 162), the glands forming a continuous layer crossed only by narrow interlayers of the collagen fiber bundles.

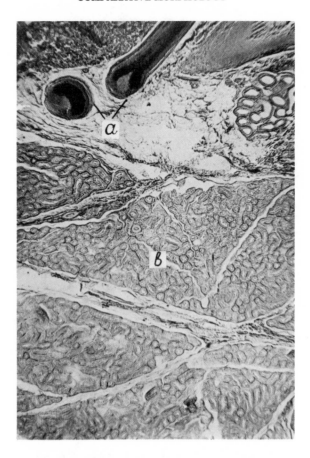

Fig. 162. *Sus scrofa.* Carpal (tubular) gland: *a,* hair bulbs;
b, tubular glands.

The interdigital gland is formed of massive accumulations of sweat glands in the skin (fig. 163). The diameter of their secretion cavities is 120μ. The secretion cavities form larger and more numerous glomi than on the body. The sum of the diameters of the secretion cavities (1 cm of the preparation) is 9.6 mm (against 2.2 to 4.6 mm on the body). The sebaceous glands in the hoof-cleft skin are not found.

Arrectores pilorum muscles are found only at bristles, where they are big (0.7 to 0.9 in diameter) and run from the caudal side of the lower part of the hair bursae (a little above the bulbs), in a few portions crosswise, upward to the posterior bulbs of common hairs and to the epidermis.

There are three categories of hair on the withers: bristles, pile, and fur (table 109). All grow singly. The bristles are straight and taper gradually from the base to the tip, which is split into several points. In cross-section,

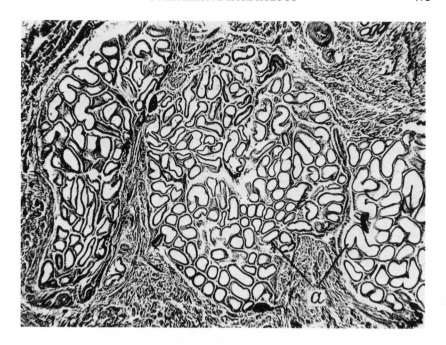

Fig. 163. *Sus scrofa*. Interdigital gland: *a*, glomi of tubular glands.

the shape is irregularly trapezoid. There is no medulla, and the pigment granules are irregular. The cuticle cells have a curved outline and are only 9μ high. There are three to five cells across the bristle. There is an average of 83 bristles per 1 cm². Both pile hairs and bristles taper from the base to a clover tip. The borders of the cuticle cells are strongly sinuate. Their height is 8μ. There are three to five cuticle cells across the hair. The medulla is absent or only slight (fig. 164). In the latter case, the hair's center may have a chain of small alveoli limited by wavy walls. The cortical layer in the pile hairs is very compact, with a peculiar wavy pattern (fig. 164). In section, the pile hairs are indistinguishable from the fur hairs, so an aggregate figure is given, 916 hairs per 1 cm². The fur hairs are long, thin, and almost uniform down the length. There are dashes of medulla. The cuticle cells are mostly ring-shaped with wavy outlines, the cells being higher than in the bristles of pile hairs (19μ).

In late autumn, in specimens (3) and (4), the epidermis is thick and the stratum corneum is relatively thin (table 108). The inner surface of the epidermis is alveolar. The alveoli are some distance apart, their height reaching 220μ on the withers skin and 50 to 110μ on the breast. The thin stratum granulosum contains one to two cell rows. No stratum granulosum is found in the breast epidermis of specimen (3). The stratum lucidum is only present in the withers epidermis. The stratum corneum consists of many rows of

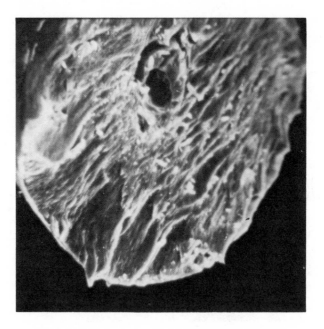

Fig. 164. *Sus scrofa.* Cross-section of withers pile hair.

scales. The stratum basale is lightly pigmented. The structure of dermis and subcutaneous fat tissue is the same as in the summer boar.

In August, before rut, the so-called kalkan is developed in males, a thickened connective layer of skin at the sides of the lower breast and shoulder blades (Dinnik, 1914; Sokolov, 1959; Geptner et al., 1961). The kalkan, which reaches from the shoulder blades backward to the last rib, is thickest (2 to 3 cm) at the ribs and it gradually levels off beyond and toward the neck. In summer, the kalkan is comparatively thin. In a fresh section it resembles a fibrous cartilage or callosity. It is elastic, very viscous, and difficult to cut with the sharpest knife. In histological structure, the kalkan of November *S. scrofa* is a typical structure of skin with very compact plexuses of collagen fiber bundles in the dermal layer. Unlike other body areas, hair bursae, arrectores pilorum muscles and glands are rare in kalkan.

The structure of the ear is very peculiar. On the outside skin of the ear is a thick layer (1.5 to 1.8 mm) of subcutaneous fat tissue not present on the inner side. The sebaceous glands of specimen (3) are smaller in the breast skin than in the withers. The reverse is true in specimen (2). The sweat glands are not as well developed in the breast as in the withers.

There are four hair categories in the withers: big bristles, small bristles, pile, and fur hairs (table 109). In the first three categories, the tips are split into threadlike parts. All the hairs grow singly. The big bristles are straight

TABLE 109

Sus scrofa: hair measurements

	Length (mm)					Thickness (μ)				
	Large bristle	Small bristle	Pile	Intermediate hair	Fur	Large bristle	Small bristle	Pile	Intermediate hair	Fur
Withers										
(1) June	148	—	21.4	—	45.7	433	—	126	—	36
(3) November	115.8	71.0	54.9	—	44.2	44.2	209	125	—	90
Breast										
(3) November	70.7	—	46.8	44.6	38.3	257	—	138	93	60
Withers										
(5) February	155	—	84	—	67	300	—	180	—	40
(6) February	105	—	97	—	39	140	—	80	—	20

and bent slightly backward, round in section (sometimes polygonal with rounded angles), and have no medulla. In a cross section of the hair, pigment granules can be seen arranged like papillae with their apexes toward the center. The cuticle cells are very short (5μ). There are four to six cells per hair, their outline being very sinuate. The small bristles are similar in shape to the big ones, but are shorter and thinner. Their cuticle cells are taller than in the first category, but the cell shape is the same. The pile hairs are still shorter, some straight, some slightly wavy. The cuticle cells are the same in shape as in the other categories, but they are taller. The slight medulla is absent at the base. The fur hairs are not true fur hairs in that they are very thick. Their shafts have four to five gentle waves and a well-developed medulla, which ends in the lower third or half of the shaft. The cuticle cells are 9μ high, with two or three around the hair. There are 41 big bristles and 833 small bristles, pile, and fur hairs per 1 cm².

On the breast, four categories can be distinguished: bristles, pile hairs, intermediate, and fur hairs (table 109). They grow singly and taper gradually from base to tip. The bristles are much smaller than on the withers and have no medulla. The cross section is irregularly round with pigment papillae pointing to the center. The pile hairs are slightly wavy, 46.8 mm long and 138μ thick, with a medulla only in the upper third of the shaft. The intermediate hairs are smaller and more curly than the pile hairs, with a very slight medulla. The fur hairs are as curly as the intermediate hairs, but are shorter and thinner with a very poorly developed medulla, reduced to occasional dashes in some hairs. There are 21 bristles, and 416 pile, intermediate, and fur hairs per 1 cm².

In February, in specimens (5), (6) and (7), the structure of the skin is similar to that of the November specimens. Indexes are given in table 108. Large cavities occur in the sebaceous glands (up to $440 \times 880\mu$). Apparently these are the lobes of glands that have stopped secreting. The sweat glands are very well developed. The secretion cavities, which have a fairly small diameter, are strongly twisted into glomeruli surrounded by fat cells.

There are three categories of hair: bristles, pile, and fur (table 109), 40 bristles per 1 cm², with 150 pile hairs and 1,000 wool hairs on the withers of the female, and 60 bristles, 60 pile hairs and 2,160 fur hairs on the yearlings.

FAMILIA TAYASSUIDAE

Tayassu tajacu

Material. Skin from withers and breast (table 108) of an adult specimen that died in the Leningrad Zoo.

The surface of the skin is irregular, with folds and hollows. The stratum malpighii consists of three to four cell layers. There is no stratum granulosum or stratum lucidum. A large numer of corneous layers form a rather loose stratum lucidum. The epidermis is not pigmented. The bristle bulbs in the withers almost touch the subcutaneous muscles. Consequently, the papillary layer (3.14 mm) is much thicker than the reticular layer (0.5 mm). On the breast, the border between papillary and reticular layers is level with the lower section of the sweat glands, and papillary and reticular layers are of almost equal thickness (1 mm).

The bundles of collagen fibers run mainly at an angle throughout the withers dermis skin, forming a compact network. The dermal collagen bundles are mostly horizontal, and sparse elastin fibers are mainly in the upper and middle papillary layer. The sebaceous glands are small, usually anterior and posterior to the bristle roots. At the roots of common hairs are two very small sebaceous glands (40 × 70μ). The secretion cavities of the sweat glands, highly twisted into glomi, lie deep in the dermis on the withers, a little above the bristle bulbs but under them in the breast skin. The ducts stretch along the roots of the bristles and open into their funnels. On the back of the peccary (about 20 cm anterior to the tail base) is a specific gland in both males and females (7.5 cm long, 3 cm wide, and 1.1 cm thick) formed of sebaceous glands with tubular glands arranged around them (Brinkmann, 1908; Houy, 1910; Frechkop, 1955; Epling, 1956; Schubarth, 1963). The secretion of tubular glands is apocrine and holocrine (Schubarth, 1963). The glands open into the primary cisterns adjoining a larger secondary cistern that opens outward. The secretion has an odor resembling that of skunk. The arrectores pilorum muscles, which are very powerful, comprise several portions and appear to run steeply upward and backward from the anterior side of the bristle root, right to the very epidermis. On the withers the hairs grow in groups of two or three bristles (usually three) and two or three common hairs (one hair beside each bristle), with 250 bristles and 250 common hairs per 1 cm².

Familia Hippopotamidae

Hippopotamus amphibius

Skin thickness in midback of an adult animal is 2.2 cm; in the middle of the side, 3.5 cm; in midbelly, 1.2 cm (Luck and Wright, 1964) (apparently this does not include the thickness of the subcutaneous tissue—V. S.). The epidermis on the back is 65 to 100μ thick, that on the belly is 85 to 130μ. Its inner surface makes up the alveoli, into which dermal papillae rise. The stratum granulosum is not pronounced. The stratum lucidum is well marked. The stratum corneum is thin, solid, and smooth exteriorly. Pigmentation varies greatly. The granules are few and far apart in the belly epidermis and numerous on the back. The sebaceous glands are not found. The sweat glands are large but far apart, opening into the hair funnel or right onto the skin surface.

SUBORDO RUMINANTIA

Ruminants have thinner skin than nonruminants, with little or no subcutaneous fat tissue and smaller skin glands. Usually the pelage is highly developed. The hairs have a thick medulla. A few of the hair categories are noteworthy. The pile hairs are usually without granna.

FAMILIA CERVIDAE

SUBFAMILIA MOSCHIDAE

In winter there are numerous fat cells through the whole of the papillary layer from right under the epidermis, but none in the reticular layer, with sebaceous and sweat glands being absent. In summer, this is reversed. There are no fat cells in the dermis, while sebaceous and sweat glands are well developed. The pile hairs are very long and thick, with a strong medulla (more developed than in other *Artiodactyla*) and a short, thin, medulla-free club. Fur hairs are very few. The hairs grow singly. There are no preorbital glands, but a characteristic abdominal specific gland exists in the males.

Moschus moschiferus

Material. Skin from withers, sacrum, and breast of specimen (1), an adult male (body length 88 cm), that died at the zoo center at Planernaya, Moscow region (table 110), and specimen (2), a summer male taken in the Sayans (table 110).

In winter, a well-developed, multilayered, loose stratum corneum is in an unpigmented epidermis, and the stratum granulosum and stratum lucidum are absent. In the dermis, the papillary layer is distinct from the reticular. Bundles of collagen fibers in the papillary layer run in varius directions to form a loose network with fat cells, ranging from 10×11; $22 \times 34\mu$ to $84 \times 130\mu$, through the whole thickness right to the epidermis, in large groups, forming the background of the papillary layer. There are no fat cells in the reticular layer, which has a very compact collagen fiber plexus (fig. 165) of mainly horizontal bundles. The papillary layer is thinner than the reticular. Not even the ducts of sweat or sebaceous glands are found. Preorbital skin glands and glands at the extremities are also absent, but there is a tail gland, appearing externally as a longitudinal fold on the tail, its excretion smelling strongly of goat. It consists of a combination of superficial sebaceous glands with apocrine type tubular glands under them (Schaffer, 1940).

A male circumcaudal gland is formed largely of sebaceous glands, with only rare tubular glands (Schaffer, 1940). There are anal glands (Campbell, 1837). The females have a medullary area of naked skin in front of the geni-

TABLE 110

MOSCHUS MOSCHIFERUS: SKIN MEASUREMENTS

| Indexes | Skin (mm) | Thickness | | | | | | Arrectores pilorum muscles (diameter) (μ) |
		Epidermis (μ)	Stratum corneum (μ)	Dermis (mm)	Papillary layer (mm)	Reticular layer (mm)	
Winter							
Withers	1.25	39	34	1.21	0.44	0.77	not found
Sacral region	1.25	*		1.20	0.50	0.70	44
Breast	0.83	*		0.83	0.38	0.45	20
Summer							
Withers	2.92	41	*	2.88	1.36	1.52	41
Breast	2.67	20	*	2.65	1.03	1.62	41
Sacral region	2.61	20		2.59	1.03	1.56	61

*Poor condition.

Fig. 165. *Moschus moschiferus.* Withers skin. Network of collagen fiber bundles in reticular layer of dermis (winter).

tal opening, which is abundant in sebaceous glands (Flower, 1875). Adult males have a musk gland at the lower abdomen between navel and preputium (Flower, 1875; Garrod, 1877; Pocock, 1910; Sokolov, 1959; Geptner et al., 1961; and others). This is a saccular intrusion of skin (6 cm by 3 cm), with glands secreting an odorous jellylike secretion. The sac has two openings: one on the belly surface, one in the preputial cavity. The sac walls contain large sebaceous and strongly twisted tubular glands. As this gland is present only in males it probably has a sexual function, though neither the odor nor the quantity vary with the season, so that it might have a different function (perhaps marking territory). Males also have small glands that are largely accumulations of tubular glands around the thighs (Schaffer, 1940).

There are two categories of withers hairs, pile and fur, with perpendicular roots. The pile hairs are long (46.1 mm), slightly wavy, with a thin club (fig. 166). The thickest part (248μ) is just above the club, beyond which they taper gradually toward the tip. In cross section they are round, and the peripheral cells of the well-developed (97.6%) medulla adjoining the cuticle are clearly distinguishable. The cells of the medulla differ in size and shape (fig. 167). Their length increases from periphery to center, the shape growing less regular. The cells of the medulla have rather a smooth surface with

Fig. 166. *Moschus moschiferus.* Withers skin. Pile hairs (summer): *a,* hair peduncle.

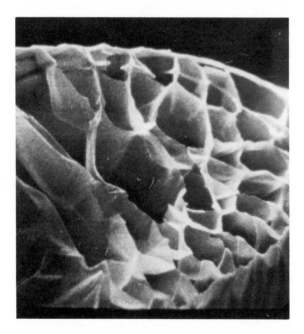

Fig. 167. *Moschus moschiferus.* Medullary cells of pile hair.

occasional folds and ridges. The cuticle cells, which are not ring-shaped, are 19μ high on the pile and 8μ on the fur hairs. The fur hairs are rather long (35.5 mm) and very thin (13μ). The hairs do not form groups or tufts, and pile hairs are much more numerous than fur hairs. There are 1,250 pile and only 83 fur hairs per 1 cm² (a ratio of 15:1).

There are two categories of breast hairs, pile (very numerous) and fur hairs (very sparse). The pile hairs, 750μ per 1 cm², 38.3 mm long, 10μ thick (medulla 98.0%) have a thin club and taper distally all the way. In cross section they are round and the peripheral cells of the medulla adjoining the cortical layer are clearly distinguishable. The fur hairs are very small (1.5 mm long, 10μ thick) and straight, thickening gradually toward the apex. There are only 166 fur hairs per 1 cm², mostly growing right beside a pile hair. Fur hairs do not form groups.

In summer there are hairs in growth and not growing. The stratum malpighii is the same as in winter. In the dermis, the main differences are the absence of fat cells, and the presence of bundles of collagen fibers running in various directions in the papillary layer, forming a rhomboid plexus in the reticular. Elastin fibers are found only in the papillary layer and there are no fat cells in the subcutaneous tissue. The sebaceous glands are saccular (fig. 137) and rather large (withers 72 x 154; 51 x 123μ; sacrum 72 x 215; 61 x 185μ; breast 62 x 226; 51 x 185μ), two glands at every hair. The ducts open into the hair bursae near the surface. The sweat glands have twisted secretion sections (46 to 62μ in diameter) which lie in the middle part of the papillary layer (fig. 168). The arrectores pilorum muscles are thin. The hairs grow singly. There are 903 hairs on the withers, 880 on the sacrum, and 1,500 on the breast per 1 cm².

Subfamilia Cervinae

The skin of the red deer (*Subfamilia Cervinae*) and the roe deer and elk (*Subfamilia Odocoileinae*) are similar enough for a general description. Skin and epidermis are moderately thick, with stratum corneum taking up most of the epidermis. The stratum lucidum and stratum granulosum are usually absent. The epidermis may have pigment. In the papillary layer, the collagen bundles run in various directions. In the reticular layer they are mainly horizontal. Elastin fibers abound in the papillary layer; subcutaneous fat tissue is rare.

The skin glands are well developed and vary seasonally in size. Most of the sebaceous glands are unilobal (some multilobal) in two or three layers at a hair. The sweat glands are tubular and usually very twisted. A metatarsal specific gland is characteristic. The interdigital glands do not normally have pockets or intrusions. The preorbital glands may or may not form intrusions. The arrectores pilorum muscles are well developed.

There are more fur than pile hairs (one to three orders) on the body. The

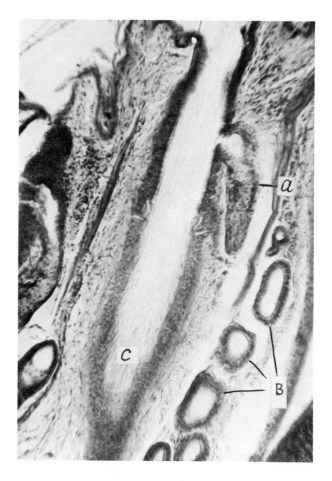

Fig. 168. *Moschus moschiferus*. Breast skin. Structure of outer
parts (summer): *a*, sebaceous glands; *b*, sweat glands; *c*, hair root.

pile hairs are closely similar in all these species. They have no granna, are
slightly wavy, and taper gradually toward the distal. The nongrowing hairs
have a short, thin, medulla-free club. The rest of the medulla is well developed. The fur hairs sometimes grow in tufts of three to eight. Seasonal variations in the hair are considerable.

Cervus nippon

Material. Skin from breast and withers of two adult males taken in winter
in the Khopper reservation (table 111).

In *C. nippon* (7 adult females) from the British Isles (season unknown),

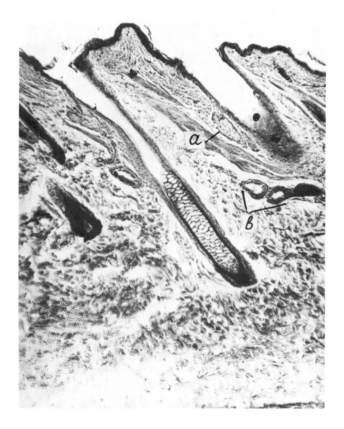

Fig. 169. *Cervus nippon*. Withers skin, general structure: *a*, sebaceous
glands; *b*, sweat glands.

the thickness of the epidermis on the body ($22.07 \pm 0.76\mu$), the size of the
sebaceous glands ($112.21 \pm 3.93 \times 351.10 \pm 9.52\mu$) and sweat glands ($75.86 \pm
3.31\mu$) (Jenkinson, 1972) differ somewhat from our data. On the withers,
the hair follicles are resting. On the breast some are active, some are resting.
The epidermis is not pigmented. The border between papillary and reticular
layers is level with the lower sections of the sweat glands. On the withers the
reticular layer is considerably thicker than the papillary. In the papillary
layer the collagen bundles are mainly horizontal. In the reticular layer they
form a solid rhomboid plexus, which becomes a loopy plexus deeper down.
A pair of single and two-lobed sebaceous glands is at every hair (fig. 169).
The moderately twisted, sausage-shaped sweat glands are large (diameter of
their secretion cavity 82 to 102μ) and open into the hair sheaths. In *Cervus
nippon* from Askania-Nova, the sweat glands number 596 to 729 per 1 cm^2
of the neck in the autumn-winter season. The sebaceous glands number
2,596 to 4,635 per 1 cm^2 (Katzy, 1972). The strong arrectores pilorum muscles are 113 to 205μ thick.

TABLE 111

CERVUS NIPPON: SKIN MEASUREMENTS

Indexes	Thickness						Sebaceous glands (μ)
	Skin (mm)	Epidermis (μ)	Stratum corneum (μ)	Dermis (mm)	Papillary layer (mm)	Reticular layer (mm)	
Specimen (1)							
Withers	6.95	41	*	6.91	1.61	5.30	102 x 205 61 x 133
Breast	4.61	215	82	4.40	2.10	2.30	102 x 369 82 x 256
Specimen (2)							
Withers	6.98	92	31	6.89	1.33	5.56	62 x 226 62 x 133
Breast	5.33	185	51	5.15	4.10	1.05	72 x 513 50 x 205

*Poor condition.

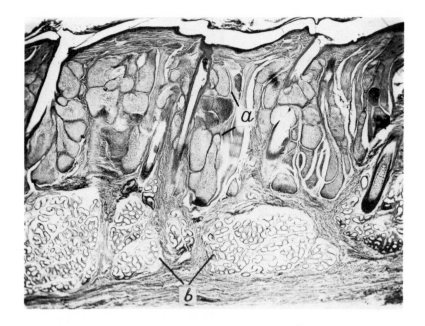

Fig. 170. *Cervus nippon.* Tarsal gland: *a*, sebaceous glands; *b*, sweat glands.

The hairs grow singly in curved rows, numbering 710 per 1 cm² on the withers and 350 on the breast. The carpal gland is formed of hypertrophied, multilobal sebaceous glands (336 × 930μ) and of very large sweat glands (diameter of secretion cavity 90μ) under them. The tarsal gland consists of superficial sebaceous glands (314 × 2,128μ) with huge sweat gland glomi (diameter of the secretion cavity 134μ) underneath (fig. 170). The interdigital glands on the front and hind feet have the same structure. The sebaceous glands are 448 × 1,344μ, and the sweat gland glomi are 762 × 1,900μ on the front feet, 504 × 728μ on the hind. The secretion cavities are 146μ in diameter. The caudal gland on the underside of the tail comprises a large group of very twisted sweat glands and a few pile hairs with big, multilobal glands.

Cervus elaphus

Material. Skin from withers, sacrum, and breast (table 112) of specimen (1), an adult male taken in September in the Caucasus; specimen (2), female (body length 179 cm); specimen (3), female (body length 169 cm); specimen (4), female (body length 198 cm); specimen (5), female (body length 206 cm); and specimen (6), female (body length 193 cm). All taken November 15-16 in the Crimea.

TABLE 112

Cervus elaphus: skin measurements

Indexes	Skin (mm)	Epidermis (μ)	Thickness Stratum corneum (μ)	Thickness Dermis (mm)	Thickness Papillary layer (mm)	Thickness Reticular layer (mm)	Sebaceous glands (μ)	Sweat glands (diameter) (μ)
Withers								
(1) Caucasus (IX)	3.5	28	11	3.47	2.27	1.2	66 × 190	120
(1) Crimea (XI)	3.6	45	28	3.55	1.85	1.7	120 × 550	143
(3) Crimea (XI)	3.3	34	17	3.26	1.86	1.4	80 × 130	80
(4) Crimea (XI)	5.7	*	*	5.7	3.5	2.2	60 × 660	110
(5) Crimea (XI)	5.0	39	17	4.96	2.36	2.6	44 × 190	44
(6) Crimea (XI)	3.9	50	34	3.85	1.75	2.1	66 × 165	80
Sacral region								
(1) Caucasus (IX)	3.8	39	17	3.76	2.66	1.1	33 × 440	187
(2) Crimea (XI)	4.7	45	22	4.65	2.85	1.8	66 × 220	165
(3) Crimea (XI)	4.3	34	17	4.26	2.36	1.9	50 × 440	90
(4) Crimea (XI)	5.0	28	12	4.97	2.77	2.2	50 × 330	80
(5) Crimea (XI)	5.2	39	17	5.16	1.86	3.3	55 × 330	80
Breast								
(1) Caucasus (IX)	3.8	39	11	3.76	1.96	1.8	130 × 495	100
(2) Crimea (XI)	2.5	45	22	2.45	1.75	0.7	120 × 440	120
(3) Crimea (XI)	2.3	25	7	2.27	1.17	1.1	110 × 440	90

*Not preserved.

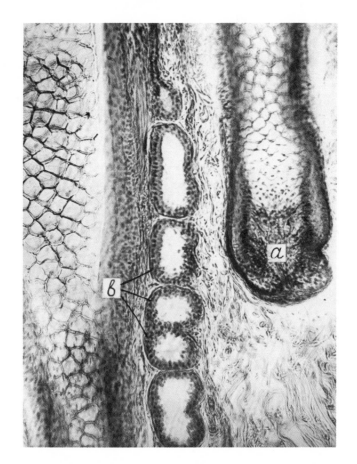

Fig. 171. *Cervus elaphus.* Withers skin: *a,* bulb of growing hair;
b, sweat gland.

In *C. elaphus* (10 adult females and adult male) from the British Isles
(season unknown), the thickness of the epidermis on the trunk (21.38 ±
0.48μ, the size of the sebaceous glands (76.42 ± 2.00 × 328.97 ± 8.00μ) and
sweat glands (92.14 ± 2.83μ) (Jenkinson, 1972) are close to our data.

On the September male, the stratum corneum is compact and the epidermis is slightly pigmented. In the dermis the papillary layer is thicker than the
reticular layer. The horizontal collagen bundles form a compact plexus. A
few horizontal bundles of elastin fibers are found in the dermis close to the
subcutaneous muscles. The sebaceous glands are mostly very small and are
not found at many hairs. The largest and most twisted sweat glands are in
the sacrum skin. They are smaller on the withers (fig. 171). The total sweat
gland diameter per square cm is estimated at not more than 3.6 mm on the
sacrum and 2.7 mm on the breast. In Askania-Nova red deer in the autumn-

winter season, there are 433 to 503 sweat glands and 3,041 to 3,263 sebaceous glands (Katzy, 1972) per 1 cm^2 of the neck skin. The arrectores pilorum muscles are well developed (160 × 220μ).

The hair follicles in specimen (1) are active. The hairs have medullae in the roots and no clubs. There are three size orders of pile hairs and one of fur hairs on the withers. The pile hairs are 49.6, 27.0, 16.9 mm long, respectively. The fur hairs are 0.8 mm. The first order pile is 91.8%. The pile size orders are not clearcut, there being many intermediate sizes. Their thickness decreases distally all the way. Both pile and fur hairs are wavy. The first order of pile hairs has eight to ten waves; the second order has four to six; the third order has three to four; and the fur hairs have two or three. The pile hairs are oval in section with lateral septa perpendicular to the hair surface . The cortical layer is 3μ thicker laterally than at the anterior and posterior. The cuticle cells are 18μ high. There are 333 first order pile hairs per 1 cm^2. The cuticle cells are the same shape on the second and third orders, but smaller (15μ high). There are 250 second and third order pile hairs per 1 cm^2. On the fur hairs the cuticle cells average 10μ in height, the cells being mostly ring-shaped. There are 2,508 fur hairs per 1 cm^2.

There is only one order of pile hairs on the breast, 8.8 mm long, 225μ thick, medulla 92.1%. The fur hairs are 5.6 mm, 26μ, and 47%, respectively. There are many growing hairs with only the tip showing. All the hairs are slightly wavy (two to four waves) and the pile hairs have clubs. The pile hairs are oval in section, but rounder on the withers. In section, the lateral septa of the outer row of medulla cells is clearly seen. The cortical layer of the hair is laterally only 1μ thicker than at the anterior and posterior. The pile hairs (500 per 1 cm^2) do not grow in tufts or groups. Some fur hairs are straight, some slightly wavy. Their very characteristic medulla, proximal and middle, may or may not be continuous. The fur hairs grow in tufts of three to five, 833 per 1 cm^2.

November specimens (2) to (6) are similar to the September animal. The stratum malpighii is two or three cells deep. The sebaceous glands are fairly large, especially on the breast, and the sweat glands (better developed on the breast than the back) have a relatively short secretion cavity with only two or three bends.

The specific skin glands of *C. elaphus* are numerous. There is a subcaudal gland on the underside of the tail (Schaffer, 1940) formed of elongated sebaceous gland lobes, five to ten of them opening into the same hair sheath. These glands are 0.8 mm long and lie in niches between the muscles. Muscular contraction expresses the secretion. There is a layer of apocrine tubular glands beneath the hair bulbs. Their 75 to 140μ diameter secretion cavities are firmly twisted into large glomi. The circumcaudal glandular organ of deer was known in Aristotle's time. It is formed by large accumulations of tubular glands (80μ in diameter) (Rapp, 1839; Zeitschmann, 1903; Schaffer, 1940).

The tarsal gland has so far been described only in *Odocoileus virginianus* (Hershkowitz, 1958; Quay, 1959) and in less detail for some other deer (Zeitschmann, 1903; Pocock, 1923). The tarsal glands of specimens (2) and (3) are on the outer, upper tarsus. The skin thickens and the hair forms a peculiar "eddy" pattern. The sebaceous and sweat glands are much bigger here; the sebaceous glands are 0.44 × 1.1 mm. The diameter of the secretion cavity of the sweat glands is 200μ. The sweat gland secretion cavities equal 6.6 mm per cm of the preparation against 0.6 to 1.9 mm on the body. The glands open into a hair bursa. There is a gradual but clear transition from the common glands around the glandular region to large glands inside the region. Large sebaceous glands are closely packed in a smaller area than the sweat glands, whose glomi are level with the hair bulbs. The structure of this gland in the red deer is the same as in *Odocoileus virginianus* (Hershkowitz, 1958; Quay, 1959).

Unlike some other deer, *C. elaphus* has no skin intrusions or sacs between the digits, but specimens (3) and (4) have large accumulations of huge (0.6 × 1 mm or even 0.9 × 0.9 mm) multilobal sebaceous and tubular glands between the digits of the front foot (fig. 172). They grow in clusters, with several ducts opening together into a hair bursa. Up to 110μ secretion cavities of the tubular glands are strongly twisted into glomi of up to 0.7 × 0.9 mm. The ducts open into hair bursae. The structure of the interdigital gland of

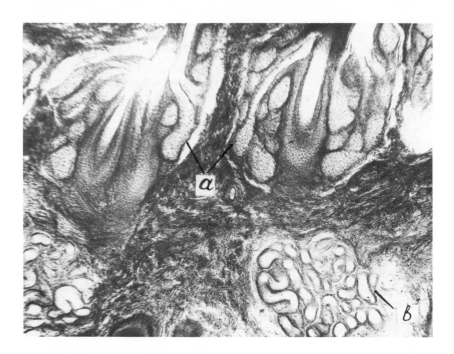

Fig. 172. *Cervus elaphus*. Interdigital gland: *a*, sebaceous gland; *b*, sweat glands.

C. elaphus is similar to that of other species of deer (Schaffer, 1940). The hind foot interdigital glands of specimens (3) and (4) are similar, large alveolar and tubular gland accumulations, though the front foot sebaceous glands are smaller (0.44 × 0.80 mm), and the sweat glands form the main glandular tissue. The secretion cavity diameter reaches 120μ; the tubular gland glomi are 0.8 × 1.0 mm. The gland diameters per 1 cm of the preparation add up to 8 mm.

In specimens (3) and (4) the preorbital gland is cutaneous intrusion into the lachrymal bone fossa. The cutaneous sac is 3.8 cm long, 1.27 cm wide, and 1.9 cm deep. Its anterior is covered with sparse, short hairs, with bulbs at about 1.1 mm below the surface. Numerous alveolar sebaceous glands form regular clusters that reach down almost to the hair bulbs (fig. 173). A bi- or trilobal sebaceous gland of up to 0.9 × 1.1 mm, or two or three separate glands, open together into the hair bursa. The frequent tubular glands are small, though some have a diameter of up to 110μ.

The arrectores pilorum muscles are well developed (110 to 460μ). There are two hair categories on the withers: two orders of pile and one of fur. The first order of pile hairs is 45.8 mm, 199μ thick, medulla 91.4%; the second order is 23.1 mm, 93μ; the fur hairs are 13.6 mm, and 15μ. None have clubs, and all are round or slightly oval in section. The lateral peripheral cells of the medulla, perpendicular to the hair surface, are clearly distinguishable. The cuticle cells, two to four across the wider side of the hair, are not ring-shaped. There are 166 first order pile hairs per 1 cm². On second order pile hairs, the cuticle cells (height 15μ) vary; some are ring-shaped, some are not. There are 416 second order pile hairs per 1 cm². The hair bulbs are 0.8 to 1.1 mm below the surface. Some of the fur hair cuticle cells (8μ high) are ring-shaped, and there are 4,160 fur hairs per 1 cm², which form no clearcut groups or tufts. However, the pile hairs give an impression of irregular rows, and the fur hairs suggest tufts of three to eight. The fur hairs are grouped to the sides, posterior to the pile hairs.

There are two categories of breast hair; pile and fur. These are 17.5 and 10.7 mm long, and 276 and 14μ thick, respectively, with 92.9% medulla in the pile hair and no clubs. Oval in section, the outer medulla cells are clearly seen adjoining the cortical layer, which is 3μ thicker laterally than at the anterior and posterior. On average, there are 583 pile hairs per 1 cm² growing in irregular rows, and 1,166 of very curly fur hairs growing in tufts of two or three.

SUBFAMILIA ODOCOILEINAE

Capreolus capreolus

Material. Skin from withers, breast, and sacrum (table 113) of two adult males; specimen (1) (body length 96 cm) and specimen (2) (body length 121 cm) taken in October 1960, near Nalchik in the Caucasus. Skin from various

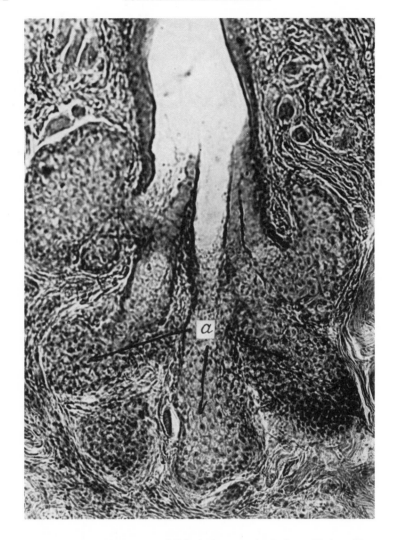

Fig. 173. *Cervus elaphus: a,* multilobed sebaceous gland of preorbital specific gland complex.

body areas of twenty-one Siberian roes (five females and sixteen males) of different ages taken in June-August 1973; in March-April, June-August, and November, 1974.

In *C. capreolus* (eleven adult females) from the British Isles (season unknown), the thickness of the epidermis on the body ($31.66 \pm 1.38\mu$), and the size of the sebaceous glands ($121.10 \pm 3.17 \times 377.79 \pm 8.14\mu$) and the sweat glands ($40.07 \pm 1.24\mu$) (Jenkinson, 1972) are close to our data.

A considerable sex dimorphism is apparent, with age-specific and seasonal changes of skin thickness in males in the head area and in the front

TABLE 113

Skin values of Capreolus capreolus

Indexes	Thickness					
	Skin (mm)	Epidermis (μ)	Stratum corneum (μ)	Dermis (mm)	Papillary layer (mm)	Reticular layer (mm)
Withers						
(1)	4.3	28	6	4.27	2.17	2.10
(2)	2.45	17	4	2.43	1.78	0.65
Sacral region						
(1)	3.7	17	3	3.68	2.08	1.60
(2)	2.7	17	5	2.68	1.98	0.70
Breast						
(2)	2.5	17	5	2.48	1.48	1.00

part of the body. Skin measurements of an adult male and a female taken in summer 1973, and in autumn 1974, are presented in figs. 174 and 175. The skin of females (fig. 174, *a-b*) does not show sharp seasonal differences in the thickness of different body areas, the maximum being within 2.5 mm. In females and males, the thinnest skin is in the groin area, on the interior side of the foot (proximal part), and on the belly (up to 1.0 mm).

The skin on the head and neck of adult males is two to five times thicker than that on the back and on the sides, forming a sort of shield over the front part of the body reaching the scapulae and even farther along the middle of the back. The maximum skin thickness (up to 10.5 mm) is 5 to 9 cm posterior to the horns. In females the skin shield is not formed at all.

The skin in the head and neck area in adult males taken in November is two to three times thinner (fig. 175*b*), and in March adult males three to five times thinner than in adult males taken in summer (fig. 175*a*). By March, the skin thickness resembles that of females (fig. 175*a, b*).

The stratum malpighii of roe specimens (1) and (2) is two or three cells deep. The stratum granulosum and stratum lucidum are absent, and the stratum corneum is thin and compact. The epidermis is not pigmented. The hair follicles are active. The hair roots are long and dilated, so the papillary layer takes up over 50% of the dermis. Horizontal elastin fibers are numerous in the upper papillary layer. They are almost nonexistent in the reticular layer, except where it borders on the subcutaneous muscles.

Small sebaceous glands occur singly, posterior to every hair, 120 x 440μ on the breast and 55 x 380; 100 x 440μ on the back.

The firmly twisted sweat glands running along the hair bursae up to the bulbs are well developed (the diameter of secretion cavities is 66 to 80μ) and larger than on the breast (the diameter 55μ).

The absolute thickness of the epidermis on the head and neck of Siberian

Fig. 174a. *Capreolus capreolus.* Skin of forehead of adult female
(summer): *a,* sebaceous glands; *b,* sweat glands.

roe males (summer) is slightly greater, and the relative percentage of skin
thickness is less than in winter (table 114). The thickness of stratum corneum of the epidermis in winter and in summer is about the same.

Thicker skin on the head and neck in males (summer) is due to the increase in the dermal reticular and papillary layers (table 114). The thickness
of the reticular layer increases more than that of the papillary layer. This
occurs despite the fact that the length of the hair roots in summer increases
due to their growth. The sweat glands attain great development, their secretion cavities being deeper than the hair bulbs.

The network of collagen fiber bundles of the dermis reticular layer in the
head and neck skin of males (summer) becomes more compact and inter-
meshed than in winter (mostly horizontal in winter). Collagen bundles in the
head and neck skin reach 70 to 74μ in summer and 26 to 34μ in winter
(mean values).

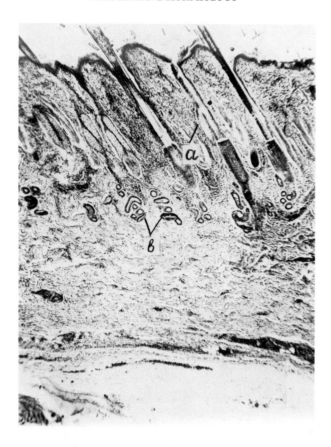

Fig. 174*b*. *Capreolus capreolus*. Skin of forehead of adult female
(autumn): *a*, sebaceous glands; *b*, sweat glands.

Trunk skin samples (males) do not vary greatly in size or structure seasonally (table 114). Head and trunk skin in females is almost similar in thickness and structure in winter and summer (table 115).

The network of collagen fiber bundles in the dermis of the forehead and top of the head in winter and in summer is predominantely horizontal and less compact than in males. The thickness of the collagen fiber bundles of the dermis in the head of females varies scarcely at all seasonally.

The skin shield in Siberian roe males begins to develop early. In 2-3-month-old male calves examined in July-August 1973, 1974, the thickening of the skin in the head and neck are noticeable, reaching a maximum (1.7-2.0 mm) between and behind the ears as in adult males. It may be noted that the skin of a 9-10-month-old roe male taken on March 13, 1974, was almost half as thin as that in 2-3-month-old male calves in summer (not exceeding

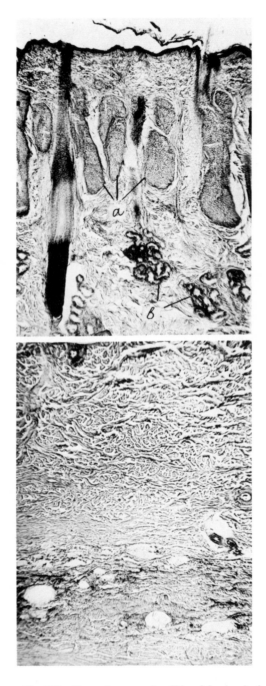

Fig. 175a. *Capreolus capreolus*. Skin of forehead of
adult male (summer): *a*, sebaceous glands; *b*, sweat
glands.

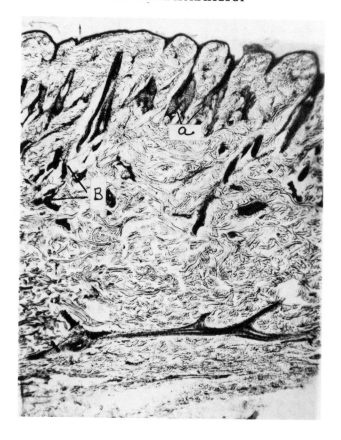

Fig. 175*b. Capreolus capreolus.* Skin of forehead of adult male
(autumn): *a*, sebaceous glands; *b*, sweat glands.

1.0 mm). The skin shield is well pronounced in a 13-14-month-old male
(subadult) taken on July 21, 1973. The skin on the head and on the neck is
about two to five times thicker than in a 9-10-month-old Siberian roe male
taken in March. But the shield is less thick than in adult males taken in
June-August.

Topographic changes in the skin measurements and structure of subadult
males are close to those in adult males, and those of juvenile males are similar to adult females (table 104).

The development of the skin shield in Siberian roe males coincides with
the increase of the size of their testicles. The greatest weight of the testicles
(54.3 g) is recorded in July. In November and March the weight of the testicles is one-fifth to one-eighth of that in July.

Three specific skin glands were studied: the metatarsal, the interdigital of
the front feet, and the interdigital of the hind feet (Sokolov and Danilkin,
1976).

TABLE 114

COMPARATIVE INDEXES OF THE SKIN OF SIBERIAN ROES (JULY, AUGUST, AND NOVEMBER)

Sample	Season	Sex and age	No.	Skin (mm)	Epidermis (μ)	Epidermis (% skin thickness)	Stratum corneum (μ)	Dermis (mm)	Papillary layer (mm)	Papillary layer (% dermis thickness)	Reticular layer (mm)	Bundles of collagen fibers in middle part of reticular layer M	Lim
Forehead	Summer	Male juv.	1	1.9	51	2.6	17	1.80	0.9	50	0.90	32	23-40
		Male subadol.	1	4.5	55	1.2	29	4.40	2.3	52	2.10	50	40-80
		Male adol.	1	6.8	76	1.1	38	6.70	2.0	29	4.70	40	34-57
		Male adol.	2	5.0	57	1.1	29	4.90	2.3	46	2.60	46	40-57
		Female adol.	1	2.3	50	2.1	25	2.20	1.2	54	1.00	25	21-29
		Female adol.	2	2.1	54	2.5	13	2.00	1.5	75	0.50	25	21-29
	Winter	Male juv.	1	1.2	40	3.3	29	1.16	0.8	68	0.36	23	17-29
		Male adol.	1	2.3	46	2.0	29	2.25	1.25	55	1.00	34	17-57
		Male adol.	2	2.4	40	1.7	29	2.36	1.2	50	1.16	34	29-40
		Female adol.	1	2.6	57	2.2	40	2.50	1.5	60	1.00	24	17-29
Top of head	Summer	Male adol.	1	7.5	57	0.8	23	7.40	2.3	31	5.10	62	57-86
		Male adol.	2	6.6	68	1.0	34	6.50	2.2	33	4.30	74	57-97
		Male adol.	3	7.5	47	0.6	17	7.45	2.0	27	5.45	57	46-68
		Female adol.	1	2.2	40	1.8	11	2.16	1.4	63	0.76	26	23-29
		Female adol.	2	2.0	54	2.7	17	1.90	1.1	57	0.80	27	23-34
	Winter	Male juv.	1	2.0	34	1.7	23	1.97	1.4	71	0.57	25	23-29
		Male adol.	1	3.2	57	1.8	46	3.10	1.5	48	1.60	26	23-29
		Male adol.	2	3.2	40	1.2	29	3.16	1.8	56	1.36	34	23-51

TABLE 114—continued

Sample	Season	Sex and age	No.	Skin (mm)	Epidermis (μ)	Epidermis (% skin thickness)	Stratum corneum (μ)	Dermis (mm)	Papillary layer (mm)	Papillary layer (% dermis thickness)	Reticular layer (mm)	Bundles of collagen fibers in middle part of reticular layer	
												M	Lim
Cheek	Summer	Male adol.	3	4.4	103	2.3	21	4.30	1.9	44	2.40	44	31-62
	Winter	Female adol.	2	1.7	72	4.2	21	1.60	1.0	62.5	0.60	35	30-41
		Male juv.	1	1.0	40	4.0	29	0.96	0.7	72	0.26	23	17-29
		Male adol.	1	1.6	51	3.2	34	1.50	1.0	66	0.50	33	23-40
		Male adol.	2	2.2	46	2.1	34	2.15	1.0	46	1.15	26	23-34
		Female adol.	1	1.3	40	3.1	29	1.26	0.7	55	0.56	26	23-29
	Summer	Male adol.	3	5.6	51	0.9	—	5.50	2.2	40	3.30	62	51-72
		Female adol.	2	2.6	38	1.4	—	2.55	1.9	74	0.65	38	21-51
	Winter	Male adol.	1	2.6	29	1.1	17	2.57	1.7	66	0.87	34	29-46
Withers	Summer	Male juv.	1	1.8	23	1.2	6	1.77	0.7	38	1.00	28	23-34
		Male subadol.	1	2.0	80	4.0	11	1.90	0.9	47	1.00	30	17-34
		Male adol.	2	2.8	108	3.8	17	2.70	1.3	48	1.40	33	23-40
		Male adol.	3	3.0	68	2.3	29	2.90	1.9	65	1.00	34	28-46
		Female adol.	1	1.9	46	2.4	17	1.85	1.2	63	0.65	26	17-34
		Female adol.	2	2.0	57	2.8	29	1.90	1.3	68	0.60	29	17-34
	Winter	Female adol.	1	1.7	46	2.7	29	1.65	0.8	48	0.85	27	17-40
Shoulder	Winter	Male adol.	1	3.0	40	1.3	29	2.96	1.8	60	1.16	26	17-34
		Female adol.	1	1.9	40	2.1	29	1.86	0.9	48	0.96	30	23-45
Side	Summer	Male adol.	3	2.1	41	1.9	—	2.05	1.3	63	0.75	18	13-21
		Female adol.	1	2.3	39	1.3	17	2.25	1.3	58	0.95	19	13-25

TABLE 115—Measurements of specific

Sample	Month	Sex and age	N	Sebaceous glands (μ)				
				Length of gland		Thickness of gland		Number of lobes
				M	Lim	M	Lim	
1	2	3	4	5	6	7	8	9
Front interdigital gland	July-August	Males adult	4	536	461-611	255	174-337	2-6
		Female adult	1	343	—	241	—	4-5
	November	Males adult	2	625	479-771	270	226-315	3-6
		Male juv.	1	440	—	133	—	2-3
		Female adult	1	334	—	168	—	3-6
	March	Male adult	1	678	—	213	—	3-6
Middle interdigital gland	July-August	Males adult	4	907	798-984	620	470-753	up to 15
		Male adol.	1	931	—	665	—	over 10
		Male juv.	1	—	—	—	—	up to 10
		Females adult	2	785	772-798	432	425-439	over 10
	November	Males adult	2	785	754-816	519	426-612	8-10
		Male juv.	1	887	—	665	—	up to 10
		Female adult	1	1,064	—	466	—	8-10
	March	Male adult	1	975	—	461	—	—
Distal parts: Metatarsus	July	Male adult	1	519	—	160	—	2-3
	March	Male adult	1	479	—	160	—	—
Metacarpus	July	Male adult	1	341	—	59	—	1-2
Metatarsal gland	July-August	Males adult	4	635	540-770	184	120-231	1-6
		Male adol.	1	652	—	172	—	1-4
		Male juv.	1	342	—	137	—	1-4
		Female adult	1	657	—	154	—	1-5
	November	Males adult	2	408	390-427	112	87-137	1-2
		Male juv.	1	692	—	154	—	1-2
		Female	1	715	—	229	—	1-4
	March	Male adult	1	530	—	188	—	1-4

SKIN GLANDS OF SIBERIAN ROE

Sweat glands (μ)									
Outer diameter		Depth		Length of secretory glomerulus		Thickness of secretory glomerulus		Sum total of outer diameters of secretion cavities per 1 cm of preparation (mm)	
M	Lim	M	Lim	M	Lim	M	Lim	M	Lim
10	11	12	13	14	15	16	17	18	19
64	62-69	898	798-1,064	880	758-1,130	330	239-426	3.45	3.30-3.71
69	—	798	—	998	—	279	—	3.83	—
46	46	1,130	665-1,595	886	842-931	253	239-267	2.74	2.56-2.91
46	—	532	—	931	—	186	—	2.60	—
54	—	879	—	887	—	275	—	3.00	—
46	—	1,463	—	1,108	—	408	—	2.21	—
74	57-110	1,230	1,117-1,463	909	732-1,143	475	352-645	4.47	3.44-4.72
41	—	1,489	—	665	—	279	—	3.10	—
41	—	931	—	625	—	332	—	2.52	—
64	62-67	1,196	1,117-1,276	994	991-997	302	299-305	3.28	3.05-3.52
41	26-56	1,144	931-1,356	800	736-864	266	239-293	2.20	1.80-2.60
44	—	1,330	—	1,108	—	213	—	1.68	—
36	—	1,915	—	957	—	359	—	—	—
62	—	931	—	1,084	—	386	—	4.00	—
62	—	1,596	—	1,197	—	399	—	3.84	—
67	—	1,463	—	1,214	—	266	—	3.84	—
54	—	1,383	—	683	—	293	—	2.70	—
82	69-92	1,077	878-1,197	1,096	825-1,212	385	292-521	4.04	3.80-4.27
46	—	1,064	—	1,046	—	284	—	3.11	—
47	—	931	—	486	—	147	—	3.40	—
51	—	1,197	—	360	—	355	—	3.46	—
51	51-52	1,489	1,463-1,516	935	807-1,064	275	266-284	4.28	3.93-4.63
41	—	1,330	—	892	—	253	—	1.62	—
68	—	1,463	—	842	—	390	—	3.50	—
51	—	1,250	—	896	—	293	—	2.62	—

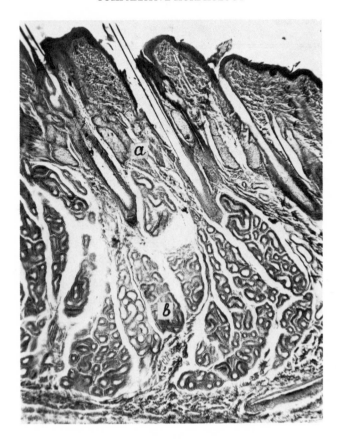

Fig. 176. *Capreolus capreolus.* Metatarsal gland: *a,* sebaceous glands;
b, sweat glands.

The metatarsal gland (fig. 176) is made up of a skin thickening covered
with long hair usually glued together at the base with solidified secretion.
The gland is exterior to the proximal part of the metatarsus (32 x 27 cm).
The metatarsal gland is formed of large, frequently multilobal sebaceous
glands. These are level with the middle of the hair roots, with ducts opening
into the hair bursae and with huge sweat gland glomi set deeper than the
hair bulbs in an almost continuous layer (fig. 176). Sweat gland ducts open
into the hair bursae. A comparison of males and females of different ages
taken in July, August, March, and November has shown equally developed
sebaceous and sweat glands (table 115). Only the March adult male and
underyearling November male had less developed sweat glands.

The interdigital gland of the front extremities is a small cavity with some
very large, multilobal sebaceous glands in its skin that open into the hair

Fig. 177. *C. capreolus.* Interdigital gland of front foot: *a,* sebaceous glands; *b,* sweat glands.

bursae (fig. 177). Sweat gland glomi, also opening into the hair bursae, are set deeper than the hair bulbs. A seasonal and sex comparison of the size of sebaceous and sweat glands in different age animals has not revealed any considerable differences (table 115). Sweat glands are less developed only in the March adult male.

The interdigital gland of the hind extremities is a deep pocket with multilobal sebaceous glands in its skin. They are larger than on the front extremities, and the sweat gland glomi are similar in size to the latter (fig. 178). Sweat glands of the November roes are less developed than in the July-August and March males (table 115).

Well-developed sweat glands are found in the skin of the distal parts of the metatarsus of the front and hind extremities, where they form a network similar in structure to the specific gland. The sebaceous glands in this area are relatively less developed than those between the fingers.

In adult males in summer, a higher development of sebaceous and sweat glands is found on the head and neck compared with other areas of the body (table 116) (fig. 175). Specifically, large sweat gland glomi function in the adult roe males on the neck in summer. The size of the glands on the head and neck in November and March is several times less than in summer. Their secretion cavities are evidently inactive (fig. 175*b*). In winter, the glands are also less developed in the skin of other areas of the body (table 116).

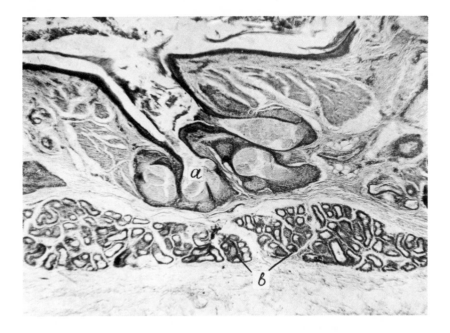

Fig. 178. *C. capreolus.* Interdigital gland of hind foot: *a,* sebaceous glands; *b,* sweat glands.

The females, unlike Siberian roe males, display no differences in the development of sweat and sebaceous glands in the skin of the head, neck, and other parts of the body (table 116). It should be noted, however, that the glands of the females are much better developed in summer than in winter. The development of sebaceous and sweat glands in the skin of 1-2-month-old males is much less than in adult animals. Their size in a one-year-old is similar to that of adult males.

Our data on Siberian roe specific skin glands have proven to be largely in agreement with those for European roe (Meyer, 1968). The interdigital specific glands of the hind feet resemble a deep sac with a narrow excretion opening, and those of the front feet appear to be gradual intrusions. On the second specimen, the front foot interdigital gland was studied. Highly developed tubular glands reach into the dermis far below the hair bulbs. The diameter of the secretory cavities is no larger than in the body skin but is much longer and very twisted. Diameters of the tubular glands cover 9.8 mm per 1 cm^2 of the preparation in the interdigital gland, but only 1.1 mm on the withers, 1.2 on the sacrum, and 0.6 on the breast.

The tubular gland ducts open into the hair funnels. The sebaceous glands are big (0.13 x 0.9 mm) and multilobal. The epidermis in the interdigital gland area is thick, a three-to-four-cell layer stratum malpighii, a single-cell-layer stratum granulosum, a thin stratum lucidum, and a multilayer stratum corneum. There are small processes and hollows on the inner side of the epidermis (our data are on the whole consistent with Solger, 1878; Schaffer, 1940; and Arvy and Bommelear, 1960). *C. capreolus* also has circumanal (Pocock, 1923) and metatarsal glands, these latter being small areas of thickened skin at the inner side of the metatarsus covered with longer hairs stuck together at the base. They excrete a blackish, odorless secretion onto the skin. The preorbital glands are inactive and rudimentary. The powerful arrectores pilorum muscles are 80 to 165μ thick.

There are two categories of withers hairs, pile (two orders) and fur. The first order pile hairs are 23.7 mm long, 286μ thick; the second order are 16.1 mm, 143μ, and 16.0 mm long and thick; the fur hairs are 15μ. The pile hairs medulla is 95.7%. Separate 65 mm long hairs that are almost straight protrude from the general pelage (Flerov, 1928), though no such hairs are found in our sample. The hairs have no clubs; the pile hairs, which grow in neither groups nor tufts, average 583 per 1 cm^2. They are almost round in section, with a pronounced peripheral layer of medulla cells adjoining the cortical layer. The first order cuticle cells (not ring-shaped) are 17μ high. The fur hairs predominate in quantity, there being 1,000 per 1 cm^2 growing in groups of two or three, but not in tufts. The 10μ high cuticle cells are mostly ring-shaped. The breast hairs also fall into two categories. The pile hairs are 25.3 mm long and 179μ thick; the fur hairs are 14.7 mm and 14μ; the pile hair medulla occupies 95.2%. Without club, the pile hairs are round in section with a pronounced peripheral layer. The fur hairs, straight, grow in vague tufts of four to eight, mostly behind or beside the pile hairs. There are 583 pile hairs and 1,749 fur hairs per 1 cm^2.

Alces alces

Material. Skin from the withers, sacrum, breast, shoulder, the side of the body below the sacrum, groin, knee, the base in the front hoof-cleft, hock, and skin at the hind hoof of six adult elks (table 117) taken near Chashnikovo (Moscow region): specimen (1), female, late November; specimens (2) and (3), female, and specimen (4), male, December; specimen (5), female, June; and specimen (6), female, July.

The November and December specimens will be treated as a winter group, and the June and July specimens will be treated as a summer group. Our data are largely consistent with the descriptions of elk skin by Lopukhova and Kulikova (1958).

In most of the samples the stratum malpighii comprises two to three cell layers (Sokolov, 1964); the stratum granulosum and stratum lucidum are sometimes present, while the stratum corneum is usually pronounced. Sometimes the epidermis is pigmented by individual pigment cells and pigment granules in the epidermal cell cytoplasm. The collagen bundles in the papillary layer run in various directions to form a compact plexus (the same

TABLE 116—Measurements

				Sebaceous glands (μ)				
				Length of gland		Thickness of gland		Number of lobes
Sample	Month	Sex and age	N	M	Lim	M	Lim	
1	2	3	4	5	6	7	8	9
Forehead	July-August	Males adult	4	779	502-1,150	249	181-352	1-3
		Male adol.	1	612	—	241	—	2-4
		Male juv.	1	275	—	113	—	1-2
		Females adult	2	410	393-427	178	174-181	1-2
	November	Males adult	2	250	209-291	62	55-68	1-2
		Male juv.	1	174	—	41	—	1
		Female adult	1	274	—	58	—	1-2
	March	Male adult	1	198	—	41	—	1-2
Top of head	July-August	Males adult	4	606	367-736	191	128-275	1-3
		Females adult	2	292	277-308	82	77-87	1-3
	November	Males adult	2	270	267-273	84	66-102	1-2
		Male juv.	1	291	—	68	—	—
	March	Male adult	1	256	—	41	—	1
Cheek	July-August	Males adult	3	696	594-854	232	160-277	1-4
		Females adult	2	320	280-359	89	82-96	1-2
	November	Males adult	2	250	192-308	72	67-77	1-2
		Male juv.	1	110	—	36	—	1
		Female adult	1	195	—	59	—	—
	March	Male adult	1	239	—	55	—	1-2
Neck	July-August	Males adult	3	489	359-718	207	133-318	1-3
		Females adult	2	327	231-423	67	56-77	1-2
	November	Male adult	1	298	—	65	—	1
Withers	July-August	Males adult	3	563	475-728	189	171-198	1-3
		Male adol.	1	696	—	109	—	1-2
		Male juv.	1	259	—	77	—	1
		Females adult	2	391	308-474	74	60-89	1-2
	November	Female adult	1	150	—	82	—	1
	March	Male adult	1	103	—	46	—	1
Side	July-August	Males adult	3	467	407-544	138	124-161	1-2
		Female adult	2	477	405-549	82	77-87	1-2
	March	Males adult	1	97	—	82	—	—
Shoulder	November	Female adult	1	168	—	82	—	—

OF SKIN GLANDS OF SIBERIAN ROE

Sweat glands (μ)									
Outer diameter		Depth		Length of secretory glomerulus		Thickness of secretory glomerulus		Sum total of outer diameters of secretion cavities per 1 cm of preparation (mm)	
M	Lim	M	Lim	M	Lim	M	Lim	M	Lim
10	11	12	13	14	15	16	17	18	19
74	62-82	1,423	1,117-1,782	637	572-830	287	168-359	1.78	0.85-2.15
56	—	1,436	—	501	—	182	—	0.92	—
49	—	585	—	232	—	126	—	1.01	—
51	51	932	931-934	425	—	279	—	1.40	—
44	37-51	1,010	984-1,037	302	272-331	145	138-151	1.27	1.23-1.31
31	—	665	—	190	—	113	—	1.04	—
51	—	931	—	406	—	160	—	1.49	—
31	—	691	—	190	—	100	—	0.67	—
70	51-81	1,270	973-1,356	868	718-975	221	172-273	2.07	1.94-2.21
51	45-57	812	811-814	308	256-359	138	113-164	1.30	1.08-1.53
23	21-25	1,010	957-1,064	359	—	93	—	—	—
41	—	798	—	356	—	96	—	0.57	—
35	—	638	—	381	—	133	—	1.19	—
95	92-100	993	891-1,117	860	751-931	284	259-310	2.64	2.11-3.40
58	54-62	644	571-718	359	333-385	104	103-105	1.61	1.49-1.74
33	31-35	636	540-731	265	256-274	110	103-118	0.82	0.70-0.94
—	—	421	—	—	—	—	—	—	—
47	—	532	—	272	—	97	—	1.04	—
31	—	763	—	423	—	180	—	1.43	—
96	92-103	851	798-957	1,188	1,024-1,496	330	252-372	3.09	2.87-3.32
62	51-72	865	732-998	454	359-550	128	79-177	1.12	0.87-1.37
37	—	1,330	—	590	—	72	—	0.38	—
87	82-92	779	691-904	771	745-791	244	213-263	2.69	2.40-2.85
62	—	851	—	638	—	154	—	1.90	—
51	—	532	—	103	—	36	—	0.74	—
51	51	771	692-851	444	441-446	88	80-87	0.86	0.64-1.08
51	—	479	—	444	—	92	—	0.62	—
51	—	293	—	180	—	46	—	0.35	—
82	82	663	559-718	631	578-683	165	142-186	1.77	1.39-2.01
75	72-78	638	532-745	511	496-525	134	126-142	1.12	0.98-1.27
31	—	266	—	239	—	51	—	—	—
47	—	792	—	466	—	120	—	0.76	—

TABLE 117

ALCES ALCES: SKIN MEASUREMENTS

Areas	Thickness						Arrectores pilorum muscles (diameter) (μ)	Sebaceous glands (diameter) (μ)	Sweat glands (diameter) (μ)	Sweat glands total diameters per 1 cm of preparation (mm)
	Skin (mm)	Epidermis (μ)	Stratum corneum (μ)	Dermis (mm)	Papillary layer (mm)	Reticular layer (mm)				
Withers*										
(1) winter	13.40	90	70	11.01	3.51	7.50	230	50 × 200	100	2.0
(2) winter	8.56	56	44	7.70	1.50	6.20	220	40 × 120	50	0.8
(3) winter	10.75	60	43	7.69	1.59	6.10	230	30 × 30	44	1.9
(4) winter	8.90	50	40	8.85	2.15	6.70	400	22 × 56	88	1.8
(5) summer	9.30	38	11	8.16	4.06	4.10	220	363 × 2,000	132	1.9
(6) summer	10.80	38	11	9.86	4.76	5.10	380	220 × 2,200	88	1.6
Sacral region										
(1) winter	8.20	100	83	8.10	1.50	6.60	350	28 × 168	88	2.8
(2) winter	7.49	56	45	7.44	1.44	6.30	500	50 × 190	60	0.8
(3) winter	7.39	90	73	7.30	1.90	5.40	550	34 × 86	100	3.4
(4) winter	6.20	84	73	6.11	1.01	5.10	440	40 × 56	70	1.2
(5) summer	7.70	39	11	7.66	3.16	4.50	250	440 × 1,540	130	3.5
(6) summer	6.60	50	17	6.55	2.85	3.70	300	300 × 1,650	120	2.5
Shoulder										
(1) winter	—	—	—	4.1	0.8	3.3	110	45 × 224	70	0.3
(2) winter	4.3	67	57	4.4	0.6	3.8	80	22 × 84	28	0.2
(4) winter	4.6	72	66	4.5	1.1	3.4	220	28 × 84	90	1.1
(5) summer	3.23	28	5	3.2	2.1	1.1	220	220 × 800	100	1.1
(6) summer	3.33	28	6	3.3	2.2	1.1	220	220 × 800	100	1.2
Flank										
(1) winter	7.8	73	67	7.7	2.2	5.5	190	55 × 660	77	1.6
(2) winter	6.4	73	67	6.3	0.8	5.5	110	absent	55	0.6

TABLE 117—continued

Areas	Thickness						Arrectores pilorum muscles (diameter) (μ)	Sebaceous glands (diameter) (μ)	Sweat glands (diameter) (μ)	Sweat glands total diameters per 1 cm of preparation (mm)
	Skin (mm)	Epidermis (μ)	Stratum corneum (μ)	Dermis (mm)	Papillary layer (mm)	Reticular layer (mm)				
(4) winter	7.2	67	57	7.1	1.6	5.5	160	absent	77	1.3
(5) summer	4.63	28	6	4.6	2.0	2.6	220	270 × 880	110	1.3
(6) summer	5.33	28	6	5.3	2.2	3.1	110	380 × 880	150	1.4
Breast										
(1) winter	6.1	56	42	6.04	1.24	4.8	130	44 × 330	90	1.7
(2) winter	7.9	61	39	7.8	3.3	4.5	220	160 × 550	120	3.7
(4) winter	4.8	67	50	4.7	1.33	3.4	160	132 × 400	100	3.5
(5) summer	4.9	84	11	4.8	2.01	2.8	130	330 × 1,200	132	2.2
(6) summer	4.7	56	17	4.64	2.54	2.1	120	132 × 1,500	110	2.0
Groin										
(1) winter	5.84	39	22	5.8	2.7	3.1	110	80 × 660	77	2.6
(2) winter	5.7	61	51	5.6	1.8	3.8	100	77 × 550	80	2.2
(4) winter	5.24	45	25	5.2	2.5	2.7	110	55 × 220	100	2.2
(5) summer	5.34	45	12	5.3	2.0	3.3	160	460 × 920	100	5.4
Carpal joint										
(1) winter	3.2	146	22	3.1	1.6	1.5	55	85 × 380	70	3.0
(2) winter	3.8	84	28	3.7	2.2	1.5	66	55 × 330	50	1.7
(4) winter	2.3	78	22	2.2	1.3	0.9	77	110 × 440	60	1.8
(5) summer	2.5	134	33	2.38	1.5	0.88	55	160 × 660	80	4.6
(6) summer	4.2	220	33	4.0	2.9	1.1	110	160 × 710	110	5.4
At hoof of front foot										
(5) summer	3.4	190	22	3.3	2.4	0.9	55	280 × 900	90	4.6
Hock										
(2) winter	4.4	62	39	4.3	1.0	3.3	50	55 × 330	60	2.4

*Subcutaneous fat tissues: (1) 2.3 mm; (2) 0.81 mm; (3) 3.0 mm; (5) 1.1 mm; (6) 0.9 mm.

is indicated by Maslov, 1967). Elastin fibers, horizontal, are most numerous on the border with the subcutaneous muscles. There is subcutaneous fat tissue in some of the skin samples, the fat cells being much the same size throughout the tissue. The interlayer of collagen bundles is horizontal, and elastin fibers stretch along the interlayers. A dense network of capillaries envelops almost every cell. Both sweat and sebaceous glands vary seasonally in size. The sweat gland ducts open into the pile hair funnels. The well-developed arrectores pilorum muscles approach the hair bursae. There are two categories of hair, pile and fur. The pile hairs, which have no granna, form no clearcut groups or tufts.

In the withers, the stratum granulosum and stratum lucidum are absent in winter, and the stratum corneum is very loose (fig. 179). The epidermis is extensively pigmented. The papillary layer is fairly thin, loosened by the roots of the hairs, large arrectores pilorum muscles, and sweat glands. Fat cells (22 × 40; 50 × 84μ) are grouped around the secretion cavities of sweat glands. The subcutaneous fat tissue is 0.8 to 3.0 mm thick, except in the December male. The fat cells are 73 × 84; 56 × 100μ. In the interlayer of collagen bundles and on the border with the subcutaneous muscles, elastin fibers are numerous, forming plexuses and bundles.

There are sebaceous glands at all pile and most fur hairs (in the December male specimen (4), not all the pile hairs have sebaceous glands) (fig. 180). They are small in December specimens (3) and (4). The secretion cavity

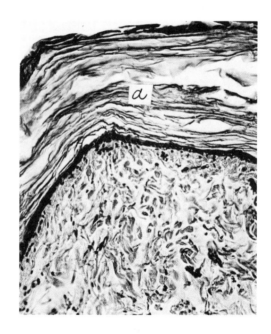

Fig. 179. *Alces alces.* Epidermis: *a,* stratum corneum.

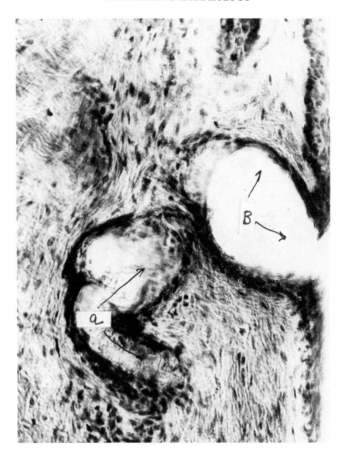

Fig. 180. *Alces alces.* Withers skin: *a,* sebaceous gland; *b,* its duct.

usually consists of several gland cells. The sebaceous gland ducts, mostly 0.3 mm from the surface, are excessively large in proportion to their ducts. The ducts are spherically dilated, tapering as they enter the hair bursa. These appear to be some form of reservoirs where secretion accumulates. The sweat glands are unevenly developed in the various specimens. The twisted secretion cavities are level with the hair bulbs. The arrectores pilorum muscles are 220 to 400μ thick.

The pile hairs are very long (158.8 mm), slightly wavy, and have a club, above which they thicken to 551μ and then gradually taper toward the tip. Most of the hairs are colorless in the proximal half or third, and some are not pigmented at all. In the pile hairs, peripheral cells of the medulla adjoining the cortical layer are pronounced (fig. 181). These are in two layers, cuboidal or slightly elongated in shape. Closer to the middle of the hair, the pattern of cell arrangement is less pronounced. Occasionally, cell walls have

Fig. 181. *Alces alces.* Section of pile hair on withers.

folds and ridges, which obviously lend them extra durability. The medulla
(94.4%) runs along the whole hair, being absent only in club and tip. The
pile hairs are round in section. The cuticle cells are rather high (30μ), three
to six across the hair. There are 350 pile hairs per 1 cm². The fur hairs are
much shorter and thinner (23 mm), very curly, and thickest in the proximal
third. They have no medulla. Some of the cuticle cells are ring-shaped, some
are not. There are never more than two cuticle cells across the hair, averaging 14μ in height. There are 916 fur hairs per 1 cm², growing in not very pronounced tufts of five to eight beside a pile hair.

In summer the hair roots are deeper and the sebaceous glands are visibly
more highly developed. The skin is about the same thickness as in winter.
The stratum malpighii is three to four cells deep. The stratum corneum is
compact. The epidermis is strongly pigmented, especially the stratum
basale. The subcutaneous fat tissue is thin (0.9 to 1.1 mm). The fat cells are
40×60; $56 \times 84\mu$. The sebaceous glands are very large and multilobal, two
or three glands at up to three different levels at every pile hair (fig. 182). The
glands lie anterior and posterior to the hair, the posterior being much larger.
Every fur hair has a sebaceous gland with one or two lobes. The secretion
cavities have about the same diameter as in winter, though the glands themselves are much longer without being more twisted (fig. 183). The deeper
hair roots have increased their distance from the sweat glands, although

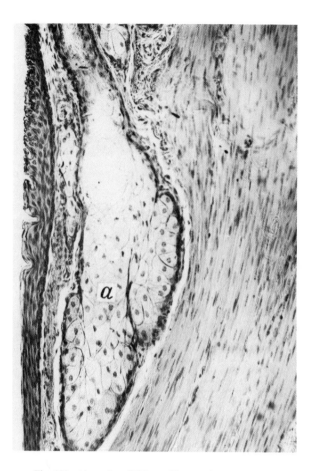

Fig. 182. *Alces alces.* Withers skin: *a,* sebaceous gland.

these are 2.7 mm from the surface. Longer, thicker, colorless winter hairs are found among the short, black summer pile hairs (about one to eight summer hairs). The pile hairs are 67.5 mm long, 259μ thick, and slightly wavy. They have no club, are round in section, with a medulla (86.8%) running down the whole length. The cuticle cells are much the same shape as on winter hair (five to eight cells across the hair). The pile hairs grow singly, 166 per 1 cm². The fur hairs are short (3.2 mm) and thin (26μ), highly pigmented, straight, thickest at the base, tapering gradually toward the tip. Some cuticle cells are ring-shaped, some are not (a maximum of two cells across the hair). They form loose tufts of four to seven, growing behind and sometimes beside pile hairs, at 500 hairs per 1 cm².

Sacrum: In winter, the skin is thicker than on the withers, but the epidermis and stratum corneum are thicker (table 117). There are many big plicae

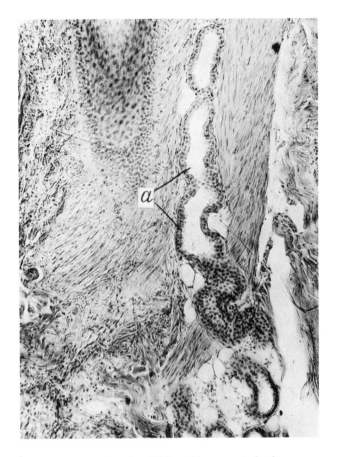

Fig. 183. *Alces alces.* Withers skin: *a,* sweat gland.

at the skin surface. The stratum malpighii is two or three cells deep. The stratum corneum is loose and multilayered. The epidermis is lightly pigmented. The dermis is proportionally thicker than on the withers. The structure of the papillary layer is much the same as on the withers. Elastin bundles, mainly horizontal, are very abundant. In the reticular layer, the largest collagen fiber bundles and the most intricate networks are found right beneath the papillary layer, where bundles rising up at various angles predominate. Nearer the subcutaneous muscles, the bundles decrease in diameter, while the number of the horizontal bundles increases. There are no fat cells in the dermis (specimen 2), but groups of them (56 x 84; 60 x 110μ) are recorded in the curves of the sweat glands. A thin, subcutaneous fat tissue (0.3 and 0.5 mm) is present in specimens (2) and (3) only. The sebaceous glands are very small. Table 78 gives maximum sizes, but most of the glands have only four or five cells. The glands are arranged on two levels at every pile hair. At most fur hairs, the secretion cavities are greatly reduced and only the sebaceous gland ducts are seen. The sebaceous gland ducts dilate spherically at both types of hair before opening into the sheath. The degree to which the sweat glands are developed is varied in the animals under study. The gland bottoms are level with the hair bulbs.

In summer the epidermis and stratum corneum are thicker than on the withers, but much thinner than on the winter elk sacrum. The stratum malpighii is four to six cells deep; the stratum granulosum has only one cell layer; the stratum corneum is compact. The epidermis is lightly or moderately pigmented.

The papillary layer is thickened as the longer growing hairs have longer roots that go down into the skin. The network of collagen bundles is the same as in summer. There are no fat cells in the dermis. The sebaceous glands are very large, but smaller than on the withers. Bilobal sebaceous glands occur even at the fur hairs.

Shoulder: In winter, a multilayered stratum corneum is very loose and the stratum basale is moderately pigmented. Only the dermal collagen bundles are oblique, changing to mainly horizontal on the inner layers of a moderately compact plexus in the papillary layer. There are numerous elastin fibers, but in the reticular layer they are found only where it borders on the subcutaneous muscles. The small sebaceous glands have single lobes, and the sweat glands are very twisted.

In summer the stratum corneum is thin and compact. The epidermis is heavily pigmented. The dermal structure is the same as in winter; large, multilobal sebaceous glands open with one duct into the hair bursae. There is a small sebaceous gland at every fur hair. The sweat glands are better developed than in winter and more twisted.

Side of body below the sacrum: In winter a multilayered stratum corneum is loose, and the epidermis is pigmented in one animal only. In the reticular layer, the collagen bundles run in various oblique directions in a rather loose plexus, while horizontal bundles predominate in the inner reticular layer. Thick elastin fibers form a compact network in the papillary layer, thinning to very few in the inner reticular layer. The sebaceous glands are shrunken, their secretion cavity absent. Only the dilated ducts remain. The sweat glands are twisted but small.

In summer the stratum corneum is thin and compact. The epidermis is slightly pigmented. The dermal structure is the same as in winter. The sebaceous glands are very large. The sweat glands are much larger than in winter animals.

Breast: In winter the skin is thinner than on the back. The outer surface is in many folds. The stratum malpighii is two to four cells deep, and in some places there is a one-layer stratum granulosum. The stratum corneum is multilayered and loose. The epidermis is lightly pigmented. The dermal collagen bundle plexus is compact. The distribution pattern of the collagen bundles in the papillary layer is the same as on the back, though the network of elastin fibers is less developed. Arrectores pilorum muscles are thinner than on the back. There are more horizontal collagen fiber bundles in the reticular layer than on the back, though elastin fibers are very rare. In specimens (1) and (3) the inner areas of the reticular layer contain groups of fat cells (34 × 56μ), concentrated around the blood vessels. The sebaceous glands are better developed than on the back, especially in specimen (2). The sweat glands are larger than on the back, reaching below the pile hair bulbs. The glands are more twisted than on the back.

There are two orders of the pile hairs, length 109.7 and 48.0 mm, thickness 527 and 512μ. The medulla of the first order is 95.3%. They all have slightly wavy clubs and taper gradually toward the tip. The lower third or half is light. The outer layer of medulla cells adjoining the cortex is pronounced. The pile hairs grow singly, 250 per 1 cm^2. The fur hairs (32.9 mm) are long, thin, and wavy, 22 mm long from the skin surface. There is no medulla, and they are thickest (21μ) at the base. They grow in loose tufts of two or three (but not in groups) 583 per 1 cm^2.

In summer the skin is thinner than in winter. The epidermis has a characteristic thin, compact stratum corneum. The stratum malpighii is three to six cells deep; the stratum granulosum one cell. There is a thin stratum lucidum, and the inner surface of the epidermis (which is not pigmented) has small processes. The dermal collagen bundle pattern is much the same as in the winter elks. The papillary layer is thicker than in the winter animals (except in specimen 2), both relative to the dermis and in absolute terms. The sebaceous glands are larger than in the skin of the elks taken in winter, but somewhat smaller than on the dorsal side. The sweat glands are as developed as in the back skin. In specimen (6) they reach deeper into the skin than the pile hair bulbs.

The hair follicles are active. Many pile hair tips are peeping out and there are still a few winter hairs. The pile hairs (29.5 mm long and 127μ thick) do not grow in groups. They are straight, have no clubs, and taper gradually toward the tip. Round in section, they have a characteristically less developed medulla (51.5%). The pile hairs grow singly, 250 per 1 cm^2. The fur hairs are tiny (length 1.3 mm) and thin, hardly showing at all. They are thickest (13μ) in the middle and have no medulla. The fur hairs grow in tufts of two to four, averaging 416 hairs per 1 cm^2.

Groin: In winter the stratum corneum is loose and multilayered. The epidermis is not pigmented. In the dermis, the mainly horizontal collagen bundles form a loose plexus. Elastin fibers are very numerous in the papillary layer (there are even small bundles). They are rare in the reticular layer except in the inner part, where they are numerous. The sebaceous glands are small. The sweat glands are large and very twisted.

In summer there is a thin stratum granulosum in the epidermis. The stratum corneum is solid and rather thick. The epidermis is not pigmented. The collagen bundles and elastin fibers in the dermis are the same as in the winter animals. The sebaceous glands are very large. The sweat glands are well developed, reaching below the hair bulbs.

Carpus: In winter, in specimens (1), (2) and (4), the thickness of the skin varies but is commonly thinner than on the back. The epidermis has internal processes that project into the dermis, increasing its thickness considerably in places. The stratum granulosum is very marked, but of a single cell layer. Specimen (2) has a thin stratum lucidum. The stratum corneum is thinner than on the body, is compact, and the epidermis is without pigment. In specimen (4) the pigmentation is light; in specimen (1), it is moderate. In the dermis, the collagen bundles are mostly horizontal (as are the few elastin fibers). The sebaceous and sweat glands are generally larger than on the body.

The pile hairs divide into three orders: 19.6, 17.0, and 9.4 mm long; 141, 117, and 82μ thick, with 71.3% medulla on the first order. All the hairs have clubs and are oval in section. There are intermediate hairs between the three orders. The medulla cells have thick walls. There are 416 pile hairs per 1 cm^2. They grow singly. The fur hairs (7.9 mm long and 25μ thick) are curly, have no medulla, and grow in tufts of three to five (mainly in fours) beside a pile hair. There are many single pile hairs and some fur hair tufts are away from pile hairs. There are 1,333 hairs per 1 cm^2.

In summer the epidermis and stratum corneum are thicker than in winter animals. Epidermal processes projecting into the dermis are marked, and there is a pronounced single-cell layer stratum granulosum. A compact stratum corneum resembles the stratum lucidum. The epidermis is moderately pigmented. The bundles of collagen fibers in the papillary layer run upward to the epidermis at several angles. The network of the elastin fibers in the dermis is loose. Collagen bundles in the reticular layers are twisted and run in various directions. The sebaceous glands are much less developed than on the body, and the sweat glands rather more.

Only one category of pile hairs is found (12.6 mm long and 125μ thick). They have no clubs and are oval in section with an underdeveloped medulla (43.5%). They grow singly, 412 fully grown and 412 growing hairs per 1 cm^2.

The base in the front hoof cleft: The skin proper was studied only in summer specimen (5). It is thinner than on the body, although the epidermis is considerably thicker. The epidermal processes are well developed. The stratum granulosum is one cell deep, and the stratum corneum resembles the stratum lucidum in structure. The stratum basale is lightly pigmented. In the papillary layer the bundles of collagen fibers are oblique, while in the reticular layer horizontal bundles predominate. Elastin fibers are few in the dermis. The sebaceous glands are very large, but are smaller than on the body. The sweat glands form large glomi, the bottoms reaching deeper than the hair bulbs.

There is only one category of hair (pile), but this is of two orders, length

14.7 and 8.5 mm; thickness 179 and 100μ. The pile hairs have no clubs and grow singly rather than in tufts or groups. The medulla is absent or poorly developed (first order 16.6%). The shafts are arched. The hairs are oval in section. There are 416 first order pile hairs and 250 second order per 1 cm^2.

There are two winter hair categories: pile (three orders) and fur. The pile hairs are 18.9, 13.1, and 9.9 mm long. The fur hairs are 7.6 mm long; thickness 180, 151, 74, and 33μ, respectively. The first order pile hairs have a 63.3% medulla. None of the hairs have clubs. Not wavy, they have an arch-shaped bend. They do not grow in groups. They form a flat oval in the medulla cells and have thick septa. In the fur hairs the medulla is usually absent, though occasionally a line of dashes is present down the whole hair. The fur hairs grow in tufts of three to five beside a pile hair. There are 333 pile and 800 fur hairs per 1 cm^2.

Hock: In winter, the skins of December specimens (2) and (3) show the following: on the inner surface of epidermis (specimen 3) there are alveoli up to 45μ thick. The stratum corneum is moderately compact. The epidermis is not pigmented in specimen (3) and lightly pigmented in specimen (2). The dermal collagen bundles are mostly horizontal, and elastin fibers are few. The papillary layer is thinner than the reticular layer. The sebaceous glands are large. The sweat glands are well developed.

The hind hoof: In winter, on December specimen (3), the skin is thinner than on the heel, the epidermis and stratum corneum being thicker. The epidermal alveoli are well developed (height 34μ). There is a discontinuous stratum granulosum and a thin stratum lucidum. The stratum corneum is rather loose, and the epidermis is highly pigmented. In the dermis the collagen bundles are predominantly horizontal. The relative thickness of the papillary layer is greater than on the heel, while sebaceous and sweat glands are about equally developed.

Samples from the tarsal gland (on the inner side of the tarsus) from a July and a December female show that the histological structure of the gland is similar. Outwardly, the tarsal gland has longer hairs oriented to form a curl. The thickness of the skin is increased to 6.2 mm. The epidermis is thick. In addition to strata basale, spinosum, and corneum, there are a stratum granulosum and patches of stratum lucidum. The bottom of the sweat glands is below the hair bulbs, level with the border between the papillary and reticular layers (4.8 mm from the epidermis). The reticular layer is 1.4 mm thick and the diameter of the arrectores pilorum muscles is as big as in the body (500 to 550μ). The sebaceous glands are large: 0.24 x 1 mm in the July specimen (1) and 0.7 x 1; 0.27 x 7 mm in the December specimen. The presence of large sebaceous glands in the winter animal at the tarsal gland skin is very peculiar when the sebaceous glands are reduced in the body skin. The tubular sweat glands are huge. Their secretion cavities are very large and very twisted, though in most of the glands the diameter of the secretion cavity is small (down to 120μ only in some). The sum of the diameters of secretion cavities (per 1 cm² of preparation) is 7.1 mm. Numerous fat cells are tucked in among the gland loops.

There is no special sac or intrusion in the hoof cleft in *A. alces*. The skin samples (June and December females) reveal no structural differences. There are extremely large accumulations of tubular glands in the skin. The diameter of the very twisted secretion cavities reaches 120μ. The sum of their diameters per 1 cm² of preparation is 8.8 mm (almost the whole skin tissue). The ducts open into hair sheaths. The sebaceous glands (0.4 x 1.1 mm) are also large and lie on either side of the hair. Unlike body sebaceous glands, they do not shrink in winter.

In elks no intrusion of skin is found in front of the eye, as in deer. The skin area of a February female (adult) shows that the skin surface is fur-covered. The skin is 5.5 mm thick, and the dermis 3.2 mm. The hair bulbs are 0.8 mm deep, level with the bottom of the very large, multilobal sebaceous glands (440 x 550; 330 x 700μ), which are arranged in pairs at every hair (fig. 184), forming an almost continuous layer near the epidermis. At a distance of 0.6 to 0.7 mm from the surface is a glandular layer formed of huge tubular gland glomi (1.1 x 1.2 mm, secretion cavity diameter up to 100μ). Their ducts open into the hair sheaths. Beneath the sweat glands is the loose reticular layer. Beneath the reticular layer is the subcutaneous fat tissue.

A skin process or pendant is found beneath the throat of both males and females, triangular, or sometimes sausage-shaped, in transverse section. It is best developed in three- to four-year-old males. Later it becomes shorter and wider. This pendant reaches 35 and even 40 cm in some deer. It is usually between 20 and 25 cm (Geptner et al., 1961).

Pendant samples from four July males (1959), one September and two December females (also 1959) have been studied. The histological structure remains the same at all seasons, even during rut. There are characteristic large hair bulbs that almost touch at the pendant center. The sebaceous glands beside the hairs are small (77 x 270μ). So are the sweat glands, with a secretion cavity diameter that does not exceed 70μ. The glands are slightly twisted. The sum of their secretion cavity diameters (per 1 cm of preparation) (long axis) is estimated at 2.2 mm, so the pendant does not appear to be a glandular formation. Its function remains uncertain.

Rangifer tarandus

Domestic and wild animals are similar, so it would seem likely that their skins are similar. This justifies including the descriptions of the skin of domestic *R. tarandus* by Bol' and Nikolaevsky (1932), Braun and Ostrovskaya (1933) and Efimov (1938, 1940).

The stratum malpighii is between one and several dozen cells deep. There is a stratum granulosum but no stratum lucidum. The outer part of the stra-

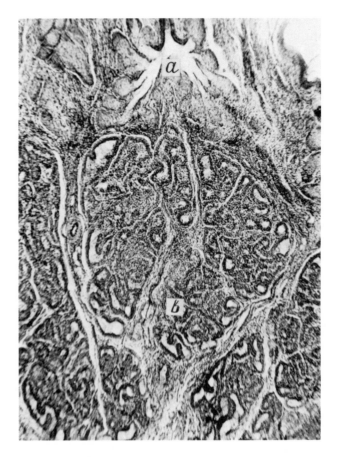

Fig. 184. *Alces alces*. Preorbital gland: *a*, sebaceous gland;
b, sweat glands.

tum corneum is loose. The inner cells of stratum malpighii are pigmented. The border between the papillary and reticular layers is level with the hair bulbs. In the papillary layer, horizontal elastin fibers, blood vessels, and lymph vessels are abundant and mostly horizontal. In the reticular layer, collagen bundles run in various directions. The subcutaneous fat tissue is made up of fine, loose connective tissue. There are abundant blood vessels and fat cells. The sebaceous glands are arranged posterior to the hairs (fig. 185). The size and number of the glands around the hairs varies in different skin areas. Sweat glands are absent in most areas (backbone, neck, sides) but well developed in others (lip). Arrectores pilorum muscles are well developed.

According to Katzy (1972), a reindeer female from Askania-Nova has 783 sweat glands and 2,908 sebaceous glands per 1 cm² of the neck skin in the

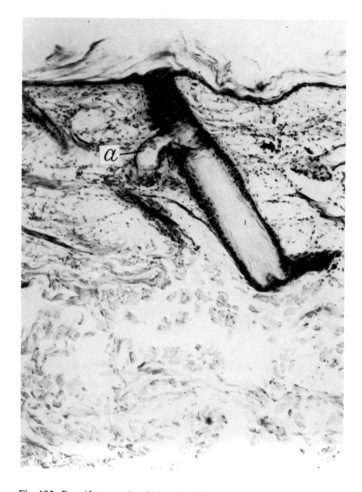

Fig. 185. *Rangifer tarandus.* Skin of middle of back: *a,* sebaceous glands.

autumn-winter season. The preorbital gland is a saccular intrusion with highly developed sebaceous glands in its walls connecting with the hair roots (two or three around every hair) and forming a continuous layer. There is an abundance of small sweat glands beneath the sebaceous glands. Quay (1955) reports that the size of the sebaceous and apocrine sweat glands of the preorbital gland are somewhat larger than in the skin near the preorbital invagination.

The gland of vestibulum nasi appears to be a saccular intrusion opening into the surface of the epidermis, with walls forming a continuous glandular layer. The interdigital gland of the hind extremities also appears as a deep, saccular, upward intrusion in the hoof cleft. Its wall differs from the sur-

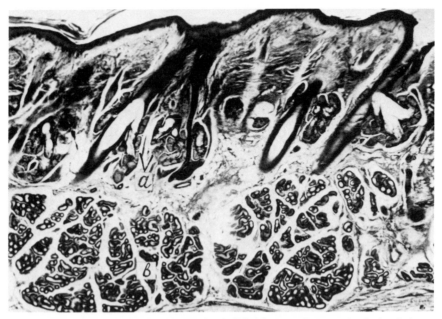

Fig. 186. *Rangifer tarandus*. Interdigital gland of hind foot: *a*, sebaceous glands; *b*, tubular glands.

rounding skin in that it has better developed sebaceous and sweat glands, the latter forming a glandular layer that comprises at least half the skin thickness (fig. 186). The interdigital gland of the front extremities differs from that of the hind extremities in that it does not have a saccular invagination and it does have developed apocrine sweat and, particularly, sebaceous glands (Quay, 1955) (fig. 187).

The tarsal gland comprises large, multilobal sebaceous glands and highly developed apocrine glands, the latter opening directly onto the skin surface with its ducts (Quay, 1955) (fig. 188). The preorbital and interdigital glands are similar in males and females, the tarsal gland being better developed in females (Quay, 1955). The caudal gland is made up of large, superficial, multilobal sebaceous glands and underlying apocrine gland glomi (Lewin and Steflox, 1967).

There are two hair categories: pile and fur, plus elastic hairs on the head (apparently a second order of pile hairs—V. S.), and intermediate hairs on the feet (apparently a third order of pile hairs—V. S.). In the first order of pile hairs, the medulla is well developed (not less than 90%). In the second order it is thin (one to two layers of vacuolized cells), and in the third order it is thin and broken or, in the fur hairs, completely absent.

The skin of the withers, back, sacrum, and dorsal neck is 0.8 to 1.8 thick in spring and 1 to 2.5 mm in autumn. The epidermis is 20μ thick in spring and 60μ in autumn. The papillary layer is 0.3 to 0.8 mm thick in spring and

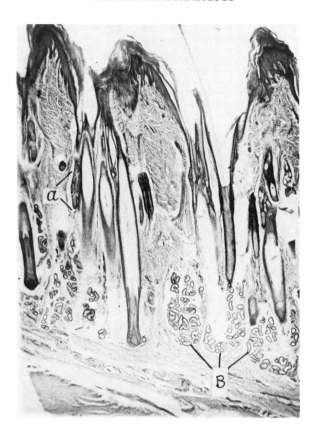

Fig. 187. *Rangifer tarandus.* Interdigital gland of front foot:
a, sebaceous glands; *b,* tubular glands.

0.5 to 1.2 mm thick in autumn. The sac-shaped sebaceous glands are moderately well developed. The sweat glands are absent. The pile hairs are 60 to
300μ thick. The fur hairs are 9 to 21μ thick. There are twice as many fur as
pile hairs (three to four times as many, according to Bol' and Nikolaevsky,
1932).

On the lateral side of the body, the skin is 0.6 to 1.3 mm thick. In autumn
it is 1.5 to 2.5 mm; the papillary layer is 0.2 to 0.4 mm in spring and 0.5 to
1 mm in summer. The sebaceous glands are round and moderately well
developed. The sweat glands are absent. The pile hairs are 150 to 500μ
thick. The fur hairs, of which there are twice as many, are 9 to 22μ thick.

The skin on the upper parts of the legs is 1.5 to 2.7 mm thick. On the
lower parts it is 4 mm. Seasonal variations are slight. The epidermis is 150μ
thick. The stratum granulosum is three cells deep. The dermal layer has
small papillae. The papillary layer on the upper parts of the legs is 0.8 to 12
mm thick, and on the lower parts 1.5 to 2 mm. The sebaceous glands on the

upper parts of the legs are moderately well developed and attain a considerable size near the phalanges in the hoof cleft. There are sweat glands with secretion cavities that are spirally twisted tubules under the hair roots. The number of sweat glands on the skin of the lower parts of the legs increases, especially near the phalanges in the hoof cleft, more on the hind than the front feet. The skin varies a great deal in different seasons.

Familia Giraffidae

Giraffa camelopardalis

Material. Skin from a seven-year-old female (Dimond and Montagna, 1975): muzzle, mane, ventral neck, back abdomen, medial thigh, upper and lower eyelids, the nictitating membrane, dorsal and ventral surface of the tail.

The epidermal thickness ranges from 100μ over most of the body to 30μ on the eyelids. The stratum malpighii is 2 to 7 cells deep. The stratum granulosum is well developed in thick epidermis and is absent in thin epidermis. The stratum corneum is 60μ thick on the back to 10μ in regions with a thin epidermis. In the pigmented skin areas, the basal epidermal layer is very heavily melanized. Pigment granules occur in the cells of all the epidermal layers. On the ventral body surface they are found only in the stratum basale cells. In the stratum basale occur dendritic pigment cells, obviously melanocytes. The dermal papillary layer attains 0.3 to 3 mm in thickness. In the reticular layer, the bundles of collagen fibers run mostly parallel to the skin surface. The elastin fibers are few in the reticular layer and abundant in the papillary (particularly at hair follicles and arrectores pilorum muscles), where they are largely horizontal.

There are few blood vessels deep in the dermis, there being many small blood vessels in the superficial dermal layers. A meshwork of acetylcholinesterase-reactive nerves is in the superficial dermal layers. No specialized nerve endings are revealed (except at some hair follicles). The sebaceous glands (small in most body areas) are uni- or bilobal. They all are situated near the hair follicles opening into the pilary canal. On the lips, nose, and eyelids they are large and multilobal. The sebaceous glands are strongly reactive for cholinesterase and for alkaline phosphatase. Most sebaceous glands are not innervated. There are many melanocytes in the sebaceous gland ducts. Every hair follicle, except those in the lips and nasal septum, is accompanied by a common, coiled tubular gland. A straight, narrow secretion duct of the sweat gland opens into the pilary canal over the sebaceous gland duct. In the skin between the nostrils, deeper than the apocrine glands, there are serous-type tubuloacinar glands. Their ducts open onto the skin surface. The secretion cells are strongly reactive for acetyl- and butyryl-cholinesterase and for alkaline phosphatase. There are large tubuloacinar

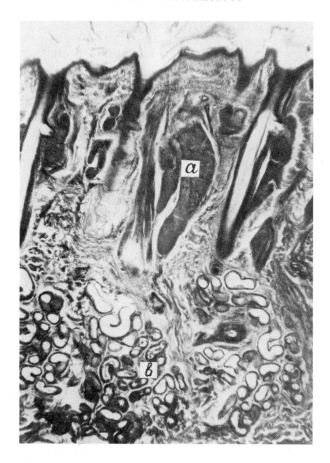

Fig. 188. *Rangifer tarandus.* Tarsal gland: *a,* sebaceous glands;
b, tubular glands.

mucous glands in the lower lip skin. The hairs follicles are situated singly. Occasionally two to three hairs form a tuft. The hairs form no groups. There are arrectores pilorum muscles.

FAMILIA BOVIDAE

Bovidae comprises a group of which many members may differ from one another greatly. The skin of the species under study varies from thin to moderately thick. The skin glands are well developed and vary little with the season. Specific glands are diversified in different members of the family. Hair divides into pile hairs or a combination of pile and fur hairs. Occasionally guard, pile, and fur hairs are found. The arrectores pilorum muscles are well developed.

SUBFAMILIA BOVINAE

The skin is moderately thick. The epidermis of the body often contains stratum granulosum and stratum lucidum in addition to the basal and corneous layers. The epidermis is pigmented. The reticular layer is a compact plexus of collagen fiber bundles which run in various directions. The sebaceous glands may be multilobal. The sweat glands are tubular and twisted. The hairs divide into three categories: sparse, straight guard hairs (they are absent in all other members of this family); more numerous, slightly wavy pile hairs; and abundant, thin, curly fur hairs. The medulla is not pronounced in the pile and is absent in the fur hairs. The cuticle cells are of a unique shape, occurring in no other *Bovidae*. The cells are low and long with the edges strongly sinuate. The hairs grow singly.

Bison bonasus

Material. Skin from withers, middle of the back, sacrum, shoulder, middle of barrels, flank, chest, belly, groin, elbow, carpus, hock, and skin at the hind hoof of four males (specimens 1, 2, 3, and 4, aged two to four years), improved by aurochs, with a blood ratio of 58/64 aurochs + 16/64 bison, taken in the Caucasian reservation, and three males (specimens, 5, 6, 7, aged two to four), seven to eight generations of absorptive aurochs breeding with bison blood traces, taken in August in the Khoper reservation.

The epidermis is relatively thin and on most specimens consists of strata basale, spinosum, granulosum, and corneum (Sokolov, 1962). Some pigment cells occur in stratum basale. The dermis is formed by a compact plexus of collagen bundles and by elastin fibers that are numerous only in the outer and the inner layers of the skin. The collagen bundles are very thin at the epidermis, becoming thicker deeper down but thin again near the subcutaneous muscles. Fat cells are usually absent in the dermis. Blood vessels are abundant in the outer dermis.

The border between papillary and reticular layers is level with the hair bulbs or the bottom of the sweat glands. In the papillary layer, the relatively thin collagen bundles are mostly horizontal, as are the thin elastin fibers. In the upper and middle reticular layer, the collagen bundles run in all directions forming a very intricate plexus. In the inner reticular layer, the collagen bundles are again mainly horizontal. Thick elastin fibers lying among the collagen bundles often form plexuses and even some loose bundles. Many of the sebaceous glands have several lobes. They are arranged posterior to both pile and fur hairs, and their ducts open into the hair follicle below the sweat gland duct.

The sweat glands have moderately twisted secretion cavities and straight ducts that open into the hair funnels posterior to the hairs, not far below the skin surface (fig. 189). Arrectores pilorum muscles at the hair roots extend

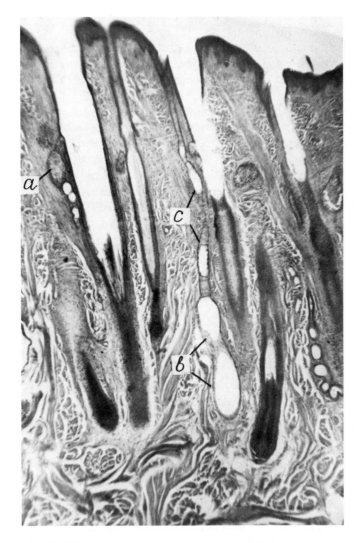

Fig. 189. *Bison bonasus.* Structure of outer parts of sacrum skin: *a,* sebaceous glands; *b,* sweat glands; *c,* their ducts.

from partway up one hair root across to the bottom of the root of another hair, posterior to the first.

Aurochs have three categories of hair: straight, thick guard hairs; slightly wavy, moderately thin pile hairs; and very thin fur hairs. The guard and pile hairs are the same length, much longer than the fur hairs (table 118). There are very few guard hairs, but abundant fur hairs. The hairs grow singly, not in tufts or groups. The guard and fur hairs are of uniform thickness throughout their length, except that they taper a little at both ends. The upper quarter of the pile hair shaft thickens slightly into a small granna. The cross section of guard and fur hairs and the lower shaft of the pile hairs is round, the granna of the pile hair being flattened to an oval. The guard and pile hairs have a medulla, which in the granna of the pile hair is 16 to 50% thick. The fur hairs have no medulla. The cuticle of the fur hairs is ring-shaped, that of the guard and pile hairs is not.

The shape of the cuticle cells does not vary with the season. The pile hair cuticle cells of these animals were compared with purebred *B. bonasus* from the Belovezhskaya Puscha and an adult male *B. bison* from the Moscow University Zoological Museum and the Department of Zoology at the Timiryazev Agricultural Academy. The cell outline and size are similar in all. The relationship between the maximum height of the cuticle cell and the thickness of the pile hair on the withers is as follows:

Species	Cell height (n=30)	Hair thickness (n=10)	Height/thickness ratio
Bison bison	17	112	1:6.5
Bison bonasus			
Khoper Reservation	15	100	1:6.6
Caucasian Reservation	15	106	1:6.7
Belovezh	13	77	1:5.8
Belovezh	15	93	1:6.1

The topographical indexes for the skin of *B. bonasus* will be found in tables 119 and 120.

Withers: In winter, in the epidermis (specimens 2, 3, and 4), the stratum granulosum consists of a single cell layer. In specimen (1), the stratum lucidum is very thin. The pigmentation of stratum basale is light and is absent altogether in specimen (3). The network of elastin fibers bordering on the subcutaneous fat tissue is very well developed: there are plexuses and loose bundles of up to 60μ thick. Specimen (4) has groups of fat cells (56 × 84; 56 ×78μ) in the inner layer of the dermis. The sebaceous glands are small. The sweat glands are small and slightly twisted with a squamous glandular epithelium.

In summer the stratum granulosum is one cell deep. There are small alveoli on the inner side of the epidermis and there is no pigment. Dermal papillae 50 to 110μ high interdigitate with the epidermal alveoli. In the inner part of the dermis, elastin fibers are abundant. The sebaceous glands are large. The sweat glands are also large and very twisted, with active glandular cells.

Middle of back: In winter, a single-cell stratum granulosum appears in patches in the epidermis (specimens 2, 3, and 4), and a very slightly developed stratum lucidum (specimen 1). Specimen (4) has slightly developed

TABLE 118

BISON BONASUS: HAIR MEASUREMENTS (WINTER AND SUMMER)

Winter

Sample	Hair	Hairs per 1 cm^2	Length (mm)	Maximum thickness (mm)
Withers	pile	500	105	0.10
	fur	1,750	57	0.04
Breast	pile	416	75	0.08
	fur	1,166	52	0.04

Summer

Sample	Hair	Hairs per 1 cm^2	Length (mm)	Maximum thickness (mm)
Withers	pile	416	68	0.12
	fur	666	34	0.04
Breast	pile	333	72	0.10
	fur	250	36	0.04

TABLE 119

BISON BONASUS: SKIN MEASUREMENTS (WINTER) (MEAN DATA FOR 4 SPECIMENS)

| Area | Skin (mm) | Thickness | | | | Sweat glands secretion cavities (diameter) ( ) | Sebaceous glands (μ) | Arrectores pilorum muscles (diameter) (μ) | Sweat glands total diameters at hair bulb level per 1 cm of preparation (mm) |
		Epidermis (μ)	Stratum corneum (μ)	Dermis (mm)	Papillary layer (mm)				
Withers	8.1	191	80**	7.9	2.4	66	116 x 376	89	1.0
Midback	8.3	118	71**	8.2	1.9	80	130 x 279	104	0.99
Rump	5.7	97	70	5.6	1.4	78	100 x 243	99	1.23
Shoulder	5.8	86	46**	5.7	1.4	64	90 x 274	87	1.43
Middle side*	9.1	67	35	9.0	1.3	82	82 x 192	76	0.41
Flank	8.5	65	37	8.4	1.3	79	77 x 241	72	1.00
Breast	11.1	208	118	10.9	3.2	106	162 x 690	72	1.58
Midbelly	10.3	88	60	10.2	2.8**	97	153 x 514	65	2.12
Groin	6.9	151**	103**	6.8**	2.3	95	142 x 379	50**	3.37
Elbow	7.2	145	98	7.1	1.8	86	96 x 346	45	1.91
Carpal joint**	5.0	96	58	4.9	1.8	102	96 x 294*	47	1.82
At front hoof**	4.9	78	35	4.8	2.8	48	187 x 503	35*	0.39
Patellar joint**	7.8	106	60	7.7	1.2	64	58 x 214	40	1.15
Hock**	5.3	96	50	5.2	1.2	73	114 x 346	57	1.25
At hind hoof**	6.3	276	120	6.0	3.4	45	164 x 548	31	0.39

*Mean for two specimens.
**Mean for three specimens.

TABLE 120

Bison bonasus: skin measurements (summer) (mean data for 3 specimens)

Area	Skin (mm)	Thickness				Sweat glands diameter (μ)	Sebaceous glands (μ)	Arrectores pilorum muscles diameter (μ)	Sweat glands total diameters at hair bulb level per 1 cm of preparation (mm)
		Epidermis (μ)	Stratum corneum (μ)	Dermis (μ)	Papillary layer (mm)				
Withers	10.6	58	15	10.5	3.3	173	127 × 411	61*	2.17
Midback	6.9	61	13	6.9	2.5	128	127 × 456	91	1.60
Rump	6.6	63	20	6.5	1.6	123	141 × 378	82	1.94
Shoulder	7.3	51	8	7.2	1.8	118	105 × 296	59	2.23
Middle of side*	5.4	40	7	5.4	1.3	123	68 × 301	55	2.46
Flank	6.9	40*	7*	6.9*	1.2*	127	75 × 137*	103	2.05
Breast	16.0	63	28	15.9	3.4*	123	209 × 730	48*	2.21
Midbelly	7.3	61	10	7.2	2.2	136	173 × 365	68*	2.13
Groin	6.6	61	11	6.5	2.1	142	155 × 365	41	5.62
Elbow*	5.4	40	10	5.4	1.7	130	109 × 253	41	2.18
Carpal joint*	3.5	55	12	3.4	1.6	116	109 × 287	54	2.08
Patellar joint	7.5	45	12	7.4	1.5	130	109 × 299	49	2.24
Hock	3.9	42	12	3.9	0.9	123	116 × 273	—	2.01

*Mean for 2 specimens.

alveoli on the inner epidermal surface. The pigmentation is the same as on the withers (specimen 3) or stronger (specimens 1, 2, 4). Specimen (4) has small dermal papillae (30μ) and some small fat cells (40 x 60; 50 x 60μ) in the inner dermis, where elastin fibers are numerous (though less so than on the withers). They do not form bundles (specimens 1, 2, 3) except in specimen (4). Sebaceous glands are small sweat glands with a squamous glandular epithelium (specimens 1, 3, 4). In specimen (2) some sweat glands are secreting.

In summer the epidermis includes a single layer stratum granulosum and, in specimen (1), patches of stratum lucidum. The inner side of the epidermis has small alveoli. Pigmentation is poor or is absent. Dermal papillae reach 50 to 110μ. Sebaceous glands are large. The sweat glands are large and twisted.

The sacrum: In winter only specimen (3) has a thin stratum granulosum. Epidermal pigmentation is moderate. Elastin fibers are abundant in the inner dermis but do not form any bundles. Specimen (4) has fat cells (40 x 70; 62 x 70μ) in the inner dermis. Sebaceous glands are poorly developed. Sweat glands are small and not twisted. In specimens (2), (3) and (4) the secretion cavity is dilated into a sac.

In summer the stratum granulosum is present in the epidermis. Specimen (5), with small (80μ) alveoli on the inner surface of the epidermis, also has a stratum lucidum. Pigmentation is poor, and is absent in specimen (7). Sebaceous glands are large. The sweat glands are smaller and less twisted than on the withers, but are very active.

The shoulder: In winter there is no stratum granulosum or stratum lucidum in the epidermis. Pigmentation is heavy in specimen (1) and very light in specimens (2), (3) and (4). Elastin fibers are numerous in the inner dermis, where they often form plexuses and bundles. The sebaceous glands are small. The sweat glands are small and slightly twisted with a squamous glandular epithelium and a secretion cavity that is narrow (specimens 1 and 3) or saccular (specimens 2 and 4).

In summer the epidermis has a slightly developed stratum granulosum and (in specimens 5 and 7) small alveoli on the inner surface. In specimen (5), the epidermis is moderately pigmented. The other two have no pigmentation. The dermal papillae are 50 to 60μ high. The sebaceous glands are medium in size. The sweat glands are slightly twisted but large with a columnar glandular epithelium.

The middle of barrel: In winter (samples from specimens 1 and 3) only specimen (3) has a stratum granulosum—it shows up in the preparation as a broken string of cells. In specimen (1), intensively pigmented areas of stratum basale occur in the epithelium, while specimen (3) has practically no pigment. The elastin fibers in the inner dermis are few. Sebaceous glands are small in specimen (1) and large in specimen (3). Sweat glands are either very small with a squamous glandular epithelium (specimen 3), or are not seen at all (specimen 1).

In summer (samples from specimens 5 and 7) in the epidermis both stratum granulosum and stratum lucidum are seen. Pigmentation is poor. The sebaceous glands are small. The sweat glands are only lightly twisted, but are large and active.

The flank: In winter the stratum granulosum occurs only in specimens (2) and (3), and then in patches. There is no pigment in the epidermis of specimen (4), and very little in specimens (2) and (3), but there are patches of strong pigmentation in specimen (1). The network of elastin fibers in the inner dermis is not very well developed. The sebaceous glands, medium in specimen (2), are small in the others. The sweat glands are small and slightly twisted with secretion cavities that are narrow (specimen 1) or dilated (specimens 2, 3, and 4).

In summer there is a single-cell stratum granulosum in the epidermis and (specimens 5 and 6) in the stratum lucidum. The pigmentation is weak (specimen 5) or moderate (specimens 6 and 7). Specimens (5) and (6) have small alveoli (up to 40μ) on the inner epidermal surface. The sebaceous glands are small. The sweat glands are large, only slightly twisted, and active.

Chest: In winter there is a single cell stratum granulosum in the epidermis and a thin stratum lucidum. Pigmentation is light. There are 110 to 230μ alveoli on the inner epidermal surface. The sebaceous glands are large. The sweat glands are only slightly twisted, but are larger than in other parts of the body. Many glands have dilated secretion cavities, though the glandular epithelium is squamous on all glands. In specimen (4) all glands are secreting.

In summer there is a thin stratum granulosum and, in specimens (5) and (6), a stratum lucidum. The inner epidermal surface has large alveoli 60 to 160μ. Pigment is absent. Sebaceous glands are very large and multilobal. The sweat glands are large and active.

The belly (*middle*): In winter specimen (2) has a pronounced stratum granulosum, but in specimens (1) and (4) it is discontinuous, and in specimen (3) is absent. The stratum lucidum occurs only in specimen (2). The epidermis is lightly pigmented (specimens 1, 3, and 4) and moderately in specimen (2). In specimens (2) and (4) the inner surface of the epidermis has alveoli up to 132μ. The sebaceous glands are large, but smaller than on the breast. In specimens (1) and (3) the sweat glands are lightly twisted and have a squamous glandular epithelium, and in specimens (2) and (4) they are more twisted than in the body areas already discussed. Most of the glands are secreting.

In summer all the animals have a single-cell-layer stratum granulosum, and specimen (5) has small alveoli 60μ on the inner epidermal surface. There is no pigment. The sebaceous glands are small in specimen (5) but rather larger in two of the others. The sweat glands are large and very twisted.

The groin: Only specimens (2) and (3) have a broken stratum granulosum. Pigment is absent. The collagen bundles form a less compact plexus in the reticular layer than in the other parts of the body and are mostly horizontal. The sebaceous glands, medium sized, are largest in specimen (3), and the sweat glands are the largest in all these animals although the epithelium in the secretion cavity is squamous.

In summer the stratum granulosum is present in all the animals. There is no pigment. The sebaceous glands are very large. The sweat glands are larger than in other samples and are very twisted and active.

Elbow: In winter (samples from specimens 1, 2, and 3) the single-layer stratum granulosum and lucidum are pronounced in all the animals. Pigmentation is slight (specimen 1) or absent (specimens 2 and 3). There are small alveoli (30μ) on the inner epidermal surface (specimen 2). The sebaceous glands are medium sized; the sweat glands are slightly twisted, mostly with a squamous epithelium in the secretion cavity (more twisted in specimen 1).

In summer (samples from specimens 6 and 7 only), there is a stratum granulosum in the epidermis, but no pigment. The sebaceous glands are small; the sweat glands are large and very twisted.

Carpus: In winter (specimens 1, 2, and 3), the stratum granulosum is clearly visible. Epidermal pigment is absent (specimens 1 and 2) or nearly so (specimen 3). In specimens (2) and (3), small alveoli are on the inner side of the epidermis (up to 160μ). The dermal network of elastin fibers is less developed than in the body. The sebaceous glands are medium in specimens (1) and (2) and large in specimen (3). The sweat glands are slightly twisted and secreting.

In summer (specimens 5 and 7), the stratum granulosum is well developed and, in specimen (2) stratum lucidum is present in patches. There is little pigmentation (specimen 7) or no pigment (specimen 5). The alveolar pattern of the epidermis is not pronounced (alveoli up to 60μ). The sebaceous glands are moderate in size. The sweat glands are large, twisted, and secreting.

Front foot hoof area: (Samples from winter specimens 1, 2, and 3 only.) There is a two-cell stratum granulosum and there are patches of stratum lucidum (fig. 190). Pigmentation is slight in specimens (2) and (3) and a little heavier in specimen (1). The inner epidermal alveolar pattern is marked, the alveoli reaching 330μ. The sebaceous glands are large. The sweat glands are very small.

Knee joint area: In winter (specimens 1, 2, and 3) the stratum granulosum and stratum lucidum are absent in the epidermis. There is little (specimen 1) or no pigment (specimens 2 and 3). The collagen bundle plexus is loose, the bundles mainly horizontal. Elastin fibers are not numerous. The sebaceous and sweat glands are very small, and their glandular epithelium is squamous.

In summer the stratum granulosum is pronounced and pigment is absent. There are small (50μ) alveoli; the sebaceous glands are medium size. The sweat glands are large and very twisted.

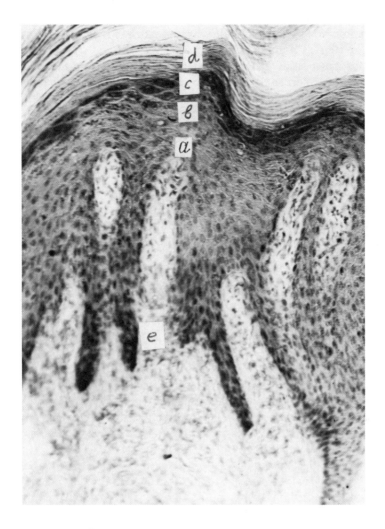

Fig. 190. *Bison bonasus.* Structure of skin near front hoof; *a*, stratum
basale; *b*, stratum spinosum; *c*, stratum granulosum; *d*, stratum corneum;
e, dermal papillae.

Hock: In winter (specimens 1, 2, and 3) there is a single-cell layer stratum
granulosum and patches of stratum lucidum. Pigment is absent (specimen 3)
or rare (specimens 1 and 2). Alveoli reach 220μ. The sebaceous glands are
medium in size. The sweat glands are small but twisted.

In summer (specimens 5 and 7) the stratum granulosum is one cell deep.
Pigment is absent in specimen (5) but plentiful in specimen (7). There are
small (80μ) alveoli in the epidermis. The sebaceous glands are medium in

size. The sweat glands are unevenly developed, medium in specimen (5), large and very twisted in specimen (7).

Hind hoof area: (Samples from winter specimens 1, 2, and 3.) The two to three layer stratum granulosum is well developed and stratum lucidum is present. Pigmentation is moderate (specimens 1 and 3) and slight (specimen 2). There are alveoli on the inner epidermal surface in all three animals. The dermal network of the elastin fibers is less developed than on the body. The dermal papillae are 260μ high. The sebaceous glands are large. The sweat glands are very small (specimen 3) or absent (specimens 1 and 2).

SUBFAMILIA ANTILOPINAE

The skin is thin as are the epidermis and stratum corneum. The strata granulosum and lucidum are absent. The epidermis is pigmented. Most of the collagen bundles in the dermis are horizontal. There is a one-lobed sebaceous gland at every hair. The sweat glands have saccular or sausage-shaped, slightly twisted secretion cavities. There is little seasonal variability in the glands. The only hairs are pile. They have a highly developed medulla and grow singly. There are numerous specific glands: preorbital, carpal, interdigital, and inguinal.

Gazella subgutturosa

Material. Skin from withers, breast, and sacral area (table 121) of specimen (1), female (body length 104 cm); specimen (2), male (body length 101 cm); specimen (3), male (body length 110 cm) taken in January; specimen (4), male (body length 120 cm) and specimen (5), female (body length 113 cm) taken in August in Badkhyz.

In summer the skin is thicker on withers and breast and thinner in the sacral region. Epidermis and stratum corneum are thickest on the breast. The stratum corneum is compact; the epidermis is not pigmented; there are large growing hair roots in the papillary layer. The collagen bundles run in all directions in the upper papillary layer to form a fairly compact plexus, while in the median and inner papillary layers they mostly line up with the hair roots. Elastin fibers are numerous and horizontal. In the reticular layer, horizontal collagen bundles predominate and form a fairly compact plexus. In this layer elastin fibers are rare. Only on the narrow outer zone (80μ) of the reticular layer are there numerous elastin fiber plexuses. Groups of fat cells (28×40; $34 \times 45\mu$) are in the reticular layer of the sacral skin. In the upper reticular layer of the breast, collagen bundles are mostly oblique. Subcutaneous fat tissue (about 0.5 mm thick) is found only in specimen (2), the fat cells being 50×90; $70 \times 80\mu$.

The largest sebaceous glands are found on the breast. Most of the sweat glands have saccular, lightly twisted secretion cavities (fig. 191) (according

TABLE 121

GAZELLA SUBGUTTUROSA: SKIN MEASUREMENTS

Indexes	Thickness						Arrectores pilorum muscles (diameter) (μ)	Sebaceous glands (μ)	Sweat glands diameter (μ)	Sweat glands total diameters per 1 cm of preparation (mm)
	Skin (mm)	Epidermis (μ)	Stratum corneum (μ)	Dermis (mm)	Papillary layer (μ)	Reticular layer (mm)				
Withers										
(1) winter	1.35	34	11	1.32	0.66	0.66	55	22 × 121	80	2.7
(2) winter	1.46	39	22	1.43	0.77	0.66	60	39 × 123	90	2.6
(3) winter	1.35	34	20	1.32	0.66	0.66	50	44 × 140	100	2.4
(4) summer	4.12	28	6	4.10	2.10	2.00	110	80 × 330	187	2.4
Sacral region										
(1) winter										
(2) winter	1.13	33	20	1.10	0.66	0.44	44	39 × 140	88	1.8
(3) winter										
(4) summer	1.56	28	11	1.54	1.10	0.44	60	66 × 330	120	3.4
(5) summer	1.87	17	5	1.86	1.20	0.66	44	55 × 165	130	3.3
Breast										
(1) winter	1.07	34	17	1.04	0.66	0.38	40	56 × 145	90	2.6
(2) winter	1.26	61	44	1.20	0.82	0.38	45	50 × 224	100	2.0
(3) winter	1.62	84	62	1.54	0.88	0.66	40	55 × 275	121	1.6
(4) summer	3.00	56	39	2.95	1.65	1.30	40	120 × 330	110	2.2
(5) summer	—	—	—	1.90	1.10	0.80	35	—	110	

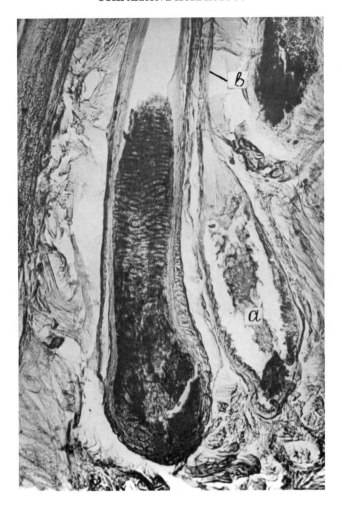

Fig. 191. *Gazella subgutturosa.* Withers skin showing sweat gland
(summer): *a,* secretion cavity; *b,* gland duct.

to Kolesnikov (1948), they are ampulla-shaped and usually straight). Usually the long axis of the secretion canal runs parallel to the hair bursae. On
the breast, the sweat glands loop gently to the hair bulbs. The sweat glands
are most abundant in the sacral region.

There are three orders of pile hair on the withers (fig. 192), 11.7, 6.6, and
3.4 mm long; 124, 90, and 47μ thick. They are similar in shape but different
in size. The thickest point of the hair is in the lower third. All three orders
have a well developed medulla (85% in the first order). The hair shafts arch
backward. They are not wavy and do not grow in tufts or groups. The hairs
grow in loose rows across the body, merging with neighboring rows. The

pile hairs (first order) are most common. They are convex or straight in section, with the convex side toward the head. The cortical layer is laterally thicker than in front and back. There are two to four cuticle cells (15μ high) across the hair. Their outline in second and third order hairs is the same as in the first order. Their height is 9μ. There are 3,499 hairs of all orders per 1 cm^2.

The breast hairs do not divide into categories or orders. The hairs are similar, but hairs in growth have no club and those with completed growth have one. Their length is 6.9 mm, thickness 149μ. The hairs taper at base and tip, being thickest in the lower third or half of the shaft. The medulla (84.8%) runs down the whole hair except the base (in hairs not in growth) and tip. The medullary cells are large and of similar size through the entire hair thickness (fig. 192). The cell walls have small surface folds and small digital processes protruding into the cell cavity. The cortical layer is laterally somewhat thicker than anteriorly and posteriorly. The shape in section is oval with one flattened (sometimes even slightly concave) side. There are no groups or tufts. The number of hairs per 1 cm^2 averages 1,333 fully grown and 500 growing hairs.

In winter the stratum malpighii is two or three cells deep. The stratum corneum is loose. The epidermis is not pigmented. In the withers dermis, the papillary layer is the same thickness or a little thicker than the reticular. In the sacral and breast skin the papillary layer is thicker than the reticular.

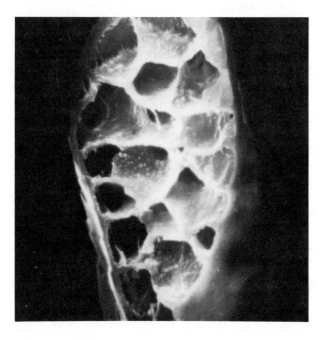

Fig. 192. *Gazella subgutturosa*. Withers: cross-section of pile hair.

The dermal structure is similar to summer animals. Subcutaneous fat tissue (0.75 mm thick) is found only on the breast of specimen (3). The skin glands are the same as in summer.

In the carpal gland in two males, specimens (2) and (3), taken in January, and specimens (4) and (5) (one male and one female) taken in August, show no seasonal or sexual variations. Outwardly, the gland can be noticed from the dark, coarser, longer hairs glued together at the base by the gland secretion. The skin thickens here to 2.9 mm as against the 1.9 mm of the surrounding skin. The bulge extends above the knee at 4.5 cm and across it at 2.2 to 3 cm. In the neighborhood of the gland, the hair bulbs may be as deep as 0.8 mm from the skin surface. The arrectores pilorum muscles are 60μ thick; the sebaceous glands reach $110 \times 330\mu$; the sweat glands are small. Two zones can be distinguished in the gland: central and marginal. In the central zone the hair bulbs are 2.2 mm below the surface. Almost the whole skin tissue gives way to huge, multilobal sebaceous glands (1.02×2.4 mm) (fig. 193), which stretch almost to the subcutaneous muscles. In the marginal zone, the roots of the hairs are shorter and the sebaceous glands are smaller (0.9×1.6 mm); there are a few tubular glands which form glomi. The arrectores pilorum muscles are very large, up to 350μ in diameter.

In the inguinal gland, peculiar skin pockets are found in the male and female groin areas (Dinnik, 1914, reports erroneously that they are found only in males). In specimen (4) (female) taken on August 7, the pockets are 5 cm deep, 3 cm in diameter. In the males, 2 to 3.4×2.6 cm (the opening of the pocket 7 mm). The samples are taken from female specimen (1) and male specimen (2) (January) and female specimen (4) (August). No seasonal or sexual differences are found. The inner surface of the inguinal sac is covered with sparse hair. Large sebaceous glands with twelve to fifteen lobes (fig. 194) open into the hair bursae. The size of the glands increases toward the middle of the sac, decreasing again at its bottom, where they give way to sweat glands. The sebaceous glands are the largest: 660×900; $660 \times 1,100\mu$. Around the middle of the sac, the sebaceous glands touch. The tubular glands are much more developed at the bottom of the sac (where they form gigantic, twisted glomi, 0.91×1.3 mm, packed close together, with a secretion cavity diameter of 70μ) than at the edges of the gland.

In the preorbital gland, closer to the median line from the inner edge of the gland, are small cutaneous intrusions (0.8×1.5 cm) in female specimen (5), August 7; (2.4×2.9 cm) in male specimen (4), August 2; (1×2 cm) in male specimen (2), January 23. Their inner surface is covered with sparse hair, mostly glued by the thick secretion. The samples are from specimen (1) (January) and specimens (4) and (5) (August). Sex dimorphism is not revealed. Well-developed tubular glands are characteristic. They have large, very twisted glomi. Their secretion cavity diameter reaches 110μ. In summer the tubular glands are slightly larger (130μ). Their ducts open into hair fun-

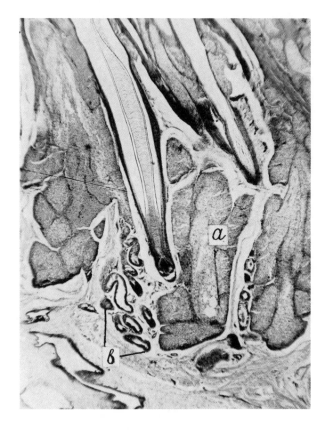

Fig. 193. *Gazella subgutturosa.* Carpal gland: *a,* sebaceous gland;
b, sweat glands.

nels. The sebaceous glands are huge. They have a peculiar, tree-shaped structure and many lobes with large ducts joining into one duct that opens into a hair sheath. The whole sebaceous gland is about 1.5 x 2.2 mm and is larger in summer.

The anterior interdigital glands are thin-walled skin sacs 1.5 to 2.4 mm long and 1.5 to 2.0 mm wide. Their inner surface is covered with sparse hair. Small sebaceous glands with several lobes (overall size 165 x 330μ), and tubular glands twisted into small glomeruli (secretion cavity diameter up to 60μ) open into the hair sheaths. Here there are no large glandular accumulations like the orbital glands. In the winter animal, the size of the sebaceous glands increases to 400 x 500μ, while the tubular glands are smaller (diameter of secretion cavity 45μ).

The posterior interdigital glands (length 1.5 to 2.1 mm, width 2.0 to 2.6

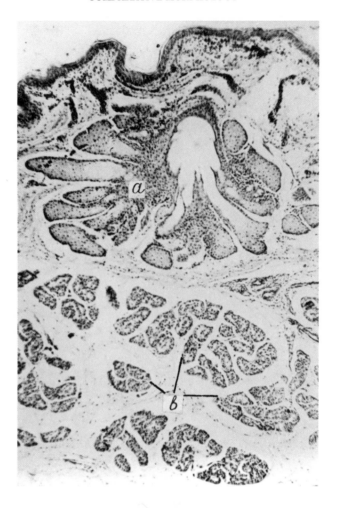

Fig. 194. *Gazella subgutturosa*. Interdigital gland: *a*, sebaceous
glands; *b*, sweat glands.

mm) differ only in size. The sebaceous glands in the summer animal are
larger than in the skin of the anterior interdigital organs ($220 \times 550\mu$). The
tubular glands form glomi $400 \times 550\mu$ (secretion cavity diameter up to 60μ).
In the winter animal, the sebaceous glands are $330 \times 550\mu$. The tubular gland
glomeruli are 250 to 660μ, their secretion cavity diameter reaching 60μ.

The arrectores pilorum muscles are highly developed. They start almost
at the hair bulb and run obliquely to the upper section of the hair bursa,
posterior to the upright hair. The single hair category, pile hairs, divides
into four size orders; length 38.8, 22.9, 12.3, and 3.5 mm; thickness 128, 81,
41, and 28μ. The hairs do not grow in groups or tufts but in ill-defined

rows. The medulla is highly developed in all, 88.7% in the first size order. The shape of the cuticle cells is the same in all size orders. Most of the first order hairs have a small thickening in the upper shaft and are bean-shaped in section—anterior convex, posterior slightly concave. The medulla runs through the whole hair except for tip and club. There are two to four cuticle cells around the hair. The second order hairs are the same shape, also with two to four cells around the hair, the third order having two to three. There are 6,000 hairs per 1 cm².

Only one category of pile hairs (three size orders) is found on the breast; length 43.9, 24.8, and 23 mm; thickness 279, 153, and 42μ. The hairs are thinner at base and tip, thickening in midsection. The thickest point is one-third to halfway up the shaft. The medulla is well developed in all the hairs (first order 91.1%). In section, the hair is bean-shaped, caudally concave, with the lateral parts of the cortical layer being thicker than the anterior and posterior parts. There are 2,333 hairs per 1 cm².

Gazella dorcas

Material. Five specimens taken in Darfur and Sudan's northern province (Ghobrial, 1970) (season and sex unknown) (table 122).

The epidermis is two to four cells thick. The dermal reticular layer is compact, 0.5 to 1.7 mm thick. The papillary layer is 0.37 to 0.50 mm. There are one to three sebaceous glands at every hair follicle. Apocrine type sweat glands are well developed. There are arrectores pilorum muscles.

SUBFAMILIA CAPRINAE

The *Caprinae* species animals under study belong to four different genera: *Saiga, Rupicapra, Capra,* and *Ovis.* Some authors treat the last two as a subfamily and put *Saiga* into the *Antilopinae* subfamily (Frechkop, 1955). Some view them as a separate subfamily (Bannikov et al., 1961).

A study of the skin reveals a close similarity between *Rupicapra* and *Capra* and differences from *Ovis.* But similarities in the hairs in *Rupicapra, Capra* and *Ovis* indicate a close relationship. In skin structure *Saiga* is so similar to *Gazella subgutturosa* that, were one judging only by the skin, the *Saiga* would seem to belong to subfamily *Antilopinae,* where Frechkop (1955) puts it. The skin is medium thick or thin. A thick bulb of skin on the breast is characteristic of *Capra* and *Ovis.* The epidermis may be pigmented. The sebaceous glands usually have single lobes and are not very large, though they are more developed in *Ovis.* The sweat glands are tubular and twisted (*Rupicapra, Capra*) or saccular, slightly or not at all twisted (*Saiga, Ovis*).

The number of specific glands varies a great deal. In *Rupicapra* and *Capra,* caudal and occipital glands are active only in season, while *Saiga*

TABLE 122

GAZELLA DORCAS: SKIN MEASUREMENTS

Area	Epidermis (μ) M (lim)	Stratum corneum M (lim) (μ)	Dermis (mm) M (lim)	Papillary layer (mm) M (lim)	Reticular layer (mm) M (lim)
Neck at top	20(15-30)	5(1-8)	2.39(1.0-3.35)	0.48(0.2-0.8)	1.7(0.67-2.0)
Neck at bottom	25(12-38)	4(3-5)	1.96(1.14-2.68)	0.42(0.2-0.67)	1.27(0.6-2.0)
Withers	21(15-27)	6(3-7)	1.32(1.07-1.65)	0.47(0.27-0.8)	1.02(0.54-1.3)
Sacrum	21(15-26)	4(3-7)	1.29(0.85-2.01)	0.5(0.4-0.6)	0.93(0.4-1.75)
Breast	23(15-30)	6(3-9)	0.84(0.67-1.2)	0.39(0.3-0.54)	0.48(0.4-0.9)
Belly	21(15-30)	6(1-10)	1.0(0.8-1.14)	0.37(0.2-0.5)	0.84(0.6-1.34)
Forelimbs (proximal):					
exterior	29(18-45)	7(3-23)	1.25(1.21-2.35)	0.46(0.27-0.94)	0.9(0.34-1.61)
inside	26(18-38)	8(1-14)	1.77(1.5-2.0)	0.45(0.2-0.67)	1.34(0.94-1.95)
Hind limbs (proximal):					
exterior	20(14-30)	5(3-10)	1.67(1.21-2.14)	0.40(0.34-0.47)	1.06(0.87-1.34)
inside	27(15-36)	7(5-10)	2.12(1.4-3.15)	0.45(0.25-0.67)	1.54(0.94-2.48)

and *Ovis* have well developed orbital, interdigital, and inguinal glands. Pile and fur hairs are distinguished, with more fur hairs in all cases, though in *Saiga* no categories can be distinguished. The medulla is developed in the pile hairs. Except for *Saiga,* it is characteristic of this subfamily that fur hairs grow in tufts. The shape of the cuticle cells varies in all the hairs and no pattern is distinguishable.

Saiga tatarica

Material. Skin from withers, shoulder, breast, middle back, side, belly, sacral region, flank, groin, knee, hock, and hoof of hind foot of animals taken in Kalmyk steppes: specimens (1) and (2), two adult males (January) and specimens (3) and (4), two adult females (July) (table 123).

The skin is thickest on the withers in the sacral region and on the breast (Sokolov, 1961), and thinnest on sides, belly, and limbs. In winter the belly skin is thickened by an increase in the subcutaneous fat layer. The epidermis is very thin all over the body. It includes a stratum basale, which is a layer of large cells with big, oval nuclei whose long axes are vertical, a stratum spinosum of one to three cell layers without round or oval nuclei (long axis of the nucleus is horizontal), and a thin stratum corneum. The lower parts of the limbs have a simple stratum granulosum and a thin stratum lucidum. The epidermis is thicker on the back and feet than on the other parts of the body, where the stratum corneum is thicker and the stratum spinosum has more cell layers, though the winter thinning of the epidermis is largely due to a thinner stratum spinosum, and the relative thickness of stratum corneum is greater. In summer pigment is found in the basal layer of the epidermis of all samples from one animal and on the limbs of the other. In winter the epidermis is not pigmented.

The dermis is thickest on the dorsal side of the body. The border between the papillary and reticular layers is level with the hair bulbs and the bottom of the sweat glands. In the papillary layer, the collagen fiber bundles form a very loose plexus. Elastin fibers are numerous and mainly horizontal. There are few blood vessels and they are small. The fat cells are absent. The papillary layer in most parts of the body is relatively thinner in winter than in summer. The reticular layer is a fairly compact network of collagen bundles, horizontal in most body areas with irregular bundles only on back and limbs. Elastin fibers are very rare in the reticular layer. Blood vessels are larger than in the papillary layer and few in number. Fat cells are absent. The relative thickness of the reticular layer is greater in winter in most body areas. Subcutaneous fat tissue is absent in almost all body areas (except the belly, where it reaches 0.1 to 1.1 mm) in summer, but in winter it is a thick (0.3 to 5.0 mm) accumulation of fat cells with very occasional thin interlayers of collagen bundles. The fat cells are the same size throughout the subcutaneous fat tissue.

The sebaceous glands are arranged in the upper papillary layer 300 to

TABLE 123

SAIGA TATARICA: SKIN MEASUREMENTS

Indexes	Thickness						Sebaceous glands (μ)	Sweat glands diameter (μ)
	Skin (mm)	Epidermis (μ)	Stratum corneum (μ)	Dermis (mm)	Papillary layer (mm)	Reticular layer (mm)		
Withers								
(1) winter	2.5*	16	4	1.8	0.9	0.9		50
(2) winter	1.5	16	6	1.5	0.8	0.7		80
(3) summer	3.0	28	8	3.0	1.0	2.0	84 x 136; 84 x 224	100
(4) summer	1.5	35	6	1.5	1.0	0.5	56 x 220	70
Shoulder								
(2) winter	1.6*	16	10	1.0	0.4	0.6		
(3) summer	1.3	22	5	1.3	0.8	0.5	84 x 170	100
(4) summer	1.2	26	4	1.2	0.8	0.4	84 x 186	100
Breast								
(1) winter	5.4*	28	16	1.5	1.0	0.5	84 x 224	85
(2) summer	1.4*	20	12	0.8	0.4	0.4	56 x 168	100
Midback								
(3) summer	1.4	28	5	1.4	1.2	0.2	80 x 170; 80 x 280	80
Midside								
(4) summer	1.0	35	5	1.0	0.7	0.3	77 x 220	95
Midbelly								
(4) summer	1.0	22	4	0.9	0.7	0.2	122 x 122	100
(3) summer	1.2	33	10	1.2	0.8	0.4	168 x 336	90
(4) summer	2.2*	22	4	0.9	0.8	0.1	110 x 300	100
Sacral region								
(1) winter	4.9*	11	5	1.8	1.0	0.8		60
(2) winter	2.3*	16	7	1.5	0.5	1.0		70

TABLE 123—continued

Indexes	Thickness						Sebaceous glands (μ)	Sweat glands diameter (μ)
	Skin (mm)	Epidermis (μ)	Stratum corneum (μ)	Dermis (mm)	Papillary layer (mm)	Reticular layer (mm)		
(3) summer	2.6	28	6	2.5	1.7	0.8	70 x 330; 110 x 220	120
(4) summer	2.6	16	3	2.5	1.9	0.6	55 x 160	100
Flank								
(1) winter	6.7*	22	12	1.6	1.1	0.5		60
(2) winter	2.2*	16	7	1.6	0.9	0.7		60
(3) summer	1.7*	22	5	1.3	0.8	0.5	80 x 170	100
(4) summer	1.6	22	4	1.6	0.9	0.7	55 x 160	90
Groin								
(1) winter	1.9*	30	20	1.5	0.9	0.6	30 x 110	70
(2) winter	—	—	—	—	—	—	56 x 160	80
(3) summer	1.6*	33	12	1.0	0.7	0.3	168 x 336	90
(4) summer	1.2*	22	4	1.0	0.7	0.3	60 x 220	100
Patellar joint								
(4) summer	1.1	28	6	1.1	0.9	0.2	—	80
Hock								
(3) summer	1.0	22	5	1.0	0.7	0.3	110 x 220	60
(4) summer	1.4	33	11	1.4	1.1	0.3	—	—
At digit of hind foot								
(3) summer	1.6	56	12	1.5	1.0	0.5	60 x 224; 80 x 224	70

*From subcutaneous fat tissue.

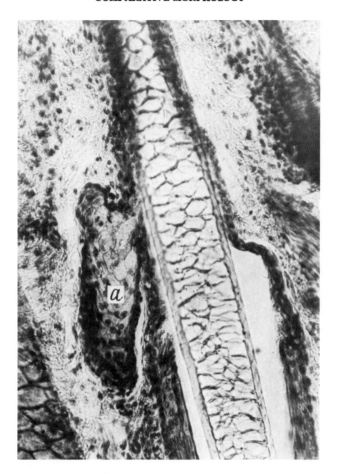

Fig. 195. *Saiga tatarica*. Sacrum skin: *a,* sebaceous gland.

800μ from the skin surface. They are shaped like elongated sacs along the hair bursae into which their ducts open (fig. 195), one posterior to every hair bursa. In winter the sebaceous glands are absent on the sides and back, and are much smaller on feet and belly. The slightly twisted sweat gland secretion cavities lie in the dermal papillary layer level with the hair bulbs (fig. 196). Their ducts open into the hair bursae under the sebaceous gland ducts. The glands are very active in summer; their number decreases in winter, the diameter of the secretion cavities being less, and the cells squamous. The sweat glands are especially developed in the inguinal region (Kolesnikov, 1948).

The interdigital gland has been studied in winter specimens (1) and (2) and summer specimen (3). It presents a thin-walled (0.4 mm) cutaneous sac 22 x 45 mm with a wide entrance (14 mm). It is well developed on both fore and hind limbs (the above measurements are taken from a single gland) (Sokolov, 1961). Inside the sac are occasional hairs with roots running through the whole thickness of the skin. Bundles of collagen fibers, especially elastin fibers, are few. There are very few small sebaceous glands (up to $100 \times 130\mu$). Very numerous twisted tubular glands are the basal structure of the inner part of the skin of this organ. The diameter of the sweat gland secretion cavities reaches 70 to 80μ. Seasonal and sexual variations in the interdigital gland structure are not recorded.

The carpal gland (studied only in the summer animals) has a skin thicker (up to 2.7 mm) than the 1.3 mm in neighboring areas. It is an accumulation of huge, multilobal sebaceous glands (fig. 197), with individual lobes reaching $220 \times 550\mu$. The sweat glands are better developed than on the body,

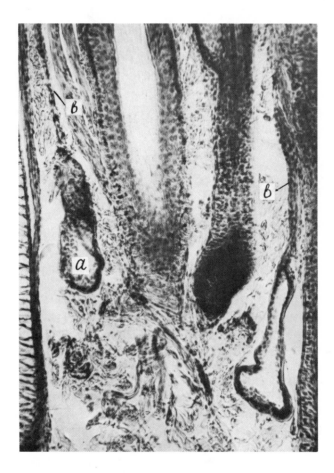

Fig. 196. *Saiga tatarica*. Sweat gland of sacrum skin: *a*, secretion cavity; *b*, gland duct.

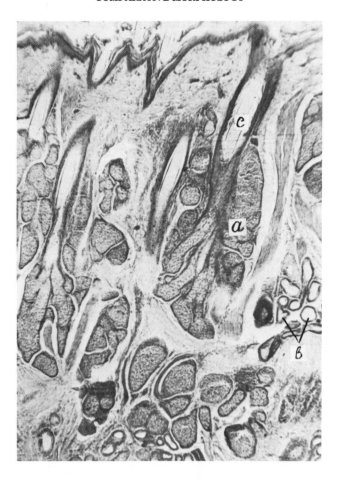

Fig. 197. *Saiga tatarica*. Carpal gland: *a*, sebaceous glands; *b*, sweat glands; *c*, hair roots.

and their secretion cavities (up to 100μ in diameter) form glomeruli in spots. The lower parts of the sebaceous glands lie characteristically much deeper in the skin than on the body, reaching to the hair bulb level.

There are small orbital glands secreting an odorous substance that glues together the bases of the hair, giving the area a grayish tint (Sokolov, 1959b). These glands are intrusions of skin, visible only from the inner side as longitudinally-oval small sacs about 10 mm in diameter containing a creamy, yellow-whitish substance with a distinct odor. On the surface of the muzzle, the glands open with a round aperture (into which a thin probe enters with difficulty) to a depth of 8 mm. There are also funnel-like inward pouches of skin about 4 cm deep on the sides of the mammary gland or scrotum.

Arrectores pilorum muscles are well developed on trunk and limbs, one end fixed anterior to the distal section of the hair bursa, the other to the lower end (just above the bulb) of the neighboring hair bursa. These muscles are somewhat thinner in winter (33 to 80μ) than in summer (30 to 120μ).

No hair categories are distinguishable in *Saiga*. This does not match the data given by Adolf (1959), who distinguishes guard, pile, and fur hairs. All the withers hairs are alike in winter. Most have a club, their shaft tapering and flattening gradually toward the apex. They are 35.7 mm long, 132μ thick, medulla 78.4%. The hair shaft is not wavy but is slightly bent. In section the hairs are convex, with the convex side toward the head. The cortical layer is laterally 6μ thicker than the anterior and posterior. The hair medulla has large cells (fig. 198). The cell walls may have a peculiar venation pattern. They have small ridges protruding into the cell cavity. The hairs do not form groups or tufts. There are 916 growing and 3,166 grown (with clubs) hairs per 1 cm². A small number of apexes of growing hairs peep out of the skin. The withers hairs are the same thickness summer and winter. There are growing and fully grown hairs, the latter having a club. They do not form groups or tufts. The hairs are 10.3 mm long and 63μ thick. They are plano-convex in section, with the convex side toward the head. In cross section, at the largest point the medulla takes up 89.8% of

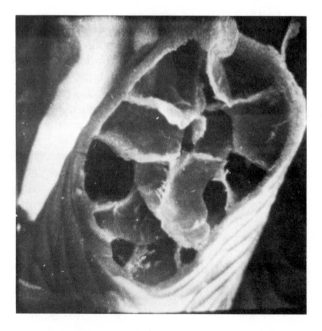

Fig. 198. *Saiga tatarica.* Withers: cross-section of pile hair
(scanning photomicrograph).

the hair. There are about 1,083 growing and 2,166 grown hairs per 1 cm^2 (2,695 hairs in all, according to Adolf, 1959).

Three size orders of hairs can be distinguished on the breast of winter animals (length 40.8, 19.0, and 5.9 mm; thickness 154, 151, and 110μ) with transitions between. The hairs thin toward base and apex, being of even thickness in the middle section. The medulla is well developed in all the hairs (83.6%, first order). The section of the first order is almost oval. The first order is most common; the third order is least common. There are about 1,500 hairs per 1 cm^2. The hairs do not form groups or tufts. All the summer breast hairs have clubs, and they do not fall into size orders. Their length is 5.4 mm, thickness 132μ, medulla 90.7%. The shaft is thickest in the middle, the hair tapering at both ends. The hairs are nearly oval in cross section. There are 1,583 hairs per 1 cm^2 which do not grow in groups or tufts but in ill-defined rows.

Rupicapra rupicapra

Material. Skin from withers and breast (table 124) of two adult females taken in July in Teberda.

The stratum malpighii epidermis has two to three cell layers; the stratum granulosum and the stratum lucidum are absent. The stratum corneum is compact on the withers and looser on the breast. The epidermis is not pigmented. The papillary layer of the dermis is rather thick owing to a high development of growing hair roots. Its collagen bundles run in various directions. In the reticular layer, the collagen bundles form a loose plexus. They are mainly horizontal on the withers but run in various directions on the breast.

The single-lobed sebaceous glands vary in size, being larger on the withers than the other members of *Caprinae* under study, but smaller on the breast. The sweat glands have long, highly twisted tubular secretion cavities. Orbital glands are absent. The interdigital glands on the front and hind legs are small inward pouches covered inside with sparse hairs. The interdigital glands open with triangular apertures at the back of the feet above the hoof. Inguinal glands are absent. Both males and females have peculiar occipital glands. In addition to *R. rupricapra,* these are typical only of *Oreamnos americanus.* These glands are usually areas of thickened, folded skin, posterior to the hairs. During the reproduction season, the glands become bulging, tightly adhering, red outgrowths. The surface is crossed by a few deep grooves rather like the folds of the brain (Hessling, 1855; Schick, 1913; Pocock, 1910, 1918; Schaffer, 1940).

In this period the animals have a distinctive odor caused by the occipital gland secretion (Sokolov, 1959). The occipital gland is made up of large sebaceous glands. There are large arrectores pilorum muscles (diameter 100 to 165μ on the withers and 50 to 70μ on the breast). The coarse hairs divide clearly into pile and fur hairs (Dinnik, 1910; Sokolov, 1959; Geptner et al., 1961). The winter fur reaches 8 to 9 cm, as many as 15 to 17 cm on back and withers. The pile and fur hairs are slightly wavy.

The withers hairs of the animals under study divide into pile and fur. The pile hairs vary in length, the shortest appearing to be still in growth. The intermediate forms between the longest and shortest hairs are easily found. Measurements taken of the longest hairs are: length 25.9 mm, thickness 149μ, medulla 79.1%. They have no club or granna, are slightly wavy, and are almost uniformly thick throughout the length. They thicken slightly toward the distal quarter then taper gradually toward the tip. Their section is an irregular oval, flattened more on one side than on the other. The cortical layer is laterally 6μ thicker than antero-posteriorly. Two to three cuticle cells, 14 mm high and not ring-shaped, surround the hair. The hairs grow singly and average 1,000 per 1 cm². The fur hairs, 4.1 mm long and 13μ thick, have one to one-and-a-half waves, no clubs, and no medulla. The cuticle cells are ring-shaped. The fur hairs grow in tufts of two to four, though they occasionally grow singly, and average 2,333 per 1 cm². The hairs do not form groups.

The breast hairs divide into two categories: pile and fur. They have no granna, and all except the fur hairs grow singly. The first order pile hairs are thicker (220μ) than the others; all the tips are broken off, so the measured 23.1 mm is less than the full length. The hairs have clubs, their shafts being slightly wavy front to back and thickest at the point where they are broken. They are oval in section. The club, the highly developed medulla (95.4%), and the broken tips indicate that these are winter hairs. The second order pile hairs also have clubs and are much the same shape, being thickest in the upper third. These, too, appear to be winter hairs. Their length is 24.5 mm, thickness 151μ. The third order pile hairs are similar in shape but have no clubs, are 22.1 mm long and 149μ at the thickest point, which is halfway up the hair. They are almost round in section. The medulla is less developed than in the two other orders. They appear to be summer hairs. The fourth order is much shorter (6.0 mm) and thinner (62μ) and is thickest halfway up the shaft. There are also many tips in the skin. In horizontal section, the orders of pile hairs are indistinguishable. The total number per 1 cm² is 1,833. The fur hairs are short (3.9 mm), thin (13μ), with little dashes or no medulla, growing in tufts of two to five.

Capra aegagrus

Material. Skin from withers, sacrum, and breast (table 124) of adults taken near Firjuza (Kopet Dag), adult males specimens (1) and (2), late February, and female specimen (3), July 29.

In winter, the skin from withers and sacrum shows that most of the stratum malpighii is two cells deep, while the stratum granulosum and the stra-

TABLE 124

Rupicapra rupicapra, Capra aegagrus, Capra sibirica, Capra caucasica: skin measurements

| Indexes | Skin (mm) | Thickness | | | | | Sebaceous glands (μ) | Sweat glands diameter (μ) | Sweat glands total diameter per 1 cm of preparation (mm) |
		Epidermis (μ)	Stratum corneum (μ)	Dermis (mm)	Papillary layer (mm)	Reticular layer (mm)			
Rupicapra rupicapra (summer)									
Withers (1)	—	28	6	—	1.6	—	165 × 475	110	1.8
Withers (2)	—	22	5	—	1.4	—	220 × 550	90	—
Breast (1)	4.4	45	23	4.4	1.7	2.7	110 × 380	121	2.0
Breast (2)	3.3	56	22	3.2	1.8	1.4	110 × 450	132	1.5
Capra aegagrus (winter)									
Withers (1)	1.21	112	95	1.1	0.4	0.7	56 × 140	88	0.7
Withers (2)	1.91	112	100	1.8	0.7	1.1	45 × 112	40	0.3
Withers (3)	4.85	56	22	4.8	1.3	3.5	84 × 350	121	1.7
(summer)									
Sacral region (2)	1.77	78	61	1.7	0.3	1.4	34 × 140	40	0.2
(winter)									
Sacral region (3)	3.04	40	11	3.0	1.3	1.7	90 × 190	90	0.8
(summer)									
Breast (3)	10.46*	583	286	3.5	1.6	1.9	88 × 275	—	—
(summer)									
Capra sibirica (summer)									
Withers (1)	3.45	56	17	3.4	1.5	1.9	88 × 220	143	2.2
Withers (2)	3.35	56	17	3.3	1.5	1.8	90 × 190	132	1.8

TABLE 124—continued

Indexes	Thickness						Sebaceous glands (μ)	Sweat glands diameter (μ)	Sweat glands total diameter per 1 cm of preparation (mm)
	Skin (mm)	Epidermis (μ)	Stratum corneum (μ)	Dermis (mm)	Papillary layer (mm)	Reticular layer (mm)			
Sacral region (1)	2.62	28	6	2.6	1.2	1.4	80 × 190	100	1.6
Sacral region (2)	2.72	28	6	2.7	1.2	1.5	55 × 300	100	1.7
Breast (1)	11.53*	484	176	4.4	2.7	1.7	190 × 440	88	0.3
Breast (2)	13.80*	803	363	3.8	3.0	0.8	132 × 440	240	2.2
Breast (3)	8.97*	415	220	3.0	2.0	1.0	150 × 440	110	0.6
Capra caucasica (summer)									
Withers (1)	5.24	40	17	5.2	2.1	3.1	100 × 220	110	0.7
Withers (2)	3.33	34	11	3.3	1.5	1.8	44 × 275	132	1.2
Breast (1)	18.39*	495	253	5.2	3.0	2.2	132 × 450	80	0.7
Breast (2)	15.03*	330	143	5.2	3.0	2.2	110 × 660	143	1.0

*Including subcutaneous fat tissue.

tum lucidum are absent. The stratum corneum is very thick and loose. The epidermis is not pigmented. In the papillary layer, collagen bundles run in various directions to form a compact plexus. Elastin fibers are rather numerous and horizontal. In the reticular layer, loosely interwoven, oblique collagen bundles predominate. Elastin fibers are present only in small numbers in the inner part of the layer.

The sebaceous glands are small but have large ducts that are spherically dilated just before they enter the hair bursae. Only the ducts are visible at the fur hairs. The sweat glands are small with narrow, slightly twisted secretion cavities. The sweat glands are developed least in the withers of specimen (2), where the secretion cavities are flattened, nonfunctioning glandular cells. In the sacrum skin, the glands are shrunken. The arrectores pilorum muscles are 33 to 55μ in diameter. The hairs divide into elastic, slightly wavy, comparatively long hairs and thin fur underhairs (Sokolov, 1959). The winter pile hairs are five to six cm long. In summer the fur hairs are absent. The pile hairs are elastic and lie close to the skin.

The withers hairs on all the animals divide into pile and fur. The pile hairs are long (35.4 mm), thin, (109μ), with a small granna. In section they are slightly oval (fig. 199). The cortical layer is laterally 5μ thicker than anteriorly and posteriorly. The height of the cuticle cells is 27μ; their outlines are sinuate. The cells are strongly twisted around the hair, one to three cells

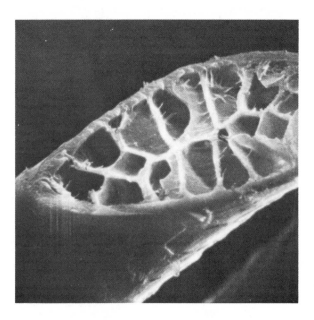

Fig. 199. *Capra aegagrus*. Cells of medulla: cross-section of
pile hair (scanning photomicrograph).

per hair. There are 670 pile hairs per 1 cm². The fur hairs are very long (38.7 mm) and thin (11μ), curly, even crinkly, so that despite their greater length, the fur hairs do not stretch to the pile hair tips. The fur hairs have no medulla. Their cuticle cells are 11μ high and ring-shaped. The fur hairs grow in tufts of five to twelve behind the pile hairs. There are 4,166 fur hairs per 1 cm². There are ill-defined groups of two or three pile hairs and two or three tufts of fur hair.

The summer skin is considerably thicker than in winter, though the withers epidermis is thinner (especially stratum corneum). However, the stratum malpighii, three to four cell rows, is somewhat thicker. The stratum granulosum and stratum lucidum are absent. The stratum corneum is compact. The epidermis is heavily pigmented. The breast epidermis, especially stratum corneum, is very thick. Epidermal processes (up to 220μ high) project into the dermis. The stratum granulosum has four to five cell rows. The stratum lucidum is thick, and the stratum corneum is very compact. In some places the stratum basale is lightly pigmented. The papillary layer is formed of thick hair roots. On the withers the collagen bundles are much the same as in winter. The collagen bundles in the reticular layer are interwoven at various angles to the surface into a compact plexus. The elastin fibers form a 1.2 mm network that borders on the subcutaneous muscles. The breast papillary layer is almost as thick as the reticular, with collagen bundles running in various oblique directions in the papillary and upper and middle reticular layers, but mainly horizontally in the inner reticular layer. Elastin fibers are not numerous in the papillary layer and are rare in the reticular layer. Groups of fat cells occur in the inner reticular layer. The thick subcutaneous fat tissue forms 63.4% of the breast skin. The sebaceous glands are larger than in winter, especially on the breast, with two at every pile hair (fig. 200). Their ducts are 250 to 300μ apart. The sweat glands are larger and more twisted than in winter (fig. 201).

There are two categories of hair on the withers: pile and fur. The pile hairs have no club and only a slight granna. They are 11.6 mm long, 143μ thick, medulla 85%, and oval in section. The medulla is absent only from tip and base. The fur hairs are spindle-shaped and with a sizable thickening in the middle, short (1.5 mm) and 55μ thick, with no medulla in the long club or the tip, though it is well developed in the thickening (85%). The fur hairs grow in tufts of two to eleven behind the pile hairs, forming loose groups of two or three pile hairs and two or three fur tufts. There are on the average 666 pile and 2,833 fur hairs per 1 cm².

The breast hairs also divide into pile and fur, neither of which has clubs. The pile hairs are 12.0 mm long and 127μ thick, straight with a small granna, broken tip, and slightly oval to almost round in section with a slight medulla (62.4%). The cortex is laterally 10μ thicker. The pile hairs grow singly, 666 per 1 cm². The fur hairs are 7.0 mm long and 32μ thick, medulla-free, though some fur hairs do have a few dashes of medulla. The fur hairs

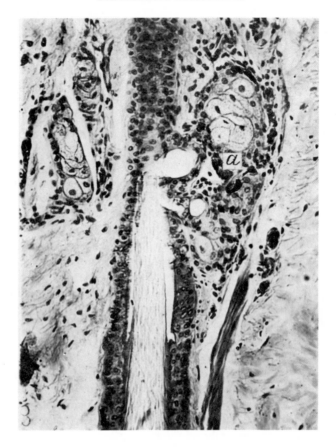

Fig. 200. *Capra aegagrus.* Withers skin: *a*, sebaceous glands
(summer).

mostly grow in tufts of two or three; there are 666 per 1 cm². Both pile and
fur hairs grow in loose rows.

Because Schaffer (1940) reports the presence of glandular formations in
the region of the occiput in *Capra*, a sample from this part of the head was
investigated in the *C. aegagrus* taken in summer. Both sweat and sebaceous
glands are developed normally and are no different from the body glands.

Capra sibirica

Material. Skin from withers, sacrum, and breast (table 124) of three adult
females, taken August 17 and 18 in western Tien Shan. Specimen (1) (body
length 130 cm); specimen (2) (126 cm); specimen (3) (112 cm).

The skin is thicker on the withers than on the sacrum. The epidermis and stratum corneum are thin. The stratum malpighii is three to four cells deep. A one- or two-cell-layer stratum granulosum (withers) is conspicuous. The stratum lucidum is absent on the withers and breast. The stratum granulosum is absent on the sacrum. The stratum corneum is compact. The epidermis is moderately pigmented. The epidermis and stratum corneum are moderately thick on the breast area (especially specimen 2). The epidermis has processes (220μ high) jutting into the dermis. The stratum granulosum is four to five cells deep. The stratum lucidum is thick (up to 200μ).

The epidermis is not pigmented (except for stratum basale, which is pigmented in patches). A well-developed papillary layer is characteristic. The collagen bundles run in various directions, elastin fibers being numerous on the withers and rare on the breast. In the reticular layer, almost all the collagen bundles run at various angles to the surface, the horizontal predominating only in the inner layer. Elastin fibers are almost absent. The subcutane-

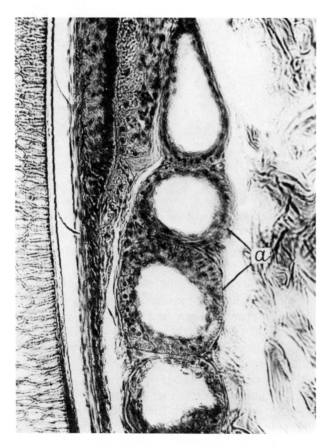

Fig. 201. *Capra aegagrus.* Withers skin: *a,* sweat gland (summer).

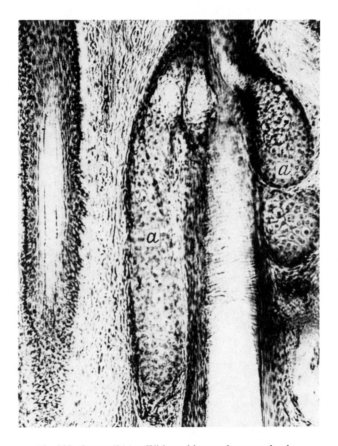

Fig. 202. *Capra sibirica.* Withers skin: *a,* sebaceous glands.

ous fat tissue is absent on the withers, but is 5.5 to 9.2 mm thick on the breast. The fat cells are 55 × 80; 60 × 65μ. The sebaceous glands are rather large (fig. 202), with two (anterior and posterior) at some pile hairs. The glands have single lobes, but are sometimes found one above the other at the same hair. Sweat glands are rare and moderate in size. The arrectores pilorum muscles are 33 to 55μ thick.

It is difficult to distinguish hair categories on the withers of the *C. sibirica* specimens under study as the hairs are in the state of growth on all specimens but two. Three size orders of pile and fur hairs are distinguished, all without clubs. The numerous first order pile hairs are the longest at 20.6 mm, thickness 156μ, medulla 88.9%. They thicken gradually from the base to the distal third of the shaft, then taper toward the tip. The shaft is slightly wavy from front to back, the cross section being a flattened oval with a slightly concave posterior and an almost straight anterior wall. The cortex is laterally 2μ thicker. The cuticle cells are low (13μ) with sinuate outlines, one to three cells around the hair. The 833 first order pile hairs per 1 cm² grow singly. The second order pile hairs are shorter and thinner and much less frequent (length 12.9 mm; thickness 68μ). They resemble the first order in outline, except that they have only one to one-and-a-half large waves (front to back) and one to three 11μ high cuticle cells around the hair. There are an average of 250 second order hairs per 1 cm². The third order hairs are much like the first two in shape, except that their thickening in the upper third is more pronounced. They are shorter, thinner (6.5 mm, 47μ), and fewer, their 9μ high cuticle cells mostly ring-shaped, though hairs occur with two cells around them. The fur hairs are of even thickness (15μ) almost their whole length (6.6 mm), tapering only toward the tip. The shaft has one to one-and-a-half large waves. Some fur hairs have a medulla 35%. Most of the cuticle cells are ring-shaped, averaging 9μ in height. The fur hairs grow in tufts of two or three, often in groups of up to nine hairs. There are 2,916 hairs per 1 cm². On the breast there are pile hairs only, with two size orders, neither of which has a granna or club.

The first order pile hairs (12.8 mm long, 48μ thick, medulla 43.6%) have broken tips and are more numerous. They are round in section, grow singly, 916 hairs per 1 cm². The second order hairs (11.2 mm long, 48μ thick) are similar in shape and have a well-developed medulla, though it is thin and broken in some hairs. They grow singly, 583 per 1 cm².

Capra caucasica

Material. Skin from withers, sacrum, and breast (table 124) of two adult males (body length 150 and 140 cm) taken July 26 in the Caucasus.

The skin thickness on the withers differs in the two specimens. The epidermis and stratum corneum are the most alike in thickness. The stratum malpighii is two to three cells deep; the stratum granulosum appears in patches and the stratum lucidum is absent. The epidermis is moderately pigmented. The pigment cells show up clearly. The breast skin is thick. The epidermis and stratum corneum are thick, and there are 220μ high alveoli on the inner epidermal surface. The stratum granulosum appears in patches. The strata corneum and lucidum are very similar in appearance. The stratum basale is moderately pigmented. Growing hairs ensure a highly developed papillary layer with bundles of collagen fibers running in various directions. In the reticular layer (withers) oblique collagen bundles predominate. In the reticular layer the collagen bundles run at various angles and form a compact plexus.

There are many groups of fat cells in the inner reticular layer. The subcutaneous fat tissue is thick on the breast (9.5 to 12.7 mm). The sebaceous glands, singly posterior to the pile hairs, are large on the breast and small on the withers. The sweat glands are saccular. They are not twisted on the withers but are tubular and slightly twisted on the breast. The males have specific glands near the tail which function in the reproductive season and give the goats their distinctive smell (1959). The arrectores pilorum muscles are 50 to 132μ thick.

The withers hairs divide into pile (two orders) and fur. In addition, numerous apexes of growing hairs are just coming through the skin. None of the hairs has clubs. The pile hairs (first order) (45.1 mm long, 132μ thick, medulla 84.8%) have no granna, are slightly wavy (front to back) and slightly flattened. The cuticle cells are an average of 25μ high, two to four around the hair. These seem to be the residual winter hairs. The second order are much shorter (20.0 mm). These are summer hairs. They have a front-to-back, flattened granna (maximum width being 159μ) in the upper third, oval in section with a posterior invagination. The cortical layer is twice as thick laterally as in the antero-posterior (12μ as compared with 6μ). The height of the cuticle cells (24μ) is almost the same as in the first order (two or three cells around the hair). There is an average of 588 hairs of the first two orders per 1 cm². The third order is very small (length 1.3 mm, thickness 32μ) with a well-developed granna. The medulla is absent at base and tip and present only in the granna. These hairs grow in tufts (or more exactly in small groups) of two or three. There are 1,750 per 1 cm².

The fur hairs are very sparse. They are long (55.8 mm), thin (23μ), and curly. They seem to be left over from the winter pelage. Their cuticle cells reach 21μ in height (which almost equals the thickness of the hair), are all pile, and divide into three size orders with some in between. They have no granna or club and grow singly in ill-defined rows. The first order pile hairs (22.9 mm long, 132μ thick, medulla 73.9%) thicken slightly in the upper part. They bend back slightly but are not wavy. Most have broken tips. In section they are irregularly round. There are 583 first and second order hairs per 1 cm². The second order pile hairs are shorter (16.8 mm) and thinner (78μ) than the first, but are similar in shape. The third order are 12.4 mm long and 78μ thick, slightly wavy (1.5 to 2.5 waves) with a well-developed medulla. There are 416 hairs per 1 cm².

Ovis ammon musimon

Material. Skin from withers, sacrum, and breast (table 125) of female (body length 107 cm) taken on November 19, 1960, in the Crimea.

The stratum malpighii is complex, three to five cell layers deep; the stratum granulosum and stratum lucidum are absent. The layers of stratum corneum are not pronounced. The epidermis is not pigmented.

The papillary layer is thicker than the reticular. The collagen bundles run in various directions to form a compact plexus. Elastin bundles are numerous in the upper and middle reticular layer. The collagen bundles in the upper reticular layer form an intricate plexus and are mainly horizontal in the inner part. The networks of elastin fibers border on the subcutaneous muscles. The sebaceous glands, small with single lobes, are posterior to the hair roots but are not visible at fur hairs. The sweat glands have saccular, slightly twisted (withers) or straight (sacrum) secretion cavities with a large diameter. The bottom of the glands does not reach the hair bulbs. These glands are rather sparse.

The withers hairs divide into pile and fur. They do not grow in groups. The pile hairs are 28.3 mm long, 160μ thick, medulla 87.5%. Some have clubs, some do not. They have no granna. They are slightly wavy and thicken from base to tip; they are oval or occasionally bean-shaped in section (Ivanov, Belekhov, 1929). The cortical layer is slightly thicker laterally. The 18μ high cuticle cells are longer than on other sheep (there are three to four cells around the hair). Their height is one-ninth of the hair thickness. The pile hairs grow singly, 583 pile hairs per 1 cm². The fur hairs are short (22.1 mm), thin (14μ), and have no medulla. The 12μ high cuticle cells are mostly ringed, almost as high as the hair is thick. The fur hairs grow in tufts of two to seven, and average 2,416 per 1 cm² (according to Ivanov and Belekhov (1929), in Askanian *O. a. musimon* 96 to 97% of the pelage is pile hairs).

Ovis ammon cycloceros

Material. Skin from withers, sacrum, and breast (table 125) of specimen (1), female (body length 120 cm); specimen (2), female (body length 118 cm); specimen (3), male (body length 121 cm), all taken between January 31 and February 2, 1961, in Firjuza, Kopet Dag; specimen (4) female (body length 120 cm) taken on January 24, 1961, near the Badkhyz reservation; specimen (5), female (body length 131 cm) taken on July 28, 1961, in Firjuza, Kopet Dag; specimen (6), male (body length 140 cm) taken August 6, near the Badkhyz reservation.

In winter specimens (1) to (3) from Kopet Dag, both withers and breast skin are thick. The stratum malpighii takes up the largest part of the epidermis, on withers and sacrum (Sokolov, 1964). The stratum malpighii is only one to three cells deep. The stratum granulosum and stratum lucidum are absent. In the sacral epidermis, the stratum granulosum is sometimes a patchy string of cells. The epidermis is not pigmented. The epidermis and stratum corneum are very thick on the breast. The stratum malpighii has many cell layers. In places, a simple stratum granulosum and stratum lucidum can be seen. The stratum corneum is very compact.

In specimens (2) and (3) there is no pigment in the epidermis, and in specimen (1) the epidermis is slightly pigmented. There are alveoli on the inner surface of the epidermis, up to 330μ in height. The bundles of collagen fibers in the papillary layer form a compact plexus. They run in different directions. Horizontal elastin fibers are numerous in the upper and middle

TABLE 125

Ovis ammon: skin measurements

Area	Thickness						Sebaceous glands (μ)	Sweat glands diameter (μ)	Sweat glands total diameter per 1 cm of preparation (mm)
	Skin (mm)	Epidermis (μ)	Stratum corneum (μ)	Dermis (mm)	Papillary layer of dermis (mm)	Reticular layer of dermis (mm)			
Withers									
Ovis ammon musimon (November)	2.8	45	17	2.8	1.7	1.1	50 × 190	165	1.6
Ovis ammon cycloceros									
(1) winter	1.4	34	17	1.4	0.9	0.5	45 × 140	165	4.2
(3) winter	1.3	50	40	1.2	0.5	0.7	34 × 84	100	1.8
(4) winter	1.0	28	17	1.0	0.4	0.6		121	2.2
(5) summer	2.6	17 17	4	2.6	1.6	1.0	55 × 275	100	0.8
(6) summer	3.0	34	17	3.0	1.7	1.3	55 × 350	121	2.2
Ovis ammon karelini									
(7) summer	3.5	17	5	3.5	2.3	1.2	132 × 275	132	1.1
(8) summer	3.9	28	6	3.9	2.3	1.6	80 × 280	140	1.7
Sacral region									
O. ammon musimon (November)	2.5	17	6	2.5	1.4	1.1	45 × 224	154	2.2
O. a. cycloceros									
(1) winter	1.5	34	17	1.5	0.8	1.7	34 × 151	132	2.2
(2) winter							48 × 90	66	0.9
(2) winter									
(3) winter	1.2	45	37	1.2	0.5	0.7	40 × 112	55	0.5
(5) summer	1.8	28	6	1.8	1.1	0.7	45 × 196	110	1.0
(6) summer	2.6	22	6	2.6	1.5	1.1	45 × 336	154	2.0

TABLE 125—*continued*

Area	Thickness Skin (mm)	Epidermis (μ)	Stratum corneum (μ)	Dermis (mm)	Papillary layer of dermis (mm)	Reticular layer of dermis (mm)	Sebaceous glands (μ)	Sweat glands diameter (μ)	Sweat glands total diameters per 1 cm of preparation (mm)
Ovis ammon karelini									
(7) summer	3.0	28	6	3.0	1.8	1.2	44 × 275	132	1.5
(8) summer	3.2	28	6	3.2	2.1	1.1	100 × 220	176	1.5
Breast*									
O. a. cycloceros									
(1) winter	7.7	671	308	2.6	1.0	1.6	112 × 252	187	2.6
(2) winter	11.6	118	84	2.9	1.8	1.1	110 × 300	187	2.0
(3) winter	10.2	299	143	2.3	1.1	1.2	165 × 330	121	1.3
(4) winter	7.8	335	165	2.5	1.6	0.9	209 × 440	165	1.0
(5) summer	23.3	330	165	3.0	1.8	1.2	100 × 440	175	1.7
(6) summer	9.1	106	61	2.9	1.8	1.1	110 × 407	132	0.4
O. a. karelini									
(8) summer	18.3	1,155	550	4.5	2.9	1.6	110 × 385	110	0.6
(9) summer	29.9	343	165	6.1	3.1	3.0	110 × 682	198	1.4

*Including subcutaneous fat tissue.

part of the layer. In the reticular layer, thin bundles of collagen fibers, mostly horizontal and longitudinal (especially in the inner parts of the layer), form a loose plexus. Elastin fibers are present only on the border with the subcutaneous muscles, where they form networks in a layer about 110μ thick. The subcutaneous fat tissue is well developed in the breast skin and is 4.5 to 8.6 mm thick. The fat cells are 40×85; $55 \times 65\mu$. The sebaceous glands are larger on the breast than the back, with two at some hairs opening into the front and back of the hair bursa. Some glands have two or three lobes, and specimen (2) has two layers of glands.

The withers hairs (specimen 1) are pile and fur. They do not grow in groups. The pile hairs, 30.9 mm long, 189μ thick, medulla 95.5%, have clubs. They are curly and have many of the tips broken off. There is no granna. They are oval in section. The cortical layer is laterally thicker. The cuticle cells are high (31μ) and short (four or five cells around the hair). The pile hairs grow singly, 1,333 per 1 cm². The fur hairs are thin (11μ), rather long (23.3 mm), and very curly. The ring-shaped cuticle cell height (19μ) exceeds the hair thickness. The fur hairs grow in tufts of 3 to 12, about 400 per 1 cm².

The breast hairs also divide into pile (two orders) and fur. The pile hairs grow singly, the first order hairs (20.0 mm long, 214μ thick, medulla of 89.1%) have clubs; they are straight or slightly wavy, with broken tips. Oval in section, there are 916 first order pile hairs per 1 cm². The second order hairs are much shorter (9.5 mm) and thinner (95μ), and the thickest point is in the upper two-thirds of the shaft. They have a thick medulla. There are 333 pile hairs (second order) per 1 cm². The fur hairs are slightly thickened in the distal part, have a characteristic long club, are 7.2 mm long, 52μ thick, with a well developed medulla. They average 1,666 per 1 cm², growing in tufts of two to four.

In summer specimen (5) from Kopet Dag, the withers skin is thicker than in winter, and the epidermis and stratum corneum are much thinner. The stratum granulosum and stratum lucidum are absent. The epidermis is not pigmented. The breast skin is much thicker than in the winter animals, though the epidermis and stratum corneum are the same. The stratum granulosum, not very well defined, is three to four cells deep. There is no pigment in the epidermis. The papillary layer is much thicker than the reticular. The networks of the bundles of collagen and elastin fibers in the dermis are much the same as in the winter animals. There is no subcutaneous fat tissue on the withers, but in the breast skin it is thick (20.0 mm). The size of the sebaceous glands is the same, but the sweat gland diameter is smaller than in winter animals.

The arrectores pilorum muscles are 55 to 80μ thick. There are three hair categories: pile (two orders), intermediate, and fur. The first order pile hairs (10.4 mm long, thickness 144μ, medulla 88.9%) are slightly wavy, with broken tips. Most have clubs. They are oval in section. The cuticle cells reach around the hair and are 13μ high, with two to four cells around the hair. The second order pile hairs are shorter (9.2 mm) and thinner (103μ), with similar shaped cuticle cells 14μ high. There are one to three cells around the hair. The medulla is well developed. The (few) intermediate hairs are 8.2 mm long and 28μ thick. Their medulla is broken and very poorly developed. The fur hairs are short (4.7 mm) and thin (12μ), with no medulla. The cuticle cells are 10μ high (almost the same as the hair thickness). It takes one (sometimes two) cells to reach around the hair.

The fur hairs grow in loose tufts of two to four hairs and form no groups. There are 1,750 hairs of the first two categories and 5,250 fur hairs per 1 cm². All the hairs on the breast are worn almost down to the root. In the sample under investigation, only eight hairs with broken tips are found which have no waves and clubs. Their average length is 5.6 mm, thickness 96μ, and their section is oval. In cross section, the long axis of the medulla is 69.7% of the hair. The cortex is laterally somewhat thickened. Horizontal section reveals hairs with a larger diameter which may be pile hairs; those with the smaller diameter are evidently fur hairs. The latter grow in tufts of two. There are 500 of the larger and 333 of the smaller hairs per 1 cm².

In the winter specimen (4) from Badkhyz, the withers skin is thinner than in the Kopet Dag animal. The epidermis is also thinner, but the stratum corneum is the same. The stratum malpighii of the epidermis is two to three cells deep. The stratum granulosum and stratum lucidum are absent. The stratum corneum is compact. Their breast skin layers relate in the same way as in the animals from Kopet Dag. The epidermis is not pigmented. The plexus of collagen bundles in the papillary and reticular layers are the same as in the Kopet Dag'animals. The sebaceous glands are smaller. Genuine glandular cells are absent, their being instead accumulations of small cells with nuclei highly stained with hematoxylin.

The sweat glands have saccular, slightly twisted or straight secretion cavities level with the hair bulbs. The arrectores pilorum muscles are 35 to 110μ thick. As in the Kopet Dag animals, the withers hairs divide into two categories: pile and fur. The pile hairs, which have no granna, taper gradually toward the tip. They are slightly wavy and have clubs, are 33.7 mm long, 141μ thick, medulla 92.9%. Many have lost their tips. They are oval in section. Peripheral medulla cells adjoining the cortical layer are clearly visible. The cortex is laterally somewhat thicker (2μ). The cuticle cells are rather tall (average 29μ) and narrow (there are three to five cells around the hair).

The pile hairs grow singly, 1,750 per 1 cm². The fur hairs are long (20.6 mm), thin (9μ), and very curly, have no medulla, and have ring-shaped cuticle cells higher (13μ) than the hair thickness. The fur hairs grow in tufts of two to five, averaging 5,177 per 1 cm². The withers hairs form no groups. The breast has pile and fur hairs. The pile hairs (22.4 mm long, 232μ thick, medulla 92.7%) are oval in section, grow singly, averaging 1,666 per 1 cm². The fur hairs are 9.5 mm long and 50μ thick, thickening slightly in the distal section. They have a long club and their medulla is well developed. The fur hairs grow in tufts of two to seven, averaging 6,083 per 1 cm^2.

In the summer specimen (6) from Badkhyz, the withers skin is three times as thick as in the winter animal; the epidermis is slightly thicker than in winter; the stratum corneum the same. The stratum granulosum and stratum lucidum are absent. The stratum corneum is compact. The epidermis is not pigmented. The breast skin is thicker than in winter, though the epidermis and stratum corneum are thinner and the alveoli of the epidermal breast skin are very poorly developed. The stratum granulosum and stratum lucidum are absent. The epidermis is not pigmented.

The dermis is the same in winter and summer. The subcutaneous fat tissue on the breast is thick (6.1 mm). It is crossed by a horizontal layer of striated muscles, under which there is more subcutaneous fat tissue. If this layer is included, the whole subcutaneous fat tissue is 14.3 mm. The breast sebaceous glands and the diameter of the sweat gland secretion cavities are somewhat smaller than in winter. On the withers, the single-lobed sebaceous glands, though not large, are larger than in winter. There is a small sebaceous gland at every fur hair which sometimes consists of just a few secretion cells. The sweat glands on the withers are slightly larger than in winter, still twisted, with the same diameter as the secretion cavity. Only their length is greater. The arrectores pilorum muscles are 33 to 110μ thick.

There are two categories of withers hairs: pile (two orders), and fur. The pile hairs (first order) are 12.5 mm long and have broken tips. Some have clubs, some do not. They are thickest (178μ) in the distal third of the shaft and have an 88.2% medulla. In section they are flattened oval. The cuticle cells (12μ high) stretch close to the hair (two to three cells around). There are 833 first order pile hairs per 1 cm^2. They do not form groups or tufts. The second order are fewer, similar in shape, and show transitions. They are 10.4 mm long, 94μ thick, and their medulla is well developed. Most of these hairs have clubs. The cuticle cells are similar to the first category, 10μ in height, one or two cells around the hair. There are 83 hairs per 1 cm^2 growing in neither tufts nor groups. The fur hairs are shorter (7.2 mm) and much thinner (11μ) than the pile hairs, and are medulla-free. The shaft has two to four big waves. The cuticle cells are ring-shaped, 9μ high. The fur hairs grow in tufts of three to twelve, averaging 4,833 hairs per 1 cm^2.

The breast hairs also divide into two categories: pile hairs (two orders) and fur. Most of the first order pile hairs have a club and a straight shaft. They are 8.5 mm long, 172μ thick, medulla 69.6%, many with broken tips. Their section is oval. The second order hairs are longer (9.1 mm) but much thinner (86μ), with a well-developed medulla. The thickest point is in the distal third of the shaft. There are 583 first and second order pile hairs per 1 cm^2, all growing singly. The fur hairs are shorter (5.2 mm) and thinner (43μ), and have a small thickening in the distal third, which contains a medulla. The hairs of this category grow in tufts of two to four, 833 fur hairs per 1 cm^2.

Ovis ammon karelini

Material. Skin from withers, sacrum, and breast (table 125) of specimen (7), male (body length 160 cm); specimen (8), male (body length 163 cm); specimen (9), male (body length 161 cm), taken August 20, 1961, in Western Tien Shan.

In summer the back skin is thicker than on any other sheep (Sokolov, 1964). The epidermis and stratum corneum are thin. The stratum malpighii is two to three cells deep. The stratum granulosum and stratum lucidum are absent. Specimen (8) has a lightly pigmented epidermis. Skin, epidermis, and stratum corneum are considerably thicker on the breast than on dorsal areas, (12.7 and 23.5 mm; if the layer of muscles under which the fat cells lie is taken into account, this rises to 14.3 to 27.8 mm). The stratum granulosum is five or six cells deep. The stratum lucidum is up to 100μ thick. The stratum corneum is compact. The epidermis is slightly pigmented in spots. The epidermal alveoli are up to 462μ high. The papillary layer is thicker than the reticular layer. The collagen plexus and the elastin network are the same as in the back and breast of other sheep. So is the summer development of the skin glands. There are two sebaceous glands anterior and posterior to every pile hair.

The withers hairs divide into pile (two orders) and fur hairs (Butorin, 1935, distinguishes in addition intermediate hairs, but we find no such category; Chagirov (1954), does not subdivide the pile hairs into orders; Chagirov gives no indication of subspecies, season, or sex). The pile hairs (first order) are wavy and taper gradually from base to tip. They are 35.8 mm long, 207μ thick, medulla 87.9%, and oval in section. The cuticle cells are 19μ high, two to four cells around the hair. The pile hairs grow in inconspicuous rows, 833 hairs per 1 cm². The second order hairs are very sparse and are much like the first in shape, 32 mm long, 69μ thick, with a well-developed medulla. The cuticle cells are 13μ high. The second order pile hairs average 83 per 1 cm². The fur hairs (5.4 mm long and 29μ thick) are wavy, with no medulla. The cuticle cells are 11μ high. The fur hairs grow in tufts of three, six, or nine (three to seven, according to Chagirov, 1954), usually lateral or posterior to pile hairs, and averaging 2,499 per 1 cm².

The breast hairs, too, are divided into pile (two orders) and fur. The pile hairs (first order) have no clubs or waves, but do have tips. They are 15.6 mm long, 207μ thick, medulla 87.9%, and are oval in section. The cortex is laterally thicker. The second order pile hairs are much the same in shape as the first, are 9.6 mm long, 98μ thick. Together they average 660 per 1 cm². The fur hairs are short (5.4 mm) and thin (29μ) with a slight medulla that may be broken. The fur hairs grow in tufts of two to four, averaging 1,000 per 1 cm².

Orbital gland: In *Ovis musimon,* 13 x 16 mm; in specimen (2), 25 x 28 mm; in specimen (3), 25 x 45 mm; in specimen (4), 14 x 20 mm. There are no seasonal differences in structure. The skin surface is covered with sparse hairs. The thickness of the skin is about 2.9 mm. The hair bulbs are 1.4 mm deep. A thin (about 0.1 mm) layer of collagen plexus is under the epidermis. Sebaceous glands are beneath it, with numerous lobes adhering to one another and to the lobes of adjoining glands. The sebaceous glands (0.6 × 1.3 mm) form a continuous 1.3 mm layer. The secretion cavities of looped (a few loops) tubular glands are arranged beneath the sebaceous glands. Our data for the preorbital gland are similar to *O. a. musimon* (Schaffer, 1940). There is a reticular layer of longitudinal collagen bundles beneath them. The preorbital glands in mountain sheep are almost the same as the orbital glands in domestic sheep (Schaffer, 1940).

Inguinal gland: The size of the inguinal glands (pockets): in specimen (2), 14 × 34 mm; in specimen (3) 22 × 60 mm) in specimen (4), 20 × 30 mm; in specimen (7), 20 × 60 mm. The skin (1.4 to 1.6 mm thick) is covered with sparse hairs. On the sides of the hair roots, large, multilobed sebaceous glands (0.5 × 1.0 mm; 0.4 × 1.1 mm) stretch horizontally (fig. 203) with large tubular gland glomi some way beneath. The size of these glomi is difficult to determine, for they join to form a continuous glandular zone. The diameter of the secretion cavities reaches 130μ. Closer to the opening of the inguinal pockets, the sebaceous glands grow larger, while the sweat glands grow smaller. No sharp seasonal differences are noted, though there is an impression that the sweat glands are more highly developed in summer than in

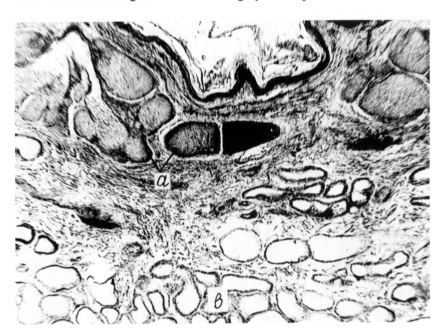

Fig. 203. *Ovis ammon karelini.* Groin gland: *a*, sebaceous glands; *b*, sweat glands.

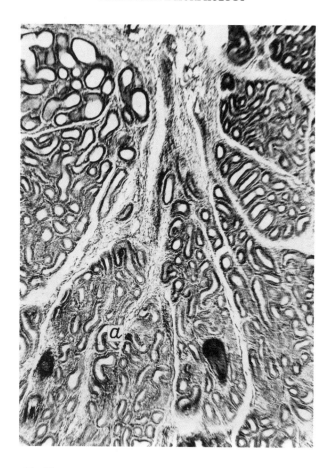

Fig. 204. *Ovis ammon karelini*. Interdigital gland of front feet:
a, tubular glands.

winter. The inguinal gland in the mountain sheep under study is similar to that of the domestic sheep (Schaffer, 1940).

The interdigital gland: On the front feet this is a bent tubular formation, 30 to 47 mm long and 3 to 7 mm in diameter. No seasonal differences are noted. The thickness of the skin ranges from 1.2 to 1.5 mm. The skin is covered with sparse hairs. Near the opening of the gland are small sebaceous glands, nearly disappearing into the end of the gland. Almost the whole skin is filled with large tubular gland glomi (fig. 204), the borders between individual glomi being indistinguishable. The diameter of the secretion section reaches 100μ. The ducts open into hair sheaths. In *O. a. musimon*, the structure of the interdigital glands of the front feet is the same, though the diameter of the secretion cavity is smaller (up to 70μ).

On the hind feet, the interdigital gland structure is the same as on the front feet. Its length is 25 to 42 mm, diameter 2.2 to 8 mm. The diameter of the tubular gland secretion cavities reaches 90μ (80μ in *O. a. musimon*). No seasonal differences are revealed. The interdigital glands of mountain sheep are similar to those of domestic sheep (Schaffer, 1940).

Domestic sheep (and evidently mountain sheep) have numerous other specific skin glands: caudal, formed by tubular glands (Schaffer, 1949); circumanal, formed by sebaceous and tubular glands (Siedamgrotsky, 1875); preputial, formed by sebaceous and individual large tubular glands (Schaffer, 1940); planum nasolabiale and meatus acousticus internus (Schaffer, 1940).

III

ADAPTATIONS OF MAMMAL SKIN
TO ENVIRONMENT

Every animal organ, and group of organs, has more than one function—this multiplicity of function has been noted by Darwin, Dorn, Severtsov and many others. The skin is undoubtedly one of the most versatile organs in this respect.

The skin provides protection against water loss, keeps out dangerous foreign bodies, shields against injury and wear, regulates the body temperature, receives sensory impressions, suckles the young, defends passively (through protective color, spines, osseous or corneous cover) and actively (through claws, hoofs, horns), gives chemical signals (the smell of secretion), functions optically (acting as mirror in *Artiodactyla*) and acoustically (the tail needles of the porcupine and back needles of *Hermicentetes semispinosus*), stores food, excretes waste, acts as auxiliary in locomotion (hoofs, skin pads of foot soles, claws for climbing [arboreal], fins [aquatic], flying membrane [gliding or flying species]), regulates the metabolism in respiration, and more.

This multiplicity of functions enables the organism to perform with great economy and versatility using a relatively small number of organ systems. It makes for highly flexible organization—in changed conditions, a function that was auxiliary can become a main function. The capacity for change in the structure and function of organs is a vital part of evolution. Severtsov (1949) speaks of a whole system of phylogenetic organ changes. Skin, perhaps the most diversified and variable organ, provides an excellent example of such phylogenetic changes. Severtsov distinguishes two groups of phylogenetic changes: quantitative changes, which exaggerate but do not alter the

573

main function; and qualitative changes, which cause revolutionary alterations in primary function.

Some patterns of evolutionary change involving skin fall into the first group. The basic working principle in quantitative change is that, in the descendants, the function of a particular organ or group of organs becomes progressively greater than in the ancestors, either through progressive change in the structure of the cells or through increase in numbers of the constituent parts. The adaptation of pelage to colder climates illustrates this principle. There is no qualitative change: the pelage becomes longer and more dense; in some species the medulla thickens. A. N. Severtsov's example, the development of the mammary glands from small tubular glands, does not illustrate this principle as effectively, for qualitative changes are also at work in the evolution of the mammary gland. Qualitative changes occur in the chemical composition of the secretion, so that his example properly belongs in the second group of phylogenetic changes.

A reduction in the number of functions is not infrequent in the evolution of mammal skin. When one function of an organ is augmented, another may drop away in the descendants. For example, the epidermis of the land ancestors of recent *Cetacea* and *Sirenia* provided bodily protection against water loss. When *Cetacea* and *Sirenia* took to the sea, their epidermis lost this function. Another example is the modification of fur into spines. Here heat insulation is sacrificed to protection.

The second group of phylogenetic alterations, qualitative changes, may be associated with an increase in functions. That is, one or several new functions may be added. In all mammals the skin has an added function—regulating temperature when heterothermal forms give way to homothermal.

During functional change, the original central function is reduced as an erstwhile auxiliary function gradually takes over as the major, central function. The organ itself changes as a result of altered function. As examples, mammary glands and specific skin glands can be adduced. The primary function of their precursors was lubricating the skin surface and fur. With shifting functions, changes occurred in glandular morphology leading to qualitative changes in secretion. In the mammary glands, the secretion turned to milk; in the specific glands, the secretions became chemical signals and odors.

In substitution of functions, the function of a given organ is replaced by a function of similar biological significance in another organ. Here Severtsov (1949) uses *Cetacea* as an example. In *Cetacea,* the subcutaneous fat tissue takes over the insulation function from the pelage of the terrestrial ancestors. In the water, fur provides poor insulation and becomes atrophied.

In mammals all phylogenetic skin variations are associated with environment. The structure of common integument has adapted successfully to a wide variety of environmental demands. The skin adaptations of mammals reflect diverse habitats and modes of life—terrestrial, arboreal, subterranean, and aquatic habitats, and flying, climbing, burrowing, and swimming modes of life. Despite a basically similar skin structure, mammals show a tremendous diversity in skin morphology. Seasonal variations, as an adaptation to changed climatic conditions, represent another instance of qualitative change. Specific skin glands exemplify the specialization of one of the organs of the common integument system (for instance, in transmission of information through chemical signaling).

SKIN ADAPTATIONS IN TERRESTRIAL MAMMALS

Land dwellers have relatively thin skins. Even the elephant's skin is thin in relation to his size. This skin and poorly developed subcutaneous fat deposits are associated with high body motility. The subcutaneous fat between the subcutaneous muscles and the muscles of the body is rarely a continuous, thick layer. It tends instead to concentrate fat deposits in specific areas. Most mammals have subcutaneous fat deposits in the following areas: a cervical pair to the anterior on the sides of the neck; a humeral pair at the elbow; interscapular; near the backbone, extending in narrow strips down both sides of the vertebrae in the thoracic and lumbar regions; a lateral pair beside the posterior edges of the thoracic region; and sacral area in the groin reaching almost to the anus (Kuznetsov, 1962). These fat deposits, of small importance as insulation, provide food storage. Consequently, they vary in size and quality with the season. Fat animals have fat cells in the subcutaneous tissue, distributed throughout the body and attaining great thickness on the sacrum.

The skin of land animals is thickest on the dorsal side, thinning toward the belly and becoming thinnest in the groin and armpits. The dorsal skin is the least motile and elastic but is more durable. In big mammals, the reticular layer of the dermis has a compact plexus of thick collagen bundle fibers in peculiar rhomboid loops.

The epidermis of land animals tends to be thin (especially in species with thick fur), with fully cornified outer epidermal layers. The division between strata malpighii and corneum is distinct except during molt. Where hair is sparse, the epidermis has a greater productive function, thereby tending (in absolute terms) to be thicker. This inverse relation between the thickness of pelage and epidermis is well illustrated in the contrast between the heavily furred *Artiodactyla* and the sparsely furred wild boar. In the former, the withers epidermis ranges from 17 to 112μ. In the latter, it ranges from 209 to 396μ. The number of hairs per cm^2 is 874 to 1,190 and 2,333 to 6,916, respectively. Relative epidermal thickness is, however, no greater in the boar. In some cases, it may actually be less than in other *Artiodactyla*.

Limbs, especially soles, get the hardest wear, so here the epidermis tends to be thickest. *Odocoileinae* (summer) have an epidermis of 38 to 50μ on the withers. On the tarsal joint it is 134 to 220μ, and near the hoofs it is 190μ. The structure of the epidermis also changes, developing an alveolar pattern on the inner surface near the hoofs. The alveoli inderdigitate with dermal papillae, giving the outer parts of the skin great strength.

On animals with a sparse pelage, the inner surface of the epidermis is alveolar throughout the body. The alveoli reach considerable height. In *Elephantidae,* the epidermis is 0.48 to 1.60 mm thick on the trunk and 5.7 mm near the hind foot. The alveoli are 0.45 to 1.80 mm high. The stratum corneum, which provides major mechanical protection, is much thicker where the fur is thin. In animals with sparse pelage, it thickens a great deal and the whole epidermis is more compact. The elephant, for instance, has the thickest stratum corneum of all mammals (0.11 to 0.55 mm). On his hind foot it reaches 3.86 mm.

Some body areas are subject to more wear than others. Sheep and goats, because they lie down on rocks, have a thickened epidermis on the chest with alveoli on the inner surface. The wild goat (winter) has a withers epidermis 112μ thick, with stratum corneum being 95μ. This contrasts with the breast, where the figures are 583 and 286μ. The corneous layer is notably more solid. On the feet, especially the harshly treated soles, the epidermis is very thick. For instance, polar bear epidermis is 54 to 214μ on the back, as compared with 5 to 7 mm on the soles. The epidermis has an alveolar inner surface and a very compact, thick stratum corneum. Strata granulosum and lucidum are also present. They are rarely found in the trunk skin.

Where the pelage is well developed, the papillary and reticular layers are usually clearly distinguishable in the dermis. The papillary layer contains hair roots, sebaceous glands, and sweat glands. Only in species with coarse, bristling pelage, where the roots pierce the whole dermis (boar, peccary), are these layers almost indistinguishable. The border between these layers is much deeper during molt than in the state of rest.

The dermis provides mechanical protection, so it is toughest and thickest in the areas where this is most needed. In autumn, in the adult wild boar, the dermis on the sides of the breast comprises small but very compact collagen fiber bundles, toughening the skin so that it can withstand the fangs of other boars. In summer, a peculiar skin shield develops on the head and neck of roe males (Sokolov and Danilkin, 1967). It is found at the sites most exposed to horn stabs, and it is most highly developed by the time of the fighting season. In contrast to the wild boar, the skin shield in the roe is formed by the thickening of both papillary and reticular layers of the dermis. In the wild boar, the reticular layer is predominantly thickened. In the adult male *Oreamnos americanus,* the dermis thickens greatly to form a kind of shield on the posterior quarter of the body, especially on the hindquarters and backside (around the anus) (Geist, 1967). This is associated with the way *O. americanus* males fight—they stand side by side, head to tail, butting each other's backsides with their horns. *Aepyceros melampus* males fight head on, and the mature *A. melampus* has a dermal shield on the most vulnerable body parts (the upper neck, the shoulders, and around the horn) (Jarman, 1972).

Subcutaneous fat tissue also has some protective value. The combination of fat cells and collagen bundles provides excellent shock absorption. Thick collagen bundles run in all directions in the skin of the soles, interspersed with large groups of fat cells. This combination makes a shock-absorbing cushion that can withstand the pressures of walking, running, and jumping.

The structure of the skin varies with its functions in different parts of the body, the dermis being the most variable both in structure and thickness. The dermis also needs relative motility for its mechanical functions. The skin is obviously always slightly stretched—where a piece is cut from a live animal, the edges contract as they would on a close-fitting glove stretched on the hand. This natural stretch is greater in the areas with a compact network of elastin fibers. It is greatest where the skin is thin but elastin fibers are abundant. The stretch is greater in young animals than in old (Bonninger, 1905; Evans et al., 1943). There is a set direction to the stretch controlled by a definite system (Pinkus, 1927). The stretchability of skin should not be confused with elasticity, which enables the skin to go back into its original position when the force that pulled it ceases to act. Elastin fibers are very important to skin elasticity. The skin properties of durability, motility, stretch, and elasticity vary according to the mechanical functions of given parts of the body.

In most land mammals the back skin possesses the greatest durability and thickness and, correspondingly, the least motility and elasticity. The thinnest, most motile, and most elastic skin on land animals is in the armpit, the groin, and on the ventral side of the body. In the most durable areas of skin, the dermal reticular layer is a very compact plexus of thick bundles of collagen fibers running vertically, diagonally, and horizontally. The weakest collagen plexus in the reticular layer is found in areas with very thin, motile skin, where collagen bundles are loosely packed and horizontal.

Land mammals invariably have well-developed sebaceous glands, which usually open high up on the hair bursae. The sebum lubricates the skin surface and hair. Sebaceous gland secretion helps the hair to remain elastic and increases the insulation properties of the epidermal stratum corneum (Kuznetzov, 1957; Yazan, 1972).

The sweat glands vary considerably and are sometimes absent. There has been very little research on their physiological function in wild land mammals. *Marsupialia, Insectivora, Carnivora, Perissodactyla,* and *Artiodactyla* that live on land all have sweat glands. There is reason to believe that sweat glands are to some extent involved in temperature control in all these mammals. Experiments on the sweat glands of the dog are of interest here (Aoki and Wado, 1951; Nielsen, 1953). It is commonly accepted that dogs have no sweat glands and can lose heat only through hyperventilation of the lungs. This view has been proven wrong, for there are well-developed sweat glands over the whole body. It has been shown that the sweat glands of dogs are a protective mechanism against sharp local temperature increase.

Of the *Artiodactyla* under study, the sweat glands are best developed in the wild boar. This appears to be associated with difficulty in losing heat through the skin due to a thick subcutaneous fat layer. This suggestion counters the view that the sweat glands of the pig do not respond to heat (Ingram, 1967). The wild boar's sweat gland secretion cavity is long and tortuous, twisting into large glomi. All other *Artiodactyla* have smaller sweat glands (though they are invariably present in the body skin (except in *Moschus moschiferus*). The shape may vary, depending on the systematic position of the animal.

Opinion differs on the functional significance of *Artiodactyla* sweat glands, but published data are concerned only with domestic species. Many authors believe that the sweat glands in *Artiodactyla* do not regulate body temperature, claiming that the sweat glands of cattle and sheep lack any capacity for this (Venchikov, 1934; Nemilov, 1946; Klimov, 1937; Kashkarov, 1938; Miropolsky, 1939; Findlay, 1953; Badreldin and Ghany, 1954; and others). It is believed that panting is the principal method of evaporative cooling in sheep. Sweating, though of lesser importance, is advantageous to the shorn sheep. Sweat glands seem to play an insignificant part in red deer thermoregulation (Johnson, Malloiy, and Bligh, 1972).

There is evidence that sweat glands in all members of the *Bovidae* family become active during hot weather (Robertshaw, 1971), though their role appears to vary in different species. For example, southern zebu cattle achieve much greater heat loss through sweating than does the gnu (Taylor et al., 1969). Numerous reports on perspiration and temperature regulation in sheep (Kolesnik, 1949; Alekseeva, 1962; Raushenbakh, 1950-1957; Raushenbakh and Erokhin, 1967), produce evidence that, among cattle in the southern regions, the sweat glands are actively involved in temperature control. Raushenbakh does claim that in sheep these glands cannot be essential for temperature regulation as they are so few. Experiments on cows bred in the Tashkent region (as well as with Dutch, Bushuevsky, and Scwiz breeds) show that Tashkent cows have the largest number of sweat glands per square centimeter and that Friesian cows have the fewest (Chizhova, 1958a). Moisture is secreted onto the skin by sweat glands, not by transpiration, and the sweat glands of cows are active in temperature regulation. It has been found that moisture evaporates twice as fast from fur as from a smooth surface, proving that fur does not prevent the evaporation of sweat. The primary role of perspiration in cooling domestic horses with a rapid formation of sweat has been established (Tangl, 1956), and for asses as well, evaporation of sweat plays the major role in heat loss (Schmidt-Nielsen, 1964).

Some authors believe that heat loss is achieved by secretion of body moisture almost exclusively through the surface of the planum nasolabiale (Venchikov, 1934; Chizhova, 1958b and others). Our data point to many regularly functioning sweat glands in the skin of ruminant *Artiodactyla*. These cannot be treated as organs that merely produce specific odors, because all *Artiodactyla* have special glands for this. Sweat does help to lubricate the pelage ("fat-sweat" in sheep). The diffusion of moisture through the epidermis is slow and may not be adequate for moistening the whole corneous layer where it is thick (Kuno, 1959). The demand seems to be met by the regular secretion of sweat. There are some interesting observations on the drying and cracking of the epidermis in extreme cold (low temperature decreases the activity of sweat glands). It seems that the reason for relatively well-developed sweat glands in many mammals in winter (the season when stratum corneum is at its thickest) is to moisten the epidermis. We believe that sweat glands have more than one function. They regulate temperature, lubricate and moisten the pelage and stratum corneum, create a specific odor, and (partially) excrete. All these functions are vital for land animals, though their relative importance varies in different species and at different seasons (e.g., in lubricating the pelage, production of "fat-sweat" is more important for domestic sheep than for other animals).

When ambient temperature approaches or exceeds the skin temperature, heat loss by radiation, convection, and conduction drops to minimum values and even absorption of heat by the body is recorded. Under such conditions heat loss is achieved only by evaporation. Many mammals do not have sweat glands capable of changing the rate of evaporation from the skin very much and they avoid getting overheated in hot weather in various ways. Their main protection is to get out of the sun and to find a cool place. For most mammals, a big rise in the temperature of their environment (so long as it is not higher than their body temperature) results in a rise in their skin temperature. This is a mechanism for increasing heat loss. In most mammals the rate of respiration also rises if the temperature is high, regulating body temperature through evaporation from the respiratory tract surfaces. This is one of the most effective heat-loss mechanisms, though it has nothing to do with the skin.

There are a number of mammals that have no sweat glands and do not use polypnea for heat regulation. Many such animals (e.g., rodents) salivate copiously in high temperatures. The saliva wets breast and belly, and the animal spreads it over its body with its paws. The saliva evaporates, cooling its whole body. Elephants have a similar way of keeping cool. They suck up water with their trunks, splashing it over their heads, backs, and sides. If the elephant cannot find water, he puts his trunk in his mouth and uses saliva (Kuno, 1959). In addition, elephants lose a considerable amount of heat through the ear surface. This is accomplished by flapping (Buss and Estes, 1971).

It is interesting to note that, unlike most placental mammals, in *Setonyx brachyurus* (which has sweat glands) neither polypnea nor sweating are as important as salivation and spreading saliva over the skin (Bartholomew, 1956).

In temperate and cold climates, the bodies of land mammals cool down, especially in winter, which leads to the growth of pelage for insulation. The corneous matter in the pelage is only a fair heat conductor, and the fur alone would not be adequate for insulation. This is achieved by the air between the hairs and in the medulla of the hairs proper. But the most important job of the hairs is to keep out the convection currents in the air immediately around the body.

A number of authors have demonstrated that the heat conductivity of a material does not affect its qualities as an insulator. They have tested this by filling a space with material. Regardless of which material is used, cotton or even fine steel wool, the insulation remains the same unless the density becomes too low. It makes little difference how good a heat conductor the filling is. The important factor is that these fibers prevent air motion and thus create still air, even inert air.

Inert air differs from still air in that convection currents are totally absent. The pelage insulation is proportional to the thickness of the inert air it holds. The more still air the pelage holds, the better the insulation. The essential thing is to keep away outer air that may create air currents.

The air held inside the medulla is the ideal inert air, so a thick medulla makes for better insulation. But a thick medulla weakens the hair. Animals whose mode of life creates heavy wear on their fur cannot afford a thick medulla. In such mammals, the main insulator is the air between hairs. They usually have numerous, very thin hairs. But large numbers of thin hairs are liable to tangle into a tight mass, reducing their insulating properties. The pelage protects itself against this by differentiating. The thin fur hairs that hold the air are interspersed with rather fewer pile and guard hairs that are long and elastic and protect different parts of the body. The distribution is usually greatest on the back, becoming less down the sides and on the belly (e.g., on the withers, the elk has pile hairs up to 158 mm long in winter and 1,166 hairs per 1 cm², while on the breast the length is 109 mm and the number per 1 cm² is 833).

The reason for this distribution is that land animals spend a lot of time in winter on their bellies or rolled up into a ball, exposing the dorsal side. Rolling into a ball reduces considerably the body surface exposed to the ambient environment (by 44% in mink, Segal and Ignatov, 1975) and gives essential protection to feet and muzzle, which are most subject to heat irradiation. In cold weather, the polar bear curls up. In hot weather, he lies belly up on his back (Ristland, 1970). On a moving animal, the back is more exposed to air currents than the belly. Many animals have fluffy tails which they curl around themselves to cover belly and head. In tropical land animals, such features are less pronounced. Many land animals have well-developed arrectores pilorum muscles to enable them to raise or lower their hairs to change the thickness of the air layer held between the hairs.

Despite many common features in hair structure, various groups of land mammals manifest structural peculiarities associated with their biology and evolution. *Carnivora* are more subject to fur damage and to injuries than other mammals, so their fur tends to be more hard-wearing. Even their fur hair has a slight granna (27 to 68% of hair thickness). Nevertheless, many cold climate carnivores have pelage that gives a high degree of insulation (especially their many fine fur hairs). Because they depend mainly on the air layer between hairs, the hairs grow in tufts and groups. Their hair differentiation is marked, and their arrectores pilorum muscles are well developed.

Carnivores vary considerably in their need for insulation. Tropical carnivores have the poorest fur. The mongoose has relatively little short fur hair. The mongoose and civet have only 6,660 hairs per 1 cm² on the withers as compared with the weasel's 24,000, and their fur hairs are much thicker (35 to 51μ) than, for instance, are the marten's.

A correlation of hair density in polar fox, fox, and sable to the relative air humidity and the wind rate in winter reveals that specifically dense hair is characteristic of animals inhabiting areas of greater humidity and wind rate (Pichugin, 1974). It is associated with the fact that skin cooling increases with the relative humidity and the air motion rate. For protection against cooling, an increase of hair density is necessary (table 126). Arctic species have about nine times the insulation of tropical mammals (Irving and Krog, 1955; Irving et al., 1955). All *Lagomorpha*'s hairs have a characteristic medulla development in every hair category. The longest hairs are ventral, evidently because the resting hare lies on his belly on the earth or snow. *Artiodactyla* have varied pelage. Nonruminant *Artiodactyla* have poorly insulating pelage. The hair is very sparse, and the fur hairs are medulla free. As size prevents them from finding shelter easily, they are more subject to the vagaries of weather (the wind, for instance) than small mammals and need better skin insulation. Most ruminant *Artiodactyla* have numerous pile

TABLE 126

POLAR FOX: GEOGRAPHIC VARIATION OF HAIR DENSITY (THOUS. PER 1 CM²) (PICHUGIN, 1974)

	Polar foxes				
	Pechora	Obdor	Enisei	Yakutia	Kamchatka
Hair density	29.25	27.71	21.60	22.49	30.50
Relative humidity (%)	85	80	70	65	85
Wind speed (m/sec)	7.0	7.4	5.5	5.0	7.5

hairs with a highly developed medulla. Holding inert air, these provide excellent insulation capable of resisting even strong winds.

Some interesting data are reported by Ignatov (1958, 1971). In an 11 m per sec wind, the deer pelage gives much better protection (he does not state which species) than the coypu's, although the latter has a much denser pelage. The level of development of this kind of insulation (through air-bearing pile hairs) varies in different ruminant *Artiodactyla*. It is the most pronounced in musk deer, whose pile hairs are long and slightly wavy, which holds them closer together. They are round in section with a 98% medulla—more than any other *Artiodactyla* except the caribou. The thick medulla makes their hair very brittle, so a thin, medulla-free, sturdy club permits the hair to bend without breaking. The musk deer's fur hairs are so sparse that their insulation value is negligible (the ratio of pile to fur hairs is 15 to 1). Other deer are similar to musk deer in this. All have long, thick pile hairs and 90% medulla (on the body hair), though their fur hairs are also rather well developed—they have more fur hairs than pile (winter ratio 3:1).

Caribou have better fur insulation than other deer. Their winter pile hairs are five to six cm, and eight cm on the belly; the cortex is thin and sometimes indistinguishable. The medulla is very highly developed (up to 98% in winter). The fur hairs consist only of cuticle and cortex (Bol' and Nikolayevsky, 1932; Efimov, 1938; Akayevsky, 1938; Kuznetsov, 1945).

The pelage of *Gazella subgutturosa* and *Saiga tatarica* is very peculiar. The steppe and desert winds demand reliable insulation. They have no fur hairs at all, only several orders of pile with a well-developed medulla. The medulla is thinner than that of the caribou and musk deer, for these live in colder climates. The medulla is usually under 90%, which still gives adequate protection against heat loss. *S. tatarica* lie in high, unsheltered places in the coldest wind and weather (Sludsky, 1963).

G. subgutturosa are more thermophilic than *S. tatarica*. Unlike *S. tatarica,* they often seek shelter against the wind, a characteristic apparently determined by this more southern species' evolutionary heritage. Morphological evidence indicates that the gazelle's pelage gives poorer insulation than the saiga's, even though some orders of the gazelle's pile hairs are longer than those of the saiga. The gazelles have a larger number of hairs, for the uneven length of their pelage prevents it from creating a continuous insulating integument around the body. *G. subgutturosa* has four different lengths of pile hairs. *S. tatarica* has only one order of pile hairs, all of equal length.

Sheep have well-developed pile hairs with a thick medulla, though they derive more insulation from the fur hairs than the animals previously discussed. In sheep the relationship between the pile and fur hairs is about the same as in deer (1:3), but in absolute numbers, sheep have many more fur hairs (4,000 to 5,000 hairs per 1 cm² against, for instance, the elk's 916). Fur hairs are even more important to *Caprinae*. In the wild goat, the ratio between pile and fur hairs is 1:6, relatively few pile hairs, and the medulla takes up only 84% of the thickness. Their fur hairs are long (longer than the pile hairs), abundant, thin (11μ), and very curly, without medulla. So goats are protected mainly by the air layer between hairs.

Bison bonasus is not exposed to very low temperatures, for in the forest he can always find shelter against the wind. So his pelage is about two-thirds sparser than that of other ruminant *Artiodactyla*. He has fewer pile hairs than fur. The medulla is less developed in his pile hairs than in those of other ruminants. The fur hairs are medulla-free and thick. The large size of the aurochs favors heat retention. Most ungulates are affected by cold less ventrally than dorsally, so the breast and belly pelage are shorter and less dense. But the yak lies on his belly in the snow, so he has a long, dense fur edging on his sides.

Pelage also insulates against the heat—notably in large ungulates who live in the deserts and are subject to temperatures above body temperature. This was demonstrated experimentally with the dromedary (K. Schmidt-Nielsen, 1964). The summer pelage of the gazelle and saiga is also an adequately developed insulator.

The perissodactyl Asiatic wild ass closely resembles the ruminant *Artiodactyla* in pelage. His insulation is largely dependent on long, thick pile hairs with a well-developed medulla (about 90%). The pile hairs are numerous, only slightly outnumbered by fur hairs (on the withers there are 3,000 pile hairs and 5,000 fur hairs per 1 cm², and on the sacrum 2,000 and 3,000, respectively). No fur hairs are recorded on the belly. The wild ass is insulated by air held in the pile hair medulla, and, to a lesser extent, by air between the hairs. His abundant fur hairs grow in tufts. This suggests that fur hairs also play a part in insulation. Well-developed arrectores pilorum allow the regulation of pelage insulation.

Animals of the same species may have different needs for insulation if their habitat is different. The members of three subspecies of mountain sheep: *Ovis ammon musimon, O. a. cycloceros,* and *O. a. karelini,* were taken at Kopet Dag near Firjuza and in Badkhyz. Their pelage differs in the number of orders of pile hairs, in length and thickness of pile and fur hairs, in the medulla development in the pile hairs, and in the number per square centimeter of skin.

The pelage is the least dense in *O. a. musimon* (29,999 hairs per 1 cm² in winter). *O. a. cycloceros* has 5,333 to 6,916, and summer *O. a. karelini* has 3,415. *O. a. musimon* has a shorter pelage than the other sheep. The winter withers pile hairs of *O. a. musimon* are 28.3 mm; his fur hairs are 22.1 mm. For *O. a. cycloceros,* the figures are 30.0 to 33.7 mm and 20.6 to 23.3 mm. For summer *O. a. karelini,* the figures are 35.8 mm and 29.5 mm. The mouflon's pile hairs are 160μ thick. In *O. a. cycloceros,* pile and fur hairs are 141 to 189μ thick, respectively, and *O. a. karelini*'s pile is 184μ. Mouflon's pile hairs have a less developed medulla (85.7%) than other sheep (*O. a.*

cycloceros, 92.9 to 95.5% and *O. a. karelini,* 96.0%). The mouflon and
O. a. cycloceros have the thickest fur hairs (14μ on the withers). *O. a. musimon's* pelage gives the poorest insulation, for the mouflon lives in a warmer
area than other sheep. The differences between sheep from Kopet Dag and
Badkhyz are slight. Sheep from Tien Shan live at a higher altitude in summer. The temperature is lower, so their pile hairs and their medulla are better developed than in any other sheep.

Tropical mammals need less insulation and so tend to have sparser fur
(e.g., elephant, rhinoceros). Other parts of the skin are also involved in
insulation. A number of mammals have a thick, loose stratum corneum in
winter with many air cavities between the scales for insulation. There are
some interesting data on the relationship of thermophily in mice and the
structure of their pelage (Herter, 1952). When normal thermophilic (white)
and cryophilic (gray) mice are compared, gray mice are found to have a
much denser fur than white (the ratio of hairs is 1.55:1). Keeping them in
warm or cold conditions does not alter hair density or length. However, the
epidermis of the mice kept in warm conditions grows thinner while the other
group's skin thickens. The layer that becomes thicker or thinner is stratum
corneum.

The dermis is less concerned with insulation. Only *Moschus moschiferus*
has large numbers of fat cells in the dermal papillary layer in winter (when
they fill the whole of it, crowding right under the epidermis). There are no
fat cells in the reticular layer. However, an extra, outer insulation layer is
important to the musk deer, possibly because it is smaller than the other
ungulates.

A few land mammals use subcutaneous fat only as insulation. The wild
boar with his sparse pelage has a covering of subcutaneous fat all over his
body, which is almost totally responsible for insulation function. The structure of the ear skin is characteristic. Ears with a large surface give off a lot
of heat, especially the outer surface. The wild boar has a thick layer of subcutaneous fat on his ears.

The skin is important for losing heat. There are four physical methods of
losing heat: radiation, convection, conduction, and evaporation. Radiation
transfers energy from skin to environment through electromagnetic waves—
spread as straight lines at the speed of light. This is expressed as radiation
energy per second per 1 cm². If an object is warmer than its ambient temperatures, radiation takes form in infrared rays. The range of infrared radiation from the human skin is 5 to 20μ, the most frequently occurring value
being 9μ (Hardy and Muschenheim, 1934). It is notable that radiation heat
losses do not appear to vary with skin color. At normal temperatures, radiation accounts for about 60% of the total heat loss, varying with the difference of skin and air temperature and with the size of the radiating surface.

Heat convection is energy transfer through the motion of air. The air
around an animal is normally colder than its skin and so will take some of

its heat. Where the air is in motion, new cool air will keep on taking up heat. Where it is not, the increased warmth of the air nearest the skin will cause that air to rise, making way for cooler air. Of the total heat loss from a naked person placed in a calorimeter at 22.7°, 66% is due to evaporation and only 15% through convection (Du Bois, 1937). When a fan is put on inside the calorimeter, radiation decreases by 41%, convection rises to 33%, and evaporation rises to 26%. A decrease of heat radiation results in this case from lowering of the skin surface temperature through convection heat loss. Heat is conducted from one medium to another with which it is in direct contact. In contrast to convection, air motion does not in this case affect heat conductivity.

Heat is lost by skin evaporation because when water turns to vapor it takes heat from the body surface. Evaporation of moisture from the skin and the respiratory tracts usually accounts for at least 20% of the total heat loss in man. In normal conditions, with only moderate perspiration the heat loss through the evaporation of perspiration accounts for 23 to 27% (Du Bois, 1927), about one-third of the total water loss being from the respiratory tract. The remaining two-thirds is from the skin surface.

Experimental data on heat radiation in wild animals are so few that they add nothing here. Mammals that require major heat loss usually have a sparse, short pelage. Many terrestrial mammals do not have adequate shelters and their pelage often has protective coloring. Pelage also plays an important part in protecting the skin against injury. The dense pelage of most terrestrial mammals acts as a kind of shock absorber. The hair shaft grows at an acute angle (very acute to almost 90°) to the skin, the dense pelage making a protective, shock-absorbing cushion.

The protective function of the pelage limits its insulating capacity somewhat, for while a well-developed medulla gives the best insulation, it also reduces hair strength. When the pelage is likely to keep brushing up against external objects as the animal moves, the medulla tends to be less developed. Greater fur density provides the main insulation. For instance, the fur of carnivores is more at risk than that of herbivores and omnivores. The relatively small medulla of carnivores increases the durability of the pelage. The breaking point of pile hairs expressed in grams averages 19.0 g in the marten, and only 12.0 g in the caribou (Kuznetsov, 1952), although the latter's hairs are much thicker (160μ) than the marten's (41μ). This is because the marten has a thinner medulla (60%) than the caribou (about 90%).

Nonmedulla hair tips, which get the hardest treatment, help to reduce wear and tear on the pelage, so the tips tend to be strongest. The longer guard and pile hairs are subject to more wear than the short fur hairs and so are the more durable. On the rump of the adult winter rabbit, the guard hair breaking point is 19.1 g, pile 5.7 g, and fur only 3.3 g (Tserevitinov et al., 1948).

The pile hairs of most *Artiodactyla,* which have a thick medulla, contain

TABLE 127

RUMINANTIA THICKNESS OF CORTEX IN DIFFERENT PARTS OF THE PILE HAIR (IN μ)

Species	Anterior	Posterior	Lateral
Moschus moschiferus	4	4	4
Cervus elaphus	8	8	11
Capreolus capreolus	5	5	5
Alces alces	11	11	19
Gazella subgutturosa	7	7	8
Saiga tatarica	11	11	17
Rupicapra rupicapra	11	11	17
Capra aegagrus	6	6	11
Capra sibirica	8	8	10
Capra caucasica	6	6	12
Ovis ammon musimon	6	6	11
O. a. cycloceros	3-6	3-6	5-7
O. a. karelini	6	6	6

an interesting device that ensures their staying evenly in place and reduces wear on the pelage. The cortical layer of the hair is antero-posteriorly thinner and laterally thicker. In almost all the ruminant *Artiodactyla* (except *Moschus moschiferus, Capreolus capreolus, Ovis ammon karelini*) the cortical layer in the pile hair is 1 to 6μ thicker laterally than antero-posteriorly (*G. subgutturosa* 1μ, wild goat 5μ, *S. tatarica* and *Capra caucasica* 6μ) (table 127). This is more pronounced in winter. The hair sections with little or no medulla have a higher breaking point. Medulla-free sections enable the hair shaft to bend without breaking, so they are often found in pile hairs at the base and before the granna. The thin medulla-free club in the thick, brittle pile hairs of *Artiodactyla* is typical.

In most species of *Artiodactyla,* the septa are thicker in the peripheral row of medulla cells, strengthening the pile hairs. This is especially pronounced in the pile hair on the limbs (e.g., in elk, *Alces alces*). The pelage of some *Artiodactyla* (*G. subgutturosa, S. tatarica*) are bean-shaped in cross section with a convex anterior and concave posterior. This gives the pelage greater resistance. The fact that the fur hair medulla is completely different at base and tip on most mammals is noteworthy. At the tip the cortex and the cell walls of the medulla are thicker, and small cortical processes run into the medulla to strengthen the tip. Since the best protection is to reduce wear and tear, the roots of the hairs are abundantly innervated so that the lightest touch on the tip of a hair is felt, signaling the animal to avoid the obstacle.

The differences of the pelage on body and legs in *Artiodactyla* provide a good example of increased strength where it is most needed. *Artiodactyla*'s body hairs have a thick medulla that makes them very brittle. The pelage of deer and sheep wears out fast since the pile hairs have a thick medulla. The medulla on the withers is 86.8 to 98.0%. The pile hair is thinner and shorter on the limbs. The pile hairs (summer pelage) on the elk withers are 67.5 mm long, 12.6 mm on the knee, and 14.7 mm above the hoof (their thickness being 259μ, 141μ and 179μ). The thickness of the medulla varies most of all. On the withers, the pile hairs of the summer elk have an 86.8% medulla, while on the tarsus it is down to 43.5%, and near the hoof it is 16.6%, with an inverse increase in strength.

The hairs grow at an angle to the skin surface, and all hairs in a given area grow in the same direction. The back hairs of most mammals point toward the tail, with the hair "currents" pointing more downward than backward on the sides but toward the middle line on the belly. In this way, the fur does not hamper the animal's movements through thickets.

In some animals the pelage loses most of its insulating function and increases its protective function. Pigs and peccaries provide examples of adaptation to living in dense thickets. Their long (115 to 155 mm) bristles, which protrude well beyond other hairs, give major protection against scratches. They are very thick (430 to 464μ) and have practically no medulla. The bristles are firmly attached in the skin; their roots reach right through the dermis, and the bulbs form peculiar dermal tubercles reaching into the subcutaneous fat tissue.

In *S. scrofa,* the arrectores pilorum muslces have no heat insulating function. They only protect against injury, contracting to raise the bristles for shock-absorption. The pile hairs, adding a second protective layer, are situated under the bristles. They are much more numerous than the bristles, and are long and thick (being 21 to 97 mm on the withers and 80 to 180μ thick). Their resistance is very high since, like the bristles, they have no medulla. The fur hairs are the most abundant, and they do insulate to some extent but their protective function dominates. They are also very thick (36 to 90μ) while the medulla is small or absent.

Many terrestrial mammals need protection against the sun. The long-wave sun rays are known to be harmful to vertebrates. They can harm various organs and the organism as a whole (Kazbekova, 1941; Kurlyandskaya, 1941 and others), and mammal skin endeavors to keep them out. The pelage is the most essential for this function, with pigment as a second line of defense. Of the ruminant *Artiodactyla* under investigation only aurochs, elk, summer wild goat, and Siberian mountain goat have a pigmented epidermis. Elk and wild goats have stronger pigmentation in summer. The ventral side of the body of *Artiodactyla* is less strongly pigmented.

The pattern, seasonal variations, and distribution of body pigment suggest the role of pigment as a shield against the sun in *Artiodactyla*. This role is attributed to pigment by a number of authors who have worked with cattle in the tropics. Their findings reveal a light-colored pelage and an intensely pigmented epidermis, with the ventral surface of the body being pigmented more lightly than the dorsal (Bonsma, 1949; Yang, 1952; Badreldin and Ghany, 1952). The absence of pigment in southern steppe and desert dwellers (e.g., *G. subgutturosa*) is due to their dense, short, summer pelage, which seems to supply adequate protection (*G. subgutturosa* spends summer days in the shade). *S. tatarica* has a lightly pigmented skin, while the wild ass (*Ordo perissodactyla*) is heavily pigmented. Both live in open, sunny places, shielded by skin pigment. Elephants in sunny areas seek daytime shade, while their thick skin with its solid stratum corneum provides excellent protection. The pelage of the elephant is too sparse to be much protection, but the skin is well pigmented. Some members of *Hyracoidea,* which dwell in steppe and semisteppe habitats, are protected from penetrating sun rays by dense pelage and epidermal pigment.

SKIN ADAPTATIONS IN ARBOREAL MAMMALS

Many mammals have adapted to life in the trees. Their pelage differs, of course, from that of aquatic mammals, but the skin and pelage of arboreal and terrestrial mammals are similar in structure. Life on the ground or in trees has much in common as it affects the skin.

Many arboreal mammals have thicker skins and longer fur in the sacral region. No special features are noted in the structure of their epidermis, dermis, subcutaneous fat tissue, or skin glands, and they do not differ from the terrestrial species of taxonomic groups to which they belong. There is no common pattern in the structure of their pelage. For hair density, they fall into two categories: dense and sparse. In tree shrews, tarsier, lemur, sloth, and tamandua, the number of the hairs per 1 cm^2 (withers) ranges from 1,332 to 9,461. The other mammals investigated here range between 17,600 and 34,000. Most of the arboreal species have coarser pelage on the ventral side, where it rubs against the tree bark.

Some arboreal mammals show the beginnings of a flying membrane. The squirrel has a small skin fold on the sides of the body. In the flying squirrel and flying opossum, the flying membrane is highly developed. In the flying lemur, the tail is a part of the flying membrane.

The membrane's structure if similar in all three species. It comprises the dorsal and ventral skins doubled together. The flying membrane, unlike that of bats, has horizontal bundles of cross-striated muscles in the middle. Very abundant elastin fibers form thick bundles that run through the middle of the membrane.

The characteristic arrangement of the hair in the sloth, and the abundance of vibrissae in many arboreal mammals, are common knowledge. On several arboreal species the fur of the pelage is parted in the middle (squirrel, flying squirrel, tree shrew).

SKIN ADAPTATIONS IN FLYING MAMMALS

Only *Chiroptera* are genuine flying mammals. The skin membrane on the wings and their intercostal membrane represents a remarkable adaptation of the skin. The evolution from gliding arboreal forms to animals with flapping wings can be assumed. However, the process of formation of wings during the course of phylogenesis is still not quite clear. The flying membrane is a doubled-up skin, the dorsal and the ventral skins. The membrane is covered with sparse hair, more abundant on the inner side. The epidermis of the membrane is rather thick (thicker in proportion to the whole skin than on the trunk) which correlates with the sparse pelage on the wings. The epidermal pigmentation is peculiar, for neither stratum corneum nor the outer rows of the cells in stratum spinosum are pigmented. The outer surface of the flying membrane carries more pigment than the inner.

In fruit bats, the pigment in the epidermis of the wings protects the body from the sun (*Megachiroptera* rest in the open, clinging to tree branches). The function of the skin pigment in bats is puzzling, for bats are normally not in the sun. Pigmentation of the bat's flying membrane is all the more interesting, because their body skin contains no pigment.

Numerous elastin fibers are found in the dermis of the flying membrane, some of which form thick bundles passing in various directions to create a continuous network. This gives needed elasticity to the flying membrane and tightens it when the wings are being folded. A few large sebaceous glands surround the roots of the sparse hair like wreaths, two or five in the late, great bat *Vespertilio murinus,* and six or seven in *Plecotus auritus.* They are needed to lubricate the big epidermal surface of the wings. The shape of the sweat gland differs in the various bat species. For instance, in the common noctule the secretion cavity is shaped like a long, very twisted tube with a short duct (Zabusov, 1910). In the noctule, the secretion cavity is sac-shaped and the duct is long (Schöbl, 1871).

The sweat glands in the flying membrane skin are mainly important for temperature regulation—the sweat evaporating from the membrane can have a considerable cooling action. Sweat also helps to moisten the membrane's epidermis. Some bats lack sweat glands on the flying membrane. We found none in *Pipistrellus kuhli,* and they appear to be absent in *Rhinolophidae, V. murinus,* and members of *Molossidae* (Jobert, 1872). Many such differences in closely related species can be explained by differences in physiology. The almost naked surface of the flying membrane easily loses excess heat. The total surface is much greater than the fur-covered body surface: 4.5 to 8 times in bats and 11 times as great in fruit bats. The membrane contains large arteries and veins. In flight, when the whole chiropteran body produces a great deal of heat, all the blood vessels on the limbs are filled with blood. The flying membrane radiates much heat, and the heat loss is further increased by the action of the wings in flight. While resting, when heat loss must be reduced, the wings are folded so that the area radiating heat is much smaller. Blood supply to the flying membrane is sharply reduced by peculiar capillary sphincters and numerous anastomoses (Nicoll and Webb, 1946).

Many bats have complicated skin formations on the muzzle (*Rhinolophidae* and *Phyllostomatidae*) and around the mouth and nostrils, shaped like leaves or rosettes. They focus sounds the bat makes into a narrow beam, to produce an accurate echo. There are no special peculiarities in the structure of the epidermis of the chiropteran body. The epidermal pigment of the fruit bat's body provides extra protection against the sun. Bats that shelter in dark caves in the daytime have unpigmented body epidermis. The chiropteran dermis contains a compact plexus of elastin fibers which provides the high degree of elasticity needed in the skin of flapping wings.

The sebaceous glands in the body of *Chiroptera* are largest in the fruit bats. Their secretion may have a double function—protecting *Megachiroptera*'s epidermis and pelage from the tropical damp or, in arid areas, protecting the epidermis from drying out.

Sweat glands are present only in *Microchiroptera*. Their absence in *Megachiroptera* appears to be a secondary characteristic (indicated by their specific glands and their sweat glands). In bats, the sweat glands are an additional organ of heat radiation. We happened to catch bats resting in the daytime under house roofs heated by strong sunshine. They were wet with sweat. Heat loss through sweating appears to be limited to the rest period. In *Megachiroptera,* excess heat is given off from the flying membrane by means of polypnea and by wetting the breast and wings with saliva.

The difference between the flying membrane surface and the fur-covered body surface is much greater in fruit bats than in ordinary bats, making it easier for them to lose surplus heat. Sharp rises in temperature (up to about 40°) increase the fruit bat's rate of respiration. They breathe through an open mouth, and they fan themselves with half-opened wings. If this is not enough, they salivate and lick their breast and wings, spreading the saliva (Leither, Bartholomew, and Nelson, 1963). The evaporating saliva cools the body.

On the other hand, it is noteworthy that the pelage of *Megachiroptera* seems to be more involved in temperature regulation that that of bats. Only fruit bats have developed arrectores pilorum muscles capable of adjusting the air as cushion in the pelage. This seems to be why the fruit bat can regulate its temperature better than *Microchiroptera* (Burbank and Young, 1934). Fruit bats also have a developed medulla in the pelage.

In temperate zones, where warm days alternate with cold nights, the insects that bats feed on are not flying. This has led to poikilothermy in these bats. When the air temperature is low, the bat's body temperature drops down to the temperature of the environment, as in *Megachiroptera torpid*. This anabiosis enables the bats to survive the cold periods. Obviously these poikilothermic bats do not need pelage with good insulation properties or a powerful source of heat radiation in the flying membrane. Indeed, their pelage is short and without medulla (the medulla is recorded only in *Megadermatidae* and *Rhinopomatidae,* Benedicts, 1957). However, in some circumstances (as during rapid flight) bats employ the pelage as insulation to protect their inner organs from excessive cooling. For this reason the abdominal surface, which is more exposed to head winds, has denser pelage than the dorsal surface. When taking off, the bat produces heat by sharp movement (Young, 1963). The pelage helps it to accumulate heat. In most of the bats, the fur hairs have peculiar cuticle cells with lateral protuberances that make for better cohesion.

SKIN ADAPTATION IN BURROWING MAMMALS

Burrowing mammals are halfway between terrestrial and subterranean mammals. Their skins are thicker dorsally or on their sides. The epidermis has a developed stratum corneum. The dermis of shallow burrowers is usually rather loose, containing great quantities of fat by autumn. Many shallow burrowers hibernate through the winter. By autumn they have accumulated a vast, continuous subcutaneous fat layer. This fat differs sharply from that of terrestrial and arboreal animals in its low congealing temperature. This prevents congealment during hibernation, when the body temperature drops down to 1 to 3°C. This character derives from a high content of unsaturated acids in the fat, of which the high iodine count is indicative. The physical and chemical properties of fat in hibernating mammals (e.g., bear and badger) with organ systems that keep going normally, resemble those of the fats of mammals active in winter (Goosen, 1941; Kuznetsov, 1962):

Hibernating mammal	Iodine count	Congealing t°C
Sable	57.0	28-30
Fitch	65.7	18
Bear	73.1	21
Badger	70.0	22
Marmot	170.9	—
Tawny ground squirrel	109.6	0
Little suslik	110.2	2-0

Most shallow burrowers have short, loose, rather coarse pelts. Wintering in warm burrows (warm even in northern latitudes) requires a pelage resistant to the wear of rubbing against burrow walls. Other shallow burrowers have warm pelts because they spend much time on the surface. The fur hairs of many rodents are abundant and thin, with a sizable air-carrying medulla of not less than 47% and often as much as 80%. Thus, rodents carry insulation both as a layer between the fur hairs and in the hairs themselves. The guard and pile hair medulla is also well developed, providing better insulation than, for instance, carnivores would have, but the hairs of the burrowers are not as hard wearing. Most rodents have guard hairs with a long, thin medulla-free tip. This is absent in squirrels, mice, and a few species in other families. These thin but durable tips protect the fur and pile hairs. The differentiation of the pelage is pronounced in rodents. Their hairs grow in tufts and form groups, with arrectores pilorum muscles in the skin.

Fur insulation may differ even in closely related burrowing mammal species. The midday gerbil is an example, with several subspecies whose pelage differs markedly in morphology, such as *M. m. meridianus* Pall. and *M. m. nogaiorum* Hept. (Kalabukhov, 1955, 1957, 1958, 1959, 1950; Kalabukhov and Pryakhin, 1954; Kalabukhov, Mokrievich, and Petrosyan, 1959; Kalabukhov and Kazakevich, 1963). Midday gerbils living on the West Volga bank (*M. m. nogaiorum*) have much in common with northern and mountain mammals living in colder climates. Investigation of the hair of these two gerbil subspecies confirms the previously published data on other organ systems. *M. m. nogaiorum* has more abundant fur than *M. m. meridianus* (the former has 35,916 hairs per 1 cm^2 in winter; the latter has 30,874). The relationship remains much the same in summer (*M. m. nogaiorum* 27,999; *M. m. meridianus* 21,966). No essential differences are found in length, thickness, or medulla.

Those mammals that need to lose a lot of heat by radiation usually have short, sparse hair. The long-clawed ground squirrel is perhaps the most vivid example. In winter, when he needs good insulation (he is active through the winter in Kara Kum, staying in his burrow only for a few days when the temperature is very low) (Sokolov, 1959a; Sapozhenkov and Sokolov, 1960), his pelage includes guard, pile, and two orders of fur hairs. The pelage is very dense and the hairs are thick. The guard and pile hairs have a highly developed medulla. These factors combine to give good winter thermoinsulation.

Unlike all other desert animals, the ground squirrel is not nocturnal in the summer, so he faces very high temperatures. It is essential for this animal to ensure rapid losses of excess heat, so his summer pelage is much shorter (down to half or a quarter the length) and very sparse. The hairs themselves are of completely different shape. They are very flat, broad hairs with a thin club and with no medulla. They are not divided into categories, though there are three size orders. The medulla in all these hairs is well developed. This short, sparse pelage allows for easy heat loss from the skin surface. When feeding in the sun, the ground squirrel retires to its burrow for a while or takes shelter in the shade of saxaul trees. Presently it comes back into the open and resumes feeding. A very curious cooling technique has been observed in a young *S. leptodactylus.* After being out in the sun it runs into the shadow of saxaul trees, shovels away the hot surface sand, and lies on its belly in the hole (Sokolov and Naumova, 1962). After a while, the animal runs out into the sun again.

Most shallow burrowers have coarse pelage on the ventral side, where the skin is barely covered by fur. The direction of the tips is pronounced because the animals move head first in the burrow. Many shallow burrowers have a peculiar topography of the pelage: the longest hairs are on the sides of the body, and the tail is not very fluffy.

Burrowers that hibernate molt only once a year. Some species of burrowers have adaptations for passive defense, such as spines and armor. Despite apparent similarity, spiny animals vary not only in skin structure but also in spine structure. Spine structure differs even in *Erinaceidae* and *Tenrecidae,* which belong to the same order. In the spine, medulla cells have very thick septa. These form peculiar beams that strengthen the spine. This is absent in the more primitive spine of the tenrec, whose strength depends on a thickened cortical layer. *Echidnidae* raises his spines differently from *Erinaceidae, Tenrecidae,* and *Hystricidae.* The spiny anteater has no arrectores pilorum muscles, though bundles of cross-striated muscles rise high into the skin. The spines are raised by contracting these muscles. *Tenrecidae, Erinaceidae,* and *Hystricidae* have powerful arrectores pilorum muscles at the spines, which usually consist of several sections.

It is interesting to note that both *Tenrecidae* and *Echidnidae* have a dense network of cross-striated muscles in the inner layers of the skin. The long, relatively heavy spines must have roots, which are also long and reach deep into the skin, so that the skin of all spined animals is thick. For instance, in *Echinosorex gymnurus,* which has no spines, the withers skin is only 0.4 mm, while *Erinaceidae* is 2.0 to 3.1 mm. *Echidnidae*'s skin is still thicker at 6.3 mm. The porcupine's skin reaches 15.7 mm (the thickest of all rodent skins under investigation). The epidermis and stratum corneum are thicker because of the sparse pelage. In *Echinosorex gummarus,* which has rather a dense pelage, the withers epidermis is only 9μ and the stratum corneum is 5μ, whereas the hedgehog's epidermis reaches 196μ and the stratum corneum 185μ. In the porcupine, the epidermis and stratum corneum are the thickest of all rodents at 330 and 296μ. The spines on the porcupine's tail (gobetlike) and in the middle of the tenrec's back (a bulge at the base), are most diverse. Both send acoustic signals (Gould and Eisenberg, 1966).

The skin of many shallow burrowers shields the body from the harmful effects of sun-rays while the animals stay on the surface. Hair being removed, skin permeability increases many times in all the animals in question except *Spermophilopsis leptodactylus.* A certain role in skin permeability is played by its thickness, specifically the thickness of epidermis and stratum corneum. Among other functions, skin pigment protects the organism against long-wave rays in the sun spectrum (even if only partially). In all the mammals under investigation, the body skin pigment (excluding the pigment in the bulbs of the growing hair) is noted in some *Rodentia, Artio-* and *Perissodactyla, Hyracoidea,* and *Insectivora.*

Among rodents, the skin of *Spermatophilopsis leptodactylus,* ground squirrel, *C. pygmaeus, C. suslicus, C. undulatus, Marmota caudata, M. bobac,* and *H. indica hirsutricostris* is pigmented. *S. leptodactylus* differ biologically from other desert rodents in that they are active even in the hottest summer daytime hours. Since the pelage of *S. leptodactylus* is sparse in summer, this animal needs extra protection against the sun. This demand is met by intensive skin pigmentation. The number of melanocytes in the skin of *S. leptodactylus* is considerably higher than in other rodents. There are large numbers of pigment cells in the epidermis, mostly in the stratum basale. Sometimes small groups of melanocytes occur in the outer dermis. Individual variations are recorded in the intensity of pigmentation, but no seasonal variations are recorded. The skin of a young *S. leptodactylus* (three months) is more strongly pigmented than the adult's.

Like *S. leptodactylus,* the ground squirrel is active in the daytime. This appears to be the reason for strongly pigmented skin. The pattern of pigment distribution is the same as in *S. leptodactylus. Sciurus vulgaris* has no skin pigment. *S. vulgaris* lives in the forest, always in the shade, and needs no extra protection from the sun. So skin pigmentation is not common to all the members of *Sciuridae* (pigment is also absent in *Sciurus palmarum* and *Callosciurus locroides). S. anomalus* lives in much sunnier areas than *S. vulgaris* and has a slightly pigmented epidermis. *S. leptodactylus, C. pygmaeus, C. suslicus, M. caudata,* and *M. bobac* have pigment cells in their skin, though many fewer than in *S. leptodactylus* and *Atlantoxerus getulus;* they are mostly found in the stratum basale. In summer, susliks and marmots tend to be active in the cool hours of morning and evening, staying in their burrows in the middle of the day, so they are little affected by the sun. *Rhombomys opimus* is also active only in early morning and evening hours in the summer. All have only light skin pigmentation. It is noteworthy, however, that their pigment distribution pattern is peculiar. Large melanocytes are found in the dermis but not in the epidermis. The pattern may depend on the systematic status of the animal. In this investigation, none of the other nocturnal mammals or rodents that live out of the sun have pigment in the skin.

Some interesting features are noted in the pattern of body pigmentation of *Spermatophilopsis leptodactylus.* On back and sides, the skin is pigmented with equal intensity. The scrotum skin is highly pigmented. Melanocytes are distributed through the whole thickness of the epidermis to the stratum corneum but are more numerous in the inner epidermal layers, where they form a continuous black background. The scrotum skin proves to be more impermeable to light than the body skin, where the sparse pelage

TABLE 128

PERMEABILITY OF HAIR AND SKIN OF THE BACK OF VARIOUS RODENTS
TO LONG-WAVE RADIATION IN SUMMER (%)

Species	Haired skin			Hairless skin		
	Wave length (m)					
	650	700	750	650	700	750
Spermophilopsis leptodactylus	5.5	7.3	9.3	8.3	11.0	12.7
Sciurus vulgaris	4.5	5.5	7.5	22.0	32.0	38.5
Citellus pygmaeus	3.0	3.0	3.7	14.7	18.0	21.3
C. suslicus	3.0	3.5	3.5	22.0	26.0	30.5
Rhombomys opimus	3.5	4.0	4.4	21.0	25.7	32.0
Meriones meridianus	3.0	3.0	3.7	55.5	60.0	61.5
Skin of scrotum of *S. leptodactylus*	5.0	6.3	8.5	5.0	6.3	8.5
Recently sacrificed *R. opimus* (November)	2.7	2.7	3.0	49.0	53.0	55.0

makes no difference at all (table 128). High pigmentation on the scrotum gives the testicles better protection, and scrotal pigmentation is recorded in many mammals. An analogous formation protecting the sexual glands from the sun is found in some daytime reptiles (Sekerov, 1912; Krüger, 1929; Dinesman, 1949) with black bellies (the belly of *S. leptodactylus* has no pigment).

Since most organs perform more than one function, the hypothesis on the shielding function of skin pigment is not inconsistent with pigmentation of foot soles of most mammals or on the body skin of some nocturnal animals (e.g., the porcupine and hedgehog), for here the presence of pigment is associated with some other aspect of tissue metabolism.

Skin plays an important role in water metabolism. It is responsible for significant water loss both in sweat and by means of so-called imperceptible perspiration. Water losses through the skin may be great and may play a decisive part in water balance. Even in animals without sweat glands, water loss through the skin can be considerable. Chew (1955) published a work on *Peromyscus maniculatus sonoriensis* trained to put their heads in special rubber collars that separated and isolated two chambers during the experiment. The experiments revealed that water loss through the skin of rodents sitting quietly averages 46% of the total loss from skin and lungs. It is interesting to note that the amount and rate of skin water loss increases much more with stimulation than does lung water loss.

The problem of maintaining the water metabolism at a permanent level is vital for desert mammals. Deserts are large and very hot in summer, and have dry air and no free water. This makes it difficult to maintain the water metabolism at a certain level. The possibilities of water loss are greater than the opportunities for water intake.

We studied the water metabolism of some mammals in the Repetek region in Kara Kum. Most of Repetek's mammal inhabitants are nocturnal, avoiding the heat of day, for night temperatures are much lower in the desert (e.g., on June 13, 1960, at an altitude of 160 cm above the surface, the temperature was 12.4° at 3 A.M. and 34.8° at 4 P.M. (Sokolov and Polozhikhina, 1964). On July 29, 1959, the surface temperature of the soil was 54° at 4 P.M., and on July 30 at 7 A.M., it was 28° (Sokolov, 1960e). Moisture in the air also increases considerably at night. In fact, on June 13, 1959, the relative moisture of the air at soil surface was 29.0% at 7.30 A.M., against only 8.0% at 5.30 P.M. (Sokolov and Varzhenevsky, 1962a). So night is relatively favorable to desert animals, minimizing water loss through the skin. With few exceptions, desert animals spend the daytime in the burrows. The climate within the burrow is more or less stable during the day, so in summer it is cooler and damper than on the surface. We developed instruments for distant sensing of temperature and air moisture in burrows over long periods (Sokolov, 1964b; Sokolov and Varzhenevsky, 1962). With a surface temperature of 54°, the temperature 60 cm down in the great gerbil's burrow is only 33.0°. A surface moisture of 17% changed to 90% in a burrow 80 cm down. Our data are consistent with the results of Strelnikov (1932, 1955), Rall (1932, 1939), Vlasov (1937), Latyshev and Sidorkin (1947), and Schmidt-Nielsen B. and Schmidt-Nielsen K., (1950a, b, c). Consequently, even when the surface temperature and air dryness are very high the microclimate in burrows is favorable for animals, ensuring a minimal moisture loss through skin.

The study of desert animal skin has failed to reveal any peculiarities helping to reduce skin water losses (although corneum and skin blood systems are known to affect transepidermal water loss (Jarret, 1969; Hattingh, 1972, 1973). Apparently the reduction of water loss in desert animals is not a result of morphological skin adaptations. It is true that some mechanisms of sparing water consumption, unrelated to skin, have been found in desert animals, such as a special kidney structure allowing for a high concentration of urine (Peter, 1909; Sperber, 1944).

The concentration of urea and creatinine in desert animal urine is much higher than is possible in nondesert species (Schmidt-Nielsen K. et al., 1948; Schmidt-Nielsen K. and Schmidt-Nielsen B., 1951, 1952; Schmidt-Nielsen B., 1950a, 1950b; Schmidt-Nielsen B. et al., 1956; Schmidt-Nielsen K. et al., 1957; Sokolov and Skurat, 1962c). Desert mammals can excrete very dry feces. They also have other mechanisms.

Some authors claim that the skin of desert mammals contains less water

TABLE 129

DESERT MAMMALS: SKIN WATER CONTENT (%)

Species	Season	Number of animals	M	Lim
Rhombomys opimus	summer	5	65.2	57.4-68.2
	winter	9	56.8	41.6-66.6
Spermophilopsis leptodactylus	summer	6	65.7	62.4-69.7
	winter	7	61.6	56.8-77.2
Meriones meridianus	winter	3	61.3	59.5-62.4
Lepus capensis tolai	summer	4	65.5	59.2-72.3
	winter	2	66.2	59.1-73.3
Canis vulpes	summer	1	65.4	—
Mus musculus severtzovi	summer	1	59.6	—
Erinaceus auritus	summer	1	72.7	—
White rat (after Winn and Haldi, 1944)	?	?	64.7 males	—
			58.1 females	—
Man (after Nadel, 1932, and Urbach, 1928)	?	3	62.8	58.9-66.0

than the skin of other mammals (Khalil and Abdel-Messeih, 1954). We collected skin from newly killed animals in Repetek. The hair was removed from the samples, which were then dried at a temperature of 105° until the dry weight was obtained. The water content was estimated as a percentage of the initial weight. The data, in table 129, show that desert mammals' skin has much the same water content as that of other mammals. A lower water content in the skin of winter animals (great gerbil, long-clawed ground squirrel) is noteworthy. This appears to be due to a thick stratum corneum in winter. No difference in the water content of the skin of desert animals was found.

Despite the absence of morphological adaptations, many authors have demonstrated that skin and lung water loss is considerably lower in desert mammals than in others (Tennent, 1946; Schmidt-Nielsen B. and Schmidt-Nielsen K., 1950; Schmidt-Nielsen B., 1950; Lindeborg, 1952, 1955; Scheglova, 1952; Chew, 1955; Butcher, 1954; German, 1962). This is accounted for by the properties of water metabolism regulation, mostly by the action of an antidiuretic hormone.

Some desert rodents (Pygeretmus zhitkovi, Salpingotus crassicauda) store fat in the tail. A similar concentration of fat is recorded in the proximal half or third of the tail of the North American Microdipodops. The fat cells appear to have belonged previously to the subcutaneous fat tissue, though they do not differ morphologically from the inner dermis (Quay, 1965a).

SKIN ADAPTATIONS IN SUBTERRANEAN MAMMALS

Subterranean animals have rather thin skins, though this varies in different parts of the body. The areas where the skin often rubs against the soil or is subject to wear while digging is usually thicker. According to the Gambaryan classification (1960), *Talpidae* have a characteristic wedge method of digging, so that the chest sustains the hardest wear. Mole skin is thickest on the breast, and the reticular layer is thicker than the papillary, a relationship also found in gophers (table 130).

TABLE 130

TALPIDAE, SPALAX, GEOMYIDAE: SKIN MEASUREMENTS

| Species | Area | Thickness (μ) | | |
		Skin	Papillary layer of the dermis	Reticular layer of the dermis
Talpa europaea	Withers	320-693	144-239	162-440
	Breast	890-913	224-232	638-660
T. europaea altaica	Withers	451-528	151-224	286
	Breast	627-784	176-337	420-431
Talpa caeca	Withers	224-285	139-242	34-73
	Breast	1,456-2,035	813-1,102	935-1,120
Talpa micrura robusta	Withers	1,210-1,485	321-990	495-880
	Breast	1,980-2,310	745-1,189	770-1,540
Talpa micrura wogura	Withers	1,133	374	748
	Breast	1,749	907	792
Spalax microphthalmus	Withers	557-661	292-330	220-275
	Breast	292-397	220-275	10-66
Geomys bulbivora	Withers	600	600	20
	Breast	2,230	330	1,870
Geomys floridanus	Withers	476	330	90
	Breast	2,234	220	1,980

Spalax microphthalmus use head and incisors to dig, so their extremities are less important for digging as compared with moles. Their skin is thicker on the back than on the belly (table 130).

Zokors use head and claws to dig, so their skin is thicker on the back than on the breast, as in the mole rat. The reticular layer of the dermis in mole rats and zokors is better developed than in related surface dwellers, but it is thinner than the papillary.

Zokors have an interesting adaptation to digging: the tips of their noses are protected by a thick, highly cornified epidermis with a pronounced, two-layered stratum granulosum up to 600μ thick. The stratum corneum is 370μ, and alveoli up to 120μ high are found on the inner side of the epidermis. The nose skin can take considerable loads.

The sebaceous glands are strongly developed in subterranean mammals, providing necessary lubrication to the pelage where it continually rubs against the burrow walls.

Subterranean mammals have a peculiar pelage. Usually it is short and soft, uniform in length and density all over the body. Direction is less pronounced than in surface animals, making it possible to move both head first and tail first, with the fur bending easily upon movement. The difference in the length of different hair categories is characteristically slight. On the common mole's withers, the guard hairs are 7.0 mm and the fur hairs are 6.6 (*Talpa caeca* 7.1 and 6.4 mm, *Altai mole* 7.2 and 6.3 mm, *Talpa micrura robusta* 6.8 and 6.2 mm, *Spalax microphthalmus* 11.1 and 9.8 mm, *M. psilurus* 18.8 and 14.6 mm, mole vole 10.6 and 7.7 mm, *Prometheomys schaposchnikowi* 13.9 and 11.5 mm).

The medulla is somewhat less developed in subterranean mammals than in their surface relatives, making the hairs more resilient. The mole has the least developed medulla. *Uropsilus soriceps,* a surface dweller, has 71% medulla in the guard hairs of the withers compared with our moles, which varied from 36 to 51%. Gophers have a better developed medulla (withers guard hair 89% to 94%) than other subterranean mammals. But in their fur hairs subterranean mammals have a rather better developed medulla than in the guard hair; the guard and pile hairs protect them from above, so they can afford to sacrifice a good deal of resilience to better insulation. In *Spalax microphthalmus,* however, the fur hair medulla at 33.3% is less developed than in our other rodents. Compensatory molts characteristically restore the fur worn out in the burrow of subterranean mammals.

In structure, subterranean mammal skins usually have much in common with related land animals. All moles have specific ligatures and longitudinal turns peculiar to most *Insectivora. Heterocephalus glaber* is distinguishable from all other subterranean mammals. The outer surface of the skin of *Heterocephalus glaber* is irregular and is covered with large folds. The hairs grow singly and far apart, and though they do not have blood lacunae at the roots like vibrissae, they undoubtedly perform a sensory function. There is no medulla. The skin is thicker on the breast than on the back, which (as in *Talpidae*) seems related to their habit of digging with their paws. The epidermis is thick, up to 56μ, of which up to 34μ is taken up by stratum corneum. The fur is sparse, so the epidermis is alone responsible for protecting the skin from injury. Despite its thickness, the epidermis has no alveoli or processes on its inner surface. The dermis is a compact plexus of mainly horizontal collagen bundles. The inner dermis has large accumulations of fat cells, apparently for insulation, and these are very well developed. Three-lobed sebaceous glands are found high on the hair bursae. Pigment cells are numerous in the outer dermis of the withers—their function in a subterranean animal is obscure.

SKIN ADAPTATIONS IN AQUATIC MAMMALS

There are several systematic groups of aquatic mammals. Their level of adaptation varies, as is manifest in the adaptation of their skins. There are two distinct groups of water mammals, amphibian and aquatic. Amphibians may spend considerable time on land, while the aquatic mammals never leave the water.

We have studied the skins of *Ornithorhynchus paradoxus, Chironectes minimus, Desmana moschata, Galemys pyrenaicus, Limnogale mergalus, Potamogale velox, Micropotamogale lamottei, Neomys fodiens, Castor fiber, Ondatra zibethica, Arvicola terrestris, Myocastor coypus, Thalarctos maritimus, Mustela lutreola, Mustela vison, Lutra lutra, Eumetopias jubatus, Callorhinus ursinus, Odobaenus rosmarus, Phoca vitulina, Cystophora cristata,* and *Mirounga leonina.*

Among these mammals, skin evolution has taken two roads, leading to a dense or a sparse pelage. Those with dense pelage form the largest group (amphibian mammals), including all the above species from the orders *Monotremata, Marsupialia, Insectivora, Rodentia, Carnivora* and *C. ursinus* of *Pinnipedia.* Animals that dive into water for relatively short spells carry an air layer in their pelage. The water forms a surface tension film around the body, so the animal's fur does not get wet when it is diving. A smaller number of amphibian mammals (*Thalarctos maritimus, Eumetopias, Odobaenus rosmarus, Phoca vitulina, Pusa sibirica, Phoca groenlandica, Cystophora cristata,* and *Mirounga leonina*) have pelage that is more or less sparse, some with only a few separate hairs. These mammals do not have an air cushion when diving—their skin gets wet.

In densely furred amphibian mammals, the skin is rather thick and durable. The epidermis and dermis may, as distinct from terrestrial species, be thicker on the belly than the back (e.g., *N. fodiens, Mustela lutreola,* and *Lutra lutra*). There are no differences in the structure of the epidermis. A high development of the papillary layer is typical, and the dermal collagen bundles are mostly horizontal. In *C. fiber* on the abdominal side of the body, and in *O. zibethica* on the dorsal and abdominal sides, numerous bundles of smooth muscle fibers unconnected with hairs are found in the papillary layer of the dermis. Most of them are horizontal. They may have an adaptive significance, since they increase the motility of the skin to facilitate swimming. The subcutaneous fat tissue is somewhat more developed than in terrestrial mammals, though it is much less so than in amphibian mammals with sparse pelage or in aquatic mammals. It usually consists of an accumulation of fat cells around certain centers. As regards chemical composition and the high solidification temperature, the densely furred amphibian's subcutaneous fat is nearer to that of the terrestrial mammals than to that of amphibians with sparse pelage or aquatic mammals. This is associated with their rather high skin temperature. The sebaceous and sweat glands are commonly large (rodents have no sebaceous glands on the body). Their secretion lubricates the pelage and is largely responsible for keeping it dry and water-repellent.

The densely furred amphibian's pelage is vital for insulation both on land and in the water. The hair categories (guard, pile, and fur) are more pronounced than in terrestrial mammals. Even in insectivores (*N. fodiens, Desmana moschata,* and *G. pyrenaicus*), these three types are clearly distinguishable; as in terrestrial species of this order (e.g., in pygmy shrews and white-toothed shrews), the fur hairs are similar to pile hairs. Hair categories are distinguishable in fur seals *Callorhinus ursinus,* whereas in other Pinnipedia with a sparse pelage the fur is homogenous. The granna of the pile hairs is usually more thickened and flattened than in terrestrial forms in this group of mammals, as is indicated by the ratio of the cross section of grannas (table 131).

The fur hairs of the densely furred amphibian species have more waves and are relatively thinner than in terrestrial mammals. In fact, the thickness of fur hairs in the tiny common shrew (15 to 17μ) is greater than in the muskrat (10 to 13μ), which is many times as big. The difference in the length of fur and pile hairs in densely furred amphibians is also greater. For instance, in shrews these hairs are the same length, while in muskrats the pile hairs are 3.4 mm longer. The granna of most of the pile hairs is longer than that of the fur hairs, the shaft being bent at the base of the granna so that the hair leans backward. Pile hairs in *Ornithorhynchus paradoxus* have a short, fat granna with a medulla in the upper third; in *N. fodiens* the granna with a thick medulla is relatively unpronounced; in *D. moschata* it is big and without a medulla; in *C. ursinus* the medulla is highly developed in the lower part of the shaft of the pile hair but is absent in the granna.

In densely furred amphibians, the medulla development in the pile hair granna (table 131) is slighter, giving the hair greater durability and flexibility when swimming. The hair's specific weight nears one.

The pelage of densely furred amphibian mammals grows in pronounced tufts and groups. These are found even in *D. moschata, G. pyrenaicus, P. velox,* and *M. lamottei,* although in all other *Insectivora* (and in *N. fodiens,* which is less adapted to life in the water) the hairs grow singly. The tufts make the pelage more fluffy and ensure a more regular hair distribution. The arrangement of the tufts in *Carnivora* and *Pinnipedia* is peculiar. The pile hair grows behind the fur, bending backward when the animal is swimming. The pile hair covers the fur hairs, which promotes air retention. Densely furred amphibians have many more hairs than most terrestrial mammals of their size. *S. araneus* has 27,000 hairs per 1 cm² on the back

TABLE 131

AMPHIBIAN AND LAND MAMMALS: COMPARISON OF WITHERS HAIR

Species	Granna to pile section axes ratio	Number of wavelike curves		Thickness of medulla in pile granna as % of granna thickness
		Pile	Fur	
Ornithorhynchus anatinus	1 : 6	absent	7-9	16
Tachyglossus sp.	1 : 3	1.5-2	absent	absent
Chironectes minimus	1 : 3	10	14	57
Galemys pyrenaicus	1 : 3.6	4	9	absent
Desmana moschata	1 : 4	5-6	14	absent
Talpa europaea	1 : 1.5	3	3	65-73
Neomys fodiens	1 : 2	3	2-3	57-60
Sorex araneus	1 : 2	1-1.5	1-2	70
Marmota caudata	1 : 2.5	absent	5	72
Ondatra zibethica	1 : 2.6	same	7	57
Arvicola terrestris	1 : 2	same	8	42
Microtus arvalis	1 : 3	same	3-5	82
Myocastor coypus	1 : 2.5	same	12	76
Lutra lutra	1 : 2.3	same	11	49
Mustela lutreola	1 : 2	3-5	7	46
Martes foina	1 : 1.7	absent	6	58
Castor fiber	1 : 2.3	absent	7	46

and 14,000 on the breast; the shrew has 40,000 and 22,000; *D. moschata* has 22,000 and 37,000; *Ondatra zibethica* has 11,000 and 12,000; the otter has 35,000 and 51,000; *Ornithorhynchus paradoxus* has 57,000 on the breast; and *Chironectes minimus* has 39,000 on the back. It is noteworthy that, unlike the terrestrial mammal, the amphibian belly is protected from over-cooling by long, dense fur. The density of hair differs much less from north to south in amphibians than in terrestrial forms, for water temperatures are much less varied in different latitudes.

All these properties (a sharp category division, long, flattened pile hairs with little or no medulla in the granna, and very dense fur) are associated with the need for the pelage to carry an air layer while the animal is swim-ming. Wide, flat grannas or pile hairs bend over and cover the fur hairs parallel to the body, forming a tile pattern under the water's surface ten-sion, retaining air in the fur (Gudkova-Aksenova, 1951; Sokolov, 1961; Bel-kovich, 1962). The air layer appears to be present even in migrating fur seals that have spent several months at sea. This peculiar external cover in amphibian mammals (a combination of air and pelage) probably has

damper properties, due to its elasticity, and is capable of reducing turbulence and lessening water drag while the animal is swimming. The fur hairs
provide good insulation. A high density of wavy fur hairs ensures an inter-
layering of air. Further, the medulla is well developed in the fur hairs (unlike that in the guard and pile hairs).

In amphibian mammals the molt is gradual and slow, for the animals continue to swim during the molt and need to keep their fur. The characteristic
dark color on the dorsal side and light color on the ventral side appears to
have protective value. Heat loss, when the body overheats in the water,
occurs mainly in those areas of the body that are bare. In densely furred amphibian forms, the active heat loss area is usually the tail, which is large and
is sparsely covered with hair or completely bare. It has a highly developed
blood circulation, and heat loss is by conduction. In the air, excessive heat is
lost by convection, radiation, sweat evaporation, or, where sweat glands are
absent, by polypnea (e.g., in coypu). In the fur seal, heat loss while swimming is achieved mainly from the surface of the fins, which are very large
and bare (Irving et al., 1942; Bartholomew and Wilke, 1956). In the air,
they may become overheated. In fact, death from heat stroke has been
known to occur at a temperature as low as 10°C when the densely furred
C. ursinus is kept ashore by hunters (Bartholomew and Wilke, 1956). On
hot days in the breeding ground, *C. ursinus* may suffer considerably from
the heat. However, the skin of both sets of limbs of *C. ursinus* has large and
abundant sweat glands and has large blood vessels approaching the skin.

Effective vasomotor control of the limbs is typical of the skin of all *Pinnipedia*. In the breeding grounds and on warm days, fur seals will flap their
hind limbs incessantly, cooling their surface both by radiation and by sweat
evaporation. Flapping the hind limbs is especially characteristic of the males,
who are most active on land, and is almost never observed in pups, whose
fur provides much poorer insulation. So this subgroup of animals has many
skin adaptations in common; although their systematic status differs, every
species under study has a number of specific properties. This testifies to the
peculiar roads of evolution taken by various species in adapting to aquatic
life.

The second group, the amphibian mammals with sparse fur, includes species in which the pelage is developed to varying degrees. Most of these mammals stay in the water longer than those of the first group. *T. maritimus*
shows no morphological adaptations to the water, and its skin is little different from the skin of terrestrial *U. arctos*. In swimming, a major adaptation
is to the cold temperature of the water. As is the case with many aquatic animals, the morphology of the common integument in the polar bear protects
against overcooling. In the water, however, it is not the pelage that provides
insulation. Although its density is uniform throughout various body areas
(except on the groin), the pelage is not adequate to retain an air layer in the
fur while the polar bear is in the water. For instance, *C. fiber*, whose fur

holds an air layer while swimming, has 30,900 hairs per 1 cm^2 on the back and 34,000 on the belly. *T. maritimus* has only 1,100 to 2,300 on the back and 1,000 to 1,800 on the belly. Water penetrates the fur easily, wetting the epidermis, with the result that the skin provides poor insulation in the water (and in the air) (Scholander et al., 1950).

The subcutaneous fat tissue in *T. maritimus,* an excellent insulation, varies considerably with the season and is absent altogether in seasons when food is scarce. Thus, the adaptation of *T. maritimus* to water appears to be physiological and biochemical rather than morphological (Scholander et al., 1950, 1950*a*; Scholander, 1955). For instance, the peripheral tissues of the polar bear overcool at low temperatures to reduce heat loss. When the bear dives into cold water, the difference in temperature between skin surface and subcutaneous fat tissue may be as high as 10 to 14° (Ristland, 1970). Polar bears appear to be able to reduce the temperature of the limbs below that of the trunk, as is observed in many northern animals (Scholander et al., 1950; Irving, 1951). This mechanism compensates for the poorer insulation of skin and limbs (as compared with body insulation) through thinner subcutaneous fat tissue and shorter and sparser pelage. This mechanism, however, does not appear to be found in all northern mammals. In fact, in mink the temperature of the extremities does not drop below 14.5°, even in hard frosts (Segal and Ignatov, 1975). The chemical composition of the subcutaneous fat on the limbs of *T. maritimus* must also be different from the rest of the body, for it prevents this fat from congealing when the temperature of the limbs is lowered. This property has been found in a number of polar animals (Irving et al., 1957).

In the sparsely furred subgroup of *Pinnipedia,* the hair is at various stages of reduction. The relatively coarse hairs are divided into pile and fur, with the two being little different in shape. The air layer retention function is lost in the water. The main function of the pelage is to provide protection against injury and better movement in the water and, among seals, over the ice. The animal slides forward more easily on snow and ice, and the elastic hairs prevent it from sliding back. In the water the pelage may modify turbulence.

Experiments on models covered with the skins of *Phoca vitulina* and *Pusa sibirica* show a reduction in friction drag as compared with rough, bare skin. Hence the damping role of pinniped pelage during swimming (Mordvinov and Kurbatov, 1972). A hydrodynamic tube experiment with a dummy having a Baikal sealskin pasted over it reveals that hair is conducive to an increase in the boundary layer thickness, to a reduction of the speed pulsation level within 1 to 10 khz, and to a reduction of drag. The flow-around rate is 8 to 10 m/sec (Romanenko et al., 1973). Amphibian mammals with sparse pelage have much shorter, thicker hairs, which are increased in strength by the total absence of the medulla. The number of hairs is smaller, with the hairs continuing to grow in tufts. The number of fur hairs is sharply reduced to five per hair tuft in *P. vitulina,* three in *Eumetopias jubatus,* and one to three in *O. rosmarus.* Arrectores pilorum muscles are absent. In newborn *C. ursinus* cubs, the subcutaneous fat tissue is poorly developed, the hair is dense, and the fur hairs are marked. The cub does not yet swim. Many authors have reported the accumulation of fat under the skin during the transition to an aquatic mode of life in a seal cub. The puppy fur gives way to an adult pelage.

As the pelage of animals in this group is sparse, the epidermis is very thick to protect against wear and injury. The level of pelage development in pinnipeds correlates well with the thickness of the epidermis and the stratum corneum. *P. vitulina* has the densest pelage. Its epidermis and stratum corneum are thinner than in *E. jubatus* and in the walrus. In most cases, all five layers are found in the epidermis. A solid stratum corneum is especially well developed. The walrus epidermis, the thickest, has the alveolar pattern at the inner surface. Because it has the greatest reduction in pelage among pinnipeds, the epidermis is the sole protection of the walrus against injury. Its alveoli interdigitate with dermal papillae to strengthen the skin. In summer, *Pinnipedia* need protection against the sun, for their relatively sparse pelage gives them none. Pigment in the pinniped epidermis appears to act as an additional shield.

This subgroup of pinnipeds has a greatly thickened dermis, thickest in the reticular layer. The collagen bundles form intricate plexuses in the dermis. Pinniped skin is permanently exposed to wear and injury both on the land and in the water. The skin on the bonnets (*O. rosmarus*) is the thickest, the collagen plexus being most compact there. These bonnets are highly developed in the male thoracic area, where they protect males from fang wounds during fighting. A network of smooth muscle bundles in the dermis is not connected to hairs. This may play a part in altering the configuration of the skin, when swimming, to reduce drag. Powerful elastin fiber networks in the reticular layer of the dermis increase the elasticity of the skin. These skin formations are least developed in *O. rosmarus,* a slow swimmer that does not consume mobile food. The fast-swimming ichthyophages—*C. ursinus, E. jubatus, P. vitulina*—have the most highly developed network of smooth muscle fibers and elastin bundles.

The insulating function is taken over by thick subcutaneous fat tissue. This tissue is thicker, both in absolute and relative terms than in terrestrial or densely furred amphibian mammals. In *O. rosmarus,* it comprises 63.1% of skin thickness on the back and front of the forelimbs; in *E. jubatus,* 79.0%; in *P. vitulina,* 85.0%. The differences in the relative thickness of the subcutaneous fat tissue in the above animals is explained by the smaller animal's (*P. vitulina*) need for better insulation.

It is characteristic of the seal that almost all body fat is in the subcutaneous tissue (over 92%), though over 25% of body weight is fat (Shepeleva, 1971). In the adult harp seal, 37.4 to 48.3% of adult body weight is fat; in the Caspian seal, 53.8 to 54%; and in *P. vitulina,* 46%. Sealskin provides excellent insulation, both in air and in icy water. In *C. fiber* and *T. maritimus,* however, the quality of insulation deteriorates sharply in the water (Scholander et al., 1950). The skin and all its layers are thickest on the ventral side of the pinniped. On land, the abdomen gets the hardest wear. Furthermore, greater insulation is needed for lying on cold stones or ice on the abdomen, and for swimming, because the internal organs are not protected from the cold by as thick an abdominal muscular layer as on the back.

The pinniped ability to cool the skin deeply and gradually reduces heat loss (Belkovich, 1964). The high iodine count and the lower congealing temperature of fat in *Pinnipedia* are associated with this ability. While swimming in icy water, the skin temperature at 36° is lower than the body temperature (Hart and Irving, 1959).

The skin glands (sweat and sebaceous) are well developed. The sebaceous secretion of most mammals is soluble in water (Rothman, 1954), but the absence of cholesterol and other soluble substances in the common seal's sebum appears to make its skin waterproof (Montagna and Harrison, 1957). This needs confirmation, because cholesterol has been recorded in *Mirounga leonina's* sebaceous glands (Ling, 1968, 1970). The sebaceous secretion penetrates between the cells of the whole of stratum corneum. This secretion does not appear to create water drag even though it makes the skin water repellent (Kobets, 1968).

In the water, pinnipeds with sparse pelage dispose of excess heat by direct radiation. Here, the blood supply of the skin is very significant (A. Sokolov, 1959). Even a negligible increase in peripheral circulation can cause a big heat loss (Scholander et al., 1950).

Aquatic mammals (*Cetacea* and *Sirenia*) have many skin adaptations. Their skin is fairly smooth to reduce water drag; their pelage is minimal. *Sirenia* have a few rows of vibrissae on the body and muzzle. Toothed whales and freshwater dolphins have some on the muzzle. In toothed whales, vibrissae grow on the mandible and maxilla and are abundant on the anterior edge of the lower jaw (Japha, 1911; Yablokov and Klevezahl, 1964), with many sensitive nerve endings in the hair bursae (Nakai and Shida, 1948). These hairs may have a sensory function when catching mass and small plankton, which is the toothed whale's main diet. Some scientists believe that *Cetacea's* head hairs determine the direction of currents or indicate water pressure when the whale dives. These suggestions do not appear very sound for they would render inexplicable the disappearance of the toothed whale's pelage. If the hairs actually acted as sensors for finding accumulations of plankton, they would be useless for catching single prey and would be reduced in the toothed whale. However, most toothed whale embryos have some hairs on the upper lip in a set pattern, the number being much the same for various species. The embryo hairs are well developed (Kükenthal, 1889; Weber, 1886; Japha, 1911), and have a hair papilla and a blood sinus between the outer and inner layers of the hair membranes. Only the absence of the sebaceous glands, nerves, and muscles is indicative of their rudimentary nature. In some dolphins (e.g., *T. truncatus*) vibrissae are not lost until one or two months after birth. In the white whale and the narwhal no such hairs are recorded, even in the embryo.

The skin glands (sweat and sebaceous) are absent in *Cetacea* and *Sirenia*. The hydrophilic properties of the stratum corneum of the epidermis accounts for the absence of a complete epidermal cornification. A high phospholipid content is likewise conducive to watertightness in the cetacean corneous layer (Spearman, 1972). In the absence of pelage and a genuine stratum corneum, the need to lubricate the skin with the secretion of sebaceous glands disappears, and the glands are reduced in number. In water, the sweat glands are unable to fulfill their main function of regulating temperature and are also lost. Despite the absence of hair and glands, the skin of *Sirenia* and *Cetacea* is structured on the same principle as in terrestrial mammals. The skin is made up of three layers: epidermis, dermis, and subcutaneous fat tissue. *Cetacea* has a much thicker skin than any land mammals.

Water is much denser than air. When the animal is moving, its skin must withstand pressure from the surrounding water. In thousands of years, whales have developed a bodily and external-organ shape that produces the least drag in the water. The skin also has become smooth, without any formations that might cause drag. Moving in a denser medium, the skin of *Sirenia* and *Cetacea* must withstand a much greater load than the skin of a land mammal. This results in a rather thick epidermis. Different parts of the body are affected differently by swimming. The "working areas" (i.e., the anterior edges of the fins and the anterior head) that cut through the water are bound to take a greater load and to have a thicker epidermis than such areas as, for instance, the side of the body or the dorsal surface of the fins (table 132). This factor also influences the structure of the stratum corneum; it is thicker in the working areas (table 133).

The alveoli on the inner epidermal surface of *Cetacea* and *Sirenia* and the dermal papillae with which they interdigitate help to resist high water pressure. It is interesting to note that if the external layers of the epidermis are desquamated and if the whale is towed, the layers over the apexes of the dermal papilla level are desquamated. The epidermal cells are distributed in regular archlike rows, imparting more resistance to the epidermal layer. Finally, according to Kükenthal (1889), a complex system of tonofibril strands occurs between the epidermal cells of stratum germinativum in *Cetacea*. The tonofibrils form a peculiar frame that increases the epidermal layer's elasticity.

All *Cetacea* dive to a particular depth. Some do not go very deep; others prefer great depths for catching their food. Only the lungs change in volume with water pressure. The air in the lungs reduces in volume until the air

TABLE 132

CETACEA SKIN: THICKNESS OF EPIDERMIS (MM)

Species	Tail flukes		Flipper			Frontal fat cushion	Back
	Anterior edge	Dorsal surface	Anterior edge	Dorsal surface			
Physeter catodon	4.3	3.1	3.3	3.2		6.3	5.8
Lagenorhynchus obliquidens	1.6	1.2	1.6	1.1		—	—
Balaenoptera physalus	—	—	3.4	3.3		—	—
B. acutorostrata	2.4	1.4	3.0	2.4		—	—

TABLE 133

CETACEA AND DUGONG DUGON SKIN: STRATUM CORNEUM THICKNESS (μ)

Species	Tail flukes		Flipper		Frontal fat cushion	Back
	Anterior edge	Dorsal surface	Anterior edge	Dorsal surface		
Physter catodon	36	18	—	—	45	24
Lagenorhynchus obliquidens	69	39	99	49	—	36
B. physalus	—	—	144	54	—	60
B. acutorostrata	405	180	345	300	—	120
Dugong dugon	—	—	41	—	—	31

pressure equals the pressure of the water. Of course, this pressure on the lungs is resisted by skin, muscles, and skeleton, so the reduction in volume is not as great as would be expected from the Boyle-Mariotte law, though the volume is bound to decrease the deeper the whale dives. The volume of the air in the lunbs (and, consequently, the volume of the lungs) must also shrink to increase the solubility of the air in the blood at high pressures. The hydrostatic pressure on the lungs is bound to transfer the gases rapidly from the lung alveoli into the blood. The lungs of deepdiving whales are under such pressure that the remaining air is pushed into thick-walled bronchi and bronchioli (Scholander, 1940). Four to five extra atmospheric pressures are enough for all the gas in the lungs to dissolve (Kreps, 1941).

An interesting formation in the skeleton of the whale prevents damaging pressure on the thorax. *Cetacea* has up to seventeen pairs of ribs, but only one to eight are anteriorly attached to the sternum. The sperm whale, a wonderful diver, has only three of ten rib pairs anteriorly attached to the sternum. This gives the thorax much greater mobility. On every dive, the belly and part of the chest are somewhat constricted (a powerful trunk and muscles withstand water pressure on the back). This squeezing affects the whale's whole body, causing slight skin resistance, which is reflected in the skin structure. In deep divers, the dermal papillae rise higher into the stratum corneum than in shallow divers (table 134).

TABLE 134

CETACEA DERMAL PAPILLAE HEIGHT (% OF THE EPIDERMAL THICKNESS)

	Back	Side	Belly, breast
P. catodon	75.8	81.3	—
B. bairdii	66.6	57.1	59.2
L. acutus	60.0	54.6	53.5
B. physalus	60.0	57.5	58.6

The stratum corneum plays no great part in decreasing the pressure on the dive: it is thinner in deep divers than in shallow divers. In deep-diving whales, the transverse epidermal septa descend to the lower level of the longitudinal epidermal septa (fig. 86), so even in the lowest layers of the epidermis, alveoli interdigitate with the dermal papillae. As distinct from those of most shallow-diving cetaceans (except *Eubalaena australis*), the epidermal alveoli are round or polygonal in cross section. The sperm whale epidermis is under pressure from all sides. This determines the shape of the epidermal alveoli. Large, peculiarly shaped processes protrude from the walls of the epidermal alveoli into the dermal papilla tissue and pass through the length of the papilla. In the tangential section, it can be seen that the proximal part of the process is much thinner than the distal part, which forms a bulge. This creates a still tighter cohesion between dermis and epidermis. In addition, the processes are much bigger on the inner productive surface of the stratum germinativum.

Comparisons of the shapes of the epidermal alveoli of the shallow-swimming *Lagenorhynchus acutus* and *Lagenorhynchus obliquidens* with the deeper-diving *Tursiops truncatus* reveal no great differences. Swimming capability does not appear to manifest itself in this aspect of epidermal structure. The skin of the first three species has other adaptations to fast swimming, explaining their similarity in structure.

The sperm whale has rough skin on back and sides. This would seem to support the thesis of M. M. Sleptsov (1955), who believes that the durability of the skin protects the body from underwater pressures. In the hollows between the bulges, ingrowths of epidermis form a peculiar epidermal prominence. In this place, the epidermis (5.8 mm thick) is 2.2 mm thicker than on the top of the projection (3.6 mm). However, it is unlikely that this outgrowth substantially strengthens the outer layers of the skin.

Balaenopteridae have, on throat and belly, numerous cutaneous folds divided by deep grooves. These grooves probably perform the same function in all species, increasing the volume of the oral cavity during feeding (Kükenthal, 1889; Sleptsov, 1955). They appear also to augment skin resistance to water pressure. When the blue whale dives to great depths, the water exerts strong pressure on its belly, and the bands are pushed together. The bands appear to be highly elastic and to resist the increasing pressure. The elastic properties of the bands are manifested in several ways: (a) when adjoining bands are pushed together under pressure, the hollows of one band fit the prominences in the next, (b) the epidermis of the groove between the bands is much less developed than at the apexes of the grooves (at the apexes, the epidermis is $3,945\mu$ thick; in the hollow between grooves it is only 168μ). This suggests interconnection of the bands, so that groove skin is protected against water pressure (otherwise a thicker epidermis like that on the apexes of the bands would develop in the grooves), and (c) the dermal papillae in the apexes of the bands are patterned to resist the kinds of pressure the bands may encounter.

As is known, the elastin fibers (being elastic) pull tissue back into place after stretching or shifting. The thick elastin bundles in the skin of the blue whale passing along the body of the animal, and the elastin fibers covering the cross-striated muscles in the subcutaneous fat tissue apparently restore bulged or squashed tissue on the underside of the body to its normal position.

Many *Cetacea* swim very fast. The following reliable data are available on this: a 210 cm dolphin swims at 33.3 km per hour; *Orcinus orca* 8 m long swims at between 38 to 50 km per hour for 20 min; large groups of *T. trun-*

catus (200 to 500 head) swim at 23 to 33 to 36 km per hour (different observations) for 8 to 25 min. Slightly wounded sei whales (12 to 15 m long) spurt off at 55 km per hour. The fact that large groups of animals swim alongside one another at high speeds should be especially noted (the turbulent eddies formed by neighboring dolphins should handicap their movement). Harpooned whales have been known to tow whaling boats at 20 to 25 km per hour. The great length of whale migration routes is well known, and whales cover these long distances remarkably quickly.

Cetacea can jump high out of the water (dolphins can jump up four to five meters) even with a small running start, which is possible only at high speeds. Gray (1936) estimates that to develop their common speeds, dolphins must have muscles at least seven times as powerful as terrestrial mammals in terms of the friction equivalent met by an inflexible projectile of equal size. If the dolphins' muscles are more powerful, the dolphin would be expected to have a higher oxygen consumption than similar terrestrial mammals. However, the dolphin's oxygen consumption is no different from that of terrestrial mammals. Gray (1936) demonstrates that if the water layer around the dolphin has a laminar flow, the dolphin's muscle power will correspond to that of other mammals. *Cetacea* appear to have mechanisms to decrease drag while swimming. Their body shapes are well designed. However, identically shaped dummies do not show much reduction in drag, nor do experiments on towing dead dolphins produce an explanation. Kramer (1960) reports a 50% reduction in drag by covering the surface of an underwater projectile with an especially designed flexible skin (Kramer, 1960*a*, 1960*b*). A number of mechanical dimensions on the skin of *T. truncatus* and *D. delphis* prove to be optimal for stabilizing the laminar boundary layer (Babenko, 1971).

Significant changes occur in a dolphin's skin with sudden increases in speed (Essapian, 1955). Wavy, lengthwise folds run from head to tail on his body. They are thought to be created by a contraction of the subcutaneous muscles (Surkina, 1968, 1971*a*, *b*). However, V. A. Rodionov at Moscow University demonstrates that similar changes of the skin occur to a lesser degree, possibly created by the turbulent eddies resulting from differences in the density of the epidermis, the dermal papillae, and the layers between. The pattern of dermal papillae along the body and limbs may have significance as an adaptation to swimming in *Sirenia* and *Cetacea* (Sokolov, 1955, 1962*c*; Purves, 1963, 1969; Sokolov and Kuznetsov, 1966; Sokolov et al., 1968; Surkina, 1971*a*, *b*). Experiments in a flow channel reveal that the arrangement of dermal ridges over the dolphin's body run in the same direction as the boundary layer (Sokolov et al., 1967; Purves, 1969). The structure of the dermal papillae and the epidermal alveoli differs in aquatic mammals (Sokolov, 1955, 1959, 1960, 1962*c*) and appears to be determined by their swimming speed.

In *Dugong dugon,* the epidermal alveoli are round in section. In all our other aquatic mammals (except the sperm whale and *E. australis*), they are oval or almost flat, with the narrow side facing the water flow. In some aquatic mammals, the epidermal alveoli walls are completely smooth (*D. dugon, B. physalus*) (fig. 86). In some, epidermal cells grow into the tissue of the dermal papillae (*Lagenorhynchus obliquidens*). Occasionally processes project from the epidermal septa to the dermal papillae (*L. obliquidens, B. acutorostrata*) (fig. 86), though they are small and not present at either end. In *Stenella coeruleoalbus* and *Balaenoptera acutorostrata,* the processes are larger (fig. 86). They are largest in *P. catodon* and *S. australis* (fig. 86), where they run through the dermal papillae. It is interesting that the killer whale, the fastest of them all, has the thickest epidermis and the highest dermal papillae.

It has been suggested that turbulent eddies are modified by epidermal squamae breaking off from the skin of a cetacean, just as turbulence is reduced in some high molecular compounds (Sokolov et al., 1969). In size, the epidermal squamae are commensurate with molecules of high molecular substances surrounded by the water molecules. Experiments with a saltwater suspension of a dolphin's epidermal biopsy specimen have proven unsuccessful. The suspension, however, is a rough imitation of natural epidermal desquamation.

Pershin et al. (1970) speculate that the caudal fin lobes may be partially responsible for high swimming speeds through a "hydroelastic effect" caused by a pulsation of blood in the complex blood vessels. A number of adaptations related to swimming are found in the dermis of aquatic mammals. The areas of maximum water friction (the anterior edge of the caudal fin in all whales; the frontal fat cushion in the sperm whale) have a thicker dermis (table 135).

The collagen plexus is highly compact in such places, most bundles being oblique, and there are fewer fat cells in the dermis. Where it is present, the nonfat layer of the dermis is thicker, giving the skin greater strength (table 136).

TABLE 135

CETACEA: THICKNESS OF BODY DERMIS (% OF SKIN)

Species	Back	Belly	Side	Tail flukes, anterior	Frontal fat cushion
L. obliquidens	10.1	2.7	10.4	22.7	—
P. catodon	14.7	—	13.6	75.2	51.3
B. acutorostrata	3	Bands 1.8	—	50.3	—

TABLE 136

CETACEA: THICKNESS OF THE FAT-FREE SUBPAPILLARY LAYER OF THE BODY DERMIS (% OF SKIN)

Species	Back	Belly	Side	Tail flukes (anterior)	Frontal fat cushion
L. obliquidens	3.7	0.3	2.9	11.9	—
P. catodon	4.8	—	3.8	75.2	19.1

Apparently, subcutaneous tissue does not help to reduce friction. It might be interesting to compare the structural differences in the subepidermal layers of the dorsal and anterior edges of the tail flukes. The layer of subcutaneous fat tissue decreases considerably approaching the anterior edge of the fin, until it disappears completely on the edge. On the frontal fat cushion of sperm whales, 37.8% of the skin is subcutaneous fat tissue, as compared with 79.5% on the back and 82.2% on the body side.

Thick collagen bundles interlayered with fat cells should increase the skin's resistance to high pressure. This is the function of the inner dermis and the middle and outer layers of subcutaneous fat. Very thick collagen bundles run obliquely, with large inclusions of fat cells between them. In whales that dive deep, these layers are very thick (table 137).

TABLE 137

CETACEA: THICKNESS OF THE SUBCUTANEOUS FAT TISSUE (MM)

Species	Back	Belly-Breast	Side
Orcinus orca	—	32	34
Berardius bairdii	212	139	133
Physeter catodon	81	119	95
Balaenoptera physalus	44	42	41

The table shows that the deep divers (Baird's beaked whale, sperm whale) have much thicker subcutaneous fat tissue than do the shallow swimmers, the killer whale, and the fin whale, and that they need thicker subcutaneous fat tissue for insulation against the lower temperatures at a greater depth (Yablokov et al., 1972). There is also reason to believe that Cetacea's subcutaneous fat tissue, the fat cells in particular, absorb oxygen (oxygen is five times as soluble in fat as in water (Kreps, 1941). This enables the whale to carry much more oxygen and so to stay under water longer. The network of the capillaries is highly developed through the whole subcutaneous fat layer and can easily supply all the needed oxygen to the fat. Oxygen absorption by fat is only hypothetical at the present time and requires further investigation. In the course of diving, the peripheral blood vessels almost cease to function. It is possible that these local oxygen stores are of physiological significance.

The common integument is significant in heat regulation. Whales spend a large part of their lives in very cold seas. Even in subtropical and tropical seas, where the temperature of the surface water layers reaches 28°, the water is cold beyond a certain depth. Heat conductivity of the water is twenty-seven times that of air, so to protect the body from overcooling, aquatic mammals need adequate insulation. In terrestrial mammals, pelage and subcutaneous fat tissue are primary insulators, but under aquatic conditions, the pelage, unable to fulfil this function, is minimal. The subcutaneous tissue, which contains great deposits of fat, takes over.

Cetacea have very thick subcutaneous fat tissue all over the body. Fat cells penetrate high into the dermal layer. In all whales, in winter the fat cells may fill up the dermal papillae. The fat insulator in whales is so efficient that, even after death, the body retains its normal temperature for a long time (Zenkovich, 1938). The fat content of the skin varies seasonally, being at its lowest in the period when whales migrate to warm tropical waters. However, even in tropical waters, the sperm whale has only slightly less fat, probably because the sperm whale has to dive very deep for its food, 200 to 300 meters. Where the temperature is between 2 to 3°C (Sleptsov, 1955), the sperm whale needs a permanent insulation layer.

According to Rubner, large animals radiate less heat per unit of mass than do smaller animals, so smaller whales need a relatively thicker insulation layer. This difference is evident in comparing the fin whale with the striped dolphin, where the disparity in size is great (table 138).

The chemical composition of the subcutaneous fat in *Cetacea* is charac-

TABLE 138

THICKNESS OF SUBCUTANEOUS FAT TISSUE

	Body length (m)	Ratio of subcutaneous fat tissue thickness to body length		
		Back	Belly	Side
B. physalus	15.65	0.2	0.3	0.2
L. obliquidens	2.2	0.5	0.4	0.4

teristic. As in pinnipeds, it is characterized by a considerable amount of unsaturated fatty acids; the fat has a high iodine count and a low solidification temperature (table 139). This is also characteristic of hibernating mammals. The low temperature of fat solidification is very important for *Cetacea,* because the temperature of the outer skin layers is usually little higher than the sea temperature.

TABLE 139

SUBCUTANEOUS FAT: IODINE COUNT AND SOLIDIFICATION TEMPERATURE; GOOSEN, 1941
(KIZEVETTER, 1953, KUZNETSOV, 1962); GRUSHETSKAYA ET AL., 1973

Species	Iodine count	t° solidification
Pusa sibirica	130-133	3
Balaena mysticetus	69-144	7-9.5
Balaenoptera musculus	99-136	5-8
B. physalus	107-155	3-5
L. acutus	118.6	0
Mustela vison	64.6	16
Mustela putorius	65.7	18
Martes zibellina	57	28-30
Thalassarctos maritimus	98.3	147
C. pygmaeus	118	0-2

The sweat glands, which play an important part in heat regulation among terrestrial mammals, cannot fulfil this function in water and consequently are reduced in *Cetacea.* Polypnea is not noted in *Cetacea.* Even under a great physical strain (in pursuit) the respiration rate barely increases. Nevertheless, *Cetacea* must have an efficient heat loss mechanism since their heat production is high. It has been estimated that, swimming at 10 knots, the whale's body temperature can increase by 1.65°C in ten minutes (Gawn, 1948). Accumulating heat at such a rate could kill the whale. The following pattern of heat loss in *Cetacea* appears to exist. A resting whale must minimize heat loss, and the blood circulation in the skin (body and fins) is of low intensity. The venous blood in the fins flows mainly along the thin-walled veins in the complex vessels. A peculiar pattern of veins, a rosette around a large artery (Tomilin, 1947, 1951, 1957; Sokolov, 1953) ensures heat exchange between the arterial blood flowing from body to fins and the venous blood flowing from fins to body. This greatly reduces the heat loss from the surface of the fins, where the temperature is low (Scholander and Schevill, 1955; Van Utrecht, 1958).

The complex vessels are found in the nonhomodynamic parts of the body: in the foreflippers, tail flukes, and dorsal fins. Apparently, their presence is determined functionally. The outer skin of *Cetacea* at this stage is a peculiar "poikilothermic membrane" (Belkovich, 1961; Yablokov et al., 1972), with a temperature nearly the same as the environment (Irving and Hart, 1957). The skin temperature increases gradually from the epidermis to the muscles, though it remains lower than the body temperature even in the innermost layers (Hart and Irving, 1959). The possibility of blood supply into the outer cool skin or only into the warm inner skin layers is important (Belkovich, 1961), for this can vary the level of heat radiation.

When the increase in the body temperature is negligible, the intensity of the blood flow increases more in the skin of the fins than of the body. There is circumstantial evidence for this, such as the fact that the surface temperature of the fins may vary considerably and may be much higher than the temperature of the body skin. If the fin is pricked with a needle it bleeds profusely. The body skin does not (Tomilin, 1946, 1947, 1948, 1951, 1957). The dermal layer of the fin skin contains many more large blood vessels than does the body dermis. Furthermore, the fin blood vessels show a higher vasometor function. This fact, established for seals (Irving, Scholander, and Grinnel, 1942), is apparently also true for *Cetacea*. As the circulation speeds up, the fin temperature increases as the exchange of heat between the arterial and venous blood decreases. Most of the venous blood is also channeled through other thickwalled veins, eliminating heat exchange with the arterial blood. The blood in the body skin begins to be pumped into upper layers, at which stage most of the heat is lost through the surface of the fins.

A considerable rise in body temperature results in a still more intense flow of blood and a further increase in the temperature of the fins. The blood is supplied to the highest layers of the common integument for maximum heat loss. Heat losses for a given period are higher per cm² of skin surface on the fins than on the body, though the total heat loss from the fins is much less than the total loss from the enormous surface of the body.

A considerable reduction in the peripheral blood circulation is noted during apnea while diving (Irving, 1939; Scholander, 1940), and there is reason to believe that the blood stops circulating in the skin (Belkovich, 1961). It is possible that at such times the only heat loss is from the fins. Cutaneous respiration in *Cetacea* has been suggested (Tinyakov et al., 1973), but a direct experiment (conducted by V. Dargoltz and V. Sokolov) to test the existence of gas exchange through the skin of a Black Sea *T. truncatus* in a small aquarium failed to support the suggestion.

SEASONAL VARIATIONS

Seasonal skin and pelage changes are most marked in the higher latitudes, enabling mammals to adapt to seasonal temperature changes, to reduce heat loss in winter, and to increase heat loss in summer. Investigation of the common integument of some mammals in winter and summer shows that most seasonal changes are shared. In winter, the skin is thinner and the epidermis is thicker, through the growth of the stratum corneum. The stratum germinativum is thinner. The stratum corneum, formed by numerous loose corneous squamae, contains small air cavities for increased insulation.

In early winter many animals have fat deposits in the subcutaneous fat tissue. In the winter *Moschus moschiferus* and *Odobaenus rosmarus,* fat cells are concentrated in the dermal papillary layer, while in *Cetacea* winter fat cells fill out the dermal papillae.

In most of the animals under study the sebaceous glands remain unchanged, though in ungulates they are bigger in the summer. This is especially pronounced in the body skin of the elk. In the common mole, the raccoon dog (Kogteva, 1963), the red fox, and the squirrel (Leschinskaya, 1952*a, b*), the sebaceous glands are also larger in summer. In the harp seal only half the sebaceous glands are active in March and April, though in summer all of them seem active (Belkovich, 1964*b*).

Sweat glands are better developed in summer than in winter for higher summer temperatures create a need for heat reduction. In some mammals, the sweat glands all but disappear in winter (e.g., musk deer). In walruses, an autumn atrophy of sweat glands is recorded (Sokolov, 1960). In *Cystophora cristata* the sweat glands are minuscule in March, and they are very shrunken in the April harp seal (Belkovich, 1964*b*).

Seasonal change in the pelage includes a winter increase in the number or the length of hairs and a thicker medulla to improve insulation. The thickness of the hairs does not change seasonally in all animals (some of the *Insectivora*). All categories may be thinner in winter (common mole), or the guard and pile hairs may be thicker and the fur hairs thinner (*Carnivora*), or all categories may be thicker (many ungulates).

Sharp seasonal weather changes affect terrestrial and arboreal mammals more than other ecological groups, giving rise to clear-cut seasonal pelage changes, especially in those features affecting insulation. Temperatures change less underground and underwater, so there are fewer seasonal changes in the fur of burrowing and aquatic mammals.

Seasonal skin changes are not pronounced in bats, which hibernate in winter. The sebaceous glands cease to function in bats in winter, and fat cells accumulate in the deep parts of the dermis.

SPECIFIC SKIN GLANDS AND CHEMICAL COMMUNICATION IN MAMMALS

The secretion of specific skin glands in mammals is one source of odorous (olfactory) chemical signals. These odorous signals (pheromones) are of great significance in mammal biology. Microorganisms may decompose the secretions of specific skin glands to produce some signaling substances (Albone et al., 1971). Many specific glands are in skin fossae and pouches, where conditions are conducive to the development of microorganisms. An examination of the anal sacs of *Herpestes auropunctatus* (Gorman et al., 1974), *Vulpes vulpes,* and *Panthera leo* (Albone et al., 1974; Albone, Gosden, and Ware, 1976) has revealed the bacterial nature of the pheromones.

Genetically, chemical signalization appears to be the most ancient of communicative systems. It is effective over far greater distances than are visual or acoustic communicative systems. It is effective in total darkness (which excludes visual communication), and because an odorous signal persists for a considerable time, it permits communication despite the absence of the communicant. Such communication is impossible when acoustic and tactile receptors are involved. The energy efficiency of the chemical signal is extremely high: a minute amount of chemical agent may convey important information.

It is well known that by means of odor mammals can identify individual species. In fact, a predator finds and chases prey by means of olfaction. Conversely, olfaction enables prey to smell a predator at a distance. Experiments reveal that *Peromyscus maniculatus* males can distinguish the smell of females of their species from that of *Peromyscus polionotus* females (Moore, 1965). Estral females of *Peromyscus maniculatus* likewise distinguish the smell of males of their species from that of *P. leucopus* males (Doty, 1972). *Mus musculus* (strain C57B1) distinguish the smell of their own species from that of *Peromyscus maniculatus* (Bowers and Alexander, 1967). *Meriones unguiculatus* distinguish the smell of their species from that of *Rattus norvegicus, Mus musculus, Mesocricetus auratus,* and *Dicrostonyx groenlandicus* (Dagg and Windsor, 1971). Thus, there is reason to believe that chemical composition of the secretion of skin glands is species-specific.

By thin-layer chromatography, we have studied the secretions of specific skin glands of twenty-six mammal species (five orders) as follows:

1. INSECTIVORA: *Desmana moschata;*
2. LAGOMORPHA: *Lepus capensis, Lepus timidus;*
3. RODENTIA: *Castor fiber, Arbicola terrestris, Rhombomys opimus, Meriones unguiculatus;*
4. CARNIVORA: *Martes zibellina, Mustela vison, Meles meles, Gulo gulo, Lutra lutra;*
5. ARTIODACTYLA: *Sus scrofa, Cervus nippon, Cervus elaphus, Capreolus capreolus, Rangifer tarandus, Alces alces, Saiga tatarica, Gazella thomsoni, Gazella granti, Aepyceros melampus, Connochaetes taurinus, Redunca redunca, Ovis ammon, Tragelaphus oryx.*

The chromatograms of each of the species under study differ significantly from one another. Consequently, the chemical composition of the secretion of specific skin glands is of taxonomic significance in mammal studies.

The secretion's chemical composition appears to be specific for members of the same group (herd, population, subspecies). If group members are related, they may share an inherited specificity of the pheromone's chemical composition. This would occur in small groups (families) as well as large communities (populations, subspecies). A comparison of the chemical composition of the secretions of two deer subspecies (*Cervus elaphus elaphus* and *C. e. xanthopygus*) has demonstrated qualitative differences (Sokolov, 1975). In addition, black-tailed deer (*Odocoileus hemionus hemionus* and *O. h. columbianus*) can distinguish between the odors of tarsal gland secretions of their own and alien subspecies members (Müller-Schwarze, 1974), and *Clethrionomys* males prefer the odor of females of their own subspecies (Godfrey, 1958).

In small groups of some species, the dominant male applies its secretion to other group members to provide them with a unified odor. This is done, for instance, by a dominant male of the flying phalanger, *Petaurus breviceps* (Schultze-Westrum, 1965). The alien odor elicits aggression in *P. breviceps*. In a family of great gerbils (*Rhombomys opimus*), young animals creep under the dominant male to lubricate themselves with its abdominal gland secretion. A similar behavior is recorded in Norway rats (*Rattus norvegicus*) and black rats (*R. rattus*); nondominant individuals creep over or under the dominant (Barnett, 1963; Sokolov, Lyapunova and Khorlina, 1977).

On the other hand, some mammals in small groups may merely memorize the individual odors of all the group members. This applies to black-tailed deer (*Odocoileus hemionus*). Members of the same herd in *O. hemionus* will sniff at one another's tarsal glands at a rate of one to two times per hour in the daytime and six times per hour at night. When an alien female approaches the herd in the daytime, the rate of sniffing at the tarsal glands increases up to three to four times per hour, the respective value attaining six to sixteen and even forty-two times per hour with the approach of an alien male. An alien male is never attacked before his tarsal gland has been sniffed, thereby determining his identity.

Numerous observations reveal that every individual has a specific odor. Dogs can trace people and, moreover, can distinguish between genetically identical monozygotic twins. Among gregarious mammals, such as *Callorhinus ursinus* and fur seals, females identify their young by odor. Occasionally a certain specific skin gland is responsible for individual identification (e.g., the tarsal gland in black-tailed deer) (Muller-Schwarze, 1971). In the wild rabbit, *Oryctolagus cuniculus*, excision of the inguinal gland prolongs individual identification. Lubrication of young rabbits with inguinal gland secretion from another female results in the mother's aggressive behavior toward them. At the same time, application of anal gland secretion from another female elicits no aggression on the part of the mother (Mykytowicz and Dudzinski, 1972).

Odor is highly significant in sex identification. In *Oryctolagus cuniculus,* the specific skin glands (anal, inguinal, mandibular, and orbital) are larger in size and more active in adult males than in adult females (Mykytowicz, 1965, 1966*a, b*). In wild rabbit males, anal and inguinal gland secretion odors are stronger than in females (Hesterman and Mykytowycz, 1968). In *Lepus europaeus,* the inguinal gland is larger in adult females. The anal gland size does not show sex dimorphism.

Chemical analyses of carbonyl compounds from the caudal gland secretion in the desman (*Desmana moschata*) reveal qualitative similarity in males and females (fig. 205) (Sokolov, Chernova et al., 1977). But the quantitative composition of a number of components is distinctive. As is shown in the histogram (see fig. 206), components 4, 7, 8, 12, 13, and 14 (named in the figure) do not quantitatively overlap in males and females. Their contents can thereby serve to identify the sex of a desman by the odor.

Considerable differences in the chemical compositions of tarsal gland

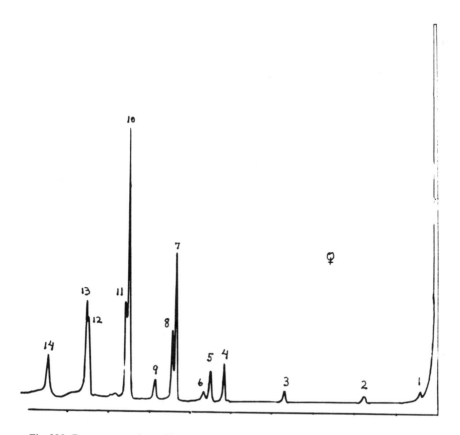

Fig. 205. *Desmana moschata.* Chromatograms of carbonyl compounds of subcaudal gland secretion in males and females.

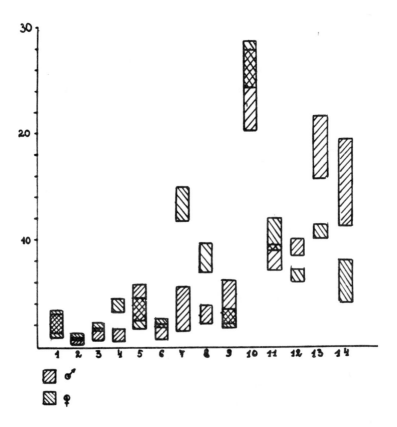

Fig. 206. *Desmana moschata*. Histogram of carbonyl compounds of sub-caudal gland secretion: *4*, pentadecanon; *7*, heptadecaenon; *8*, heptadecanon; *12*, icosadienon; *13*, icosaenon; *14*, tricosaenon; abscissae, peak numbers; ordinates, content percentage.

secretions of males and females are revealed in reindeer (Sokolov, 1977). The composition of the alcohols obtained from reindeer secretions through reduction by lithium alumohydrite is found to be qualitatively and quantitatively sex specific. The peak area on the chromatogram corresponds to the quantity of the substance (the identified alcohols are presented in the figure). It is noteworthy that the tarsal gland secretion composition in castrated reindeer is identical to that of females.

Gas chromatography data indicate that the chemical composition of the tarsal gland secretion in black-tailed deer (*Odocoileus hemionus*) (Müller-Schwarze, 1971) and of the caudal gland of *Apodemus flavicollis* (Stoddart, 1976) is sex specific.

In some instances, olfaction appears to be of primary importance for age determination. It is well known that an adult dog will not attack a cub on sniffing it. As is demonstrated in *O. cuniculus,* rabbits determine age by the odor of skin gland secretion (Hesterman and Mykytowicz, 1968). Likewise, gas chromatography of the lateral gland secretion in *Arvicola terrestris* reveals considerable differences in chemical composition between adults and young (Stoddart, 1976).

The secretions of specific skin glands play a role in identification of individual social status (Schultze-Westrum, 1965). The weights of specific skin glands (anal, inguinal, and mandibular) show a statistically significant correlation with individual rank; they are considerably larger in dominants (Mykytowicz and Dudzinski, 1972). A similar correlation is found between social rank and the histological structure of these glands; they are larger and more active in dominants (Mykytowicz, 1965, 1966*a, b*). In house mice (*Mus musculus*) dominants have a larger preputial gland (Christian et al., 1965). In dominant guinea pigs (*Cavia porcellus*), the perineal glands produce more secretion than they do in subdominants (Beauchamp, 1974). This holds true as well in *Meriones unguiculatus* (Thiessen and Yahr, 1977).

As a rule, dominants mark the territory more frequently than subdominants. This is observed in rabbits (Mykytowicz, 1965, 1967), *Petaurus breviceps* (Schultze-Westrum, 1965), *Mesocricetus auratus* (Johnston, 1972, 1975, 1977), *Rhombomys opimus* (Sokolov, Isaev et al., 1977), and *Cavia porcellus* (Beauchamp, 1974). As a subdominant takes over the dominant position, his marking rate greatly increases.

Many mammals dwell in areas partially or entirely protected by the territorial male. The territorial male marks its home range in some way, mostly chemically. To mark the home range, a number of mammals use specific skin gland secretions. Rabbits mark the home range with fecal hillocks (Mykytowicz, 1968, 1974). It is noteworthy that this involves fecal pellets specially lubricated with spontaneously produced anal gland secretions. Experimental rabbits react to pellets from other rabbit hillocks by showing anxiety and producing lubricated pellets themselves. They remain indifferent to pellets scattered over the home ranges of other rabbits. Some protruding objects found in the home range are marked with secretions from the chin gland. The secretion of this gland (imperceptible to the human nose) is used by male rabbits to mark objects impossible to mark with feces or urine, to mark the entrance to the burrow, to mark pellets of other rabbits and their own old pellets. The chin gland secretion is commonly used in their own home ranges and is not utilized in alien home ranges.

A *Rhombomys opimus* male often produces signal hillocks by throwing up a small soil mound with its front feet and smoothing it with its belly. These signal hillocks may be marked with midventral gland secretion particles, feces, and urine. In addition, the *R. opimus* male may rub his midventral gland against the soil surface. Signal hillocks and rubbins are renewed systematically as the odor left by the male disappears rapidly. At 60°C with no wind, gerbils stop perceiving odor after two hours (Sokolov, Isaev et al., 1977). The male does not mark the entire area but only those areas near inhabited and visited burrows and near permanent paths (Sokolov, 1977). The marking rate increases greatly if an alien individual appears in the colony area or if an alien male odor is introduced. Some rodents, such as the long-clawed ground squirrel (*Spermophilopsis leptodactylus*) mark their territory with fecal hillocks evidently lubricated with anal gland secretion. Marmots, such as *Marmota flaviventris* (Armitage, 1974, 1976), *M. marmota* (Koenig, 1957; Münch, 1958), *M. olympus* (Barash, 1973), *M. broweri* (Rausch and Rausch, 1971), use buccal gland secretion for marking territory.

Territorial marking is characteristic of many carnivores. Urination and defecation are commonly employed, but specific skin gland secretion appears also to be involved. In the weasel (*Mustela nivalis*) in captivity, three types of marking of the area or of objects are observed (1) rubbing against the object or substrate with the anal region; (2) rubbing with inguinal region of the belly; (3) leaving of feces (Rozhnov, 1975). The first and third types of marking involve anal glands, the second involves skin, inguinal, and abdominal glands (Neal, 1948). In the wild, the weasel uses the abdominal gland secretion to mark definite sites.

Other *Mustelidae* rub against the substrate with the anal region, specifically in the course of rut. These include the ermine (*Mustela erminea*), the Siberian weasel (*Musela sibirica*), the Siberian polecat (*Mustela eversmanni*) (Zverev, 1931), (*Mustela altaica*) (Zlobin, Shiryaev, 1975), and the mink (*Mustela vison*) (Ktitorov, 1950). A tame *Mustela eversmanni* female rubs against all the objects in the room new to it with its underbelly (Il'enko and Zagorodnyaya, 1977). The sable (*Martes zibellina*) rubs against the soil with its belly or inguinal region (Manteifel, 1934; Raevsky, 1947). The American marten (*Martes americana*) marks the area with inguinal gland secretions (Herman and Fuller, 1974). The wolverine (*Gulo gulo*) uses the highly odorous secretion of the perianal gland for marking the area and prey (Weddle, 1968). The wolverine will push the snow away from the object being marked and will rub against the object with the anal region; Pullianen and Ovaskainen (1975) report a wolverine leaving twenty-six marks over 7.5 km. The badger (*Meles meles*) in captivity, after sniffing at new objects, will touch them with its tail base (the site of the caudal gland) (Pod'yapolsky, 1933; Eibl-Eibesfeldt, 1950). Among all *Mustelidae,* feces and urine appear also to be involved in marking.

Territorial marking is common among most ungulates. Males of the Siberian roe (*Capreolus capreolus pygargus*) mark small trees and shrubs in their territories, decorticating them with the horns or rubbing against them with the frontal portion of the head, cheek, and neck (Sokolov, Danilkin, 1975, 1977) (fig. 207). Specific skin glands have not been found in these body regions, but this does not rule out their presence. Sebaceous and sweat glands are present and are better developed than in adjoining skin regions. Within their territories, Siberian roe males hoof certain sites of land, marking them with the odor of the interdigital gland secretion. European roes mark their territories in a similar way (Meyer, 1968). Siberian roe males mark their territories not only at the boundaries but throughout the whole

Fig. 207. Marking of small trees and shrubs by Siberian roe male. Marks made by rubbing with frontal portion of head.

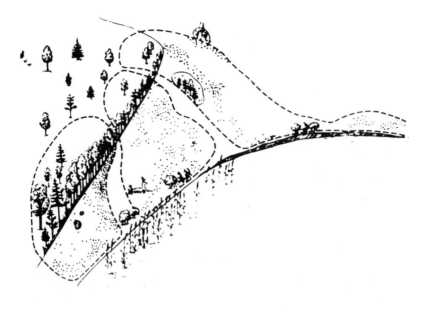

Fig. 208. Siberian roe. Home ranges. Boundaries shown by broken line, marking of trees and shrubs by dots. Ranges are located near a lake.

area (fig. 208). Normally, when the territorial male feeds on his home range with no one worrying him, the number of rubbings against trees per hour ranges from five to fifty (fig. 209). The secretion odor disappears rapidly and must be constantly reintroduced. The number of rubbings with the head increases sharply if the range boundary is trespassed by another male. The rubbing rate may increase up to thirty per minute and over four-hundred per hour.

Some ungulates mark their territories with the orbital gland. A male antelope (*Ourebia ouribi*) bites off a plant stalk at the head level and marks it with secretion by touching the tip of the remaining plant portion (Gosling, 1972). The *Damaliscus lunatus* male marks the territory in a similar way (Duncan, 1975), as does the Thomson's gazelle (*Gazella thomsoni*) (Grau, 1976) and the dik-dik (*Madoqua*). The latter appears to use dunghills for marking the territory. The repellent effect of the marking odor is commonly noted. Of equal importance is the fact that communicative odors are attractive for a territorial male, providing comfort and self-assurance, ensuring the victory of a weaker territorial male over a stronger stranger.

Some pheromones apparently communicate danger signals. A live-trap by which a rat or a mouse has been caught is avoided by other rats or mice for a long time, evidently due to the persistent danger odor (it is uncertain that skin glands alone are responsible for this odor). The danger odor (and

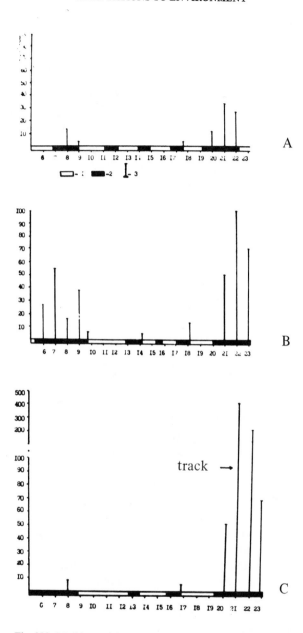

Fig. 209. Marking activity of roe male. *A,* observations at a distance (August 10, 1973); *B,* when being followed by a man (August 12, 1973); *C,* on discovery of young male's tracks within range (August 14, 1973); 1, rest; 2, activity; 3, number of male contacts with trees per hour; abscissae, time of day; ordinates, number of male contacts with trees.

possibly the odor of fleeing from danger) is evidently produced by the caudal glands of some deer (*Cervus nippon*). In danger, *C. nippon* flees, tail up. This tail posture is regarded as a visual signal, opening a white field on the posterior portions of the thighs. As noted earlier, a large part of the tail incorporates tubular gland tissue situated over and under the vertebrae. When the deer is running fast, a head wind blows over its raised tail. From the tail surface, specifically from the lower, hairless portion, secretion evaporates, creating an odorous air used as a cue by other deer. This is of particular importance in the thicket where the visual cue may be ineffective. In the damans (*Procavia* and *Heteroxyrax*), during anxiety the hairs surrounding the dorsal gland are raised to open the glandular field, the secretion evaporation being evidently augmented. In the suslik (*Citellus*), the anal gland sacs bulge when the animal is gripped. In the maral (*Cervus elaphus sibiricus*), during fright or pain the orbital glands are wide open.

Some carnivores use the anal gland secretion for protection from enemies. Polecats (*Mustela eversmanni, Mustela putorius*) produce malodorous and pungent secretions from the anal glands during danger. The same is done by *Viverra, Viverricula,* and *Herpestes*. The anal glands of some *Viverridae* and skunks (*Mephitis* and *Spilogale*) are most developed and specialized for protection (Blackman, 1911; Bourliere, 1954; Dücker, 1965; Wynne-Edwards, 1962; Goethe, 1964). Skunks can spray the pungent anal gland secretion for considerable distances (4 to 5 meters), directing it at the enemy. It is suggested that the duck-billed platypus utilizes a gland located at the limbs to produce venomous secretion during fights for territory and female (Calaby, 1968).

The specific skin gland secretion can be involved in aggressive behavior. The significance of odors in the agonistic behavior of mountain goats (*Oreamnos americanus*), musk-oxen (*Ovibos moschatus*), blacktailed deer (*Odocoileus hemionus*) (Geist, 1964), and a number of rodents (Ropartz, 1977; Fass and Stevens, 1977). In agonistic behavior, the yellow-bellied marmot (*Marmota flaviventris*) uses its anal glands (Armitage, 1974). Marking is an essential component of aggressive behavior in *Mesocricetus auratus* (Dieterlen, 1959; Johnston, 1975a), guinea pigs (*Cavia porcellus*), and some other rodents (Kunkel and Kunkel, 1964). It is part of agonistic behavior in the wild rabbit; males give up fighting to dig, immediately marking the dug-up earth and other objects within reach of the chin gland. Production of fecal pellets lubricated with the anal gland secretion is also frequently observed at intervals throughout fighting. Raising of the posterior portion of the body and wagging the tail by rabbit males during fights indicates that the odor of inguinal and anal glands bears aggressive significance. Between strength demonstrations or fights, the aggressive behavior of *Capreolus capreolus* males, in particular, may include marking small trees and shrubs with the secretions of head and neck.

Olfactory signals are of great significance for breeding. However, because these odors are largely produced not by skin glands, they are not considered here. Involvement of some specific skin glands in breeding is found in a number of mammal species. During the breeding season, such glands are especially well developed as the lateral glands in *Sorex, Neomys,* and *Arvicola terrestris,* the temporal gland in elephants, and the gland behind the horn in *Rupicapra rupicapra.* Excision of the lateral gland in males of *Mesocricetus auratus* prevents mating (Lipkow, 1954).

In camel males during rut, the temporal gland is highly developed. The camels rub against various objects, leaving a heavily odorous secretion (Pilters, 1954). *Saiga* males (*Saiga tatarica*) mark the harem females with urine and inguinal gland secretion (Sludsky, 1975). It is suggested that the musk gland of the musk deer (*Moschus moschiferus*) includes among its functions the marking of covered females (Ustinov, 1967). The male of *Litocranius walleri* and *Ammodorcas clarkei,* in the breeding season, rubs with the forehead and ocular region against the chest and the posterior portion of the female body, evidently marking it with the gland secretion (Walther, 1958, 1963).

It is suggested that the odor of the tarsal gland secretion ensures the sexual separation of the two subspecies of blacktailed deer, *Odocoileus hemionus columbianus* and *O. h. hemionus* (Müller-Schwarze and Müller-Schwarze, 1975). Similarly, the lateral gland secretion odor provides sexual separation of *Sorex vagrans* and *Sorex obscurus* (Hawes, 1976). *Microtinae* females (many or perhaps all) become estral under the effect of the male odor (the source of the odor is unknown). The induced estrus pheromone is revealed in *Microtus ochrogaster, Ondatra zibethica, Clethrionomys gapperi, Pitymys pinetorum, Microtus mexicanus* (Richmond and Stehn, 1976), *Microtus pennsylvanicus* (Clulow and Mallory, 1970) and *Dicrostonyx groenlandicus* (Hasler, 1974).

Specific gland function is conducive to finding a mate (male or female) during rut. Thus, white-tailed deer (*Odocoileus virginianus*) track down females by the odor of the interdigital glands (Moore and Marchinton, 1974).

REFERENCES

Achten, G.
  1959. Recherches sur la keratinization de la cellule épidermique chez l'homme et le rat. Arch. Biol., 70: 1-119.

Adolf, T. A.
  1959. O volosyanom pokrove saigakov (On the pelage of saiga.) Uch. Zap. Mosk. Gos. Ped. Inst. im. Potemkina, 104: vyp. 8.

Afrosin, M. F.
  1975a. Elektronno-mikroskopichesky analiz morfologii sekretornykh kletok trubchatogo otdela prianalnoi zhelezy norok (An electron microscopic study of the morphology of the tubular portion of secretory cells in the mink anal gland). Problemy Molekulyarnoi Biologii i Patologii, 80: 130-134.
  1975b. Submikroskopicheskaya struktura alveolarnoi chasti epiteliya prianalnoy zhelezy norki (The submicroscopic structure of the epithelium alveolar in the mink anal gland). Problemy Molekulayrnoi Biologii i Patologii, 80: 134-140.

Agarkov, G. B., and V. G. Khadzhinsky.
  1970. K voprosu o stroenii i innervatsii kozhnykh pokrovov chernomorskikh delfinov v svyazi s ikh zaschitnoi funktsiei (On the problem of the structure and innervation of the skin of Black Sea dolphins in connection with their protective function). Bionica, 4.

Akaevsky, A. I.
  1939. Anatomiya severnogo olenya (Anatomy of reindeer). L, Selkhozgiz.

Albone, E. S., and M. W. Fox.
  1971. Anal gland secretion of the red fox. Nature, 233: 569-570.

Albone, E. S., G. Eglinton, J. M. Walker, and C. G. Ware.
  1974. The anal sac secretion of the red fox, *Vulpes vulpes;* its chemistry and microbiology. A comparison with the anal sac secretion of the lion, *Panthera leo.* Life Sci., 14: 387-400.

631

Albone, E. S., P. E. Gosden, and G. S. Ware.
  1976. Bacteria as a source of chemical signals in mammals. *In* Chemical signals
        in vertebrates (eds. D. Müller-Schwarze and M. M. Mozell). New York
        and London: Plenum Press, pp. 35-43.
Alekseeva, G. I.
  1962. Ekologo-fiziologicheskoe napravlenie issledovanij na primere izucheniya
        biologicheskikh osobennostei karakulskoi ovtsy v uslovyakh pustyni (Eco-
        physiological approach as exemplified by biological features of Karakul
        sheep under desert conditions). Moscow: Trudy Sovesch. po Biol. Osno-
        vam Povysheniya Produktivnosti Zhivotnykh.
Allen, G.
  1916. The whalebone whales of New England. Mem. Boston Soc. Natur. Hist.,
        8 (2): 107-322.
Allen, J. A.
  1908. The North Atlantic right whale and its near allies. Bull. Am. Mus. Nat.
        Hist., 24: 277-329.
Allen, T. E., and J. Bligh.
  1968. A comparison of the patterns of sweat discharge in several mammalian
        species. J. Physiol., 196: 36-37.
Andersen, K.
  1917. On the so-called colour phases of the rufous horseshoe bat of India (*Rhi-
        nolophus rouxi,* Temm.). J. Bombay Natur. Hist. Sci., 25: 260-273.
Aoki, T. and M. Wado.
  1951. Functional activity of the sweat glands in the hairy skin of the dog. Sci-
        ence, 114: 123.
Arao, T. and E. Perkins.
  1969. The skin of primates: Further observations on the Philippine tarsies (*Tar-
        sius syrichta*). Amer. J. Phys. Anthropol., 31: 93-96.
Argyris, T. S.
  1962. Glycogen in the epidermis of mice painted with methylcholanthrene.
        J. Nat. Cancer Inst., 12: 1159-1165.
  1965. Succinic dehydrogenase and esterase activities of mouse skin during regen-
        eration and fetal development. Anat. Rec., 126: 1-13.
Armitage, K. B.
  1974. Male behaviour and territoriality in the yellow-bellied marmot. J. Zool.
        London, 172: 233-265.
  1976. Scent marking by yellow-bellied marmots. J. Mammal., 57: 583-584.
Armstrong, J., and T. Quilliam.
  1961. Nerve endings in the mole's snout. Nature, 191 (4796): 1379-1380.
Arvy, L. and D. Bommelear.
  1960   Recherches histoenzymologiques sur les glandes interdigitales chez *Capre-
  (1961). olus capreolus* L. Comp. Rend. Soc. Bio., 154 (12): 2112-2215.
Astanin, L. P.
  1958. Organy tela mlekopitayuschikh i ikh rabota (Organs of the mammalian
        body and their function). Moskva: "Sovetskaya Nauka."
Auber, L.
  1952. The anatomy of follicles producing wool fibers with special reference to
        keratinization. Trans. Roy. Soc. Edinburgh, 62: 191-254.

Azbukina, M. D.

1967. O pakhuchikh zhelezakh amerikanskoi norki (On the odorous glands of mink). *In* Sbornik Nauchno-Tekhnicheskoi Informatsii. Vsesoyuzny Naucho-Issledovatelsky Institut Zhivotnogo Syrya i Pushniny. Moscow: Ekonomika, pp. 67-71.

1970. Spetsificheskie kozhnye zhelezy schenkov *M. vison* (Specific skin glands of *Mustela vison* cubs). Sbornik Nauchno-Tekhnicheskoi Informatsii. Vsesoyuzny Nauchno-Issledovatelsky Institut Okhotnyechyego Khozyaistva i Zverovodstva, Kyrov, 30.

1972. Morfologiya muskusnykh zhelez amerikanskoi norki (Musk-gland morphology in mink). Sbornik Nauchno-Tekhnicheskoi Informatsii. Vsesoyuzny Nauchno-Issledovatelsky Institut Okhotnichyego Khozyaistva i Zverovodstva. Kirov, 37-39: 127-135.

1973. Vozrastnye osobennosti stroenya muskusnykh zhelez amerikanskoi norki (Age properties of musk-gland structure in mink). Sbornik Nauchno-Tekhnicheskoi Informatsii. Vsesoyuzny Nauchno-Issledovatelsky Institut Okhotnichyego Khozyaistva i Zverovodstva. Kirov, 90-91: 103-108.

Bab, H.

1904. Die Talgdrüsen und ihre Sekretion. Beitr. Klin. Med., Festschr. 70, Gebrtstag: 1-37.

Babenko, V. V.

1971. Nekotorye nekhanicheskie kharakteristiki kozhnykh pokrovov del'finov (Some mechanical properties of dolphin skin). Bionica, 5. Kiev: "Naukova Dumka."

Babenyshev, V. P.

1938. Znachenye mekhovogo pokrova i razmerov tela nekotorykh vidov gryzunov dlya teplootdatchi i ikh stoikost' k deistviyu vneshnikh faktorov (The significance of fur and body size of some rodent species for heat radiation and their resistance to the effect of environmental factors). Zool. Z., 17: 3.

Badreldin, A. L., and M. A. Ghany.

1952. Adaptive mechanisms of buffaloes to ambient temperature. Nature. London, 170: 457-458.

1954. Species and breed differences in the thermal reaction mechanism. J. Agric. Sci., No. 44 (pt. 2): 160.

Bailey, V.

1897. Revision of the American voles of the genus *Evotomys*. Proc. Biol. Soc. Washington, 2: 113-138.

1900. Revision of the American voles in the genus *Microtus*. N. Amer. Fauna, 17: 5-88.

1936. The mammals and the life zones of Oregon. N. Amer. Fauna, 55: 1-416.

Balo, J. and I. I. Banga.

1949. Die Zeratörung der elastischen Fasern der Gefässwand. Schwez. Z. allgem. Pathol. und Bakteriol., 12: 350.

Bannikov, A. G., L. V. Zhirnov, L. S. Lebedeva, and A. A. Fandeev.

1961. Biologiya saigaka (Biology of saiga). Moscow: "Selskokhozyaistvennaya Literatura."

Barabash-Nikiforov, I. I., S. V. Marakov, and A. M. Nikolaev.

1968. Kalan (morskaya vydra) (Sea otter). Leningrad: "Nauka."

Barash, D. P.
1973.   The social biology of the Olympic marmot. An. Behav. Monogr., 6: 173-245.
Barnett, S. A.
1963.   A study in behaviour., London: Methuen.
Bartholomew, G. A.
1956.   Temperature regulation in the macropod marsupial *Setonyx brachyurus*. Physiol. Zool., 29: 26-40.
Bartholomew, G. A., and F. Wilke.
1956.   Body temperature in the northern seal, *Callorhinus ursinus*. J. Mammal., 37: 327-337.
Bashkirov, I. S., and I. V. Zharkov.
1934.   Biologya i promysel krota v Tatarii (Mole biology and industry in Tataria). Uch. Zap. Kaz. Gos. Univers., 94 (Kniga 8): Vyp. 3.
Bear, R. S.
1952.   The structure of collagen fibrils. Adv. Protein Chem., 7: 69-160.
Beauchamp, G. K.
1974.   The perineal scent gland and social dominance in the male guinea pig. Physiol. and Behav., 13: 669-673.
Becker, S. W., T. B. Fitzpatrick, and H. Montgomery.
1952.   Human melanogenesis: cytology and histology of pigment cells (melanodendrocytes). Arch. Dermatol. Syphilol., 65: 511-523.
Belkin, A. N.
1964.   Materialy po embrionalnomu razvitiyu kozhnogo i volosyanogo pokrova ushastykh tyulenei (*Otariidae*) (Materials on embryonic development of skin and hair of *Otariidae*). Trudy VNIRO: 51. Morskie Kotiki Dalnego Vostoka.
Belkovich, V. M.
1959*a*. Nekotorye osobennosti stroeniya kozhi belukhi (Some properties of the skin structure of *Delphinapterus leucas*). Trudy I Konf. Molodykh Nauchn. Sotr. IMZH. AN SSSR: M.
1959*b*. O mekhanizme smeny okraski u belukhi (On the mechanism of coloration change in *Delphinapterus leucas*). Dokl. AN SSSR, 127: 4.
1959*c*. Nekotorye ekologo-morfologicheskie osobennosti kozhi belukhi (Some ecomorphological properties of the skin of *Delphinapterus leucas*). Tez. Dokl. III Nauchn. Molodezhn. Konf. IMZH AN SSSR. M.
1960.   Nekotorye osobennosti krovosnabzheniya i teplootdachi kozhi belukhi (Some properties of the blood supply and heat emission of *D. leucas* skin). Tez. Dokl. III Nauchn. Molodezhn. Konf. IMZH AN SSSR: M.
1961.   O fizicheskoi termoregulyatsii belukhi (On physical thermoregulation in *Delphinapterus leucas*). Trudy Soveschaniya Ikhtiol. Komissii AN SSSR, 12.
1962.   Prisposobitelnye osobennosti stroeniya kozhnogo pokrova vodnykh mlekopitayuschikh (Adaptive properties of the skin structure of aquatic mammals). Avtoref. Kand. Diss. M.
1964*a*. Kozhny pokrov i fizicheskaya termoregulyatsia (Skin and physical thermoregulation). *In* S. E. Kleinenberg et al. Belukha: Nauka, M.
1964*b*. Stroenie kozhnogo pokrova nekotorykh lastonogikh (The skin structure

in some pinnipeds). *In* Morfologicheskie Osobennosti Vodnykh Mlekopitayuschikh (ed. S. E. Kleinenberg). Moscow: "Nauka."

Benedicts, A. F.
1957.   Hair structure as a generic character in bats. Univ. Calif. Publ. Zool., 59: 285-548.

Benfenati, A., and F. Brillanti.
1939.   Sulla distribuzione della ghiandole sebacee nella cute del corpo umano. Arch. Ital. Dermatol. Siphilol. e Venerol., 15: 33-42.

Bergensen, B.
1931.   Beitrage zur Kenntnis der Haar einiger Pinnipedien. Skr. Norske. videnskaps. akad. Oslo, 5.

Berlin, L. B.
1957.   Gistologicheskie izmeneniya kozhi krolikov i morskikh svinok, vyzyvaemoe pitaniem na nikh kleschei (Histological changes of rabbit and guineapig skin caused by mite parasitism). Dokl. Akad. Nauk. SSSR, 112: 2.

Bezler, F. M.
1935.   Ob osnovnykh momentakh gistoarkhitektoniki kollagena sobstvenno kozhi svinyi i krupnogo rogatogo skota (On principal features of the histoarchitectonics of skin proper in swine and cattle). Tr. Histol. Konf.

Bielschowsky. M.
1907.   Ueber sensible Nervendigungen in der Haut zweier Insectivoren (*Talpa europaea* und *Centetes ecaudatus*). Anat. Anz., 31(6): 187-194.

Billingham, R. E., and P. B. Medawar.
1953.   A study of the branched cells of the mammalian epidermis with special reference to the fate of their division products. Philos. Trans. Roy. Soc., B(237): 151-171.

Birbeck, M. S. C., E. H. Mercer, and N. A. Barnicot.
1956.   The structure and function of pigment granules in human hair. Exp. Cell. Res., 10: 505-514.

Bizzozero, G., and G. Vassale.
1887.   Über die Erzeugung und die physiologische Regeneration der Drüsenzellen bei den Säugetieren. Virchow's Arch. Pathol. Anat. und Physiol., 110: 155-214.

Blackman, M. W.
1911.   The anal glands of *Mephitis mephitica,* Anat. Rec., 5: 497-523.

Blank, J. H.
1953.   Further observations on the factors which influence the water content of the stratum corneum. J. Invest. Dermatol., 21: 259.

Boardman, W.
1943.   The hair tracts in marsupials. Part I. Proc. Linnean Soc. N. S. Wales, 68, Part 3-4 (307-308): 95-113.

1943.   The hair tracts in marsupials. Part I. Proc. Linnean Soc. N. S. Wales, 70 (321-322): 179-202.

1949.   The hair tracts in marsupials. Part III. Proc. Linnean. Soc. N. S. Wales, 74.

1950.   The hair tracts of marsupials. Proc. Linnean Soc. N. S. Wales, 75, Part 4 (347-348): 89-95; Part 5-7 (351-252): 254-278.

1952.   The hair tracts in some American marsupials. Proc. Zool. Soc. London, 121: 845-850.

Boeke, J.
 1926. Die Beziehungen der Nervenfasen zu den Bindegewebselementen und Tastgellen. J. Mikrosk. Anatom. Forschung, 4 (H. 1): 448-509.
Bogomolova, O.
 1933. Gistologicheskoe stroenie svinoi shkury v zavisimosti ot porody, pola i vozrasta (The histological structure of swine skin depending on breed, sex, and age). Sb. Trudov. TSNIKP. Novye Vidy Kozh. Syrya, Vyp. 1.
Bol', B. K., and A. D. Nikolaevsky.
 1932. Stroenie kozhnogo pokrova severnogo olenya i ego izmeneniya po vreme-nam goda (The structure of the skin of reindeer and its seasonal variabil-ity). In Sb. po Olenevodstvu, Tundrovoi Veterenarii i Zootekhnike. Mos-cow: Vlast Sovetov.
Bollinger, A.
 1944. On the fluorescence of the skin and the hairs of Trichosorus vulpecula. Austral. J. Sci., 7: 35.
Bollinger, A., and M. D. McDonald.
 1949. Histological investigation of glycogen in skin and hair. Austral. J. Exper. Biol. and Med. Sci., 27: 223-228.
Bonin, W., and V. D. Vladykov.
 1940. La peau du marsoin blanc ou beluga (D. leucas). Naturaliste Canad., 67: 253-287.
Bonninger, M.
 1905. Die elastische Spannung der Haut und deren Beziehung zum Oedem. Z. exper. Pathol. und Therap., I: 163-183.
Bonsma, J. S.
 1949. Breeding cattle for increased adaptability to tropical and subtropical en-vironments. Agric. Sci., 39: 204-221.
Borowski, S.
 1952. Sezenowe zmiany uwlosienia u Soricidae. Ann. Univ. Mariae Curie Sklo-dowska. Sect. C, 7 (2): 65-117.
 1959. Variations in density of coat during the life cycle of Sorex araneus L. Acta Theriol. (14): 286-289.
 1968. On the moult in the common shrew. Acta Theriol. 13 (30): 483-498.
 1973. Variations in coat and colour in representatives of the genera Sorex L. and Neomys Kaup. Acta Theriol., 18 (14): 247-249.
Bourlière, F.
 1954. The natural history of mammals. New York: Alfred A. Knopf.
Bouteille, M., and D. C. Pease.
 1971. Tridimensional structure of native collagenous fibrils, their proteinaceous filaments. J. Ultrastructur. Res., 35: 314-338.
Bowers, J. M., and B. K. Alexander.
 1967. Mice: individual recognition by olfactory cues. Science, 158: 1208-1210.
Bowes, J. J., and R. H. Kenten.
 1949. The effect of deamination and etherification on the reactivity of collagen. J. biochem., 44: 142.
Brachet, B.
 1942. La localisation des acides pentosenucleiques dans les tissues animaux et les oeufs d'Amphibiens en voie de développement. Arch. Biol., 53: 207-257.

Braun, A. A.

1935. Sravnitelny analiz mikrostruktury kozhnogo pokrova loshadi i krupnogo rogatogo skota (A comparative analysis of the skin microstructure in horse and cattle). *In* Topografiya Kozhnogo Pokrova Loshadi, Vyp. 1.

Braun, A. A., and P. I. Ostrovskaya.

1933. Materialy k topograficheskoi gistologii kozhnogo pokrova. V. Severny olen' (Materials on topographical histology of skin. V. Reindeer). Arch. Anat. Histol. i. Embryol., XII (2): 33-362.

Braun, A. A. and A. I. Lebedev.

1947. Materialy k topograficheskoi gistologii kozhnogo pokrova (Materials on the topographical histology of skin. VI [anteater]). Uch. Zap. Lenin Ped. Inst. im. Gerzena, X.

Braun-Falco, O.

1956a. Über die Fähigkeit der menschlichen Haut zur Polysaccharidsynthese, ein Beitrag zur Histotopochemie der Phosphorylase. Arch. klin. und exper. Dermatol., 202: 163-170.

1956b. Zur Histotopographie der SS-Glucuronidase in normaler menschlicher Haut. Arch. klin. und exper. Dermatol., 203: 61-67.

1956c. Zur histochemischen Darstellung von Glukose-6-Phosphatase in normaler Haut. Dermatol. Wochenschr., 134: 1252-1257.

1958. The histochemistry of hair follicles. *In* The biology of hair growth (eds. W. Montagna and R. A. Ellis). New York: Academic Press, pp. 65-87.

Braun-Falco, O., and B. Rathjens.

1954. Histochemische Darstellung der Bernsteinsäuredehydrogenase in der menschlichen Haut. Dermatol. Wochenschr., 130: 1271-1276.

1955. Über die Histochemische Darstellung der Kochlen Säurenhydratase in normaler Haut. Arch. Klin. und Exper. Dermatol., 201: 73-82.

Brinkmann, A.

1908. Die Rückendruse von Dicotyles. Anat. Heft., 36: 281-307.

1912. Die Hautdrüsen der Säugetiere. Ergebn. Anat. Entwicklungs gesch., 20: 1173-1231.

Brody, I.

1959. An ultrastructure study on the role of keratohyalin granules in the keratinization process. J. Ultrastruct. Res., 3: 84-104.

Bromley, G. F.

1956. Linka kolonka i prognoz daty nachala ego promysla (Moult of the Siberian weasel and forecasting the onset of hunting). Trudy Daenevostochnogo filiala AN SSSR. Ser. Zool., 3: 6.

Bronson, F. H.

1971. Rodent pheromones: Biology of reproduction, 4 (3): 333-357.

Brosset, A.

1962. La reproduction des chiropteres de l'ouest et du centre de l'Inde. Mammalia, 26: 176-213.

Brownlee, R. G., R. M. Silverstein, D. Müller-Schwarze, and A. G. Singer.

1969. Isolation, identification and function of the chief component of the male tarsal scent in black-tailed deer. Nature, 221: 284-285.

Bryden, M. M.

1964. Insulating capacity of the subcutaneous fat of the southern elephant seal. Nature, 203: 1299-1300.

Bullough, W. S.
1948. Mitotic activity in the adult male mice. *Mus musculus* L. The diurnal cycles and their relation to waking and sleeping. Proc. Roy. Soc. B., 135: 212-233.
1962. Gross control of mammalian skin. Nature, 193: 520-523.

Bullough, W. S., and E. A. Eira.
1949. The diurnal variations in the tissue glycogen content and their relation to the mitotic activity in the adult male mouse. J. Exper. Biol., 26: 257-263.

Bullough, W. S., and E. B. Lawrence.
1964. The production of epidermal cells. *In* The mammalian epidermis and its derivatives (ed. F. J. Ebling), pp. 1-23.

Bunting, H.
1950. The distribution of acid mucopolysaccharides in mammalian tissues as revealed by histochemical methods. Ann. N. Y. Academy Sci., 52: 977-982.

Bunting, H., G. B. Wislocki, and E. W. Dempsey.
1948. The chemical histology of human eccrine and apocrine sweat glands. Anat. Rec., 100: 61-77.

Burbank, R. C. and J. Z. Young.
1934. Temperature changes and winter sleep of bats. J. Physiol., 82: 459-467.

Burstone, M. S.
1958. Histochemical demonstration of proteolytic activity in human neoplasms. J. Nat. Canc. Inst., 16: 1149-1161.

Burstone and Folk.
1956. Histochemical demonstration of aminopeptidase. J. Histochem. and Cytochem., 4: 217-226.

Buss, I. O. and J. A. Estes.
1971. The functional significance of movements and positions of the pinnae of the African elephant, *Loxodonta africana*. J. Mammal., 52: 21-27.

Butcher, E. O.
1954. Fluid loss through epidermis in the rat and factors affecting it. Proc. Exper. Biol. and Med., 87: 3.

Butorin, N. S.
1935. Sravnitelnoye izychenie stroeniga shersti dikikh baranov i kirghizskoi kurdyuchnoi ovtsy (A comparative study of the structure of the wool of wild sheep and Kirghizian fat-tailed sheep). Sovet po izycheniu prirodnykh resursor. Moscow.

Bychkov, V. A.
1968. Kharakteristika razmerov i okraski volosyanoga pokrova morskikh kotikov (Size indices and coloration of the fur of fur seals). Trydi VIVIRO, 68. Lastonoghie Tikhogo okeana, 117-133.

Calaby, J. H.
1968. The platypus (*Ornithorhynchus anatinus*) and its venomous characteristics. *In* Venomous animals and their venoms (eds. W. Bücherl et al.). New York: Academic Press, I: 15-29.

Campbell, A.
1837. Notes taken at the post mortem examination of musk deer. J. Asiat. Soc. Bengal, 6: 118.

Carus, C. G., and A. W. Otto.
1840. Erläuterungstafeln zur vergleichenden Anatomie. Leipzig: Barten, H. V. 1843, H. VI.

Caruthers, C., W. C. Quevedo, Jr., and D. L. Woernley.
1959. Influence of hair growth cycle on cytochrome oxidase and DPNH-cyto-chrome C reductase in mouse epidermis. Proc. Soc. Exper. Biol. and Med., 101: 374-376.

Cauna, N., and P. Alberti.
1961. Nerve supply and distribution of cholinesterase activity in the external nose of the mole. Zellforsch., 54(2): 156-158.

Cave, A. J. E., and D. B. Allbrook.
1959. The skin and nuchal eminence of the white rhinoceros. Proc. Zool. Soc. London, 132: 99-107.

Chagirov, I. A.
1954. Gistostruktura kozhi arkhara i ego gibridov arkharov merinosov (Histo-structure of the skin of *Ovis ammon* and its hybrid merino sheep). Trudy Inst. Exper. Biol. AN KAZ. SSR, II.

Chapsky, K. K.
1936. Morzh karskogo morya (The walrus of the Kara Sea). Moscow: Trudy Arkticheskogo Instituta, 67: 23-47.

Charles, A.
1960. Electron microscopic observation of the human sebaceous gland. J. In-vest. Dermatol., 35: 31-36.

Chevremont, M., and J. Frederic.
1943. Une nouvelle méthode histochemique de mise en evidence a fonction sulf-hydrile. Application à fonction sulfhydrile. Application à l'épiderme au poile et à la levure. Arch. Biol., 54: 589-605.

Chew, R. M.
1955. The skin and respiratory water losses of *Peromyscus maniculatus sonori-ensis*. Ecology, 36.

Chizhova, S. S.
1958a. Gistofiziologicheskaya kharakteristika potovykh zhelez raznykh porod krupnogo rogatogo skota (Histophysiological description of sweat glands of different cattle breeds). Sborn. Rabot. Aspirantov. AN Uzb. SSR., Otd. Biol. Nauk, 2, Tashkent, Izd. AN Uzb. SSR.
1958b. Rol' nosovogo zerkal'tsa korov v teplootdatche (Role of planum naso-labiale of cows in heat radiation). AN Uzb. SSR., Otd. Biol. Nauk, 2, Tashkent, Izd., AN Uzb. SSR.

Chlopin, G.
1932. Über einige Wachstums- und Differenzierungserscheinungen an der emb-ryonalen menschlichen Epidermis in Explantat. Roux' Arch. Entwick-lungmech. Organ., 126: 69-89.

Christian, J. J., J. A. Lloyd, and D. E. Davies.
1965. The role of endocrines in the self-regulation of mammalian population. Recent Progr. Hormone Res., 21: 501-578.

Christophers, E.
1971. Cellular architecture of the stratum corneum. Dermatology, 56(3): 165-169.

Clara, M.
  1929. Morfologia e svilluppo delle ghiandole sebacee nell'uomo. Recherche morfol., 9: 121-182.
Clulow, F. V., and F. F. Mallory.
  1970. Oestrus and induced ovulation in the meadow vole, *Microtus pennsylvanicus*. J. Reproduct. and Fertility.
Cnyrim, E.
  1914. Zu der Schlädendrüse und zum Lidapparate des Elefanten. Anat. Anz., 273-279.
Comel, M.
  1933. Fisiologia normale e patologica della cute umana. Milano: Treves, 1: 650-675.
Compton, A. S.
  1952. Cytochemical and cytological study of the connective tissue muscles. Amer. J. Anat., 91: 301-329.
Constantine, D. G.
  1957. Colour variation and moult in *Tadaridae brasiliensis* and *Myotis velifer*. J. Mammal., 38: 461-466.
  1958. Colour variation and moult in *Mormoops megalophylla*. J. Mammal., 39: 344-347.
Cooper, Z. K.
  1930. Histological study of the integument of the armadillo, *Tatasia novemcincta*. Amer. J. Anat., 45: 1-38.
Cowles, R. B.
  1947. Vascular changes in the wings of bats. Science, 105: 362-363.
Dagg, A. I. and D. E. Windsor.
  1971. Olfactory discrimination limits in gerbils. Can. J. Zool., 49: 283-285.
  1960. The Cutaneous Innervation (ed. W. Montagna). New York: Academic Press.
Dalquest, W. W., and H. G. Werner.
  1954. Histological aspects of the faces of North American bats. J. Mammal., 35(2): 147-150.
Dan Baran, S. E., and Glickman.
  1970. Territorial marking in the Mongolian gerbil: A study of sensory control and function. J. Compar. and Physiol. Psychol., 71(2): 237-245.
Dantschakowa, W. E.
  1910. Ueber die Entwicklung der embrionalen Blutbildung bei Reptilien. Vergleichende Anatomie G. Bruxells.
Darrow, C. W.
  1937. Neural mechanism controlling the palmar sweating. Arch. Neurol. Psychiatr., 37: 641-643.
Dehnel, A.
  1950. Bodania nad rodzajem Neomys Kaup. Univ. Mariae Curie-Sklodowska, Sect. C, 5(1): 1-63.
Deparma, N. K.
  1951. Lin'ka krotov, ee posledovatel'nost' i sroki (Molt of moles, its sequence and timing). Trudy VNII Okhot. Promysla, Vyp. 10.

Dick, J. C.
  1947. Observations on the elastic tissue of the skin with a note on the reticular
          layer at the junction of the dermis and epidermis. J. Anat., 81: 201-211.
Dieterlen, F.
  1959. Das Verhalten des syrischen Goldhamsters (*Mesocricetus auratus* Water-
          house). Z. Tierpsychol., 16: 47-103.
Dimond, R. L. and W. Montagna.
  1975. The skin of giraffe. Anat. Rec., 185: 63-76.
Dinesman, L. G.
  1949. O rasprostranenii i ekologii reptilij v svyazi s zonami solnechnoi radiatsii
          (On the distrbution and ecology of reptiles with reference to the zones of
          solar radiation). Problemi Fizicheskoi Geografii, 14.
Dinnik, N. Ya.
  1914. Zveri Kavkaza (Mammals of the Caucasus). Pts. I-II. Tiflis.
Dobson, G.
  1883-1890. Monograph of Insectivora. London.
Dobson, R. L., and W. C. Lobitz.
  1958. Some histochemical observations on the human eccrine sweat glands. IV.
          The recovery from the effects of profuse sweating. J. Invest. Dermatol.,
          31: 207-213.
Dobson, R. L., W. C. Formisano, Jr., and D. Brophy.
  1958. Some histochemical observations on the human eccrine sweat glands. III.
          The effect of profuse sweating. J. Invest. Dermatol., 31: 147-159.
Doty, R. L.
  1972. Odor preferences of female *Peromyscus maniculatus bairdii* for male
          mouse odors of *P. m. bairdii* and *P. leucopus noreboracensis* as a function
          of estrous state. J. Comp. Physiol. Psychol., 81: 191-197.
Doty, R. L. and R. Kart.
  1972. A comparative and developmental analysis of the midventral sebaceous
          glands in 18 taxa of *Peromyscus,* with an examination of gonadal steroid
          influences in *Peromyscus maniculatus bairdii.* J. Mammal., 53 (1): 83-99.
Du Bois, E. F.
  1927. Basal metabolism in health and disease. Philadelphia: Lea & Febiger.
  1937. The mechanism of heat loss and temperature regulation. Stanford: Stan-
          ford University Press.
Ducker, G.
  1965. Das Verhalten der Viverriden. Hand. Zool., 8: 38, 10 (20a): 1-48.
Duncan, P.
  1975. Topi and their food supply. Ph.D. dissertation, University of Nairobi.
Duran-Reynolds, F., H. Bunting, and G. Wagenen.
  1950. Studies on the sex skin of *Macaca mulata.* Ann. N. Y. Acad. Sci., 52:
          1006-1014.
Durward, A., and K. M. Rudall.
  1958. The vascularity and patterns of growth of hair follicles. *In* The biology of
          hair growth (ed. W. Montagna and R. A. Ellis). New York: Academic
          Press, pp. 189-217.
Dwyer, P.
  1966. Observations on *Chalinolobus dwyeri* (*Chiroptera: Vespertilionidae*) in
          Australia. J. Mammal., 47: 716-718.

Eadie, W. R.
  1939. A contribution to the biology of *Parascalops breweri*. J. Mammal., 20(2): 153-173.
  1948. The male accessory reproductive glands of *Condylura* with notes on a unique prostatic secretion. Anat. Rec., 101(1): 59-79.
  1954. Skin gland activity and pelage description in moles. J. Mammal., 35(2): 186-196.

Eales, N.
  1925. External characters, skin, and temporal gland of a foetal African elephant. Proc. Zool. Soc. London, 95: 445-456.

Ebling, F. J.
  1970. Steroid hormones and sebaceous secretion. *In* Advances in steroid biochemistry and pharmacology, 2: 1-39.

Edwards, E. A., and S. Q. Duntley.
  1939. Pigments and odour of living human hair. Amer. J. Anat., 65: 1-33.

Efimov, A. E.
  1938. Materialy po morfologii kozhnogo pokrova severnogo olenya (Materials on morphology of the skin of caribou. III. Histology of labium). Soobsch. III. Gistologiya guby. Trudy Omskogo Vet. Inst., XI: 23-31.
  1939. Gistologiya obschego kozhnogo pokrova i ego sezonnye izmenenya u severnogo olenya. Trudy Omskogo Vet. Inst., XII: 293-327.

Eggeling, H.
  1901. Über die Schläfendrüse des Elefanten. Biol. Cbl., 21: 443-452.
  1931. Hautdrüsen. Handbuch vergleichende Anatomie der Virbeltiere, Bd. 1: 663-692.

Eibl-Eibesfeldt, I.
  1950. Über die Jugendentwicklung des Verhaltens eines männlichen Dachses (*Meles meles* L.) unter besonderer Berücksichtigung des Spiels. Z. Tierpsychol., 7: 327-355.
  1957. Ausdrucksformen der Säugetiere. *In* Hand. Zool. Helnicke u. von Lengerken, 8, 10 (6): 1-26.

Eimer, T.
  1871. Die Schauze des Maulwurfes als Tastwerkzeug. Arch. Microsk. Anat., 7: 181.

Eliseev, V. G.
  1961. Soedinitelnaya tkan' (Connective tissue). Moscow: Medgiz.

Ellenberger, W.
  1906. Handbuch Vergleichender Mikroskopischer Anatomie der Haustiere, Bdl.

Ellis, R. A.
  1964. Enzymes of the epidermis. *In* The Epidermis (eds. W. A. Montagna and W. C. Lobitz, Jr.). New York: Academic Press.

Ellis, R. A., and W. Montagna.
  1958. Histology and cytochemistry of human skin. XV. Sites of phosphorylase and amylo-1.6-glucosidase activity. J. Histochem. and Cytochem., 6: 201-207.
  1962. The skin of primates. IV. The skin of the gorilla (*Gorilla gorilla*). Amer. J. Phys. Anthropol., 20: 79-94.

Ellis, R. A., W. Montagna, and H. Fanger.
  1958. Histology and cytochemistry of human skin. XIV. The blood supply of the cutaneous glands. J. Invest. Dermatol., 30: 137-145.

Epling, G. P.
  1956. Morphology of the scent gland of the javelina. J. Mammal, 37 (2): 246-248.

Eremeeva, K. M.
  1954. Sezonnaya izmenchivost' kozhnogo i volosyanykh pokrovov golubykh pestsov (Seasonal variability of skin and hair in the blue fox). Karakulevodstvo i Zverovodstvo: I.
  1956. Sezonnye izmeneniya kozhnogo i volosyanykh pokrovov golubykh pesthsov (Seasonal variations of skin and hair in the blue fox). Karakulevodstvo i Zverovodstvo: I.
  1957. Razvitie kozhnykh zhelez pushnykh zverei (The development of skin glands in furbearing mammals). Karakulevodstvo i Zverovodstvo: I.
  1967. Razvitie kozhnogo i volosyanogo pokrovov u norok (The development of skin and hair in mink). Izv. TSKHA, 2: 192-198.

Ermilov, A. G., and M. Ya. Marvin.
  1971. Stroenie mekha rechnoi vydry Lutra lutra l. (Hair structure in Lutra lutra L.). Uchenye Zapiski Uralsk. Gosud. Universiteta. Ser. Biol., 9: 10-14.

Essapian, F. S.
  1955. Speed-induced skin folds in the bottle-nosed porpoise, Tursiops truncatus. Breviola, 43: 1-4.

Evans, R., E. V. Cowdry, and P. E. Nielsen.
  1943. Aging of human skin. I. Influence of dermal shrinkage on appearance of epidermis in young and adult fixed tissue. Anat. Res., 86: 545-560.

Fadeev, E. V.
  1955. Stroenie kozhi nutrii (The structure of coypu skin). Trudy Vses. Nauchn. Issled. Inst. Okhtn. Promysl., Vyp. XV.

Fass, B., and D. A. Stevens.
  1977. Pheromonal influences on rodent agonistic behaviour. In Chemical signals in vertebrates (eds. D. Müller-Schwarze and M. M. Mozell). New York and London: Plenum Press, pp. 185-206.

Feldman, A., and O. G. Mitchell.
  1968. The post-natal development of the pelage and ventral gland of the male gerbil. J. Morphol., 125: 303-313.

Felts, W. G. L.
  1966. Some functional and structural characteristics of cetacean flippers and flukes. In Whales, dolphins and porpoises (ed. K. Norris). Berkeley and Los Angeles: University of California Press.

Fetisov, A. S.
  1931. Stroenie mekha zaitsa belyaka vostochno-sibirskogo kryazha (The structure of pelage of the blue hare of the East-Siberian mountain ridge). Ogiz, Irkutsk.

Findlay, J. D.
  1953. Farm animals and high temperature. Brit. Agric. Bull., 6 (28): 212.

Fitzpatrick, T. B., and A. B. Lerner.
  1953. Terminology of pigment cells. Science, 117: 640.

Fjelstrup, A.
   1888.   Über den Bau der Haut bei Globiocephalus melas. Zool. Anz., 269: 11-15.
Flerov, K. K.
   1928.   O sezonnykh izmeneniyakh volosyanogo pokrova u kosuli (On seasonal
           variations of pelage in roe). Dokl. AN SSSR, 22.
   1952.   Kabargi i oleni (Musk-deer and deer). Fauna SSSR. Mlekopitayuschie I,
           Vyp. 2. Izd. AN SSR. Moscow, Leningrad.
Flower, W. H.
   1875.   On the structure and affinities of the musk deer (*Moschus moschiferus
           Linné*). Proc. Zool. Soc. London, 159-190.
Ford, E.
   1934.   A note on the sternal gland of *Myrmecobius*. J. Anat., III: 346-349.
Formisano, V., and W. C. Lobitz, Jr.
   1957.   The Schiff-positive nonglycogen material in the human eccrine sweat
           glands. I. Histochemistry. A. M. A. Arch. Dermatol., 75: 202-209.
Frade, F.
   1955.   Ordre des Proboscidiens. Tégument. *In* Traité de zoologie (ed. P. P.
           Grassé). Paris: Masson et Cie., 17 (1): 717-721.
Frechkop, S.
   1955.   Ordre des Paraxoniens. *In* Traité de zoologie (ed. P. P. Grassé). Paris:
           Masson et Cie., 17.
Fular, W.
   1955.   Der Lellersatz in der menschlichen Epidermis. Morphol. Jahr., 96: 1-13.
Gabe.
   1967.   Le tégument et ses annexes.
Gabe and J. Dragesco.
   1971.   Contribution a l'histologie des glandes cutanées des Hyracoides. Ann.
           Fac. Sci. Cameroun, 7-8: 173-191.
Gambaryan, P. P.
   1960.   Prisposobitelnye osobennosti organov dvizheniya royuschikh mlekopita-
           yuschikh (Adaptive features of the organs of locomotion of digging mam-
           mals). Erevan.
Garrod, A. H.
   1877.   Notes on the anatomy of the musk-deer (*Moschus moschiferus*). Proc.
           Zool. Soc. London, 287-292.
Gawn, R.
   1948.   Aspect of the locomotion of whales. Nature, 161: 4080.
Geist, V.
   1964.   On the rutting behaviour of the mountain goat. J. Mammal., 45: 551-568.
   1967.   On fighting injuries and dermal shields of mountain goats. J. Wildlife
           Mgmt., 31: 192-194.
Geptner, V. G., and V. I. Tsalkin.
   1947.   Oleni SSSR (Deer of the USSR). Materialy k poznaniu fauny i flory SSSR,
           Novaya Seriya. Otd. Zool., 10(XV). MOIP. Moscow.
Geptner, V. G., A. A. Nasimovich, and A. G. Bannikov.
   1961.   Mlekopitayuschie Sovetskogo Soyuza. v. 1. Parnokopytnye i Neparno-
           kopytnye (Mammals of the USSR. v. 1. *Artiodactyla* and *Perissodactyla*).
           Moscow: Vyshaya Shkola.

Gerasimova, M. A.
1955. Osennyaya link'ka zaitsa-rusaka (Autumn moult of European hare). Trudy Vses. Nauchn. Issled. Inst. Okhotnychyego Promysla, XV: 457-521.

German, A. L.
1962. Izmeneniya kozhno-legochnykh poter' vody pri vodnom golodanii u nekotorykh myshevidnykh gryzunov (Dynamics of water losses in skin and lungs under water shortage in some Muridae). Biul. MOIP, ser. Biol., 67: 4.

Gersch, I., and H. R. Catchpole.
1949. The organization of ground substance and basement membrane and its significance in tissue injury, disease, and growth. Amer. J. Anat., 85: 457-521.
1960. The nature of ground substance of connective tissue. Perspectives in Biol. Med., 3: 282-319.

Ghobrial, L. I.
1970. A comparative study of the integument of the camel, *Dorcas gazella,* and jerboa in relation to desert life. J. Zool. London, 160: 509-521.

Giacometti, L.
1967. The skin of the whale (*Balaenoptera physalus*). Anat. Rec. 159: 69-76.

Gibbs, H. F.
1938. A study of the development of the skin and hair of the Australian opossum *Trichosurus vulpecula.* Proc. Zool. Soc. London, Ser. B, 108 (III): 611-648.

Giroud, A., and H. Bulliard.
1930. Recherches sur la kératinization de l'épiderme et des phanères. Paris: G. Doin.

Giroud, A., C. P. Leblond, and R. Ratsimamanga.
1935a. La vitamine C (acide ascorbique) dans la peau. Compt. Ren. Ser. Biol., 118: 321-322.
1935b. La vitamine C dans l'organisme. Accumulation-elimination. Comt. Ren. Assos. Anat. Montpelier, 1-8.

Goethe, F.
1964. Das Verhalten der Musteliden. Hand. Zool. (8): 37-10 (19): 1-80.

Godfrey, J.
1958. The origin of sexual isolation between bank voles. Proc. Roy. Phys. Soc. Edinburgh, 27: 47-55.

Goosen, P.
1941. Glaveishie konstanty zhira ne kotorykh pushnykh zverei (The major fat indices of some fur-bearing mammals). Moscow: Trudy Moskovskogo Zootekhnicheskogo Instituta.

Gorman, M. L., D. B. Nedwell, and R. M. Smith.
1974. An analysis of the anal scent pockets of *Herpestes auropunctatus.* J. Zool., 172: 389-399.

Gosden, P. E., and G. C. Ware.
1976. The aerobic bacterial flora of the anal sac of the red fox. J. Appl. Bact., 41: 271-275.

Gosling, L. M.
  1972. The construction of antorbital gland marking sites by male oribi (*Ourebia ourebia,* Zimmerman, 1783).
Gould, E., and J. E. Eisenberg.
  1966. Notes on the biology of the *Tenrecidae.* J. Mammal., 47: 660.
Grant, M. E., and D. J. Prockop.
  1972. The biosynthesis of collagen. N. Engl. J. Med., 286 (4): 242-249, 291-300.
Grassé, P. P.
  1955. Traité de Zoologie, t. 17. Paris: Masson et Cie, 1715.
Grau, G. A.
  1976. Olfaction and reproduction in ungulates. *In* Mammalian olfaction, reproductive processes and behavior (ed. R. L. Doty), 219-242.
Gray, J.
  1936. Studies in animal locomotion. J. Exper. Biol., 13: 2.
Green, L. M. A.
  1961. Sweat glands in the skin of Quokka of Western Australia. Austr. J. Exper. Biol. and Med. Sci., 39 (6): 481-486.
Gross, J.
  1949. The structure of elastic tissue as studied with electron microscope. J. Exper. Med., 89: 699-708.
  1950*a*. A study of certain connective tissue constituents with the electron microscope. Ann. N. Y. Sci., 52: 964-970.
  1950*b*. A study of the aging of collagenous connective tissue of rat skin with the electron microscope (abstract). Amer. J. Pathol., 26: 708.
Gross, J., and F. O. Schmitt.
  1948. The structure of human skin collagen as studied with the electron microscope. J. Exper. Med., 88: 555-568.
Grushetskaya, L. A., A. A. Kischinsky, and S. M. Uspensky.
  1973. Issledovanie zhira belogo medvedya (A study of the fat of the polar bear). *In* Ekologiya i morfologiya belogo medvedya. Moscow: Nauka, pp. 62-68.
Gudkova-Aksenova, N. S.
  1951. Sreda obitaniya i ee vliyanie na organizatsiyu nekotorykh vodnykh nasekomoyadnykh i gryzunov (The environment and its effect on some aquatic *Insectivora* and *Rodentia*). Uch. Zap. Gorkovsk. Gosuniversiteta, 19: 135-174.
Gustavsson, K. H.
  1949. Some protein-chemical aspects of tanning processes. Adv. Protein Chem., 5: 353-421.
  1956. The chemistry and reactivity of collagen. New York: Academic Press.
Haffner, K.
  1957. Bau, Eigenschaften und ehemalige Verwendung der Haut der seit, 1768 ausgeboronen Steller'schen Seekuh (*Rhytina stelleri* Retz). Mitt. Hamburg. Zool. Mus. und Inst., 55.
Haitlinger, R.
  1968*a*. Comparative studies on the morphology of hair in representatives of the genus *Apodemus* Kaup, 1829, found in Poland. Zool. Pol., 18 (3): 347-380.

1968*b*. Seasonal variation of pelage in representatives of the genus *Apodemus* Kaup, 1829, found in Poland. Zool. Pol., 18 (3): 329-345.

Hale, C. W.

1946. Histochemical demonstrations of acid polysaccharides in animal tissues. Nature, 157: 802.

Hambrick, C. W., and H. Blank.

1954. Whole mounts for the study of skin and its appendages. J. Invest. Dermatol., 23: 437-453.

Hammel, T. H.

1955. Thermal properties of fur. Amer. J. Physiol., 182: 369-376.

Hamperl, H.

1923. Zur Kenntnis der in der Analgegend bei Insectivoren vorkommenden Drüsen. Verhandl Anat. Ges. Ver. Heidelberg, 32: 233-242.

Hanson, G., and W. Montagna.

1962. The skin of primates. 12. The skin of the owl monkey (*Aotus trivirgatus*). Amer. J. Anat., 107: 214-230.

Hardy, J. D., and C. Muschenheim.

1934. Radiation of heat from human body; emission, reflection and transmission of infrared radiation by human skin. J. Clin. Invest., 13: 817-831.

Harmer, S.

1930. Remarques sur les lacets faits de peau de beluga. La Nature, 125.

Harrison, D. L., and D. V. Davies.

1949. A note on some epithelial structures in *Microchiroptera*. Proc. Zool. Soc. London, 119: 351-357.

Harrison, D. L., and J. D. L. Fleetwood.

1960. A new race of flat-headed bat *Platymops barbatogularis* Harrison from Kenya Colony, with observations on the anatomy of the gular sac and genitalia. Durban Mus. Novitates, 5: 269-284.

Harrison, R. G.

1910. The outgrowth of the peripheral nerve fibers in altered surroundings. Roux' Arch. Entwicklungsmech., 30: 15-33.

1938. Die Naturalleiste. Anat. Anz., Erganz, H. 85: 4-30.

Harrison, R. J., and K. W. Thurley.

1974. Structure of the epidermis in Tursiops, Delphinus, Orcinus and Phocoena. *In* Functional anatomy of marine mammals, 2 (ed. R. J. Harrison). New York: Academic Press, pp. 45-71.

Hart, J. S.

1956. Seasonal changes in insulation of the fur. Canad. J. Zool., 34: 53-57.

Hart, J. S., and L. Irving.

1959. The energetics of harbor seals in air and in water with special consideration of seasonal changes. Canad. J. Zool., 37: 4.

Hasama, B.

1929. Über den Einfluss der direckten mechanischen, thermischen und elektrischen Reizung auf die Schweiss-sowie Wärmezentren. Arch. Pathol. und Pharmacol., 146: 129-161.

Hasler, J. F.

1974. Reproduction of the collared lemming, *Dicrostonyx groenlandicus*. Ph.D. dissertation. University of Illinois, Urbana.

Hass, G., and McDonald.
1940. Studies of collagen. I. The production of collagen in vitro under variable experimental conditions. Amer. J. Pathol., 16: 525-548.

Hattingh, J.
1972. The influence of blood flow on transepidermal water loss. Acta Derma., 52: 1-6.
1973. The relationship between skin structure and transepidermal water loss. Comp. Biochem. Physiol., 45A: 685-688.

Hausman, L. A.
1920. A micrological investigation of the hair structure of the *Monotremata*. J. Amer. Anat., 27(4): 463-493.
1932. The cortical fuse of mammalian hair shafts. Amer. Nat., 66: 461-470.
1944. Applied microscopy of hair. Sci. Monthly, 59: 195-202.

Hawes, L. A.
1976. Odor as a possible isolating mechanism in sympatric species of shrews (*Sorex vagrans* and *Sorex obscurus*). J. Mammal., 57: 404-406.

Hayasaki, B.
1959. A study of measurements of dermal tissues in *Macacus cyclopis*. Nagasaki Igakkai Zassi., 43: 701-734.

Hegewald, K.
1914. Vergleichende histologische Untersuchungen über den äusseren Gehörgang der Haussäugetiere. Z. Morphol. und Anthropol., 16: 201-238.

Hellman, K.
1955. Cholinesterase and amine oxidase in the skin; a histochemical investigation. J. Physiol. (Engl.), 129: 454-463.

Herman, T., and K. Fuller.
1974. Observations of the marten, *Martes americana,* in the Mackenzie District, Northwest Territories. Canad. Field Natur., 88(4): 501-503.

Herreid, II, C. F.
1960. Comments on the odors of bats. J. Mammal. 41: 396.

Hershey, F. B.
1964. Quantitative histochemistry of skin. *In* The Epidermis (eds. Montagna and Lobitz). New York: Academic Press.

Hershkowitz, P.
1958. The metatarsal glands in white-tailed deer and related forms of neotropical regions. Mammalia, 22: 537-546.

Herter, K.
1952. Der Temperatursinn der Säugetiere. Beiträge zur Tierkinde und Tierzucht, 3.

Hessling, T.
1855. Über die "Brunftfeige" der Gämse. Z. wiss. Zool., 6: 265.

Hesterman, E. R. and R. Mykytowicz.
1968. Some observations on the intensities of odors of anal gland secretions from the rabbit *Oryctolagus cuniculus* (L.). CSIRO Wildlife Res., 13: 71-81.

Hill, W. C. O.
1951. Epigastric gland of *Tarsius*. Nature, 167: 994.
1956. Body odour in lorises. Proc. Zool. Soc. London, 127: 580.

Hill, W. C. O., H. M. Appleyard, and L. Auber.
1958- The specialized area of skin glands in *Aotes humboldt* (*Simae platyrrhini*).
1959. Trans. Roy. Soc. Edinburgh, 63, Pt. III: 535-552.

Hill, W. C. O., A. Proter, and M. D. Southwick.
1962. The natural history, endoparasites and pseudo-parasites of the tarsiers (*Tarsius carbonarius*) recently living in the Society's menagerie. Proc. Zool. Soc. London, 122: 79-119.

Hoepke, H.
1927. Die Haare. *In* Handbuch der mikroskopischen Anatomie des Menschen. Bd. 3: Teil 1, W. V. Möllendorf (Hrsg). Berlin: Springer, 67-88.

Holfreter, J.
1933. Der Einfluss von Wirtsalter und verschiedenen Organbezirken auf die Differenzierung von angelagertem Gastrulaectoderm. Roux'Arch. Entwicklungsmech., 127: 619-775.

Holmgren, H., and O. Wilander.
1937. Beitrag zur Kenntnis der Chemie und Funktion der Ehrlichschen Mastzellen. Z. Mikrosk. Anat. Forsch., 42: 242-278.

Holyoke, J. B., and W. C. Lobitz.
1952. Histologic variations in the structure of human eccrine sweat glands. J. Invest. Dermatol., 18: 147-167.

Horst, R.
1966. Observations on the gular gland of *Molossus rufus nigricans*. Anat. Rec., 154: 465.

Horstmann, E.
1952. Über den Papillarkörper der menschlichen Haut und regionalen Unterschiede. Acta Anat., 14: 23-42.
1957. Die Haut. *In* Handbuch der mikroskopischen Anatomie des Menschen, Bd. 3, Teil 3. W. V. Möllendorf (Hrsg.) Berlin: Springer, pp. 1-48.

Houy, R.
1909. Makroskopische und mikroskopische Präparate der Entwicklung der Rückendrüse von Dicotyles labiatus. Verhandl. Anat. Ges. Ver. Giesse, 23: 182-183.
1910. Über die Entwicklung der Rückendrüse von Dicotyles. Anat. Hefte, 40.

Howell, A. B.
1930. Aquatic mammals. Their adaptation to life in water. Baltimore: Springfield.

Hrabě, V.
1971. Circumanal glands of Central European *Gliridae* (*Rodentia*). Zoologické listy, 20(3): 247-258.

Huggel, H. J.
1959. La pression sanguine du système veineux autonome de l'aile de la rousette Eidilon helvum Kerr (*Macrochiroptera*). Rev. Suisse Zool., 66: 315-321.

Hurley, H. J., and W. B. Shelly.
1954. The human apocrine sweat gland: two secretions? Brit J. Dermatol., 66: 43-48.

Ignatov, Ju. V.
1958. Opredeleinie teplozaschitnykh svoistv mekha metodom regulyatornogo rezhima (Determination of heat insulation properties of fur by the regula-

tion regime technique). Trudy VNII Zhivotnovodstva, Syrja i Pushniny, Vyp. XVII.

1971. Metod otsenki i issledovannja teplozaschitnykh svoistv mekha (A method for the evaluation and study of heat insulation properties of fur). Avt. kand. diss. M.

Il'enko, A. I., and G. Ju. Zagorodnyaya.

1977. Povedenie svetlogo khorya po klichke "Marfa" pri poluvol'nom soder-zhanii v moskovskoi kvartire (Behavior of Siberian polecat named "Marfa" semifree in a Moscow apartment). *In* Upravlenie povedeniem zhivotnykh (ed. B. P. Manteifel). Moscow: Nauka, pp. 122-124.

Im, M. J. C., and W. Montagna.

1965. The skin of primates. 26. Specific and nonspecific phosphatases in the skin of rhesus monkey. Amer. J. Phys. Anthropol., 23: 131-134.

Ingram, L. D.

1967. Stimulation of cutaneous glands in the pig. J. Comp. Path., 77: 93.

Irving, L.

1939. Respiration in diving mammals. Physiol. Rev., 19: 122-124.

1951. Physiological adaptations to cold in arctic and tropic animals. Federat. Proc., 10: 543-545.

Irving, L. and S. Hart.

1957. The metabolism and insulation of seals as bare-skinned mammals in cold water. Canad. J. Zool., 35: 497-511.

Irving, L., P. F. Scholander, and S. W. Grinnel.

1942. The regulation of arterial blood pressure in the seal during diving. Amer. J. Physiol., 135: 3.

Irving, L., and H. J. Krog.

1955. Temperature of the skin in the Arctic as a regulator of heat. J. Physiol., 7: 355-364.

Irving, L., H. J. Krog, and M. Manson.

1955. The metabolism of some Alaskan mammals in winter and in summer. Physiol. Zool., 28: 173-185.

Irving, L., K. Schmidt-Nielsen, and N. S. B. Abrahamsen.

1957. On the melting points of animal fats in cold climates. Physiol. Zool., 30: 93-105.

Ito, T.

1943. Über den Golgiapparat der ekkrinen Schweissdrüsenzellen der menschli-chen Haut. Okajima's Folia Anat. Japon., 22: 273-280.

Ito, T., and K. Iwashige.

1951. Zytologische Untersuchung über die ekkrinen Schweissdrüsen in men-schlicher Achselhaut mit besonderer Berücksichtigung der apokrinen Sek-retion derselben. Okajima's Folia Anat. Japon., 23: 147-165.

1951. Studien über die Basophile Substanz (Ribonukleinsäure) in den Zellen der menschlichen Schweissdrüsen. Arch. Histol. Japon., pp. 279-287.

Ivanov, M. F., and P. P. Belikhov.

1929. Unasledovanie kachestv shersti gibridami ot muflona i merinosnykh ovets (Hair properties heredity in crosses from *Ovis ammon musimon* and me-rino sheep). Bull. Zoot. Opytn. i Plemmenoi Stantsii v Gossapovednike "Chapli" (former Askania Nova), 5: 73-89.

Ivanov, M. F., and I. G. Urazov.
  1951. Stroenie kozhnogo pokrova morskogo kotika i ego izmeneniya v protsesse proizvodstvennoi obrabotki (The structure of skin in fur seal and its changes in processing). Legkaya Promyshlennost', 9: 33-35.

Iwashiga, K.
  1952. Zytologische und histologische Untersuchungen über die ekkrinen Schweissdrüsen der Achselhaut von gesunden Menschen höheren Alters. Arch. Histol. Japon., 4: 75-90.

Japha, A.
  1905. Über den Bau der Haut des Seihwales (*Balaenoptera borealis Lesson*). Zool. Anz., 29 (14): 442-445.
  1907. Über die Haut nord-atlantischer Furchenwale. Zool. Jahrb. Ab. Anat. und Ontogenie Tierre, 4 (I): 1-40.
  1910*a*. Weitere Beiträge zur Kenntnis der Walhaut. Zool. Jahrb. Suppl., 12 (3): 711-718.
  1910*b*. Die Haare der Waltiere. Zool. Jahrb. Ab. Anat. und Ontogenie der Tiere, 32 (I): 1-42.
  1911. Die Haare der Waltiere. Zool. Jahrb. Ab. Anat., 32 (S): 537-559.

Jarman, P. J.
  1972. The development of a dermal shield in impala. J. Zool. London, 166: 349-356.

Jarret, A.
  1969. The physiology of the skin. The Practitioner, 202: 12-22.

Jenkinson, D.
  1972. The skin structure of British deer. Res. Vet. Sci., 13: 70-73.

Jobert, M.
  1872. Études d'anatomie, sur les organes de toucher. Ann. Sci. Natur. Zool., 14.

Johnson, P. L., and G. Bevelander.
  1946. Glycogen and phosphatase in the developing hair. Anat. Rec., 95: 193-199.

Johnson, K. G., G. M. Malloiy, and J. Bligh.
  1972. Sweat gland function in the red deer (*Cervus elaphus*). Amer. J. Physiol., 223: 604.

Johnston, R. E.
  1972. Scent marking, olfactory communication and social behavior in the golden hamster, *Mesocricetus auratus*. Diss. Abstr. Int., 32: 10.
  1975*a*. Scent marking by male golden hamster (*Mesocricetus auratus*). I. Effects of odors and social encounters. Z. Tierpsychol., 37: 75-98.
  1975*b*. Scent marking by male golden hamster (*Mesocricetus auratus*). III. Behavior in a seminatural environment. Z. Tierpsychol., 37: 213-221.
  1977. Sex pheromones in golden hamsters. *In* Chemical signals in vertebrates (eds. D. Müller-Schwarze and M. M. Mozell). New York: Plenum Press.

Jones, K. K., M. C. Spencer, and S. A. Sancher.
  1951. The estimation of the rate of secretion of serum in man. J. Invest. Dermatol., 17: 213-226.

Jorpes, E., H. Holmgren, and O. Wilander.
  1937. Über das Vorkommen von Heparin in den Gefässwänden in den Augen. Ein Beitrag zur Physiologie der Ehrlichschen Mastzellen. Z. Mikrosk. Anat. Forsch., 42: 279-301.

Kalabukhov, N. I.

1955. Eco-fiziologicheskie osobennosti "zhiznennykh form" gryzunov lesostepi i stepei levoberezhya Ukraini i Evropeiskoi chasti RSFSR (Ecophysiological properties of "life forms" of forest-steppe and steppe rodents of the Cis-Dnepr Ukraine and European RSFSR). Zool. Z., 34: 4.

1957. Predpochitaemaya temperatura mlekopitajuschikh i ee svyaz' s drugimi osobennostyami termoregulyatsii (The temperature preferred by mammals in relation to other thermoregulation features). Mikrobiologiya, Epidmiologiya i Immunobiologiya, 9.

1958. Osobennosti termoregulyatsii gryzunov kak odin iz faktorov ikh chuvstvitel'nosti k chumnoi infektsii (Features of rodent thermoregulation as a factor of susceptibility to plague infection). Mikrobiologiya, Epidemiologiya i Immunobiologiya, 9.

1959a. Sezonnye izmenenia v organizme gryzunov, ne vpadajuschikh v spyatchku, i ikh znachenie dlya kolebanij chustvitel' nosti k infektsii (Seasonal changes in nonhibernating rodents and their significance for variability of susceptibility to infection). In Prirodno Ochagovaya Epidemiologiya Osobo Opasnykh Infektsionnykh Zabolevanij. Saratov.

1959b. Sravnitel'naya ecologia mlekopitajuschikh, vpadajuschikh v spyachku (Comparative ecology of hibernating mammals). Usp. Sovr. Biol., 48 (3): 6.

1960. Vozniknovenie ekologo-fiziologicheskikh osobennostej blizkih form zhivotnykh kak nachalnyi etap divergentsii (Origin of ecophysiological properties of related forms as an initial stage of divergence). In Evolyutsia fiziologicheskikh funktsii. Leningrad.

Kalabukhov, N. I., and B. A. Pryakhin.

1954. Nekotorye ekolog-fiziologicheskie osobennosti peschanok-grebenschikovoi i poludennoi (Some ecophysiological properties of southern and Grebenschikov's gerbils). Zool. Z., 33: 4.

Kalabukhov, N. I., N. A. Mokrevich, and Petrosyan, E. A.

1959. Nekotorye ekologo-fiziologicheskie osobennosti raznykh vidov i geograficheskikh populyatsij odnogo vida peschanok (Some ecophysiological properties of different species and geographical populations of the same species of gerbils). Sovesch. po Ecol. Physiol. Thezisy Dokl., Vyp. I.

Kalabukhov, N. I., and D. P. Kazakevich.

1963. Sezonnye izmenenia soderzhania askorbinovoi kisloty v nadpochechnikakh nekotorykh vidov gryzunov (Changes in the ascorbic acid content in suprarenal glands of some rodent species). Ukr. Biochem. Z., 35: 3.

Kalantaevskaya, K. A.

1972. Morfologia i fiziologia kozhi cheloveka (Morphology and physiology of human skin). Kiev.

Kaletina, L. G., E. A. Korshunova, and P. N. Saunina.

1957. Sezonnye izmenenia volosyanogo pokrova golubykh pestsov (Seasonal changes of blue fox pelage). Trudy Mosk. Vet. Akad., 16, Zverovodstvo.

Kalkowski, W.

1968. Social orientation by traces in the white mouse. Folia Biol., 16 (4): 307-322.

Kano, K.

1951. Zytologische und histologische Untersuchungen über die Schweissdrüsen in Greisenaltern. Beobachtungen der ekkrinen Schweissdrüsen bei den an Krankheit gestorbenen Fällen. Arch. Histol. Japon., 3: 91-105.

Kao Van Shung.

1970. Sezonnye izmenenia teploizolyatsii i ee znachenie v podderzhanii teplovogo gomeostazisa organizma zhyvotnykh (Seasonal changes in thermoinsulation and its significance in maintaining thermal homeostasis in animals). Avt. kand. diss. L. ZIN AN SSSR.

Kashkarov, D. N.

1938. Osnovy ekologii zhivotnykh (The fundamentals of animal ecology). Leningrad: Uchpedgiz.

Kaszowski, S., C. C. Rust, and R. M. Shackelford.

1970. Determination of hair density in the mink. J. Mammal., 51 (1): 27-34.

Kattsnelson, Z. S., and I. I. Orlova.

1956. Gistologicheskoe stroenie preputsialnykh zhelez rechnogo bobra i priroda "bobrovoi strui" (Histological structure of beaver preputial glands and the nature of "beaver spray"). Dokl Ak. Nauk SSSR, 106 (3): 548-550.

Katzberg, A. A.

1952. The influence of age on rate of desquamation of the human epidermis. Anat. Rec., 112: 418.

Katzy, G. D.

1972. Morfologicheskie izmenenia zhelezistogo apparata kozhi u nekotorykh mlekopitajuschikh (Morphological changes of the skin glandular apparatus in some mammals). Sel'skokhozyaistvennaya biologia, VII (3): 401-407.

Kazbekova, E. P.

1941. K voprosu o roli nervnoi systemy v mekhanizme biologicheskogo deistvia razlichnikh spektralnykh uchastkov luchistoi energii (On the role of the nervous system in the mechanism of the biological effects of different spectral regions of radiation energy). I. Arkh. Biol. Nauk., 64: 3.

Kechijan, P.

1965. The skin of primates, 19. The relationship between melanocytes and alkaline phosphatase-positive cells in the potto (Perodicticus potto). Anat. Rec., 152: 317-324.

Kelly, D. E., and J. H. Luft.

1966. Fine structure, development and classification of desmosomes and related attachment mechanisms. VI Inter. Congr. Electron Microscop. Kyoto.

Kenyon, K. W.

1972. The sea otter. In Mammals of the sea (eds. S. H. Ridgway and C. Thomas). Illinois: Springfield, pp. 205-214.

Khalil, F., and G. Abdel-Messeih.

1954. Water content of tissues of some desert reptiles and mammals. J. Exper. Zool., 125 (3): 407-414.

Kizevetter, I. V.

1953. Zhiry morskikh mlekopitajuschikh (Lipids of marine mammals). Primorskoe Kraevoe Izd. Vladivostok.

Klaatsch, H.

1888. Zur Morphologie der Tastballen. Morphol. Jahrb., 14: 407.

Kligman, A. M.

1964. The biology of stratum corneum. In The epidermis (eds. W. Rothman and W. C. Lobitz). New York: Academic Press.

Klima, M., and J. Gaisler.
1967.   Study on growth of juvenile pelage in bats. I. *Vespertilionidae*. Zool.
Listy, 16: 111-124.
Klimov, A. F.
1937.   Anatomia domashnikh zhivotnykh (Anatomy of domestic animals). Moscow: Selkhozgiz.
Klumov, S. K.
1962.   Gladkie kity Tikhogo okeana (*Balaenoptera* of the Pacific). Trudy Inst.
Okeanol, 58.
Kobets, G. F.
1968.   Vliyanie smachivania poverkhnosti na soprotivlenie morskikh zhivotnykh
(The influence of surface wetting on drag in aquatic animals). *In* Mekhanizmy Peredvizhenia i Orientatsii Zhivotnykh, Kiev: Naukova Dumka,
11.
Koenig, L.
1957.   Beobachtungen über Reviermarkierung sowie Droh-, Kampf- und Abverhalten des Murmeltieres (*Marmota marmota* L.). Z. Tierpsychol., 14: 510-
521.
Kogteva, E. Z.
1963.   Sezonnaya izmenchivost' i vozrastnye osobennosti stroenia kozhi i volosyanogo pokrova krota, zaitsa i enotovidnoi sobaki (Seasonal variabilities
and the age properties and structure of the skin and hair of mole, blue
hare, and raccoon-dog). *In* Sb. Nauch. Statei Vses. Nauchn.-Issl. Inst.
Zhyv. Syr. i Pushniny, 2.
Kolchev, V. V., and Z. V. Morieva.
1953.   Tekhnokhimicheskij sostav sala kashalota i izmenenia ego pri obezzhirivanii (Chemical composition of Physeter catodon fat and its changes
when defatting). Trudy VNIRO, 25.
Kolesnik, N. N.
1949.   Evolyutsia krupnogo rogatogo skota (Evolution of cattle). Dushanbe. Izd.
Tadj. Fil. AN SSSR.
Kolesnikov, V. V.
1948.   Potovye zhelezy saigi, jeirana i ovtsy (Sweat glands of saiga, jeiran, and
sheep). Trudy Odesskogo Selkhoz instituta, V: 143.
Kölliker.
1868.   Neue Untersuchungen über die Entwicklung des Bindegewebes. Würzburger. Naturwiss. Z., 2: 141-170.
Kolpovskii, V. M.
1972*a*.   Nadvyinaya zheleza u amerikanskoi norki (Supracervical gland in mink).
Dissertatsia na soiskanie uchenoi stepeni kandidata biologicheskikh nauk.
Alma-Ata.
1972*b*.   Embryonal'nyj gistogenez kozhi u amerikanskoi norki (Embryonic histogenesis in mink skin). Sbornik Nauchno-Tekhnicheskoi Informatsii. Vses.
Nauchn. Issled. Inst. Okhotnich. Khoz. i Zverovodstva. Kirov, 35: 54-63.
Kolpovskii, V. M., and V. B. Zaitsev.
1972.   Submikroscopicheskaya struktura epiteliya nadvyinoi zhelezy (Submicroscopic structure of the epithelium of supracervical gland). Sbornik
Nauchn. Tekhnich. Informatsii. Vses. Nauchn. Issled. Inst. Okhotnich.
Khoz. i Zverovodstva. Kirov, 36: 51-58.

Komarova, L. G.
1968. Sezonnaya izmenchivost' volosyanogo i kozhnogo pokrova golubykh pestsov norvezhskogo tipa (Seasonal variation of pelage and skin in Norwegian-type blue fox). Sb. Nauch.-Tehn. Inf., 23.
Kott, H.
1950. Prisposobitel'naya okraska zhivotnykh (Adaptive coloration of animals). IL, Moscow.
Kozhukhovskaya, A. F.
1969. Sravnenie struktury volosyanogo pokrova nekotorykh predstavitelei podsemeistva polevok (A comparison of pelage structure of some voles). Vestnik Leningradskogo Universiteta, 21: 42-50.
Kramer, M. O.
1960a. Boundary-layer stabilization by distributed damping. J. Airspace Sci., 27: 1.
1960b. The dolphin's secret. New Scientist, 7: 181.
Kratochvil, J.
1968. Das Vibrissenfeld der europäischen Arten der Gattung Apodemus Kaup, 1829. Zool. Listy, 17: 193-209.
1974. Das Stachelkleid des Ostigels (Erinaceus coencolor roumanicus). Přirodovedné práce ústavu Československé akademie věd v Brně, 11 (8): 1-52.
Kratochvil, J., and V. Hrabě.
1967. Zur Kenntnis der Analdrüsen des Tatra-Gebirgmultiers, Marmota marmota latirostrus Kratochvil, 1961 (Rodentia, Sciuridae). Zool. Listy, 16: 31-40.
Kreps, E. M.
1941. Osobennosti fiziologii nyryajuschikh zhivotnykh (The properties of physiology of diving animals). Uspekhi Sovremen. Biol., 14: 3.
Krüger, P.
1929. Über die Bedeutung der Strahlen für den Wärmehaushalt der Poikilotermen. Biol. Zbl., 49B, H2.
Ktitorov, P. M.
1950. Iz opyta provedenia gona norok (From an experiment in organizing rut in the mink). Karakulevodstvo i Zverovodstvo, 4.
Kükenthal, W.
1889. Vergleichend-anatomische und Entwicklungsgeschichtliche Untersuchungen an Waldtiere, T.I, II, Jena.
1909. Haare bei erwachensen Delphine. Anat. Anz., 35.
Kunkel, P., and I. Kunkel.
1964. Beiträge zur Ethologie des Hausmarschweinchens, Cavia aperca, f. parcellus (L.). Z. Tierpsychol., 21: 602-641.
Kuno, Y.
1934. The Physiology of Human Perspiration. London: Churchill.
Kurlyandskaya, E. B.
1941. K voprosu o mekhanizme vozdeistvia infrakrasnykh luchei na organism (Concerning the mechanism of action of infrared rays on the organism). Bull. Exp. Biol. Med., XI: 6.
Kurosumi, K.
1961. Electron microscopic analysis of the secretion mechanism. Intern. Rev. Cytol., 2: 1-117.

Kuznetsov, B. A.
  1927. Stroenie mekha i osennyaya lin'ka belki (The structure of the fur and autumn molting in the squirrel). Zagotizdat, M.
  1949a. Stroenie volosyanogo pokrova (shkurok) lisits (The structure of skin and hair of fox pelts). Trudy Nauchn. Issled. Inst. Mekhovoi Promyshlennosti, Sb. III.
  1949b. Osnovy tovarovedenia pushnogo syrya (The fundamentals of fur-industry science). M. Gosizdat Technich i Ekonomich. Lit. po Voprosam Zagotovok.
  1952. Osnovy tovarovedenia pushnogo syrya (The fundamentals of fur-industry science). Moscow Zagotizdat.
  1962. Zhirovye otlozhenya v tele mlekopitajuschikh i ikh zavisimost' ot ekologii zhivotnykh (Fat deposits in the mammal body and their dependence on animal ecology). Izv. TSKHA, 6: 49.

Lansing, A. I., T. B. Rosenthal, M. Alex, and E. W. Dempsey.
  1952. The structure and chemical characterization of elastic fibers as revealed by elastase and electron microscopy. Anat. Rec., 114: 555-575.

Latyshev, N. I., and A. P. Sidorkin.
  1947. Letniye nabliudeniya nad norami tonkopalogo suslika v Murgabskoi doline (TSSR) (Summer observations of the burrows of the long-clawed ground squirrel in the Murgab valley [Tadjik SSR]). Biul. MOIP, otd. biol., vyp. 1.

Lavrov, N. P., and S. P. Naumov.
  1934. Stroenie mekha i lin'ka tonkopalogo suslika (*Spermophilopsis leptodactylus* Licht.) pustyni Kara-Kum (The structure of the hair and molt of *Spermophilopsis leptodactylus* Licht. in the Kara Kum desert). Zool. Z., 13: 2.

Laws, R. M.
  1959. The fetal growth rates of whales with special reference to the finwhale, *Balaenoptera physalus* L. Discovery Repts., Cambridge.

Lederer, E., and D. Mercier.
  1948. Sur les constituants de la graisse de laine IV. La peau de mouton comme organe de la biosynthese des alcools triterpeniques. Biochem. et Biophys Acta, 2: 91-94.

Lehman, E. V.
  1962. Über die Seitendrüsen der Mitteleuropäischen Rötelmaus (*Clethrionomys glareolus Schreber*). Zeitschrift für Morphologie und Ökologie der Tiere, 51 (3): 335-344.
  1966. Über die Seitendrüsen der Mitteleuropäischen Wühlmäuse der Gattung Microtus Schrank. Zeitschrift für Morphologie und Ökologie der Tiere, 56 (4): 436-443.
  1967. Die Seitendrüsen der Feldmaus (*Microtus arvalis*). Zeitschrift für Morphologie und Ökologie der Tiere, 59 (4): 436-438.
  1969. Die Rückendrüse des Europäischen Maulwurfs (*Talpa europaea*). Z. Säugetierkunde, 34 (6): 356-361.

Lehner, J.
  1924. Das Mastzellen-Problem und die Metachromasie Frage. Anat. und Entwicklungsgesch., 25 (67): 164.

Leither, P., G. A. Bartholomew, and J. E. Nelson.
  1963. Body temperature regulation in three species of Australian flying foxes. Proc. XVI Intern. Congr. Zool. Washington, 2: 55.

Leschinskaya, E. M.
1952*a*. Sezonnye izmenenia kozhnogo pokrova pushnykh zverei (Seasonal changes of the skin of furbearing animals). Trudy Mosk. Tushno-Mekh. Inst., 3.
1952*b*. Sezonnye izmenenia kozhnogo pokrova mlekopitajuschikh (Seasonal changes of mammals skin). Zool. Z., 31, Vyp. 3.

Lewin, V., and J. G. Steflox.
1967. Functional anatomy of the tail and associated behavior in woodland caribou. Canad. Field Nat., 81 (1): 63-66.

Leydig, F.
1859. Über die äusseren Bedeckungen der Säugetiere. Arch. Anat., Physiol. und wiss. Med.
1876. Über den Bau der Zehen Batrachieren und die Bedeutung des Fersenhöckers. Morphol. Jahrb., 2: 165.

Leyhauser, P.
1956. Das Verhalten der Katzen. *In* Handbuch der Zoologie (eds. von Helmcke und von Lengerken), 8 (10): 1-34.

Lindeborg, R. G.
1952. Water requirements of certain rodents from xeric and mesic habitats. Contrib. Lab. Mich., 58: 1-32.
1955. Water conservation in *Perognathus* and *Peromyscus*. Ecology, 36: 2.

Ling, J. K.
1965*a*. Hair growth and moulting in the Southern Elephant Seal, *Mirounga leonina* (Linn.). *In* Biology of the Skin and Hair Growth (eds. A. G. Lyne and B. E. Short). New York, pp. 525-544.
1965*b*. Functional significance of sweat glands and sebaceous glands in seals. Nature, 208: 560-562.
1968. The skin and hair of the southern elephant seal, *Mirounga leonina* L. III. Morphology of the adult integument. Austral. J. Zool., 16: 629-645.
1970. Pelage and molting in wild mammals with special reference to aquatic forms. Quart. Rev. Biol., 45: 1, 16-54.
1974. The integument of marine mammals. *In* Functional Anatomy of Marine Mammals. 2 (ed. R. J. Harrison). New York: Academic Press, pp. 1-44.

Lipkow, J.
1954. Über das Seitenorgan des Goldhamsters (*Mesocricetus auratus Waterh.*). Z. Morphol. Ökol. Tiere, 42: 333-372.

Lobitz, W. C., Jr., J. B. Holyoke, W. Montagna.
1954*a*. The epidermal eccrine sweat duct unit. A morphological and biological entity. J. Invest. Dermatol., 22: 157-158.
1954*b*. Response of the human eccrine sweat duct to controlled injury. Growth center of the "epidermal sweat duct unit." J. Invest. Dermatol., 23: 329-344.

Lobitz, W. C., Jr., J. B. Holyoke, and D. Brophy.
1955. Histochemical evidence for human eccrine sweat duct activity. A. M. A. Arch. Dermatol., 72: 229-236.

Loewenthal, L. J. A.
1960. The human eccrine sweat gland ampulla. J. Invest. Dermatol., 34: 233-235.
1961. The eccrine ampulla: morphology and function. J. Invest. Dermatol., 36: 171-182.

Lombardo, C.
1907. Il glicogeno della cute. Gior. Ital. malvener., 42: 448-464.
1934. Il glicogeno in alcuni derivati epidermici della cute umana. Giorn. ital. dermatol. e sifilol., 75: 185-186.

Lopukhova, V. A., and V. I. Kulikova.
1958. Sravnitel'ny gistologicheskij analiz kozhi losya i krupnogo rogatogo skota (Comparative histological analysis of the skin of elk and cattle). Sb. Stud. Rabot Mosk. Technol. Inst. Molochn. i Myasn. Prom., 5: 66-74.

Lototsky, B. V., E. F. Sosnina, and V. A. Tsvileneva.
1959. O sluchayakh glubokogo pogruzhenia iksodovykh kleschei v kozhu gryzunov (Cases of deep penetration of ixode mites into the skin of rodents). Zool. Z., 33: 3.

Luck, C. P., and P. G. Wright.
1964. Aspects of the anatomy and physiology of the skin of hippopotamus (*H. amphibius*). Quart. J. Exper. Physiol. and Cognate Med. Sci., 49: 1-14.

Lund, F.
1930. Emotions of Men. New York: McGraw Hill.

Lvov, V.
1884. Sravnitel'noe issledovanie i opisanie volosa, schetiny, igly u mlekopitajuschikh i pera u ptits (Comparative studies and descriptions of hair, bristles, and spines in mammals and feathers in birds). Uch. Zap. Mosk. Un-ta, otdel Est.-Ist., 4: 1-88.

Lyne, A. G.
1951. Notes on external characters of the barred bandicoot (*Parameles gunnii Gray*), with special reference to the pouch young. Proc. Zool. Soc. London, 121 (III): 587-598.
1959. The systematic and adaptive significance of the vibrissae in the *Marsupialia*. Proc. Zool. Soc. London, 133: I.

Lyne, A. G., G. S. Molyneux, R. Mykytowycz, and P. F. Parakkal.
1964. The development, structure and function of the submandibular cutaneous (chin) glands in the rabbit. Austr. J. Zool., 12 (3): 340-348.

Machida, H., and W. Montagna.
1964. The skin of primates. XXII. The skin of Lutong (*Presbytis pyrrus*). Amer. J. Phys. Anthropol., 22 (4): 443-451.

Machida, H., and L. Giacometti.
1968. X. The Skin. *In* Biology of the howler monkey (*Alouatta caraya*) (ed. M. R. Malinow). Basle: Karger, pp. 126-140.

Machida, H., L. Giacometti, and E. Perkins.
1966. The skin of the lesser anteater (*Tamandua tetradactyla*). J. Mammal., 47 (2): 280-286.

Machida, H., and E. Perkins.
1966. The Skin of Primates. 30. The skin of the woolly monkey (*Lagothrix lagotricha*). Amer. J. Phys. Anthropol., 24: 309-319.

Machida, H., E. Perkins, and F. Hu.
1961. The skin of primates. 35. The skin of the squirrel monkey. Amer. J. Phys. Anthropol., 26: 45-46.

Machida, H., E. Perkins, and L. Giacometti.
1966. The skin of primates. 29. The skin of pigmy bushbaby (*Galago demidovi*). Amer. J. Phys. Anthropol., 24: 199-203.

Machida, H., E. Perkins, W. Montagna, and L. Giacometti.
    1955.   The skin of primates. 27. The skin of the white-crowned mangobey (*Cero-
            cebus atys*). Amer. J. Phys. Anthropol., 23: 165-180.
Mackenzie, I. C.
    1972.   The ordered structure of mammalian epidermis. *In* Epidermal wound
            healing (eds. H. I. Maibach and D. T. Rovee). Chicago: Yearbook Publi-
            cations, pp. 5-25.
Magkour, G. A.
    1961.   The structure of the facial area in the mousetailed bat, *Rhinopoma hard-
            wickei cystops,* Thomas. Bull. Zool. Soc. Egypt, 16: 50-64.
Mamed-Zade, E. S.
    1973.   Sezonnye izmenenia kozhnogo i volosyanogo pokrovov nutrij (Seasonal
            changes of hair and skin in coypu). Trudy MOIP, Zool. Ser., 40: 191-194.
Manteifel, P. A.
    1934.   Sobol' (Sable). M-L, KOIZ.
Marvin, M. Ya.
    1958a. Stroenie mekha amerikanskoi norki (The structure of mink fur). Uralsk.
            Gosuniversitet, Sverdlovsk.
    1958b. Stroenie mekha kolonka (The structure of Siberian weasel fur). Uralsk.
            Gosuniversitet, Sverdlovsk.
    1958c. Stroenie mekha gornostaya (The structure of ermine fur). Uralsk. Gosuni-
            versitet, Sverdlovsk.
    1969a. Stroenie mekha *Muridae* (The structure of the fur of *Muridae*). Trudy
            Uralsk. Otd. MOIP, 3: 137-143.
    1969b. Stroenie mekha uralskogo sobolya (The structure of the fur of the Ural
            sable). Trudy Uralsk. Otd. MOIP, 3: 144-150.
    1969c. Sravnitel'naya morfologia volosyanogo pokrova norok (Comparative
            morphology of mink pelage). Trudy Uralsk. Otd. MOIP, 3: 151-160.
    1974.   Stroenie volosyanogo pokrova gryzunov Urala (The structure of the pel-
            age of Ural rodents). Nauchn. Trudy Sverdlovskogo Gosudarstvennogo
            Instituta. Sb., 235: 11-22.
Marvin, M. Ya., and L. V. Shumakhova.
    1958.   Stroenie mekha zaitsa belyaka (Fur structure of Lepus timidus). Uralsk.
            Gosuniversitet.
Maslov, K. I.
    1967.   K morfologii kozhno-volosyanogo pokrova losya (On the morphology of
            skin and hair of elk). Trudy Pechoro Ilych. Gosud. Zapovednika, XII:
            111-115.
Masson, P.
    1926.   Les naevi pigmentaires, tumeurs nerveuses. Ann. Anat. Pathol., 3: 417-
            452, 657-696.
Matoltsy, A. G.
    1969.   Keratinization of the avian epidermis. An ultrastructural study of the
            newborn chick skin. J. Ultrasr. Res., 29: 5-6.
Matoltsy, A. G., and P. F. Parakkal.
    1965.   Membrane-coating granules of keratinizing epithelia. J. Cell. Res., 24:
            297.
Matthews, E.
    1929.   Die Dickenverhältnisse der Haut bei den Mammalia in allgemeinen, den
            Sirenia im besonderen. Z. wiss. Zool., 134: 345-537.

Matthews, L. H.
    1938. Notes on the southern right whale, *Eubalaena australis*. Disc. Reps., 17:
           169-182.
Matveev, B. S.
    1942. Kozha, ee zhelezy i podoshvennye kozhnye zhelezistye organy u sobolya
           (Skin and its glands and sole glandular organs in sable). Zool. Z., 21: 5.
Maximov, A. A.
    1906. Über die Zellformen des lockeren Bindegewebes. Arch. mikrosc. Anat.,
           67: 680.
Mayer, B.
    1855. Über die Hautbedeckungen der Cetacean. Breslau-Bonn.
McGregor.
    1952. The sweating reaction of the forehead. J. Physiol., 116: 26-34.
Medawar, P. B.
    1950. Pigment spread in guinea pig. Zool. Res. New York, 35, 21.
de Meijere, J. C. H.
    1894. Über die Haare der Säugetiere besonders über ihre Anordnung. Morphol.
           Jahrb., 21 (3): 312-424.
Meng, M.
    1955. Das Verhalten der Pigmentzellen gescheckter Meerschweinchen während
           der Ontogenie und bei Regenerations und Transplantationsversuchen.
           Roux' Archiv. Entwicklungsmech, 148: 92-122.
Meyer, J.
    1955. A modification of Gomori's method for the demonstration of phosphami-
           dase in tissue sections. J. Histochem. and Cytochem., 3: 134-140.
Meyer, J., and J. P. Weinmann.
    1957. Occurrence of phosphamidase activity in keratinizing epithelia. J. Invest.
           Dermatol., 29: 393-405.
Meyer, K.
    1945. Mucoids and glycoproteins. Adv. Protein Chem., 2: 249-275.
    1947. The biological significance of hyaluronic acid and hyaluronidase. Physiol.
           Rev., 27: 335-359.
Meyer, K., and M. M. Rapport.
    1951. The mucopolysaccharides of the ground substance of connective tissue.
           Science, 113-159.
Meyer, P.
    1968. Territoriumsmarkierung beim Rech und Morphologie des sogenannten
           Stirnorgans. Beiheft zu Nature, Kultur, und Jagd. Beitrage zur Natur-
           kunde Niedersachens, 2-4, Hannover, 4-54.
Michels, N. A.
    1938. The mast cells. *In* Downey handbook of hematology., New York, I: 231-
           322.
Mikhailov, A. N.
    1949. Fiziko-khimicheskie osnovy tekhnologii kozhi (Physico-chemical founda-
           tions of skin technology). Moscow: Gislegprom.
Minamitini, K.
    1941a. Zytologische und histologische Untersuchungen der Schweissdrüsen in
           menschlicher Achselhaut. Über das Vorkommen der Besonderen Formen
           der apokrinen und ekkrinen Schweissdrüsen in Achselhaut von Japanern.
           Okajimas Folia Anat. Japon., 20: 563-590.

1941*b*. Zytologische und histologische Untersuchungen der Schweissdrüsen in der menschlicher Achselhaut. Zur Zytologie der apokrinen Schweissdrüsen in der menschlichen Achselhaut. Okajimas Folia Anat. Japon., 21: 61-94.

Miraglia, T., and A. J. Santos.
1971. Histochemical data on the sebaceous glands in the lips of the marmoset (*Callithrix jacchus*). Acta Anat., 78: 295-305.

Miropolsky, S. V.
1939. Issledovanie po ekofiziologii domashnikh zhivotnykh (Investigations on ecophysiology of domestic animals). Voprosy ekologii i biotseonologii, 5.

Mislin, H., and M. Kaufman.
1947. Beziehungen zwischen Wandbau und Function der Flugautvene (*Chiroptera*). Rev. Suisse Zool., 54: 240-245.

Mitchell, O. G.
1965. Effect of castration and transplantation of ventral gland of the gerbil. Proc. Soc. Exper. Biol. and Med., 119: 953-955.

Mkhitaryants, A. F.
1974. Sravnitelnaya kharakteristika kozhno-volosyanogo pokrova nekotorykh polevok, v raznoi stepeni svyazannykh so vodnoi sredoi (Comparative characterization of skin and hair of some voles associated with an aquatic environment in varying degrees). Dissertatsia na soiskanie uchenoi stepeni kandidata biol. nauk.

Möbius, K.
1892. Die Behaarung des Mammoths und der lebenden Elefanten, vergleichend untersucht. Sitzungs Berlin. Akad. Wiss., 527-538.

Mohr, E.
1952. Die Robben der europäischen Gewässer. Monographien der Wildsäugetiere. Frankfurt am Main: Paul Schöps, Bd. XII: 1-283.
1961. Schuppentiere. Die Neue Brehm-Bücherei. A Wittenberg Lutherstadt: Ziemsen Verlag.

Monakhov, G. I., and E. M. Leontyev.
1968. Pakhovaya zheleza i ee znachenie vo vnutrividovykh svyazyakh u sobolei (Inguinal gland and its significance in intraspecific communication in the sable). Sbornik nauchno-tekhnicheskoi informatsii. Vsesojuzny nauchno-issledovatel'skiy institut zhivotnogo syrya i pushniny. Kirov, 22: 52-59.

Monod, Th., and P. L. Dekeyser.
1968. Sur la glande dorsale de *Procavia ruficeps latastei* Thomas. Bull. l'IFAN. Ser. A, XXX (1): 307-317.

Montagna, W.
1949. Anisotropic lipids in the sebaceous glands of the rabbit. Anat. Record, 104: 243-254.
1950*a*. The brown inguinal glands of the rabbit. Amer. J. Anat., 87: 213-238.
1950*b*. Perinuclear sudanophil bodies in mammalian epidermis. Quant. J. Microsc. Sci., 91: 205-208.
1955. Histology and cytochemistry of human skin. VIII. Mitochondria in the sebaceous glands. J. Invest. Dermatol., 25: 117-121.
1957. Histology and cytochemistry of human skin. XI. The distribution of β-glucuronidase. J. Biophys. and Biochem. Cytol., 3: 343-348.
1960. The cutaneous innervation. New York: Academic Press.
1962. The structure and function of skin. 2d edition. New York: Academic Press.

Montagna, W., and C. R. Novack.
  1946. The histochemistry of the preputial gland of the rat. Anat. Record, 96: 111-128.
  1947. Histochemical observations on the sebaceous glands of the rat. Amer. J. Anat., 81: 39-62.
  1964. The histology of preputial gland of the rat. Anat. Rec., 96: 41-54.
Montagna, W., and J. B. Hamilton.
  1949. The sebaceous glands of the hamster. II. Some cytochemical studies in normal and experimental animals. Amer. J. Anat., 84: 365-396.
Montagna, W., H. B. Chase, and J. B. Hamilton.
  1951. The distribution of glycogen and lipids in human skin. J. Invest. Dermat., 17: 147-157.
Montagna, W., H. B. Chase, and W. J. Lobitz, Jr.
  1952. Histology and cytochemistry of human skin. II. The distribution of glycogen in the epidermis, hair collagen, hair follicles, sebaceous glands, and eccrine sweat glands. Anat. Record, 114: 231-248.
  1953. Histology and cytochemistry of human skin. IV. The eccrine sweat glands. J. Invest. Dermatol., 20: 415-423.
Montagna, W., and V. R. Formisano.
  1955. Histology and cytochemistry of human skin. VII. The distribution of succinic dehydrogenase activity. Anat. Rec., 122: 65-77.
Montagna, W., and R. J. Harrison.
  1957. Specialization in the skin of the seal (*Phoca vitulina*). Amer. J. Anat., 100: 81-114.
Montagna, W., and R. A. Ellis.
  1958a. L'histologie et la ctyologie de la peau humaine. XVI. Rapartition et la concentration des esterases carboxiliques. Ann. Histochim., 3: 2-17.
  1958b. The vascularity and innervation of human hair follicles. *In* The biology of hair growth (eds. W. Montagna and R. A. Ellis). New York: Academic Press, pp. 219-227.
  1958c. The biology of hair growth.
  1959. The skin of primates. I. The skin of the potto (*Perodictis potto*). Amer. J. Phys. Anthropol., 17: 137-162.
  1960a. Sweat glands in the skin of *Ornithorhynchus paradoxurus*. Anat. Rec., 137 (3): 271-277.
  1960b. The skin of primates. II. The skin of slender loris (*Loris tradigradus*). Amer. J. Phys. Anthropol., 18: 19-44.
  1963. New approaches to the study of the skin of primates. *In* Evolutionary and genetic biology of primates (ed. J. Buettner-Janusch). New York: Academic Press, I: 179-196.
Montagna, W., and H. Machida.
  1966a. The skin of primates. XXXII. The Philippine tarsier (*Tarsius syrichta*). Amer. J. Phys. Anthropol., 25: 71-84.
  1966b. The skin of primates. XXX. The skin of angwantibo (*Arctocebus calabrensis*). Amer. J. Phys. Anthropol., 25: 277-290.
Montagna, W., H. Machida, and E. Perkins.
  1966a. The skin of primates. XXIII. The stump-tail macaque (*Macaca speciosa*). Amer. J. Phys. Anthropol., 24: 71-86.

Montagna, W., and P. F. Parakkal.
  1974. The structure and function of skin. 3d edition. New York: Academic
          Press.
Montagna, W., K. A. Yasuda, and R. A. Ellis.
  1961a. The skin of primates. III. The skin of the slow loris (*Nycticebus coucang*).
          Amer. J. Phys. Anthropol., 19: 1-19.
  1961b. The skin of primates. XV. The skin of the black lemur (*Lemur macaco*).
          Phys. Anthropol., 19: 115-130.
Montagna, W., and J. S. Yun.
  1962a. The skin of primates. VII. The skin of the great bushbaby (*Galago crassi-
          caudata*). Amer. J. Phys. Anthropol., 20: 149-166.
  1962b. The skin of primates. VIII. The skin of Anubis baboon (*Papio doguera*).
          Amer. J. Phys. Anthropol., 20: 131-142.
  1962c. The skin of primates. The skin of the ring-tailed lemur (*Lemur catta*).
          Inter. J. Phys. Anthropol., 20: 95-118.
  1963a. The skin of primates. XV. The skin of the chimpanzee (*Pan satyrus*).
          Amer. J. Phys. Anthropol., 21: 189-204.
  1963b. The skin of primates. XVI. The skin of lemur mongoz. Amer. J. Phys.
          Anthropol., 23 (3): 371-381.
Montagna, W., J. S. Yun, and H. Machida.
  1964.   The skin of primates. XVII. The skin of rhesus monkey (*Macaca mulatta*).
          Amer. J. Phys. Anthropol., 22 (3): 307-319.
Montagna, W., J. S. Yun, A. F. Silver, and W. C. Quevedo, Jr.
  1962.   The skin of primates. XIII. The skin of the tree shrew (*Tupaia glis*). Amer.
          J. Anthropol., 20: 431-440.
Moore, R. E.
  1965.   Olfactory discrimination as an isolation mechanism between *Peromyscus
          maniculatus* and *Peromyscus pelionotus*. Am. Midl. Nat., 73: 85-100.
Moore, W. G., and R. L. Marchinton.
  1974.   Marking behavior and its social function in white-tailed deer. *In* V. Geist
          and F. Walther. The behaviour of ungulates and its relation to manage-
          ment. IUCN Publications. New Series. Switzerland: Morges, 24: 447-456.
Mordvinov, Yu. E., and B. V. Kurbatov.
  1972.   Vliyanie volosyanogo pokrova nekotorykh vidov nastoyashchikh tjulenei
          (*Phocidae*) na velichinu obschego gidrodinamicheskogo soprotivlenia
          (The effect of hair of some *Phocidae* on the magnitude of total hydro-
          dynamic resistance). Zool. Z., 51 (2): 242-247.
Moretti, G., and H. Mescon.
  1956a. Histochemical distribution of acid phosphatases in normal human skin.
          J. Invest. Dermatol., 26.
  1956b. A chemical-histochemical evaluation of acid phosphatase activity in
          human skin. J. Histochem. and Cytochem., 4: 247-253.
Moretti, G., R. Ellis, and H. Mescon.
  1959.   Vascular patterns in the skin of the face. J. Invest. Dermatol., 33: 103-
          112.
Moretti, G., and W. Montagna.
  1959.   Istologia e. citochimica della cute umana. XVII. Le Unitá Vascolari.
          Giron. Ital. Dermatol. 3: 242-254.

Müller, D., and E. Lemperle.
   1964. Objektivierung und Analyse olfaktorischer Signale der Säugetiere mit Hilfe der Gaschromatographie. Naturwissenschaften, 14 (Jg. 51): 346-347.
Müller-Schwarze, D.
   1971. Pheromones in black-tailed deer (*Odocoileus hemionus columbianus*). Anim. Behaviour, 19: 141-152.
   1974. Olfactory recognition of species, groups, individuals and physiological states among mammals. *In* Pheromones (ed. M. C. Birch). North-Holland Research Monographs. Frontiers of Biology. Amsterdam: North-Holland Publishing Co., 32: 316-326.
Müller-Schwarze, D., and C. Müller-Schwarze.
   1975. Subspecies specificity of a response to a mammalian social order. J. Chem. Ecology, 1: 125-131.
Münch, H.
   1958. Zur Ökologie und Psychologie von *Marmota m. marmota*. Z. f. Säugetierh., 23: 129-138.
Munger, B. L., and S. W. Brusilov.
   1971. The histophysiology of rat plantar sweat glands. Anat. Rec., 169 (1): 1-7.
Murariu, D.
   1971. Contributions à la connaissance du systéme glandulaire tégumentaire chez *Sorex araneus araneus* L. et *Talpa europaea* L. (*Mammalia,* Ord. *Insectivora*). Travaux du Muséum d'Histoire Naturelle Grigore Antipa, XI: 429-435.
   1972. Observations concerning the form, size, structure and spreading of the sebaceous and sudoriferous glands in the *Erinaceous europaeus* L., *Crocidura leucodon Herm.,* and *Neomys fodiens Schreb.* (Ord. *Insectivora, Mammalia*). Travaux du Muséum d'Histoire Naturelle Grigore Antipa, XII: 393-405.
   1973. Données macro- et microscopiques sur les organes glandulaires latéraux chez *Sorex araneus* L., *Neomys fodiens* Schreb., et *Crocidura leucodon* Herm. de Roumanie. *Ibid.,* XIII: 445-458.
   1974. L'histologie des glandes plantaires chez les Mammifères Insectivores de Roumanie. *Ibid.,* XV: 381-391.
   1975*a.* Les glandes de la région anale chez quelques *Soricides* (*Insectivora*). *Ibid.,* XVI: 295-302.
   1975*b.* Étude anatomo-histologique des glandes de la région anale et circumanale chez *Erinaceus europaeus* et *Talpa europaea* (*Insectivora*). *Ibid.,* XVI: 303-310.
   1976. Les glandes tégumentaires de certains *Insectivores* (*Mammalia, Insectivora*) de Roumanie. Anatomie, histologie et histochimie. *Ibid.,* XVII: 387-413.
Mykytowicz, R.
   1965. Further observations on the terrestrial function and histology of the submandibular cutaneous gland in the rabbit, *Oryctolagus cuniculus* (L.). Anim. Behavior, 13 (4): 400-412.
   1966*a.* Observations on the odoriferous and other glands in the Australian wild rabbit, *Oryctolagus cuniculus* (L.), and the hare, *Lepus europaeus,* Pt. II. The inguinal glands. CSIRO Wildlife Research, 11: 49-64.

1966b. Observations on odoriferous and other glands in the Australian wild rabbit, *Oryctolagus cuniculus* (L.), and the hare, *Lepus europaeus*, Pt. I. The anal gland. CSIRO Wildlife Research, 11: 31-47.

1967. Communication by smell in the wild rabbit. Proc. Ecol. Soc., 2: 125-131.

1968a. Territorial marking by rabbits. *In* Vertebrate adaptations. San Francisco: W. H. Freeman and Co., pp. 334-341.

1968b. Territorial marking by rabbits. Sci. Am. 218: 116-126.

1972. The behavioural role of the mammalian skin glands. Naturwissenschaften, 59 (4): 133-139.

1974. Odor in the spacing behavior of mammals. *In* Pheromones (ed. M. C. Birch). North-Holland Research Monograph. Frontiers of Biology. Amsterdam: North-Holland Publishing Company, 32: 327-343.

Mykytowicz, R., and M. L. Dudzinski.

1966. A study of the weight of odoriferous and other glands in relation to social status and degree of social activity in the wild rabbit, *Ornyctolagus cuniculus* (L.). CSIRO Wildlife Research, 11: 31-47.

1972. Aggressive and protective behavior of adult rabbits, *Oryctolagus cuniculus* (L.) towards juveniles. Behaviour, 43 (1-4): 97-120.

Naaktgeboren, C.

1960. Die Entwicklungsgeschichte der Haut des Finwales, *Balaenoptera physalus* L. Zool. Anz., 165 (5): 159-167.

Nadel, A.

1932. Chemische Studien an der menschlichen Haut. II. Untersuchungen an der normalen und pathologischen Menschenhaut. Arch. für Dermat. u. Syph., 165: 507-524.

Nakai, J., and T. Shida.

1948. Sinus-hairs of the Sei whale (*Balaenoptera borealis*). Sci. a. Rep. Whale Res., 1: 41-47.

Narkhov, A. S.

1937. Morfologia muskulatury khvostovoi oblasti *Delphinus delphis* i *Tursio tursiops* (Morphology of the muscles of caudal region in *Delphinus delphis* and *Tursio tursiops*). Zool. Z., 16: 4.

Neal, E. G.

1948. The badger. London: Pelican Books.

Nemilov, A. V.

1936. Gistologia i embriologia domashnikh zhivotnykh (Histology and embryology of domestic animals). Moscow: Medgiz.

Neuville, H.

1917. Du tégument des Proboscidiens. Bull. Mus. Nat. Hist. Natur. Paris, 22: 6.

1918. Sur quelques particularités du tégument des élephants et sur les comparaisons qu'elles suggèrent. Bull. Mus. Nat. Hist. Natur. Paris.

1923. Sur la glande iléo-caecale des élephants. Bull. Mus. Nat. Hist. Natur. Paris, 29: 3.

Nicolau, S.

1911. Recherches histologiques sur la graisse cutanée. Ann. Dermatol. et Syphil., 2: 641-658.

Nicoll, P. A., and R. L. Webb.

1946. Blood circulation in the subcutaneous tissue of the living bat's wing. Ann. N. Y. Acad. Sci., 46: 697-711.

Nielsen, S. V.
1953. Glands of the canine skin—morphology and distribution. Am. J. Vet. Res. 14 (52): 448-454.

Niethammer, J.
1970. Der Stachelwechsel der Igel (*Erinaceidae*). Zool. Anz. 33. Supplementband: 530-534.

Niu, M. C.
1947. The axial organization of the neural crest, studied with particular reference to its pigmentary component. J. Exper. Zool., 105: 79-113.

Ochi, J., and Y. Sano.
1970. Morphological attempts to solve some unsettled problems on the human eccrine sweat glands. *In* Advances in climate physiology (eds. S. Ito, K. Ogata, and H. Yoshimura), pp. 79-92.

Odland, G. F.
1964. Tonofilament and keratohyalin. *In* The epidermis (eds. W. Montagna and W. Lobitz). New York: Academic Press, pp. 237-249.

Odland, G. F., and T. H. Reed.
1967. The Epidermis. *In* Ultrastructure of normal and abnormal skin (ed. A. S. Zelickson). Philadelphia, p. 54.

Orlovskaya, G. V., and A. L. Zaides.
1952. Stroenie voloknistykh struktur soedinitel'noi tkani (argirofil'nye i kollagenovye volokna i ikh submikroskopicheskaya struktura) (The structure of the fibrous connective tissue [Argyrophilous and collagen fibers, their submicroscopic structure]). Arch. Path., L.

Ortmann, R.
1960. Die Analregion der Säugetiere. *In* Handbuch der Zoologie (ed. W. Kukenthal). Berlin: Walter de Gruyter und Co., Band 8, Lieferung 2*b*.

Osipuk, O. E.
1975. Materialy k voprosu ob embrional'nom razvitii kozhi kashalota (Data on embryonic development of the skin of the sperm whale). *In* Aktual'nye Problemy Teoreticheskoi i Klinicheskoi Meditsiny. Minsk, pp. 144-145.

Ota, R.
1950. Zytologische und histologische Untersuchungen der apokrinen Schweisdrüsen in der normalen, keinen Achselgeruch (*Osmidrosis axillae*) gebenden Achselhäuten von Japanern. Arch. Anat. Japon, I: 285-308.

Pacht, I.
1954. Dürfen die Seitendrüsen als spezifisch-taxonomisches Merkmal der Cleithrionomys, Arten aufgefasst werden (*Rodentia*). Zool. Anz., 153: 321-322.

Palay, D.
1957. Funktsional'nye izmeneniya komponentov kletki (Functional changes in cell components). *In* Strukturnye Komponenty Kletki. IL. Moscow.

Palmer, E., and G. Weddel.
1964. The relationship between structure, innervation, and function of the skin of the bottle-nose dolphin (*Tursiops truncatus*). Proc. Zool. Soc. London, 143 (4): 553-558.

Parakkal, P. F., and N. J. Alexander.
1972. Keratinization. New York: Academic Press.

Parakkal, P., W. Montagna, and R. Ellis.
  1962. The skin of primates. XI. The skin of white-browed gibbon (*Hylobates hoolock*). Anat. Rec., 143: 169-178.

Pavlova, E. A.
  1951. Sezonnaya izmenchivost' mekha sobolya i lesnoi kunitsy (Seasonal variability of the fur of sable and pine marten). Trudy VNII Okhotnichego Prom., X: 78-92.
  1955. Vozrastnaya i sezonnaya izmenchivost' mekha ondatry (Age and seasonal variability of ondatra fur). Trudy VNII Okhot. Promysla., XV: 59-92.
  1959. Sezonnaya izmenchivost' volosyanogo pokrova i sroki lin'ki gornostaya (Seasonal variability of hair and molt periods in ermine). Trudy Vses. Nauchn.-Issled. In-ta zhivotnogo syrya i pushniny, XVIIII: 147-173.

Pavlova, N. R.
  1968. Stroenie volosyanogo pokrova dlinnokhvostovoi shinshilly (*Chinchilla lanigera Gray*) (The structure of the hair of *Chinchilla lanigera Gray*). Sb. Trudov Zool. Muzeya, X. Issledovanie po faune Sovetskogo Sojuza (mlekopitajuschie), 109-120. Izd. MGU.
  1969. Stroenie kozhnogo pokrova shinshilly (Structure of chinchilla skin). VNII Syrya i Pushniny. Sb. Nauchn. Techn. Inf. Kirov, 24: 89-91.
  1970. Stroenie i izmenchivost' volosyanogo pokrova *Chinchilla laniger* (Structure and variability of the hair of *Chinchilla laniger*). *In* Transactions of the XIth International Congress of Game Biologists. Moscow, 919-922.
  1971. Stroenie i izmenchivost' kozhnogo i volosyanogo pokrova chinchilly (*Chinchilla laniger*) (Structure and variability of the skin and hair in *Chinchilla laniger*). Dissertatsia na Soiskanie Uchenoi Stepeni Kandidata Biol. Nauk. Moscow.

Pawlowsky, E. N.
  1927. Gifttiere und ihre Giftigkeit. Jena: Gustav Fischer Verlag.

Pawlowsky, E. N., and S. P. Alfeeva.
  1941. Patologo-gistologicheskie izmenenia kozhi krupnogo rogatogo skota pri ukuse klescha *Ixodes ricinus* (Pathologo-histological changes in the skin of cattle bitten by mite *Ixodes ricinus*). Trudy Voenno-Med. Ak., 25.
  1949. Sravnitel'naya patologia kozhi mlekopitajuschikh pri ukuse kleschami (Comparative pathology of the skin on mammals bitten by mites). Izv. AN SSR, Ser. Biol., 6.

Pearce, R. H., and E. M. Watson.
  1949. The mucopolysaccharides of human skin. Canc. J. Res., 27: 43-57.

Perkins, E. M.
  1960. The skin of primates. 29. The skin of Goeld's marmoset (*Callimico goeldi*). Amer. J. Phys. Anthropol., 30: 231-250.
  1966. The skin of primates. 31. The skin of the black-collared tamarin (*Tamarinus nigricollis*). Amer. J. Phys. Anthropol., 25: 41-70.
  1968. The skin of primates. 36. The skin of pigmy marmoset *Callithrix* (= *Cebuella*) *pygmaea*. Amer. J. Phys. Anthropol., 27: 20-34.
  1969a. The skin of primates. 41. The skin of the silver marmoset, *Callithrix* (= *Mico*) *argentata*.
  1969b. The skin of primates. The skin of cottontop pinché saguinus (*Oedipomidas oedipus*). Amer. J. Phys. Anthropol., 30; 13-28.

Perkins, E. M., T. Arao, and E. H. Dolnick.
  1968. The skin of primates. 37. The skin of the pigment macaque (*Macaca nemestris*). Amer. J. Phys. Anthropol., 28: 75-84.

Perkins, E. M., T. Arao, and H. Uno.
  1968. The skin of primates. 38. The skin of the red uacari (*Cacajo rubicundus*). Amer. J. Phys. Anthropol., 29: 57-80.

Perkins, E. M., and D. M. Ford.
  1969. The skin of primates. 39. The skin of the white-browed capuchin (*Cebus albifrons*). Amer. J. Phys. Anthropol., 30: 1-12.

Perkins, E. M., and H. Machida.
  1967. The skin of the golden spider monkey (*Ateles geoffroyi*). Amer. J. Phys. Anthropol., 26: 35-44.

Perry, J. S.
  1953. The reproduction of the African elephant. Philos. Trans. Roy. Soc. London. Ser. B., 643: 93-149.

Pershin, S. V., A. S. Sokolov, and A. G. Tomilin.
  1970. O reguliruemoi spetsial'nymi sosudistymi organami uprugosti plavnikov del'finov (On the elasticity of dolphin flippers regulated by specific vascular organs). Dokl. AN SSSR, 190 (3): 709-712.

Peter, K.
  1909. Die Nierenkanälchen des Menschen und andere Säugetiere. Untersuchungen über Bau und Entwicklung der Niere. Jena.

Peters, W. H. K.
  1852. Naturwissenschaftliche Reise nach Mozambique. Berlin.

Petrov, O. V.
  1954. O stroenii volosyanogo pokrova gornostaya (On the structure of hair in ermine). Uch. Zap. LGU, 181, Ser. Biol., 38.

Petskoi, P. G., and V. M. Kolpovsky.
  1970. Sheinoe zhelezistoe obrazovanie u kunyikh (A cervical glandular structure in *Mustelidae*). Zool. Z., 69 (8): 1208-1219.

Petskoi, P. G., Yu. V. Pichugin, and A. N. Sinyavina.
  1971. O morfologii i funktsii fialkovoi zhelezy u pestsa (On morphology and function of the violet gland in polar fox). Trudy Kirovskogo Selkskokhozyaistvennogo Instituta, 28: 129-139.

Pichugin, Yu. V.
  1974. Adaptivnye cherty v strukture volosyanogo pokrova nekotorykh mlekopitajuschikh Evropeisko-Aziatskoi podoblasti Palearktiki (Adaptive features in the structure of hair in some mammals of the Eurasian subarea of the Palearctic). Bull. MOIP, 89 (3): 28-31.

Pichugin, Yu. V., and V. V. Sudakov.
  1973. K voprosu izuchenia sezonnoi izmenchivosti mekha krasnoi lisitsy (On seasonal variability of the fur of red fox). Sbornik Nauch. Tekhnich. Informatsii. VNII Okhotnych. Khoz. i Zverovodtsva. Kirov, 40-41: 131-133.

Pilters, H.
  1954. Untersuchungen über angeborene Verhaltensweisen bei Tylopoden, unter besonderer Berücksichtigung der neuweltlichen Formen. Z. Tierpsychol., 11.

Pinkus, F.
1927. Die normale Anatomie der Haut. *In* Jadassohn, Handbuch d. Haut und Geschlechtskir., 1 (1): 1-378.

Pinkus, H.
1939. Notes on the anatomy and pathology of the skin appendages. The wall of the intra-epidermal part of the sweat duct. J. Invest. Dermatol., 2: 175-186.

1952. Examination of the epidermis by the strip method. II. Biometric data on regeneration of the human epidermis. J. Invest. Dermatol., 19 (19): 431-446.

1954. Biology of epidermal cell. *In* Physiology and biochemistry of the skin (ed. S. Rothman). Chicago: University of Chicago Press, pp. 584-600.

1958. Embryology of hair. *In* Biology of hair growth (eds. W. Montagna and R. A. Ellis). New York: Academic Press, pp. 1-32.

Pocock, R. J.
1910. On the specialized cutaneous glands of ruminants. Proc. Zool. Soc., London: 840-986.

1918. On some external characters of ruminant Artiodactyla. II. The *Antilopinae, Rupicaprinae* and *Caprinae*. Ann. and Mag. Natur. Hist., Ser. 9 (2): 125-144.

1923. On the external characters of Elaphurus, Hydropotes, Pudu and other Cervidae. Proc. Zool. Soc., London: 181-207.

1924. The external characters of the *Bongolis manidae*. Proc. Zool. Soc. London, 2: 707-723.

1926*a*. The external characters of the flying lemur (*Galeopterus temminckii*). Proc. Zool. Soc., London: 429-444.

1926*b*. The external characters of an adult female China pangolin (*Macius pentadactyla*). Proc. Zool. Soc., London: 213-220.

Podyapolsky, H.
1933. Barsuk, ego zhizn' i khozyaistvennoe znachenie (Badger, its life and economic significance). Moscow: "Zhizn' i Znanie."

Pokrovskaya, E. I.
1957. Patogennoe deistvie ukusov polovozrelykh kleschei *Dermacentor marginatus* na khozyaina (Pathogenic effects of mature mite [*Dermacenta marminatus*] bites on host). Zool. Z., 36: 2.

Polkey, W., and W. N. Bonner.
1966. The pelage of the Ross seal. Br. Antarct. Surv. Bull., 8: 93-96.

Porta, A.
1910. Sulle glandule faciale del Vesperugo noctula (Schreb). Zoo. Anz., 36: 186-189.

Porter, K. R., and G. D. Pappas.
1959. Collagen formation by fibroblasts of the chick embryo dermis. J. Biophys. and Biochem. Cytol., 5: 153-166.

Poulton, E. B.
1894. The structure of the bill and hairs of *Ornithorhynchus paradoxus* with a discussion of the homologies and origin of mammalian hair. Quart. J. Microsc. Sci., 2 (37): 143-200.

Priselkova, D. O.
1958. O strukture i funktsii potovykh zhelez u nekotorykh vidov selskokhozya-

istvennykh zhivotnykh (On the structure and function of sweat glands in some species of farm animals). Bull. Nauch. Tehnicheskoi Informatsii VNII Zhivotnovodstva, 1 (5).

Pullianen, E., and P. Ovaskainen.
1975. Territory marking by a wolverine (*Gulo gulo*) in Northeastern Lapland. Ann. Zool. Fenn., 12 (4): 268-270.

Purves, P. E.
1963. Locomotion in whales. Nature, 197: 334-337.
1969. The structure of the flukes in relation to the laminar flow in Cetaceans. Säugetierkunde, 34 (1): 1-8.

Quay, W. B.
1954. The dorsal holocrine skin gland of the kangaroo rat (*Dipodomys*). Anat. Rec., 119: 161-176.
1955. Histology and cytochemistry of skin gland areas in the caribou, *Rangifer*. J. Mammal., 36 (2): 187-201.
1959. Microscopic structure and variation in the cutaneous glands of the deer, *Odocoileus virginianus*. J. Mammal., 40.
1965a. Integumentary modifications of North American desert rodents (eds. A. G. Lyne and B. F. Short). New York: American Elsevier Publishing, pp. 59-74.
1965b. Variations and taxonomic significance in the sebaceous caudal glands of pocket mice (*Rodentia: Heteromyidae*). Southwestern Nat., 10 (4): 282-287.
1965c. Comparative survey of the sebaceous and sudoriferous glands of the oral lips and angle in rodents. J. Mammal., 46 (1): 23-37.
1968. The specialized posterolateral sebaceous glandular regions in microtine rodents. J. Mammal, 49 (3): 427-445.
1970. Integument and derivatives *In* Biology of bats (ed. W. A. Wimsatt). New York: Academic Press.

Quay, W. B., and P. Q. Tomich.
1963. A specialized midventral sebaceous glandular area in *Rattus exulans*. J. Mammal., 44: 537-542.

Raevsky, V. V.
1947. Zhizn' kondo-sos'vinskogo sobolya (The life of Kondo-Sosva sable). Izdatel'stvo Glavnogo Upravlenia po Zapovednikam. Moscow.

Rahm, U.
1956. Notes on pangolins of the Ivory Coast. J. Mammal., 37: 531-537.

Rall, Yu. M.
1932. K metodike izucheniya mikroklimata gnezd suslika (On a method of recording the microclimate of souslik nests). Vestnik mikrobiologii, epidemiologii i parazitologii, 11: 1.
1939. Teplovye usloviya v norakh peschannykh gryzunov i metodika ikh izucheniya (Heat conditions in the burrows of sand rodents and methods of their recording). Zool. Zhurnal., 28: 1.

Ranvier, L.
1887. Le méchanism de la sécrétion. J. Microgr., 11: 1-146.

Rapp, W.
1839. Über ein eigentümliches drüsenartiges Organ des Hirsches. Müller's Archiv., 363.

Rashek, V. A.
  1972. Okraska, volosyanoi pokrov i lin'ka *Equus hemiones onager* (Coloration, hair and molt in *Equus hemiones onager*). Bull. MOIP, 77 (4): 43-59.

Rausch, R. L., and V. R. Rausch.
  1971. The somatic chromosomes of some North American marmots (*Sciuridae*) with remarks on the relationships of *Marmot broweri* Hall and Gilmore. Mammalia, 35: 85-101.

Raushenbakh, Ju. O.
  1950. O poznanii prirody ustoichivosti sel'skokhozyaistvennykh zhivotnykh k neblagopriyatnym faktoram sredy (Identification of the nature of resistance of farm animals to unfavorable environmental conditions). Izv. AN Uzb. SSR, 3.
  1952. O prirode ustoichivosti sel'skokhozyaistvennykh zhivotnykh k spetsificheskim faktoram sredy (On the nature of the resistance of farm animals to specific environmental factors). Trudy Sovesch. po Biol. Osnovam Povyshenia Produktivnosti Zhivotnykh.
  1956. Osobennosti termoregulyatsii u zhivotnykh pri vysokoi temperature sredy (Features of thermoregulation in animals under high environmental temperature). Thez. Dokl. Nauchn. Konf. Selskhoz. Vuzov po Fiziol. Zhivotn.
  1957. O strukturnykh osobennostyakh kozhi v svyazi s termoregulyatsiei u sel'skokhozaistvennykh zhivotnykh (On structural properties of the skin with special reference to thermoregulation in farm animals). Trudy Grodnenskogo Sel'skhozinstituta, 3.

Raushenbakh, Ju. O., and P. I. Erokhin.
  1967. Znachenie razlichnykh mekhanismov termoregulyatsii dlya teploustoichivosti krupnogo rogatogo skota (The significance of various mechanisms of thermoregulation for heat resistance in cattle). *In* Physiologo-geneticheskie issledovania adaptatsii u zhivotnykh (ed. A. D. Slonim). Leningrad: Nauka.

Rawles, M. E.
  1953. Origin of the mammalian pigment cell and its role in the pigmentation of hair. *In* Pigment cell growth, New York: Academic Press, pp. 1-16.

Reed, S. H., and J. M. Henderson.
  1940. Pigment cell migration in mouse epidermis. J. Exper. Zool., 85: 409-418.

Reeder, W. C., and R. B. Cowles.
  1951. Aspects of thermoregulation in bats. J. Mammal., 32: 389-403.

Richmond, M., and R. Stehn.
  1976. Olfaction and reproductive behavior in microtine rodents. *In* Mammalian Olfaction, Reproductive Processes and Behavior (ed. R. L. Doty). New York: Academic Press, pp. 197-218.

Riley, J. F.
  1959. The Mast Cells. London: E. and S. Livingston, Ltd.

Ring, J. R., and W. C. Randall.
  1947. The distribution and histological structure of sweat glands in the albino rat and their response to prolonged nervous stimulation. Anat. Rec., 99 (1): 7-20.

Ristland, N. A.
  1970. Temperature regulation of the polar bear (*Thalarctos maritimus*). Compar. Biochem. and Physiol., 37: 225-233.

Robertshaw, D.
  1971. The evolution of thermoregulatory sweating in man and animals. Int. J. Biometeor., 15 (4): 263-267.
Robertshaw, D., and C. R. Taylor.
  1969. Sweat gland function of the donkey (*Equus asinus*). J. Physiol., 205: 79-80.
Rogers, G. E.
  1953. The localization of dehydrogenase activity and sulfhydryl groups in wool and hair follicles by the use of tetrazolium salts. Quart. J. Microsc. Sci., 93: 253-268.
  1956. Electron microscopy of mast cells in the skin of young mice. Exper. Cell. Res., 11: 393-402.
  1957. Electron microscope observation on the glossy layer of hair follicles. Cell Res., 13: 521-528.
Romanenko, E. V., V. E. Sokolov, and N. M. Kalinichenko.
  1973. Gidrodinamicheskie osobennosti volosyanogo pokrova baikal'skogo tyulenya (*Phoca sibirica*) (Hydrodynamic properties of the hair of *Phoca sibirica*). Zool. Z., 52 (10): 1537-1542.
Ropartz, P. N.
  1977. Chemical signals in agonistic and social behaviour of rodents. *In* Chemical signals in vertebrates (eds. D. Müller-Schwarze and M. M. Mozell). New York and London: Plenum Press, pp. 169-184.
Rose, A. M.
  1948. The Negro in America. New York: Harper and Row.
Roth, S. I., and W. H. Clark, Jr.
  1964. Ultrastructure evidence related to the mechanism of keratin synthesis. *In* The epidermis (eds. W. Montagna and W. C. Lobitz). New York: Academic Press, pp. 303-337.
Rothman, S. T.
  1954. Physiology and biochemistry of the skin. Chicago: University of Chicago Press, pp. 1-741.
  1964. Keratinization in historical perspective. *In* The epidermis (eds. W. Montagna and W. C. Lobitz). New York: Academic Press.
Roux, W.
  1883. Beitrage zur Morphologie der functionellen Anpassung. Arch. Ant. und Entwicklungsgesch., 4: 76-162.
Rozhnov, V. V.
  1975. Agressivnoi i markirovochnoi povedenie laski v nevole (Aggressive and marking behavior of weasel in captivity). *In* Vtoraya Vsesojuznaya Konferentsia Molodykh Uchenykh po Voprosam Sravnitel'noi Morfologii i Ekologii Zhivotnykh (Tezisy Dokl.). Moscow: Nauka, pp. 154-155.
Rubner, M.
  1883. Einfluss der Körpergrösse auf den Stoff und Kraftwechsel. Zeitschr. für Biologie, 19: 35-545.
Rudall, K. M.
  1946. The structure of epidermal protein. *In* Proc. sympos. fibrous proteins (Soc. Colourists). University of Leeds, pp. 15-32.
Rumyantsev, A. V., F. I. Bezler, and V. P. Baturov.
  1933. Mikrotopografia svinoi shkury (Microtopography of pig hide). Sb. Trudov TSNIKP. Novye Vidy Kozhsyrya, 15.

Ryder, M. L.
1960. Study of the coat of the mouflon, *Ovis musimon,* with special reference to seasonal change. Proc. Zool. Soc. London, 135: 387-408.

Rzhanytsina, N. S.
1960. K sravnitel'noi topografii i neirogistologii kozhnogo pokrova (krupny rogaty skot, ovtsy, svynyi) (On comparative topography and neurohistology of skin of cattle, sheep, and swine). Kand. diss. Troitsk. Med. Inst.

Sadovskaya, N. P.
1961. Potovye zhelezy v kozhe svinei (Sweat glands in the skin of pigs). Trudy Sib. Nauch. Techn. Inst. Zhivotnov., 15.

Sale, J. B.
1970. Unusual external adaptations in the rock hyrax. Zool. Africana, 5 (1): 101-113.

Sandritter, W.
1953. Ultraviolet-mikrospektrometrische Untersuchungen an Plattenepithelien. Frank Z. Pathol., 64: 520-530.

Sansone, G., and J. G. Hamilton.
1969. Glyceryl ether, wax ester and triglyceride composition of the nouse preputial gland. Lipids, 4: 435-440.

Sapozhenkov, V., and V. E. Sokolov.
1960. Zimnyaya fauna v okrestnostyakh Repeteka (Winter fauna in the vicinity of Repetek). Vestn. MGU, 5.

Sasakawa, M.
1921. Beitrage zur Glykogenverteilung in der Haut unter normalen und pathologischen Zuständen. Arch. Dermat. Syph., 134: 418-443.

Satyukova, G. S.
1964. Makromikroskopicheskoe issledovanie lokal'nykh osobennostei krovenosnogo rusla kozhi (Macromicroscopic study of local peculiarities of skin blood vessels). Archiv. Anat. Gist. Embryol., 9.

Saunders, R. L., and C. H. De.
1961. Micrographic studies and vascular pattern in the skin. *In* Advances in biology of skin. II. Blood vessels and circulation (eds. W. Montagna and R. A. Ellis). New York: Pergamon Press.

Saunders, R. L., C. H. De, J. M. Lawrence, and D. A. Maciver.
1957. Microradiographic studies of the vascular pattern in muscles and skin. *In* X-ray microscopy and microradiography (eds. E. V. Coslett, A. Engström, and H. H. Patter, Jr.). New York: Academic Press, pp. 539-550.

Schaffer, J.
1924. Zur Einteilung der Hautdrusen. Anat. Anz., 53: 353-371.

1927. Die Drüsen I. Teil. *In* Handbüch der mikroskopischen Anatomie des Menschen (ed. W. von Möllendorf). Berlin: Springer, 2 (I): 132-148.

1930. Zur Phylogenese der Talgdrüsen. Z. microsc. anat. Forsch., 22: 579-590.

1933. Lehrbuch der Histologie und Histogenese. Leipzig: Engelmann.

1938. Über die Gesichtsdrusen der spätfliegenden Fledermaus (*Eptesicus serotinus* Schreb.). Z. mikrosk. anat. Forsch., 43: 332-333.

1940. Die Hautdrüsenorgane der Säugetiere mit besonderer Berücksichtigung ihres histologischen Aufbaues und Bemerkungen über die Proktodaäldrüsen. Münich: Urban und Schwarzenberg.

Scheffer, V. B.
1958. Seals, sea lions and walruses. Stanford: Stanford University Press.

1962. Pelage and surface topography of the northern fur seal. Washington. US Fish and Wildlife Service.

1964. Estimating abundance of pelage fibers on fur-seal skin. Proc. Zool. Soc. London, 143: 37-41.

Scheglova, A. I.

1952. Prisposoblenie vodnogo obmena u nekotorykh vidov souslika k usloviyam zhizni (Adaptation of water metabolism to life conditions in some suslik species). Dokl. AN SSSR, 83: 5.

Scherbakov, A. N.

1959. Kabarga i ee muskus (Musk deer and its musk). Uch. Zap. Khabarovsk. Gos. Ped. Inst., IV.

Schick, F.

1913. Über die Brunftfeige (Brunftdrüse) der Gamse. Zschr. Wiss. Zool., 104: 359-387.

Schieffedeker, P.

1917. Die Hautdrüsen des Menschen und Säugetiere. Biol. Zschr. 37: 11-12.

Schiller, S.

1959. Mucopolysaccharides of the estrogen-stimulated chick oviduct. Biochem. et Biophys. Acta, 32: 315-319.

Schiller, S., and A. Dorfman.

1957. The metabolism of mucopolysaccharides in animals. IV. The influence of insulin. J. Biol. Chem., 227: 625-632.

Schliemann, H.

1970. Bau und Funktion der Hafforgane von Chiroptera und Myzopoda (*Vespertilionidea, Microchiroptera, Mammalia*). Zschr. Wiss. Zool., 181 (3-4): 353-400.

Schmidt-Nielsen, B.

1950a. Kidney efficiency and reduced evaporation in the desert rodents, *Heteromyidae*. 18th Intern. Physiol. Congr. Abstr. Communs. Copenhagen, 453-463.

1950b. Kidney physiology of the desert rat (*Diplodomys merriami*). Federta. Proc., 9: 113.

Schmidt-Nielsen, B., and K. Schmidt-Nielsen.

1950. Evaporative water loss in desert rodents in their natural habitats. Ecology, 31 (1).

Schmidt-Nielsen, B., K. Schmidt-Nielsen, T. R. Haupt, and S. A. Jarnum.

1956. Water balance of the camel. Amer. J. Physiol., 185: 1.

Schmidt-Nielsen, K.

1964. Desert animals. Oxford: Clarendon Press.

Schmidt-Nielsen, K., and B. Schmidt-Nielsen.

1951. A complete account of the water metabolism in kangaroo rats, an experimental verification. J. Cell and Compar. Physiol., 38 (2): 165-181.

1952. Water metabolism of desert mammals. Physiol. Revs., 31: 2.

Schmidt-Nielsen, K., B. Schmidt-Nielsen, and A. Brohaw.

1948. Urea excretion in desert rodents exposed to high protein diets. J. Cell. and Compar. Physiol., 32 (3): 361-380.

Schmidt-Nielsen, K., B. Schmidt-Nielsen, S. A. Jarnum, and T. A. Houpt.

1957a. Body temperature of the camel and its relation to the water economy. Am. J. Physiol., 188: 103-112.

1957b. Urea excretion in the camel. Amer. J. Physiol., 188: 3.

Schmitt, F. O.
   1959. Interaction properties of elongate protein macromolecules with particular reference to collagen (Tropocollagen). *In* Biophysical sciences. A study program (ed. J. L. Oncley). New York: Wiley, pp. 349-358.
Schneider, H.
   1930. Die Sinushaare der grossen Hufeisennase Rhinolophus ferrumequinum (Schreber, 1774). Z. Säugetierkunde, 28 (H. 6): 342-349.
Scholander, P. F.
   1940. Experimental investigation of the respiratory function in diving mammals and birds. Hvaldrets skr., 22: 1-131.
Scholander, P. F., and W. F. Schevill.
   1955. Countercurrent vascular heat exchange in the fins of whales. J. Appl. Physiol., 8 (3): 279-282.
Scholander, P. F., V. Walters, R. Hock, and L. Irving.
   1950. Insulation of some arctic and tropical mammals and birds. Biol. Bull. 99: 225-236.
Schubarth, H.
   1963. Zur makroskopische und mikroscopischei Anatomie der Rückendrüse vom Halsbandpekari (Dicotyles torquatus Cuvier, 1817). Zool. Jahrb., 80 (4): abstr. 2.
Schultze-Westrum.
   1965. Innerartliche Verständigung durch Düfte beim Gleit beutler Petaurus breviceps papuanus Thomas (Marcupialia, Phalangeridae). Z. vergl. Physiol., 50: 151-220.
Segal, A. N., and Yu. V. Ignatov.
   1975. Sokhranenie tepla i teplootdacha s poverkhnosti tela u amerikanskoi norki (Mustela vison) (Retaining of heat and heat irradiation in mink). Zool. Z. (54) 11: 1678-1687.
Selby, C. C.
   1955. An electron microscopic study of the epidermis of mammalian skin in the sections. I. Dermo-epidermal junction and basal cell layer. J. Biophys. and Biochem. Cytology, 1: 249-444.
Semenov, P. A.
   1975. *Clethrionomys rufocanus* na polyarnom urale (*Clethrionomys rufocanus* in the polar Urals). Dissertatsia na soiskanie uchenoi step. kand. biol. nauk., Sverdlovsk.
Serri, F.
   1955. Note de enzimologia cutanea. II. Ricerche bio-ed. istochimiche sull'attivita succinodeidrasica della cute umana normale. Bol. Soc. Med. Chir. Pavia, 69: 3-12.
Setoguti, T., and F. Sakuma.
   1959. On the density of hair and hair groups in *Macacus cyclopsis*. Okajimas Folia Anat. Japon, 33 (2-3): 150-170.
Shelley, W. B., S. B. Cohen, and G. B. Koelle.
   1955. Histochemical demonstration of monoamine in human skin. J. Invest. Dermatol., 24: 561-565.
Shepeleva, V. K.
   1971. Prisposoblenie tiulenei k obitaniyu v Arktike (Adaptation of seals to the Arctic environment). Novosibirsk: Nauka.

Shipman, M., H. B. Chase, and W. Montagna.
　1955.　Glycogen in skin of the mouse during cycles of hair growth. Proc. Soc. Exper. Biol. and Med., 88: 449-451.

Shööl, T.
　1871.　Die Flughaut der Fledermous, namentlich die Endigung ihren Nerven. Arch. Mikrosk. Anat., 7: 1-31.

Shubnikova, E. A.
　1967.　Tsitologia i tsitofiziologia sekretornogo protsessa (Cytology and ctyophysiology of the secretory process). Izd. Mosk. U-ta.

Siedamgrotsky.
　1875.　Über die Altereiniger Haustiere vorkomenden Drüsen. A. Wiss. Prakt. Tierheilunde, 1: 438-449.

Sisk, M. O.
　1957.　A study of the sudoriferous glands in the little brown bat, *Myotis lucifugus lucifugus*. J. Morphol., 101 (1-2): 425-458.

Skurat, L. N.
　1969.　Stroenie kresttsovykh zhelez u nekotorykh polevok (*Microtinae*) (The structure of sacral glands in some voles (*Microtinae*). Zool. Z., 48: 8.
　1972.　Stroenie i znachenie spetsificheskykh kozhnykh zhelez gryzunov (The structure and significance of specific skin glands of rodents). Cand. diss., MGU., M.

Sleptsov, M. M.
　1955.　Kitoobraznye dal'nevostochnykh morei. Vladivostok (Cetacea of Far Eastern seas).

Sludsky, A. A.
　1963.　Dzuty v evropeiskikh stepyakh i pustynyakh (Jutes in European steppes and deserts). Trudy In-ta Zool. AN Kaz. SSR, XX.
　1975.　Osobennosti povedenia kopytnykh zverei aridnykh i poluaridnykh zon (Behavioral characteristics of ungulates in aird and semiarid regions). *In* Kopytnye Fauny SSSR. Ed. V. E. Sokolov. Moscow: Nauka.

Smith, D. E., and Y. S. Lewis.
　1957.　Electron microscopy of the tissue mast cells. J. Biophys. and Biochem. Cytol., 3: 9-14.

Smith, F.
　1889.　The histology of the skin of the elephant. J. Anat. and Phys., 24: 493-503.

Smith, H. H.
　1933.　The relationships of the medullar and cuticular scales of the hair shafts of the *Soricidae*. J. Morphol., 55 (I): 137-149.

Snell, R.
　1965.　An electron microscopic study of the human epidermal keratinocyte. Z. Zellforsch., 79: 4.

Sobakina, A. I.
　1961.　Nekotorye dannye o kolichestve volos severnogo olenya (Some data on the number of hairs of the reindeer). Uch. Zap. Yakutsk. Un-ta, 11.

Sokolov, A. S.
　1959.　Osobennosti kalibra i raspredelenia nekotorykh krovenosnykh sosudov u vodnykh (tyulen') i nazemnykh (sobaka) mlekopitajuschikh (The peculiarities of the calibers and distribution of some blood vessels in aquatic [seal] and terrestrial [dog] mammals), 37: 12.

Sokolov, I. I.
1959. Kopytnie zveri. Fauna SSSR. Mlekopitajuschie I, Vyp. 3 (Ungulates. Fauna of the USSR. Mammals, I: 3). Izd. AN SSSR, M.-L.

Sokolov, V. E.
1953. Struktura kozhnogo pokrova morskikh mlekopitayuschikh (Skin structure of marine mammals). Candidate dissertation, summary p. 14. Moscow: Moscow Fur Institute.
1955. Struktura kozhnogo pokrova nekotorykh kitoobraznykh (The structure of the skin of some *Cetacea*). Bull. MOIP, 60: 6.
1959a. Struktura kozhnogo pokrova lastonogikh (Soobschenie. I) (The structure of the skin of *Pinnipedia*. I). Bull. MOIP, Otd. Biol., 44: 6.
1959b. Adaptations of the skin in marine mammal fauna of the USSR to some conditions of aquatic life. Proc. XV Intern. Congr. Zool., pp. 277-280.
1960a. The skin structure in *Pinnipedia* of the USSR fauna. J. Morphol., 107: 3.
1960b. Some similarities and dissimilarities in the structure of the skin among members of the suborder *Odontoceti* and *Mystacoceti*. Nature, 185: 4715.
1960c. Struktura kozhnogo pokrova lastonogikh (soobschenie II) (Skin structure of Pinnipedia. II). Bull. MOIP. Otd. Biol., 45: 4.
1960d. Strukturnie osobennosti kozhnykh pokrovov usatikh kitov, povyshaiuschie uprugost' kozhi (Structural peculiarities of the skin of Mysticeti that increase skin resilience). Zool. Z., 33: 2.
1961a. Stroenie kozhi saigaka (Skin structure of saiga). Ed. A. G. Bannikov, et al. Biology of Saiga. Izd. Sel'skhoz. Lit., M.
1961b. Stroenie i prichiny vozniknovenia kozhnykh narostov u japonskikh kitov (Structure and origin of skin "bonnets" in *Eubalaena glacialis*). Zool. Z., 40: 9.
1962a. The structure and seasonal variability of skin in aurochs (*Bison bonasus* L.). Acta Biol. Cracoviensia. Ser. Zool., 5.
1962b. Skin adaptations of some rodents to life in the desert. Nature, 193: 4818.
1962c. Adaptations of mammal skin to the aquatic mode of life. Nature, 195: 464.
1962d. Struktura kozhnogo pokrova nekotorykh kitoobraznykh. Soobschenie II. (Skin structure of some *Cetacea*. II.). Nauch. Dokl. Vysh. Shkoly. Biol. Nauki, 3.
1963. Skin structure in wild *Artiodactyla* of the USSR fauna. Proc. XVI Intern. Congr. Zool.
1964a. Stroenie kozhnogo pokrova i yazika losei (Skin and tongue structure of elk). *In* Biologia i promysel losya (ed. A. S. Bannikov). Sb. I. Rosselkhozizdat., M.
1964b. Ekologicheskaya morfologia kozhnogo pokrova mlekopitajuschikh (Ecological morphology of mammalian integument). Dokt. diss., MGU, M.
1964c. Stroenie kozhnogo pokrova trekh podvidov gornogo baraba (Skin structure of three subspecies of mountain sheep). Tezisy dokl. sovesch. po vnutrividovoi izmenchivosti i mikroevolyutsii, M.
1964d. Morfologia kozhi kulana (*Equus hemionus* Pallas) (Skin morphology of *Equus hemionus* Pall.). Izv. AN Turkm. SSR. Ser. Biol. Nauk, 6.
1970. Osobennosti stroenia kozhnogo pokrova podzemnykh mlekopitajuschikh (Peculiarities of the skin structure of subterranean mammals). Nauch. Dokl. Vysh. Shk., 8.

1971.   Stroenie kozhnogo pokrova nekotorykh kitoobraznykh (Skin structure of some *Cetacea*. III). *In* Morfologia i ekologia morskikh mlekopitajuschikh. Ed. V. E. Sokolov. Moscow: Nauka.

1974.   Struktura kozhnogo pokrova rechnykh del'finov (*Familia Platanistidae*) (Skin structure of freshwater dolphins (*Platanistidae*). *In* Morfologia, ekologia i akustika morskikh mlekopitajuschikh. Ed. V. E. Sokolov. Moscow: Nauka, pp. 79-86.

1975.   Khemokommunikatsia mlekopitajuschikh (Chemocommunication of mammals). Vestnik AN SSSR, 2: 44-54.

1977.   Khimicheskaya kommunikatsia mlekopitajuschikh (Chemical communication of mammals). *In* Uspekhi sovremmennoi teriologii. Ed. V. E. Sokolov. Moscow: Nauka.

Sokolov, V. E., I. Bulina, and V. Rodionov.

1969.   Interaction of dolphin epidermis with flow boundary layer. Nature, 222 (5190): 267-268.

Sokolov, V. E., O. F. Chernova, E. P. Zinkevich, and G. V. Khakin.

1977.   Spetsificheskaya podkhvostovaya zheleza vykhukholi (*Desmana moschata*) (Specific subcaudal gland of desman [*Desmana moschata*]). Zool. Z., 56 (3): 250-256.

Sokolov, V. E., and A. A. Danilkin.

1975.   Markirovka territorii samtsami sibirskoi kosuli (Marking of the territory by Siberian roe males). *In* Kopytnye fauni SSSR. Ed. V. E. Sokolov. Moscow: Nauka.

1976.   Kozhny shchit u samtsov sibirskoi kosuli (*Capreolus capreolus pygargus* Pall.) (Skin shield in Siberian roe males [*Capreolus capreolus pygargus* Pall.]). (In press).

1977.   Mechenie territorii samtsami sibirskoi kosuli (Marking of the territory by Siberian roe males). Zool. Z., 56 (10): 1543-1556.

Sokolov, V. E., S. I. Isaev, A. S. Orlov, and G. K. Pashkova.

1977.   Markirovochnoe povedenie bol'shikh peschanok v model'nykh uslovijakh laboratorii (Marking behavior of the great gerbil under model laboratory conditions). *In* Ekologia i meditsinskoe znachenie peschanok fauny SSSR. Ed. V. E. Sokolov. Moscow, pp. 213-216.

Sokolov, V., and M. Kalashnikova.

1971.   The ultrastructure of epidermal cells in *Phocaena phocaena*. *In* Investigations of *Cetacea*. Ed. G. Pillerri. Vol. III, Pt. 2: 194-199.

Sokolov, V. E., M. M. Kalashnikova, and V. A. Rodionov.

1971.   Mikro- i ultrastruktura kozhi morskoi svinji (Micro- and ultrastructure of the skin of *Phocaena phocaena*). *In* Morfologia i ekologia morskikh mlekopitajuschikh. Ed. V. E. Sokolov. Moscow: Nauka.

Sokolov, V. E., and G. V. Kuznetsov.

1966.   Napravlenie dermal'nykh valikov v kozhe del'finov v svyazi s osobennostju obtekania tela vodoi (The direction of dermal papillae in the skin of dolphins with special reference to features of the water flow along the dolphin body). *In* Tez. dokl. III Vses. sovesch. po Izucheniu morskih mlekopitajuschikh. Vladivostok: Nauka.

Sokolov, V. E., G. V. Kuznetsov, and V. A. Rodionov.

1968.   Napravlenie dermal'nykh valikov v kozhe del'finov v svyazi s osobennostju obtekania tela vodoi (The direction of dermal papillae in the skin of dolphins with special reference to features of the water flow along the dolphin body). Bull. MOIP, Otd. Biol., 73: 3.

Sokolov, V. E., K. L. Lyapunova, and I. M. Khorlina.
1977. Osobennosti povedenia serykh krys (Peculiarities of the behavior of gray rats). *In* Povedenie mlekopitajuschikh. Ed. V. E. Sokolov, pp. 84-106.

Sokolov, V. E., and E. I. Naumova.
1962. K biologii tonkopalogo suslika (Concerning the biology of long-clawed ground squirrel). Izv. AN Turk. ASSR. Ser. Biol., 2.

Sokolov, V. E., and V. F. Polozhinin.
1964. Mikroklimat gnezd ptits i nor gryzunov v pustyne (Microclimate of bird nests and rodent burrows in the desert). Nauchn. dokl. vysh. shkoly. Biol. Nauki, I.

Sokolov, V. E., and L. N. Skurat.
1966. Spetsificheskaya srednebrjushnaya zheleza peschanok (Specific midventral gland in gerbils). Zool. Z., XIV: 11.
1969. Rost i razvitie bol'shoi peschanki (*Rhombomys opimus*) (The growth and development of great gerbil [*Rhombomys opimus*]). Zool. Z., 48: 12.
1975. Spetsificheskie zhelezy serykh polevok (*Microtus*) fauny SSSR (Specific glands of the common vole [*Microtus*] of the USSR fauna). Zool. Z., 65 (7): 1066-1075.

Sokolov, V. E., O. V. Smirnova, N. B. Chernysheva, and S. B. Nikitina.
1972. Exocrinovye zhelezy i ikh rol' v zhiznedeyatel'nosti bolshoi peschanki (Exocrine glands and their role in the life of *R. opimus*). *In* Povedenie zhivotnykh. I. Vses. soveschanie. Ref. dokl. Moscow: Nauka.

Sokolov, V. E., and E. B. Sumina.
1962. Stroenie kozhnogo pokrova nasekomojadnykh (Skin structure of the *Insectivora*). *In* II. Zool. konf. Litovskoi SSR, Izd. AN Lit. SSR. Vilnius.
1974. Gistostruktura kozhnogo pokrova kaspijskogo tjulenya (The histostructure of the skin of Caspian seal). *In* Morfologia, fiziologia i akustika morskikh mlekopitajuschikh. Ed. V. E. Sokolov. Moscow: Nauka.

Sokolov, V. E., E. B. Sumina, and O. N. Dyachkova.
1977. Nekotorie osobennosti kozhnogo pokrova baikal'skoi nerpi, svyazannye so smenoi embrional'nykh volos (Some features of the skin of the Baikal ringed seal associated with replacement of embryonic hair). (In press).

Sokolov, V. E., and N. S. Varzhenevsky.
1962. Novy pribor dlya distantsionnogo izmerenia vlazhnosti v norakh zverei i gnezdakh ptits (A new instrument for remote measurement of humidity in animal burrows and bird nests). Bull. MOIP. ser. Biol., 67: 1.

Solger, B.
1878. Schweissdrüsenlager beim Reh. Zool. Anz., 1: 174-176.

Spalteholz, W.
1893. Die Verteilung der Blaugefrässe in der Haut. Arch. Anat. und Etwicklungsgesch, pp. 1-54.
1927. Blutgefässe in der Haut. *In* Handbüch der Haut und Geschlechtskrankheiten (ed. Jadassohn). Berlin: J. Springer.

Spannholf, I.
1960. Untersuchungen zur Histophysiologie und Funktion Analbeutels vom Rotfuchs (*Vulpes vulpes* L.). Forma et Functio, 1: 26-45.

Spearman, R. I. C.
1966. The keratinization of epidermal scales, feathers and hairs. Biol. Rev., 41 (1): 59-96.
1968. A histochemical examination of the epidermis of the southern elephant seal (*Mirounga leonina* L.) during the telogen stage of hair growth. Austr. J. Zool., 16: 17-26.
1970. The epidermis and its keratinization in the African elephant (*Loxodonta africana*). Zool. Africana, 5 (2): 327-338.
1972. The epidermal stratum corneum of the whale. J. Anat., 113 (3): 373-381.
1973. The integument. Cambridge: Cambridge University Press.
Spener, F., M. K. Mangold, G. Sansone, and J. G. Flomilton.
1969. Long chain alkyl acetates in preputial gland of the mouse. Biochem. et Biophys. Acta., 192: 516-521.
Sperber, I.
1944. Studies on the mammalian kidney. Zool. Bidrag. Uppsala, 22.
Sperling, F., and T. Koppanyi.
1949. Histophysiologic studies on sweating. Amer. J. Anat., 84: 335-363.
Starck, D.
1958. Beitrag zur Kenntnis der Armatschen und anderer Hautdrüsenorgane von Saccopteryx bilineata Temminck, 1838. (*Chiroptera, Emballonuridae*). Morphol. Jahrb., 99: 3-25.
Starett, A., and P. Starett.
1955. Observations on young block fish *Globicephala*. J. Mammal., 36: 3.
Staricco, R., and H. Pinkus.
1957. Quantitative and qualitative data on the pigment cells of adult human epidermis. Invest. dermatol., 24: 816-823.
Stearns, M. L.
1940a. Studies on the development of connective tissue in transparent chambers in the rabbit's ear. I. Amer. J. Anat., 66: 133-176.
1940b. Studies on the development of connective tissue in transparent chambers in the rabbit's ear. II. Amer. J. Anat., 67: 55-97.
Steigleder, G. K.
1957. Die Histochemie der Epidermis und ihrer Anhangsgebilde. Arch. klin. und exper. Dermatol., 206: 276-317.
Stein, G. H.
1954. Materialen zum Haarwechsel deutscher Insectivoren. Zool. Mus. Berlin, 30 (1): 12-34.
Stepanyan, B. G.
1969. Ultrastruktura i protsess keratinizatsii epidermisa. Arch. Anat., 56-57: 12.
Stevens, P. G.
1945. American musk. IV. On the biological origin of animal musk. Two large ring ketones from the muskrat. J. Amer. Chem. Soc., 67: 907.
Stevens, P. G., and J. L. E. Erikson.
1942. American musk. I. The chemical constitution of the musk of the Louisiana muskrat. J. Amer. Chem. Soc., 64: 144.
Stoddart, D. M.
1972. The lateral scent organs of *Arvicola terrestris* (*Rodentia, Microtinae*). J. Zool. London, 166: 49-54.

1976. Effect of the odor of weasels (*Mustela nivalis* L.) on trapped samples of their prey. Oecologia, 22: 439-441.

Strauss, W. L., Jr.

1942. The structure of the crown-pad of the gorilla and of the cheek-pad of the orangutan. J. Mammal., 23: 276-278.

1950. The microscopic anatomy of the skin of the gorilla. *In* The anatomy of the gorilla. Ed. W. K. Gregory. New York: Columbia University Press, pp. 213-226.

Strelnikov, I. D.

1932. Izuchenie mikroklimata v norakh gryzunov (A study of microclimate in burrows and nests of rodents). Trudy Vsesoyuznogo Instituta Zaschity Rastenyi, 4.

1955. Mikroklimat nor i gnezd gryzunov (The microclimate of rodent burrows and nests). Zapiski Leningradskogo Selskokhozysistvennogo Instituta, p. 9.

Stroganov, S. U.

1957. Zveri Sibiri. Nasekomoyadnie (Wild animals of Siberia. *Insectivora*). Izd. AN SSSR. M.

Stubbe, M.

1969. Die analen Markierungsorgane der Martes-Arten. Acta Theriologica, XIV (22): 303-312.

1970. Zur Evolution der analen Markierungsorgane bei Musteliden. Biologisches Zentralblatt, 89 (2): 213-223.

1971. Die analen Markierungsorgane des Dachses, *Meles meles* (L.). Zool. Garten N. F. Leipzig, 40 (3): 125-135.

Stupel, H., and A. Szakall.

1957. Die Wirkung von Waschmitteln auf die Haut. Heidelberg, Huthig.

Surkina, R. M.

1968. Stroenie soedinitel'notkannogo ostova kozhi delfina (The structure of the connective tissue framework of dolphin skin). *In* Mekhanizmy peredvizhenia i orientatsii zhivotnykh. Kiev: Naukova Dumka.

1971a. O stroenii i funktsii kozhnoi muskulatury delfinov (On the structure and function of the skin musculature in dolphins). *In* Bionika. Kiev: Naukova Dumka.

1971b. Raspolozhenie dermalnykh valikov na tele delfina-belobochki (The disposition of dermal ridges on the body of the white-sided dolphin). *In* Bionika. Kiev: Naukova Dumka.

1973. Stroenie kozhi del'finov i ee prisposobitel'nie osobennosti (The structure of dolphin skin and its adaptive features). Dissertatsia na Soisk. Uch. Step. Kand. Biol. Nauk. Odessa.

Sylvén, B.

1940. Studies on the liberation of sulfuric acid from the granules of the mast cells on the subcutaneous connective tissue after exposure to roentgen and gamma rays. Acta Radiol. 21: 206-212.

Szabó, G.

1954. The number of melanocytes in human epidermis. Brit. Med. J., I: 1016-1017.

1958. The regional frequency and distribution of hair follicles in human skin. *In*

The biology of hair growth. Eds. W. Montagna and R. A. Ellis. New York: Academic Press.

Szakall, A.
1955. Über die Eigenschaften, Herkunft und physiologischen Funktionen der die H-Ionenkonzentration bestimmenden Wirkstoffe in der verhortnen Epidermis. Arch. klin. und Exper. Dermatol., 201-231.

Szymonowicz, L.
1909. Über die Nervenendigungen in den Haaren des Menschen. Arch. Mikrosk. Anat., 74: 622-634.

Takagi, S.
1952. A study on the structure of the sudoriferous duct transversing the epidermis with fresh material by phase contrast microscopy. Japan J. Physiol., 3: 65-72.

Takeuchi, T.
1958. Histochemical demonstration of branching enzyme (amylo-1, -4-, -1, 6-transglucosidase) in animal tissues. J. Histochem. and Cytochem., 6: 208-216.

Tamponi, M.
1940. Structure Nervose della Cute Umana. Edittori Capelli I. Bologna.

Tangl, H.
1956. Die Wärmeregulierende Fähigkeit der Haustiere. Tierzucht, 10 (12): 410.

Tarasov, P. P.
1960. O biologicheskom znachenii pakhuchikh zhelez u mlekopitajuschikh (On the biological significance of odorous glands in mammals). Zool. Z., 39 (7): 1062-1068.

Taylor, C. R., D. Robertshaw, and R. Hofmann.
1969. Thermal panting: a comparison of wildebeest and zebu cattle. Am. J. Physiol., 11 (217): 907-910.

Tekhver, Yu. T.
1971. Gistologia kozhnogo pokrova domashnikh zhivotnikh (Histology of skin of domestic animals). Estonsk. Selkhoz. Akad. Tartu.

Tennent, D. M.
1946. A study of the water losses through the skin in the rat. Amer. J. Physiol., 145: 436-440.

Thiessen, D., and P. Yahr.
1977. The gerbil in behavioural investigation. Austin and London: University of Texas Press.

Thomson, M. L.
1954. A comparison between the number and distribution of functional eccrine sweat glands in Europeans and Africans. J. Physiol., 123: 225-233.

Tinyakov, G. C., V. P. Chumakov, and B. A. Sevastyanov.
1973. Nekotoroye osobennosti v mikrostrukture kozhi kitoobraznykh (Some peculiarities of the microstructure of Cetacean skin). Zool. Z. 52 (3): 399-407.

Toldt, K.
1906. Interessante Harrformen bei einem Kurschnabeligen Ameisenigel. Zool. Anz., 30: 305-319.
1928. Über die Zeithaare und den Haarkleides von Talpa europaea L., Z. Morphol. und Ökol. Tiere, 12.

Tomilin, A. G.
1946. Termoreguljatsia i geograficheskie rasy kitoobraznykh (Thermoregulation and geographical races of *Cetacea*). Dokl. AN SSSR. Novaya Seriya, 55: 57.
1947. Problemy ekologii kitoobraznykh (Problems of Cetacean ecology). Doct. diss. M.
1948. K biologii i fiziologii chernomorskih del'finov (On biology and physiology of Black Sea dolphins). Zool. Z., 27: 1.
1951. O termoregulyatsii u kitoobraznykh (On thermoregulation in *Cetacea*). Priroda, 6.
1957. Zveri SSSR i Prilegajuschikh Stran. IX. Kitoobraznye (The Mammals of the USSR and Contiguous Countries. IX. *Cetacea*). Izd. AN SSSR.

Troll-Öbergfell, B.
1930. Über einige Besonderheit der Haare des gemeinen Seehundes (*Phoca vitulina* L.). Anat. Anz., 69 (1-3): 404-415.

Tsalkin, V. I.
1938. Morfologicheskaya kharacteristika, sistematicheskoe polozhenie i zoogeograficeskoe znachenie morskoi svinji Azovskogo i Chernogo morei (Morphological features, systematic position, and zoogeographical significance of *Ph. phocoena* of the Azov and Black Seas). Zool. Z., 17: 4.

Tserevitinov, B. F., I. K. Kirsanov, and V. V. Bigman.
1948. Tovarovedenie zhivotnogo syrya (Commercial science of animal materials). Moscow.
1951. Differentsirovka volosyanogo pokrova pushnykh zverei (Differentiation of the hair of furbearing mammals). Trudy Vses. Inst. Okhot. Promysla. X. M. Gos. Izd. Techn. i Ekonom. Literatury po Voprosam Zagotovok.
1958. Topograficheskie osobennosti volosyanogo pokrova pushnykh zverei (Topographical peculiarities of the pelage of furbearing mammals). Trudy vses. inst. zhivotnogo syrya i pushniny, XVII.
1969. Stroenie volosyanogo pokrova dikikh koshek (The hair structure of wild cats). Tr. Mosk. Veter. Akad., 55.

Tsuchiya, K.
1954. Über die ekkrine Schweissdrüse des menschlichen Embryos, mit besonderer Berücksichtigung ihrer Histo- und Cyto-genese. Arch. Histol. Japon, 6: 403-432.

Tsukagoshi, N.
1951. Zur Zytologie der ekkrinen Schweissdrüsen der Tiere, mit besonderer Berücksichtigung des Vorkommens der zwei Arten Drüsenzellen und ihrer apokrinen Sekretion. Arch. Histol. Japon., 2: 481-497.

Twitty, V. C.
1949. Developmental analysis of amphibian pigmentation. Growth, 9: 133-161.

Urbach, E.
1928. Beiträge zu einer physiologischen und pathologischen chemie der Haut. II. Der Wasser; Kochsalz-, Reststickstoff-, und Fettgehalt der Haut in der Norm und unter pathologischen Verhältmissen. Arch. für Dermat. u. Syph., 156: 73-101.

Ustinov, S. K.
1967. Kabarozhya struja (Musk deer spray). Okhota i Okhotnich'e Khozyaistvo, 3: 9.

Utkin, L. G.
1969. Gistogenez kozhi i ee proizvodnykh u norok (Histogenesis of skin and its derivatives in mink). Nauchn. Trudy Nauchn. Issled. inst. pushnogo zverovodstva i krolikovodstva, VIII: 98-112.

Van Utrecht, W. L.
1958. Temperaturregulierende Gefassysteme in der Haut und anderen epidermalen Strukturen bei Cetacean. Zool. Anz., 161 (3-4): 77-82.

Van Vleck, D. B.
1965. The anatomy of the nasal rays of *Condylura cristata*. J. Mammal., 46 (2): 248-253.

Veinger, G. M.
1974. Morfologicheskaya kharakteristika kozhnogo pokrova i izmenchivosti okraski kashalota (*Physeter macrocephalus*) kak populyatsionnyi priznak (Morphological characterization of skin and variability of coloration in *Physeter macrocephalus* as a populational character). Dis. na soiskanie uchenoi stepeni kand. biol. nauk., Vladivostok.

Venchikov, A. I.
1934. Potootdelenie v gornykh klimaticheskikh uslovijakh (Sweating in mountain climates). Trudy sredneaziatskogo gos. universiteta. Ser. 18: 1.

Volzhina, N. S.
1951. Izmenchivost' volosyanogo pokrova letuchikh sobak i nekotorykh nasekomoyadnykh rukokrylykh (Variability of the pelage of kalongs and some insectivorous *Chiroptera*). Bull. MOIP. Ser. Biol., 56 (4): 21-30.

Vorontzov, N. N., and N. N. Gurtovoi.
1959. Stroenie srednebryushnoi zhelezy nastoyashchikh khomyakov *Cricetus, Cricetinae, Rodentia, Mammalia* (The midventral gland structure of true hamsters, *Cricetus, Cricetinae, Rodentia, Mammalia*). Dokl. AN SSSR, 125: 673-676.

Vrieš, V.
1930. Glandular organs on the flanks of the water rat, their development and changes during breeding season. Biol. Spisy vyx. Šk. vzěrolék. Brno, 4.

Vshivtsev, V. P.
1972. Vydra Sakhalina (The Otter of Sakhalin). Novosibirsk: Nauka.

Wagner, R.
1834. Lehrbuch der Zootomie, T. 1. Anatomie der Wirbeltiere. Leipzig.

Wallin, L.
1967. The dorsal skin gland of the Norwegian lemming, *Lemmus l. lemmus* (L.). Zeitschrift für Morphologie und Ökologie der Tiere, 59 (1): 83-90.

Walther, F.
1958. Zum Kampf- und Paarungsverhalten einiger Antilopen. Z. Tierpsychol., 15: 340-380.
1963. Einige Verhaltensbeobachtungen am Dibatog (*Ammodorcas clarkei,* Thomas, 1891). Zool. Garten (NF), 27: 233-261.

Weber, M.
1886a. Studien über Säugetiere. Ein Beitrag zur Frage nach dem Ursprung der Cetaceen. Jena, 1-252.
1886b. Der Brauchligbaum der Cetacean. Vergleichen mit den übriegen Mammalia. *In* Studien über Säugetiere. Jena, 78-87.

1894. Beitrag zur Anatomie und Entwicklung des Genus Manis. *In* Zoologisches Ergebnisse der Reise Niederl. Ostindien, Bd. 2, Leiden, 1-116.

1927-28. Die Säugetiere. Bd. 2, Jena: Gustav Fischer.

Weddel, G., G. W. Pallie, and E. Palmer.

1954. The morphology of peripheral nerve termination in the skin. Quart. J. Microsc. Sci., 95: 483-501.

1955. Nerve endings in mammalian skin. Biol. Revs., 30 (2): 159-193.

Weddle, F.

1968. The wolverine: the problems of a wilderness outcast. Defenders Wildlife News, 43 (2): 156-168.

Weiner, J. S., and K. Hellmann.

1960. The sweat glands. Biol. Revs., 35 (2): 141-186.

Werner, H. J., and W. W. Dalquest.

1952. Facial glands of the tree bats *Lasiurus* and *Dasypterus*. J. Mammal., 33: 77-80.

Werner, H. J., W. W. Dalquest, and J. H. Roberts.

1950. Histological aspects of the glands of the bat, *Tadarida cynocephala* (le Conte), 31: 395-399.

Werner, H. J., and D. M. Lay.

1963. Morphologic aspects of the chest gland of the bat, *Molossus rufus*. J. Mammal., 44: 552-555.

Wilcox, H. H.

1950. Histology of the skin and hair of the adult chinchilla. Anat. Rec., 108 (3): 385-399.

Wildman, A. B., and J. Manby.

1938. The hairs of *Monotremata,* with special reference to their cuticular-scale pattern. Trans. Roy. Soc. Edinburgh, 59 (11): 333-347.

Wilgram, G., J. B. Caufield, and E. B. Madgic.

1964. A possible role of the desmosomes in the process of keratinization. *In* The epidermis. Eds. W. Montagna and W. Lobitz. New York and London: Academic Press, pp. 275-301.

Winkelmann, R. K.

1960. Similarities in cutaneous nerve-end organs. Advances in Biology of Skin. VI. Cutaneous Innervation. (ed. W. Montagna). New York: Pergamon Press.

1961. Nerve endings in the skin of the gorilla. J. Compar. Neurology, 116: 145-155.

1962. Cutaneous sensory end organs of some anthropoid apes. Science, 136: 384-386.

Winn, W., and J. Haldi.

1944. Further observations on the water and fat content of the skin and body of the albino rat on a high fat diet. Am. J. Physiol., 142: 508-511.

Wislocki, G. B.

1928. Observations on the gross and microscopic anatomy of the sloths (*Bradypus griseus* Gray and *Choloepus hoffmanni* Peters). J. Morphol., 46: 317-377.

Wislocki, G. B., H. Bunting, and E. W. Dempsey.

1947. Metachromasia in mammalian tissues and its relationship to mucopolysaccharides. Amer. J. Anatomy, 81: 1-37.

686 REFERENCES

Wislocki, G. B., D. W. Fawcett, and E. W. Dempsey.
  1951. Staining of stratified squamous epithelium of mucous membranes and skin of man and monkey by the periodic acid-Schiff method. Anat. Rec., 110: 359-578.
Wohnlich, H.
  1951. Mucine der menschlichen Epidermis. Ihre Abtrennung von Glykogen. Biochem. Z., 322: 76-78.
Wolbach, S. B.
  1951. The hair cycle of the mouse and its importance in the study of sequence of experimental carcinogenesis. Ann. N. Y. Acad. Sci., 53: 517-536.
Wolf, J.
  1940. Das Oberflächenrelief der menschlichen Haut. Z. microsk. anat. Forsch., 46: 170-202.
Wynne-Edwards, V. C.
  1962. Animal dispersion in relation to social behaviour. Edinburgh: Oliver and Boyd, Ltd.
Yablokov, A. V., V. M. Bel'kovich, and V. I. Borisov.
  1972. Kity i delfiny (Whales and Dolphins). Moscow: Nauka.
Yablokov, A. V., and G. A. Klevezal'.
  1964. Vibrissy kitoobraznykh i lastonogikh, ikh raspredelenie, stroenie i znachenie (Vibrissae of *Cetacea* and *Pinnipedia,* their distribution, structure and significance). *In* Morfologicheskie osobennosti vodnykh mlekopitajuschikh. M.
Yang, S. H.
  1952. A method of assessing cutaneous pigmentation in bovine skin. J. Agric. Sci., 42: 465-467.
Yasuda, K., and W. Montagna.
  1960. Histology and cytochemistry of human skin. XX. The distribution of monoamine oxidase. J. Histochem. and Cytochem., 8: 356-366.
Yasuda, K., T. Aoki, and W. Montagna.
  1961. The skin of primates. IV. The skin of the lesser bushbaby (*Galago senegalensis*). Amer. J. Phys. Anthropol., 19: 23-33.
Yasuda, K., H. Furusawa, and N. Ogatha.
  1958. Histochemical investigation on the phosphorylase in the sweat glands of axilla. Okajimas Folia Anat. Japon, 31: 161-169.
Yazan, O. Ya.
  1972. Znachenie sal'nykh zhelez v termoregulyatsii standartnykh norok na yuge SSSR (The significance of the sebaceous glands for thermoregulation in standard minks of southern USSR). Sbornik Nauch. Tekhnich. Informatsii. Vsesojuznyi Nauchn. Issled. Inst. Okhotnich. Khoz. i Zverovodstva. Kirov, 36: 59-67.
Young, J. Z.
  1963. The life of mammals. Oxford.
Yuditskaya, A. I.
  1949. The chemical nature of reticulin. Chem. Abstr., 43: 6263.
Yun, J. S., and W. Montagna.
  1965. The skin of primates. 25. Melanogenesis in the skin of bushbabies. J. Phys. Anthropol., 23: 143-148.

Zabusov, N. P.
1910. K innervatsii letatelnoi pereponki u letuchikh myshei (On the innervation of the flying membrane in bats). Tr. Ob-va Estesttvoisp.. Kaz. In-ta, 43: 1-67.

Zamakhaeva, N. M.
1972a. Stroenie volosyanogo pokrova rechnogo bobra (Hair structure of the European beaver). Sborn. Nauchn. Tekhnich. Inf. Vses. Nauchn. Issled. Inst. Okhotnich. Khoz. i Zverovodstva. Kirov, 35: 71-78.
1972b. Nekotorye vozrastnye osobennosti stroenia kozhnogo i volosyanogo pokrova rechnogo bobra (Some age-related peculiarities of the structure of skin and hair in the European beaver). Sborn. Nauchn. Tekhnich. Informatsii. Vses. Nauchn. Issled. Inst. Okhotnich. Khoz. i Zverovodstva. Kirov, 37-39: 140-144.

Zenkovich, B. A.
1938. Temperatura tela kitov (Body temperature of whales). Dokl. AN SSSR, XVIII: 9.

Zharkov, I. V.
1931. Stroenie mekha i osennyaya lin'ka zaitsa-belyaka (The structure of the fur and autumn molt of L. europaeus). Raboty Volzhsko-Kamsk. Kraevoi Promyslovoi Stantsii, 1: 153-167.

Zeitschmann, E. H.
1903. Beiträge zur Morphologie und Histologie einiger Hautorgane der Cerviden. Z. Wiss. Zool., 74: 1-63.

Zietschmann, O.
1904. Vergleichende histologische Untersuchungen über den Bau Augenglider der Haussäugetiere. Grefes Arch. Ophtalmol., 58: 61-122.

Zimmermann, A. A., and T. Cornblet.
1948. The development of epidermal pigmentation in the Negro fetus. J. Invest. Dermatol., 11: 383-395.
1950. The development of epidermal pigmentation in the Negro fetus. Sci. Contrib. N. Y. Zool. Soc., 35: 10-12.

Zlobin, B. D., and V. V. Shiryaev.
1975. Nekotorye voprosy ekologii solongaya (Some problems of the ecology of Mustela altaica pallas). Sbornik Nauchn. Tekhnich. Informatsii. VNII Okhotnich. Khoz i Zverovodstva. Kirov, 47-48: 80-85.

Zubin, A. M.
1938. Izuchenie gistologicheskogo stroenia kozhi zaitsa-belyaka (A study of the histological structure of the skin of L. europaeus). N. I. Labor. Glavmekhproma. Sb. Trudov, Gosizdat, legr. prom. M. L., 1: 3-31.

Zubin, A. M., and L. P. Pchelina.
1951. Osobennosti stroenia kozhnogo pokrova morskogo kotika (Structural peculiarities of the fur seal skin). N. I. Institut Mekhovoi Promyshlennosti. Gizlegprom, 2.

Zverev, M. D.
1931. Materialy po biologii i sel'skokhozyaistvennomu znacheniu v Sibiri khor'ka i drugikh melkikh khischnikov iz semeistva Mustelidae (Materials on biology and agricultural significance in Siberia of the polecat and other

small predators of the *Mustelidae* family). Trudi po zaschite rastenij Sibiri. 1 (8). Novosibirsk: 5-48.

Zweifach, B. W.

  1959. Structural aspects and hemodynamics of microcirculation in the skin. *In* Human integument. Ed. S. Rothamn. Washington, D. C.: Amer. Assoc. Advanc. Sci.

INDEX